New Trends in Synthetic Medicinal Chemistry

Edited by Fulvio Gualtieri

WILEY-VCH

Methods and Principles in Medicinal Chemistry

Edited by
R. Mannhold
H. Kubinyi
H. Timmerman

New Trends in Synthetic Medicinal Chemistry

Edited by
Fulvio Gualtieri

WILEY-VCH

Weinheim · New York · Chichester · Brisbane · Singapore · Toronto

Series Editors:
Prof. Dr. Raimund Mannhold
Biomedical Research Center
Molecular Drug Research Group
Heinrich-Heine-Universität
Universitätsstraße 1
40225 Düsseldorf
Germany

Prof. Dr. Hugo Kubinyi
ZHB/W, A 30
BASF AG
67056 Ludwigshafen
Germany

Prof. Dr. Hendrik Timmerman
Faculty of Chemistry
Dept. of Pharmacochemistry
Free University of Amsterdam
De Boelelaan 1083
1081 HV Amsterdam
The Netherlands

Volume Editor:
Prof. Dr. Fulvio Gualtieri
Dipartimento di Scienze Farmaceutiche
Università degli Studi di Firenze
Via Gino Capponi, 9
50121 Firenze
Italy

Library of Congress Card No. applied for.

British Library Cataloguing-in-Publication Data: A catalogue record for this book is available from the British Library.

Die Deutsche Bibliothek – CIP Cataloguing-in-Publication-Data
A catalogue record for this publication is available from Die Deutsche Bibliothek

Printed on acid-free and chlorine-free paper.

Composition: Datascan GmbH, D-67346 Speyer
Printing: Betzdruck GmbH, D-63291 Darmstadt
Bookbinding: Osswald & Co., D-67433 Neustadt (Weinstraße)

Printed in the Federal Republik of Germany.

Preface

In recent years, both the number and complexity of tools available to medicinal chemists for the design, identification and investigation of test drugs have increased dramatically. Among these tools, computational methods, combinatorial chemistry, biotechnology, and high-throughput screening perhaps deserve special mention. The dominant role of synthetic chemistry has been increasingly challenged by knowledge of the structure and functions of biomacromolecules including enzymes, receptors, transporters and membrane channels, as well as by an extensive growth of information in areas such as genetics, biology and pathology.

Despite these advances, the development of new drugs still depends heavily on the ability of chemists to produce – with the benefit of simple, reliable and rapid methods – the molecules that have been designed initially. Hence, although biology is becoming increasingly important in drug development, synthetic organic chemistry continues to play a fundamental role in the research that is performed both in academia and in industry. It is imperative therefore that medicinal chemists remain aware of the current rapid progress being made in different areas of synthetic organic chemistry, and the present volume is dedicated to this topic. This book reviews those modern synthetic techniques that are of major interest to medicinal chemists, among which combinatorial organic synthesis, and in particular solid-phase synthesis, enantioselective and biocatalyzed syntheses, have had the greatest impact on drug research. As a result, each of these subjects is covered in respective chapters of the present volume. Racemate resolution, although not a synthetic technique, is of major importance and is also easier to perform than enantioselective synthesis; thus, a chapter on the subject has been included. In recent years, two classes of biopolymers – carbohydrates, nucleotides and oligonucleotides – have emerged as targets of increasing interest in modern drug research; consequently, two chapters are included on the synthesis and properties of these biopolymer groups. Furthermore, as medicinal chemists become increasingly involved not only in the synthesis of series of compounds but also in the derivation of structure-activity relationships and optimization of properties of the drug molecules prepared, a chapter introducing the subject of series design has been added.

As series editors, we would like to thank the contributors to this volume – and in particular Fulvio Gualtieri – for their co-operation in its preparation. We feel sure that those scientists who have an interest in new synthetic approaches in medicinal chemistry will, in this volume, find not only an abundance of information on the topics described, but also the guidance and encouragement to apply the appropriate techniques with success.

January 2000

Raimund Mannhold, Düsseldorf
Hugo Kubinyi, Ludwigshafen
Henk Timmerman, Amsterdam

List of Contributors

Valery Antonenko

Affimax Research Institute
3410 Central Expressway
Santa Clara
95051 California
USA

Maurizio Benaglia

Dipartimento di Chimica Organica
e Industriale
Universita' di Milano
Via C. Golgi, 19
20133 Milano
Italy

Gottfried Blaschke

Institut für Pharmazeutische Chemie
Westfälische Wilhelms-Universität
Hittorfstrasse 58–62
D-64271 Münster
Germany

Giuseppe Capozzi

Dipartimento di Chimica organica
Universita' di Firenze
Via G. Capponi, 9
50121 Firenze
Italy

Bezhan Chankvetadze

Department of Physical Chemistry
School of Chemistry
Tbilisi State University
Chavchavadze Ave 1
380028 Tbilisi
Georgia

Mauro Cinquini

Dipartimento di Chimica Organica
e Industriale
Universita' di Milano
Via C. Golgi, 19
20133 Milano
Italy

Sergio Clementi

Dipart. Chim. Organica
Universita' di Perugia
Via Elce di Sotto 10
06100 Perugia
Italy

Franco Cozzi

Dipartimento di Chimica Organica
e Industriale
Universita' di Milano
Via C. Golgi, 19
20133 Milano
Italy

Antony W. Czarnik

Illumina, Inc.
9390 Towne Centre Drive
Suite 200, San Diego
CA 92121-3015
USA

Oreste Ghisalba

Novartis Pharma AG
S-508.202A
CH-4002 Basel
Switzerland

Fulvio Gualtieri

Dipartimento di Scienze Farmaceutiche
Universita' di Firenze
Via G. Capponi, 9
50121 Firenze
Italy

Lucius Kaufhold

Johann Wolfgang Goethe-Universität
Institut für Pharmazeutische Chemie
Biozentrum
Marie Curiestr. 9
60439 Frankfurt am Main
Germany

Nicolay V. Kulikov

Affimax Research Institute
3410 Central Expressway
Santa Clara
95051 California
USA

Torbjörn Lundstedt

Melacor Therapeutics
Kungsgatan 62
SE-75321 Uppsala
Sweden

Stefano Menichetti

Dipartimento di Chimica organica
Universita' di Firenze
Via G. Capponi, 9
50121 Firenze
Italy

Reza Mortezaei

Affimax Research Institute
3410 Central Expressway
Santa Clara
95051 California
USA

Cristina Nativi

Dipartimento di Chimica organica
Universita' di Firenze
Via G. Capponi, 9
50121 Firenze
Italy

Christian R. Noe

Johann Wolfgang Goethe-Universität
Institut für Pharmazeutische Chemie
Biozentrum
Marie Curiestr. 9
60439 Frankfurt am Main
Germany

Manuel Pastor

Dipart. Chim. Organica
Universita' di Perugia
Via Elce di Sotto 10
06100 Perugia
Italy

Contents

1 Organic Synthesis and Medicinal Chemistry

Fulvio Gualtieri

1.1 Setting the Scene

Medicinal chemistry is, at its most basic level, an empirical science largely based on chemistry. However, in the past decades the tools available to the medicinal chemist to identify, design and test drugs have increased dramatically, both in quantity and sophistication. Computational methods, combinatorial chemistry, biotechnologies and high-throughput screening are among the many powerful techniques that have been harnessed to bring an element of rationality to the search for new drugs. The role of the ongoing research of human genoma on drug discovery is difficult to evaluate at the moment, but will very likely be crucial, at least for some diseases.

The dominant role of synthetic chemistry has been increasingly challenged by knowledge of the structure and functions of enzymes, receptors, channels, membrane pumps, nucleic acids and by the exponential growth of information about biology, genetics and pathology, giving paramount importance to the dialogue between chemists and biologists. According to G. Wess, who discussed the problem in a recent perspective on the challenges for medicinal chemistry [1], "The Pharmaceutical industry appears to be in transition from a chemistry-based industry to one based more on human biology and genetic information".

Nevertheless, as in the old days, the development of new chemical entities is still highly dependent on the ability of chemists to obtain, with simple, reliable, fast and possibly inexpensive methods, the molecules that have been designed. Even if it is an undisputed fact that biology has become exceedingly important in drug research, it is reasonable to imagine that chemistry, and in particular synthetic organic chemistry, will continue to play a fundamental role in academic research and in the R&D departments of drug companies of the third millennium. As a consequence, medicinal chemists, while being receptive to the progress of knowledge in biology and pharmacology, must continue to excel in synthetic organic chemistry, and keep themselves up-to-date on the progress of the discipline.

The role of organic synthesis in drug research and development nowadays and over the next decades was discussed at the 11th Camerino-Noordweijkerhout Symposium (11–15 May 1997), in a section dedicated to New Developments in Synthetic Medicinal Chemistry [2] which suggested to the editors of the *Methods and Principles in Medicinal Chemistry* series, the preparation of the present book. Of course, a book of this size cannot be, and actually does not intend to be, an exhaustive review of all new synthetic methods. Rather, our goal was that of reviewing modern synthetic methods that are likely to be of great interest for medicinal chemists. Among them combinatorial organic synthesis, the closely related solid-phase synthesis, enantioselective and biocatalyzed synthesis are probably the methodologies which have the greatest impact on drug research, and they represent the most important chapters of the book.

Even if not strictly a synthetic method, racemate resolution is an argument of great importance in academic and industrial drug research, since, when practicable, this approach is simpler and more economical than enantioselective synthesis; accordingly, a chapter dedicated to this topic was introduced.

There are three classes of oligomeric compounds, related to the biological fundamental polymers – proteins, nucleic acids and polysaccharides – that appear promising in modern drug research: modified oligopeptides, oligosaccharides, and oligonucleotides. As a consequence, unnatural amino acids, carbohydrates and nucleotides, which represent their usual precursors, have been the target of increasing interest. While unnatural amino acids and related peptides have been thoroughly studied in the past, medicinal chemists' attention to modified carbohydrates and nucleotides, and to the corresponding oligomers, is relatively recent. Two chapters are dedicated to their synthesis and properties. Last but not least, since medicinal chemists, to establish structure-activity relationships (SARs) and to optimize drug properties, are generally involved with the synthesis of series of compounds, an introductory chapter on their design has been included.

1.2 Classical and Bio-assisted Organic Synthesis

Organic synthesis has made so much progress during the past decades that it appears as a mature science, ready to face any synthetic problem, however difficult that problem may appear. The mechanistic advance in understanding organic chemistry has made prediction possible of the reactivity of organic compounds toward different reagents, and also to take on more and more complicated synthetic tasks. Retrosynthetic analysis has taught us to disconnect strategic bonds in order to dissect target molecules and identify useful synthetic strategies. As a consequence, thanks also to the explosion of organometallic chemistry, the collection of reactions, reagents and catalysts available has been growing steadily and it is a common feeling that, given enough time and money, one can obtain any organic molecule, including those of the most complex natural compounds. As a matter of fact, the synthesis of natural compounds has been the arena where the skill of outstanding organic chemists has challenged and shown the power of the discipline. An impressive review of the most important achievements in this field is reported in the book by Nicolaou and Sorensen [3].

During the past 30 years, many milestones have been attained in the synthesis of natural compounds with pharmacological interest. The synthesis of prostaglandins, started by Corey in the 1960s [4] and which continued with outstanding results for years; the synthesis of vitamin B_{12} realized by Woodward in the 1970s [5]; the synthesis of the 64-chiral-centers palytoxin, performed by Kishi in the 1980s [6], and the achievement, almost at the same time, of the synthesis of the much-sought taxol by two different teams in the present era [7–9].

From a Medicinal Chemistry point of view, it must be stressed that total syntheses, even if not of practical use for commercial production of the drugs, are however very useful in opening the way to analogous and simpler molecules that may help to establish SARs and to design easier-to-obtain compounds. There are hundreds of examples of this approach in the history of medicinal chemistry. One of the most recent regards the discovery [10] and the synthesis [11] of dynemicin A (**1**), a member of a family of natural compounds (enediyne antibiotics) that combine unprecedented molecular structure with striking biological activity and a new mechanism of action [12]. Moreover, this class of compounds represents a good example of how the mechanistic interpretation of organic chemistry allows an understanding of a drug's mechanism of action (Fig. 1).

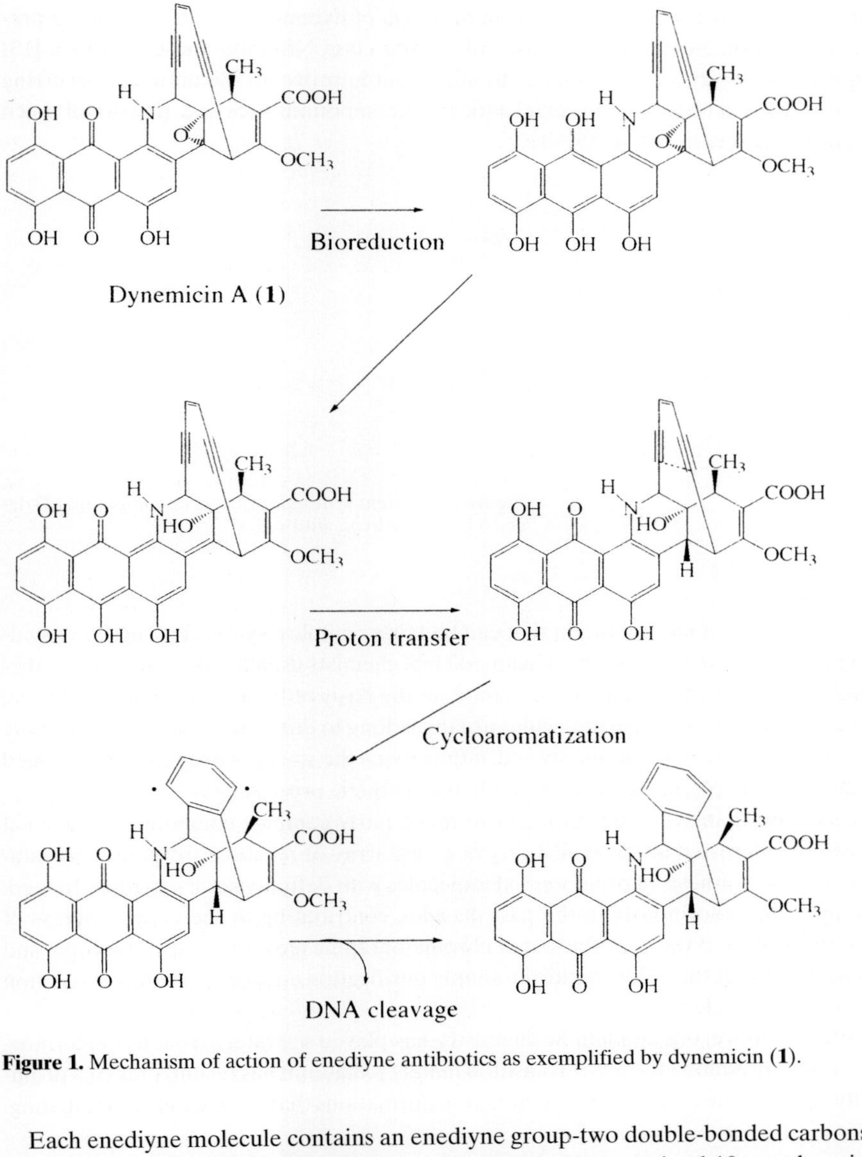

Figure 1. Mechanism of action of enediyne antibiotics as exemplified by dynemicin (**1**).

Each enediyne molecule contains an enediyne group-two double-bonded carbons flanked by two pairs of triple-bonded carbons – all embedded in a strained 10-member ring, which represents the pharmacophoric group. This group is like a spring-loaded trap which is sprung when a reaction is triggered that starts electrons flowing toward the epoxide group, as in the case of dynemicin A, thus causing it to open. The strain of the 10-membered ring is thus relieved and the terminal carbons of the system are brought closer, causing the enediyne group to cyclize. The species that actually produces biological activity (cleavage of DNA) is a highly reactive 1,4-benzenoid radical formed in the reaction which abstracts two hydrogen atoms from DNA's sugar-phosphate backbone, causing the strand to break.

Knowing the structure and the mechanism of action of dynemicin A, and exploiting prevous experience in the synthesis of compounds of this class, Nicolaou and co-workers [13] have designed a new series of compounds to mimic but improve on the naturally occurring drug, and succeeded in obtaining powerful anticancer compounds such as **2**, that are of much simpler structure and easier to synthesize.

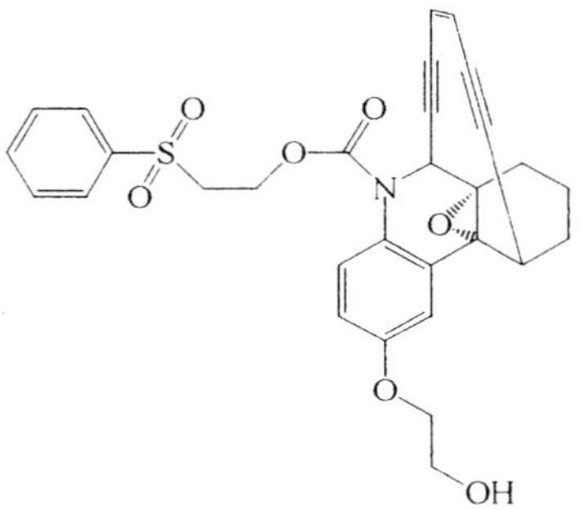

Figure 2. A potent synthetic anticancer compound (**2**) derived from enediyne antibiotics.

Unlike compounds of natural origin that can be fairly complex, synthetic drugs are usually much simpler molecules. Therefore, what medicinal chemists usually ask of organic synthesis are methods to make any structure designed on the basis of biological, pharmacological and structural information. These molecules usually belong to different chemical classes, may have strict stereochemical requirements and, if they reach the status of drug candidates, need to be produced by simple, inexpensive and reliable synthetic procedures.

Therefore, more than synthetic strategies to make fairly complex compounds, medicinal chemists will be interested in the availability of a vast array of reactions, reagents and catalysts that allow the synthesis of programmed molecules with definite steric structure. Indeed, they have increased enormously in the past decades, contributing to the great progress of organic synthesis, even if the contribution of chromatographic techniques, spectroscopic and other physical methods that allow rapid and simple purification and identification of reaction products cannot be neglected.

As mentioned above, organometallic chemistry has played a strategic role in modern organic synthesis [14]. Among the other transition metals, palladium has recently taken a prominent position due to the manifold and unique transformations that it is capable of mediating, often in a catalytic mode. A nice example of this approach is the synthesis of heliotridane reported by Bäckvall [15]. The primary amide group of **3** is the nucleophile that gives a two-fold attack in the 1,4-positions of the Pd(II)-coordinated 1,3-diene, allowing a one step synthesis of **4** which can be easily transformed into (±)-heliotridane (Fig. 3).

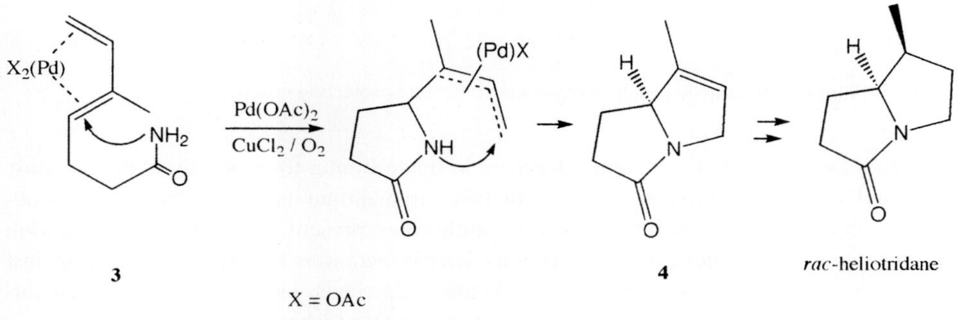

Figure 3. Palladium-catalyzed synthesis of heliotridane nucleus.

The development of protecting groups that can be introduced and removed under a variety of mild conditions has accelerated and paralleled the progress of synthetic organic chemistry. The struggle to add new entries in this class of reagents is still very great: particularly to develop new methods for removal. Indeed, multiplying the number of ways to remove protecting groups gives chemists more control over the course of a synthetic scheme and increases its flexibility. Recently, protecting groups that can be removed photolytically have been reported: o-benzoylbenzoate esters can be used to protect alcohols and mercaptans, being then removed by irradiation with a mercury vapor lamp [16] (Fig. 4).

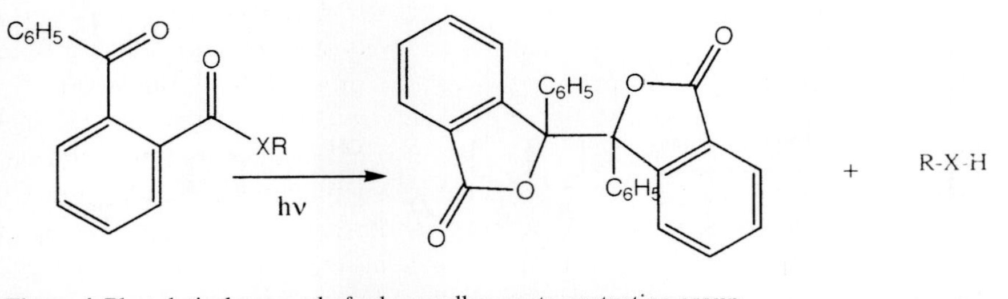

Figure 4. Photolytical removal of o-benzoylbenzoate protecting group.

Primary or secondary amines can be protected with o-hydroxy-*trans*-cinnamic acid and then restored by ultraviolet irradiation [17] (Fig. 5).

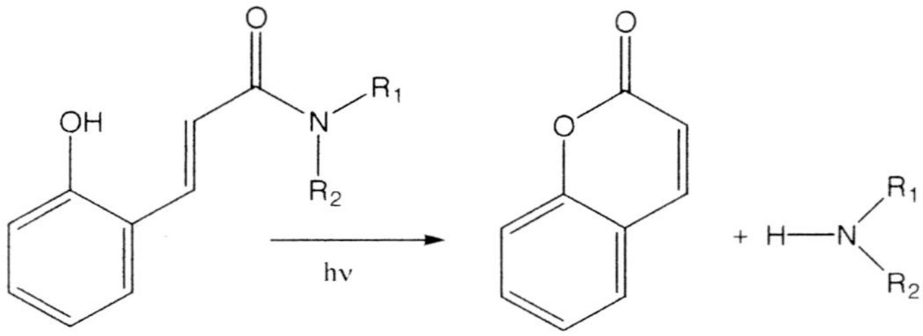

Figure 5. Photolytical removal of o-hydroxy-*trans*-cinnamate protecting group.

Even if racemic switch – the redevelopment in single-isomer form of a chiral drug already approved as a racemic mixture – apparently failed to maintain its promise because of problems in obtaining approval from regulatory authorities, present rules for drug registration strongly favor enantiomerically pure drugs as there is increasing regulatory pressure against licensing racemic drugs. As a consequence, chemists are as busy as ever in inventing new catalytic asymmetric syntheses, as well as racemate resolution techniques.

The seminal enantioselective allylic alcohols epoxidation realized by Katsuki and Sharpless [18] to which other similarly steroselective reactions soon followed (e.g., bishydroxylation, cyclopropanation, lactonization and catalytic hydrogenation) have been invaluable for this purpose. One of the most significant applications, as far as drug synthesis is concerned, of the Sharpless method from the innumerable ones which have been found in the past 20 years is the routine preparation (Fig. 6) of antipodal pairs of known chirality of β-blockers such as propranolol (**5**) [19].

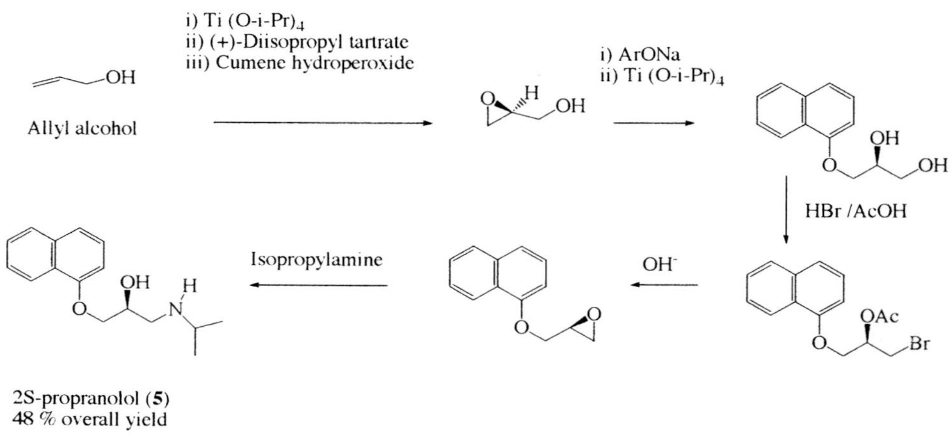

Figure 6. Routine synthesis of 2S-propranolol (**5**).

The most recent development of enantioselective epoxidation requires no transition metal and uses a catalyst derived from an inexpensive sugar [20]. Four hydroxyl groups of the pyranose form of D-fructose are protected as acetonides and the remaining one is oxidized to keto group. The chiral compound obtained is transformed by potassium monopersulfate to a dioxirane that behaves as the asymmetric catalyst and epoxidize the olefin (Fig. 7).

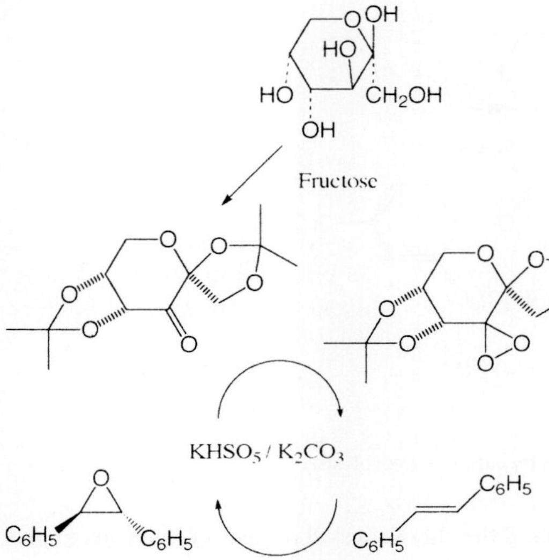

Figure 7. Fructose-based enantioselective epoxidation of olefins.

Resolution of racemates by classical separation of diastereomeric salts remains a valid alternative to enantioselective synthesis. A simple but ingenious idea has recently been proposed and shown to work successfully: the use of families of resolving agents that would work as a library on the racemate to be resolved, so that the least soluble salt would crystallize out. As a matter of fact, Broxterman [21] has shown that a mixture of acylated tartaric acids like **6a–c** of Fig. 8, gave immediate crystallization in 97 out of 100 amine racemates studied, allowing a more rapid choice of resolving agents to be made in a more suitable manner than was previously possible.

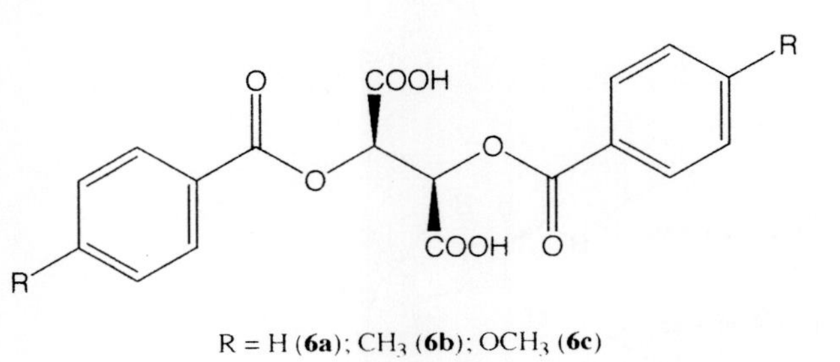

Figure 8. Acyclated tartaric acids (**6a–b**) as a family of resolving agents.

In one of the latest achievements in the field of resolution of racemates with non classical methods, Jacobsen has developed a catalyst that selectively hydrolyzes epoxides [22] (Fig. 9). The reaction can be run as a resolution process, making epoxide in high enantiomeric excess by hydrolysis of the unwanted isomer. Or it can be used as a synthetic process, in which enantiomeric excess of the glycol product is optimized.

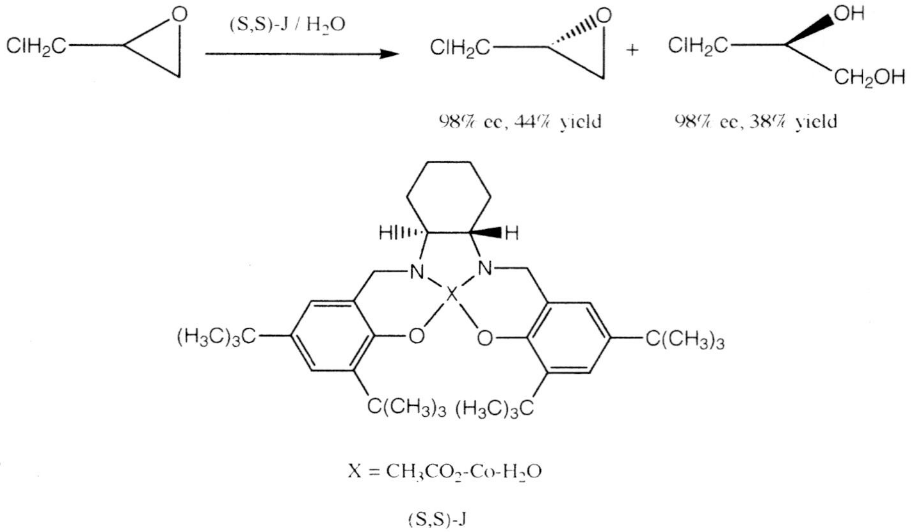

Figure 9. Kinetic resolution of terminal epoxides by catalytic hydrolysis.

For a very long time, humans have exploited the chemical skill of micro-organisms to produce or modify organic molecules. From these and other sources, many enzymes are now available that can cleanly, rapidly and inexpensively afford organic reactions on a variety of substrates. The recent progress in this field has made it possible to obtain a variety of molecules [23], usually with a definite stereochemistry, and had a tremendous impact on medicinal chemistry, especially on the industrial production of drugs. A case in point is that regarding the synthesis of 6-aminopenicillanic acid (6-APA, **7**), a fundamental intermediate in the production of semisynthetic penicillin antibiotics. In earlier processes, the removal of the side chains of penicillins G and V (Pen G, **8**; Pen V, **9**), produced in large amounts by fermentation,

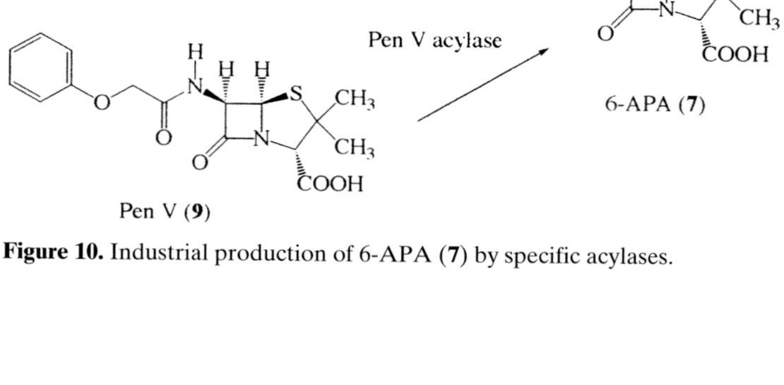

Pen V (**9**)

Figure 10. Industrial production of 6-APA (**7**) by specific acylases.

was performed via protection of the β-lactam with trialkylsilyl chloride and subsequent treatment with phosphorous pentachloride. The discovery of enzymes that specifically remove the side chains under mild aqueous conditions (Pen G acylase, from *Escherichia coli* and Pen V acylase, from *Fusarium oxysporum*) has completely replaced the earlier chemical processes (Fig. 10) [23].

A combination of the technique of organic synthesis with the power of enzymes asymmetrically to catalyze reactions, which is now within reach, can be another major improvement toward the synthesis and optimization of drugs. Moreover, it would allow the development of economically and environmentally acceptable processes for industrial synthesis. This enzyme–organic approach has proven to be particularly useful in the synthesis of oligosaccharides and other carbohydrates [24], as can be exemplified by the use of *N*-acetylneuraminic acid aldolase, both in free and immobilized form, to obtain sialic acid derivatives. A recent application of this methodology by C-H Wong [25], using a large-scale enzymatic reaction cycle on special substrates, has allowed the multikilogram synthesis of sialyl Lewis X (**10**, Fig. 11), a compound of pharmaceutical interest that is potentially useful in treating inflammation and heart attack.

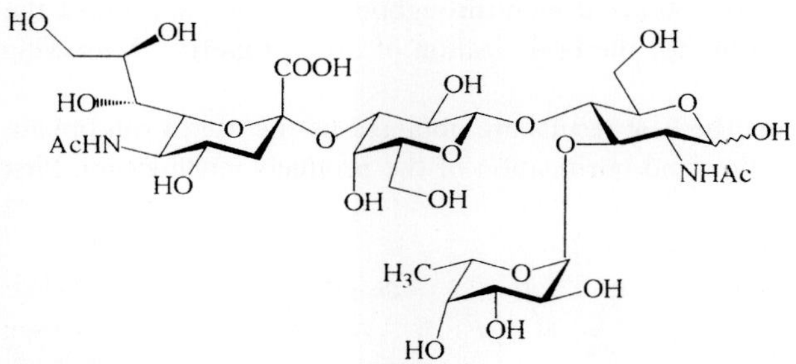

Figure 11. Sialyl Lewis X (**10**).

1.3 New Strategies

Until a few years ago, pharmacological screening was the critical step, and synthesized compounds 'piled up' waiting to be tested. Today, the development of high-throughput screening techniques has reversed this situation, allowing scientists to screen from hundreds to thousands of compounds each day. In contrast, the process of synthetic chemistry had, until recently, remained unchanged. As a consequence, the perception of the bottleneck that slackened the pace of drug discovery; the need for finding more and more hits and leads for new drugs; the pressure to reduce the time between lead discovery and clinical trials, have led to an epochal change in the philosophy of drug discovery that can be summarized as follows: it is more useful to spend time generating diversity, rather than highly pure but inactive compounds.

Originated in 1984 by Geysen [26], and brought to the attention of the scientific community by two seminal articles a decade later [27, 28], combinatorial chemistry is the cornerstone of this new strategy. Although applications of combinatorial chemistry are widespread and cover fields as diverse as material sciences, molecular recognition and development of new catalysts, the technique has captured the collective imagination of the medicinal chemistry community because of its potential for revolutionizing drug discovery. Indeed it appears as a powerful tool to increase the chemical diversity of drugs available for biological tests and to accelerate drug discovery programs and it can be useful both for lead generation and for lead optimization.

In the first case, large libraries, usually containing some 1000–100000 compounds in nanomolar quantity, for which encoding is usually required, are most suitable. They are referred to as *prospecting* or *discovery* libraries, and chemical diversity is their main property.

Lead optimization usually requires smaller libraries, containing some 100–1000 compounds in larger amounts (micro, or better, millimolar quantity). As topographical methods (parallel synthesis) are usually used, compounds do not need tagging and their structure is unambiguously identified by their co-ordinates. They are referred to as *focused* or *optimization* libraries.

Automation and miniaturization [29] have represented the necessary follow up to the introduction of combinatorial chemistry. Progress in this field is exponential: the offer of new chemi-informatic tools that provide total synthetic process control, fully automated analytical controls, efficient integration with robotized high-throughput screening and make the parallel investigation of SARs possible for the optimization of drug properties is growing every day.

Both for large and small libraries, the best results are obtained by the use of solid-phase synthesis [30] that makes the isolation and purification of the products much easier. First

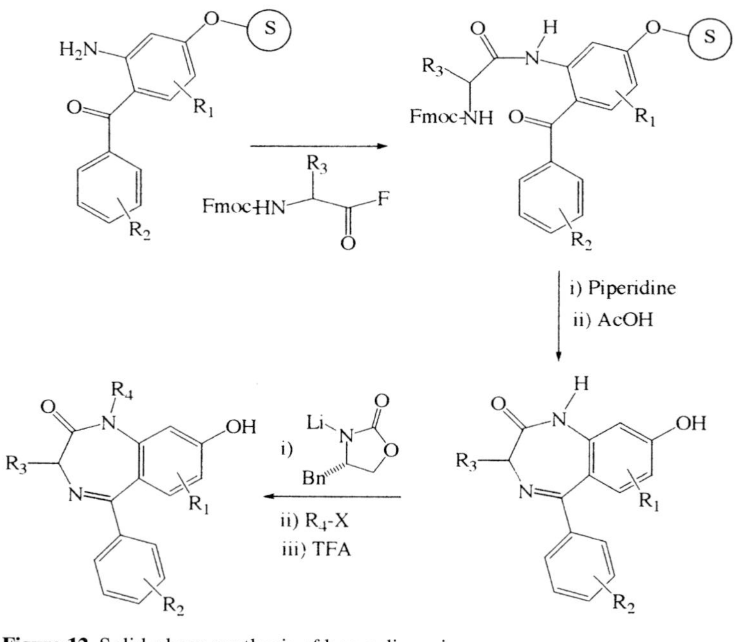

Figure 12. Solid-phase synthesis of benzodiazepines.

started as a technique for the synthesis of millions of peptides in only a few days [26], solid-phase synthesis has evolved into a widely established method for producing non-oligomeric compounds with *in vitro* biological activity [31]. In a pioneering study, J. A. Ellman [32] synthesized one of the first small molecule libraries by creating variants of benzodiazepines, a class of compounds that has been a fertile source of successful drugs (Fig. 12).

To render solid-phase synthesis widely useful, organic chemists had to solve challenging problems about the nature of solid supports, linkers and on their cleavage conditions in order to make them compatible with the synthetic conditions required by the most important organic reactions. The wide range of reactivity among a set of diverse inputs, the need for high or low temperature conditions and compatibility of the chemistry with solid-phase conditions can challenge use of this approach. In fact, a dramatic limit to the extensive use of this technique is that, of the many reactions available to the synthetic organic chemist, only a few have been successfully adapted to solid-phase synthesis [33]. As a consequence, more work to adapt the most useful organic reactions to solid phase is badly needed. It can be anticipated that much of the effort in this direction will regard stereoselective reactions.

The current interest in solid-phase synthesis has led to a renewed interest in a complementary technique in which solid-supported reagents and scavengers are used in solution-phase chemistry. These reagents have proven to be useful for a wide variety of transformations important to chemists in general, and particularly to medicinal chemists. A comprehensive and clear review on this topic has recently appeared [34]. To exemplify, in Fig. 13 is documented the use of polymer-bound 4-hydroxy-3-nitrobenophenone to synthesize a library of herbicidal amides [35].

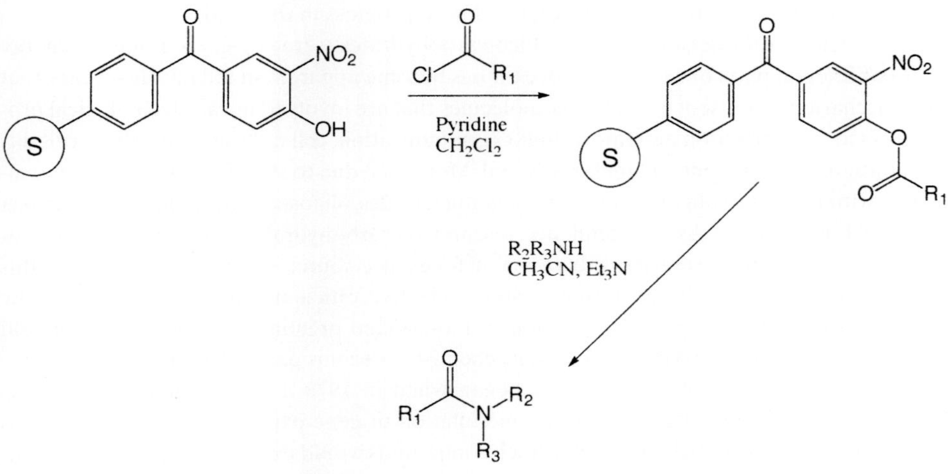

Figure 13. Synthesis of amides from polymer-bound 4-hydroxy-3-nitrobenzophenone esters.

The potential of combinatorial chemistry has been widely discussed, and one may ask where are the results in terms of new drugs. This is quite a difficult question to ask: perhaps untimely. Considering the time required for drug development and considering that combinatorial chemistry has only recently become widely used, we should reasonably wait a few years to evaluate fully the impact of the new approach on drug discovery. Indeed, many companies seem to have combinatorial chemistry-originated compounds in clinical trials and the approval of the first drug derived from this technology could come at the turn of the millennium.

1.4 Oligomers

The fundamental role of biological polymers in living organisms does not need to be emphasized. By using a few amino acids, monosaccharides and nucleotides as building blocks, nature has generated an immense variety of structures (peptides and proteins, nucleic acids, polysaccharides) that represent all chemical information and nearly all materials necessary for life. No wonder that scientists have developed a strong interest in their functions, synthesis and biological activity, and that the synthesis and pharmacological activity of small fragments thereof, such as peptides, oligonucleotides and oligosaccharides has been widely studied. However the pace of development of amino acid-, nucleotide- and carbohydrate-based pharmaceuticals has not been uniform. While the technology concerning peptides and, to a lesser extent, oligonucleotides, has advanced smoothly, the development of drugs based on carbohydrates has been slower than that of the other classes. There is a variety of reasons for this: the complexity of oligosaccharides, due to their stereochemistry and branching, is highly challenging for synthesis and analysis; there is no method to amplify oligosaccharides for sequence analysis; there is no machine available for automated sequencing or synthesis. All these problems, together with poor bioavailability which, on the other hand, is common to the other two classes of compounds, have slowed progress in this field.

Nevertheless, after decades during which carbohydrates were considered only as energy stores or integral parts of cellular matrices, it has become apparent in the past few years, that oligosaccharides represent a class of biomolecules that are involved in various biological processes such as recognition and intercellular communication, cell adhesion, infection, cell differentiation, development and metastasis [36]. Moreover, due to their higher molecular complexity with respect to oligopeptides and oligonucleotides, oligosaccharides have the greatest potential for specificity. As a consequence, research on carbohydrates is now undergoing considerable growth, and promises to be a major focus as a source of drug candidates. To this end, progress in the development of new, stereoselective, efficient and inexpensive synthetic methods is crucial. As mentioned previously, bio-assisted organic synthesis – which would combine the power of enzymes and organic chemistry – seems particularly useful.

Interest in modified oligonucleotides began when in 1978 a gene-targeted therapeutic approach was proposed, based upon the modulation of gene expression by the exogenous administration of oligonucleotides [37]. Such compounds would exploit the property of forming, *via* Watson–Crick hydrogen bonding, sequence-specific high-affinity complexes with mRNA. By binding to the mRNA, the oligonucleotide (antisense oligomer) would, by one of several mechanisms, interfere with and arrest cytosolic translation of the mRNA into proteins.

It was soon realized that natural oligonucleotides suffered from several drawbacks that severely hampered their use *in vivo* and their evolution toward therapeutically useful drugs. Natural phosphodiester oligomers are too rapidly degraded by both endo- and exonucleases, have poor bioavailability, and too many side effects to be used directly in cellular or whole animal systems, and therefore to be reliable drug candidates. Hence, intense synthetic work was started to modify the natural nucleotides by changes not only in the backbone (see Fig. 14) but also in the sugar or base moieties [38].

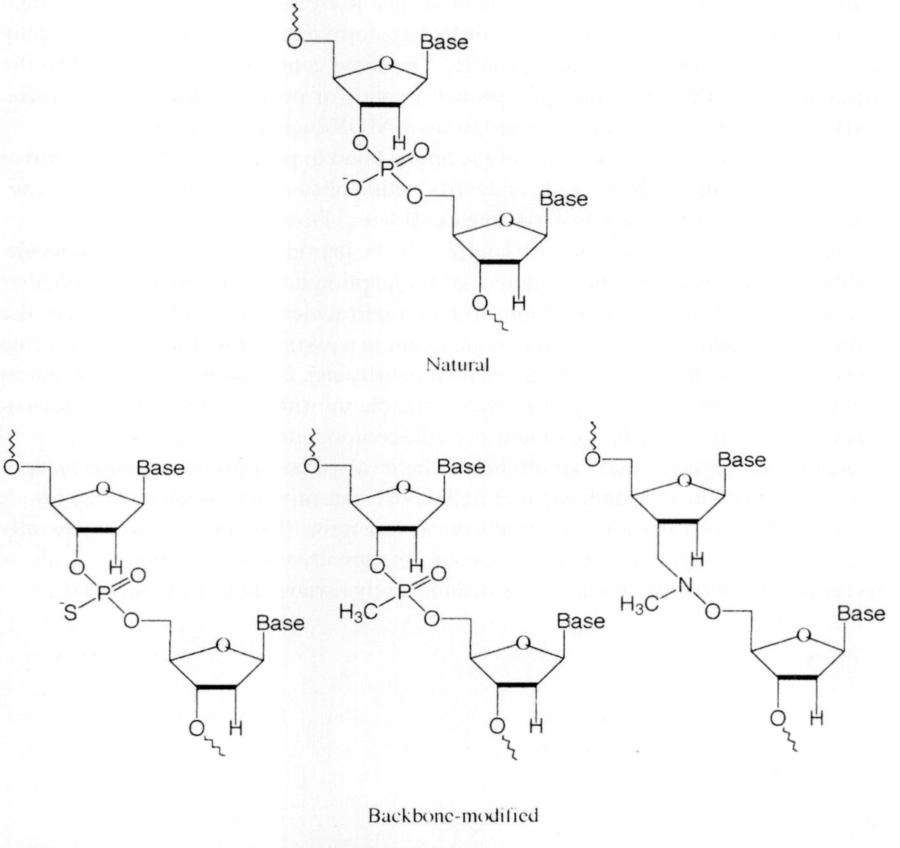

Figure 14. Natural and backbone-modified oligonucleotides.

As a result, an approach that, until a few years ago, was considered innovative but useful to generate sophisticated and precious investigational tools rather than compounds of practical use in therapy is on the way to expressing a few therapeutically useful drugs. The first compound of this class ISIS 292, a phosphorothioate oligonucleotide with the sequence GCG TTT GCT CTT CTT CTT GCG (formivirsen) [39] was approved by the NDA in August 1998 and launched by ISIS Pharmaceutical with the trade name Vitravene®. Used for cytomegalovirus (CMV) retinitis often found in HIV patients, Vitravene represents the first commercially available antisense oligonucleotide for drug therapy world-wide.

Amino acids and peptides have been on the stage for a long time, and their popularity is still untouched, as physiology, biochemistry, pharmacology and related disciplines have made it clear that peptides and proteins constitute the major classes of bioactive ligands for drug design and development. Unfortunately, many peptide hormones, neurotransmitters and other natural products are designed by nature to be rapidly degraded because they act as biological switches. Moreover, due to their chemical structure, peptides usually suffer from poor bioavailability. These problems can be overcome either by modifying the structure of the peptide sufficiently to render it stable and bioavailable, by maintaining at the same time its high affinity and selectivity for the target, or by transforming the peptide in a biologically equivalent non-peptide ligand (peptidomimetic). The same concepts can be applied to the design of inhibitors of those enzymes that process peptide or protein substrates, as was the case of HIV aspartic protease inhibitors used to treat AIDS, such as saquinavir.

In the first case, isosteric modifications of the amide bond to produce peptide bond surrogates, insertion of unnatural amino acids, and introduction of conformational constraint, represent crucial steps along the path toward drug candidates [40, 41].

In the second case, the goal is to abolish completely the peptide character of the molecule. Toward this end, the pharmacophoric moiety of the peptide has to be evidenced together with its stereochemical arrangement. A popular method to achieve this result is to reduce the conformational flexibility of the original peptide either in a particular region of the molecule (local constraint) or in the whole peptide (global constraint). Evaluation of the biological activity of the new peptide would then provide valuable information about the pharmacophore, opening the way to the design of non-peptidic compounds.

In both cases, amino acids that can modulate chemical–physical properties such as lipophilicity (like **11**, present in saquinavir and nelfinavir), can produce supplementary bonds (like **12** and **13** [42]), or can induce stereochemical constraint (like **14** [43]) are frequently used (Fig. 15). The design and synthesis of conformationally constrained amino acids as scaffolds for peptidomimetics synthesis, has been recently reviewed by Hanessian [44].

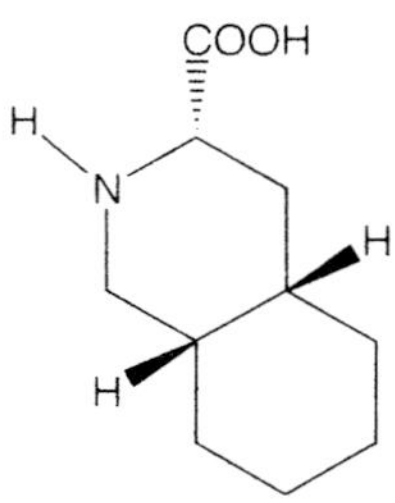

11

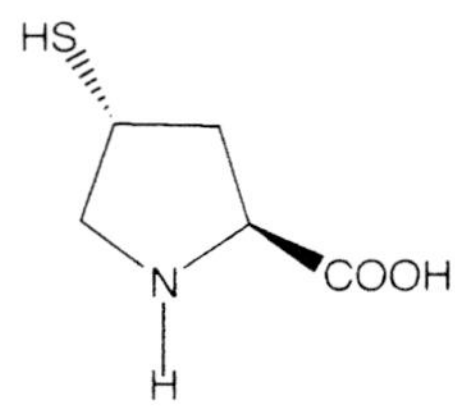

12

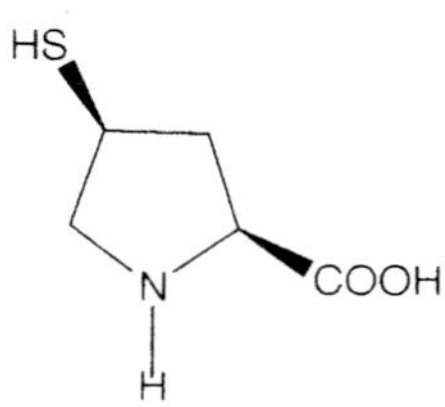

13

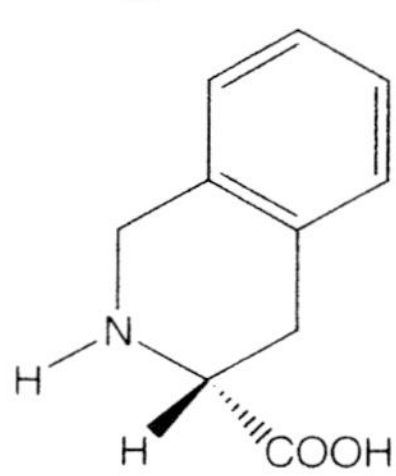

14

Figure 15. Some useful unnatural amino acids.

1.5 Conclusion

Nowadays, medicinal chemists can rely on a continuous flow of information generated by molecular biology, physiology, pathology, pharmacology and, very recently, genetics, to design drugs. A variety of strategies [45] and of computational tools [46], including the recently developed *virtual libraries* approach [47], are available to deal with the difficult problem of deciding which compounds to synthesize. Combinatorial chemistry and high-throughput screening have made it possible to synthesize and test large numbers of compounds. Nevertheless, when a drug has been reliably designed, in order to convert the virtual molecule into a solid one, to give it the proper potency and pharmacokinetics, and to put it on the shelves of the pharmacy shop, hardcore organic synthesis is needed; this was true in the old days, it is at present, and very likely it will remain in the foreseeable future. It is hoped that the present book will help medicinal chemists to cope with this challenging task.

References

[1] Wess, H. *Drug Discovery Today* **1**, 529–532 (1996).
[2] Gualtieri, F. in *Trends in Drug Research II*, Van Der Goot (ed.) Elsevier Science: Amsterdam; p. 15, 1998.
[3] Nicolaou, K.C. and Sorensen, E.J. *Classic in Total Synthesis*. Wiley-VCH: Weinheim, 1996.
[4] Corey, E.J., Andersen, N.H., Carlson, R.M., Paust, J., Vedejs, E., Vlattas, I., and Winter R.E.K. *J. Am. Chem. Soc.* **90**, 3245–3247 (1968).
[5] Woodward, R.B. *Pure Appl. Chem.* **33**, 145–177 (1973).
[6] Kishi, Y. *Pure Appl. Chem.* **61**, 313–324 (1989).
[7] Holton, R.A., Somoza, C., Kim, H.B., Liang, F., Biediger, R.J., Boatman, P.D., Shindo, M., Smith, C.C., Kim, S., Nadizadeh, H., Suzuki, Y., Tao, C., Vu, P., Tang, S., Zhang, P., Murthi, K.K., Gentile, L.N., and Liu, J.H. *J. Am. Chem. Soc.* **116**, 1597–1598 (1994).
[8] Holton, R.A., Kim, H.B., Somoza, C., Liang, F., Biediger, R.J., Boatman, P.D., Shindo, M., Smith, C.C., Kim, S., Nadizadeh, H., Suzuki, Y., Tao, C., Vu, P., Tang, S., Zhang, P., Murthi, K.K., Gentile, L.N., and Liu, J.H. *J. Am. Chem. Soc.* **116**, 1599–1600 (1994).
[9] Nicolaou, K.C., Yang, Z., Liu, J.J., Ueno, H., Nantermet, P.G., Guy, R.K., Claiborna, C.F., Renaud, J., Couladouros, E.A., Paulvannan, K., and Sorensen, E.J. *Nature* **367**, 630–634 (1994).
[10] Konishi, M., Ohkuma, H., Tsuno, T., Oki, T., VanDuyne, G.D., and Clardy, J. *J. Am. Chem. Soc.* **112**, 3715–3716 (1990).
[11] Myers, A.G., Fraley, M.E., Tom, N.J., Cohen, S.B., and Madar, D.J. *Chem. Biol.* **2**, 33–43 (1995).
[12] Smith, A.L., and Nicolau, K.C. *J. Med. Chem.* **39**, 2103–2117 (1996).
[13] Nicolaou, K.C., Dai, W.-M, Tsay, S.-C., Esteve, V.A., and Wrasidlo, W. *Science* **256**, 1172–1178 (1992).
[14] Schlösser, M. *Organometallics in Synthesis*. Wiley: New York, 1994.
[15] Andersson, P.G., and Bäckvall, J.-E. *J. Am. Chem. Soc.* **114**, 8696–8698 (1992).
[16] Jones, P.B., Pollastri, M.P., and Porter, N.A. *J. Org. Chem.* **61**, 9455–9461 (1996).
[17] Wang, B., and Zheng, A. *Chem. Pharm. Bull.* **45**, 715–718 (1997).
[18] Katsuki, T., and Sharpless, K.B. *J. Am. Chem. Soc.* **102**, 5974–5976 (1980).
[19] Klunder, J.M., Ko, K., and Sharpless, K.B. *J. Org. Chem.* **51**, 3710–3712 (1986).
[20] Wang, Z.-X., Tu, Y., Frohn, M., Zhang, J.-R., and Shi, Y. *J. Am. Chem. Soc.* **119**, 11224–11235 (1997).
[21] Vries, T., Wynberg, H., vanEchten, E., Koek, J., tenHoeve, W., Kellogg, R.M., Broxterman, Q.B., Minnaard, A., Kaptein, B., vander Sluis, S., Hulshof, L., and Koistra, J. *Angewandte Chemie – International Edition* **37**, 2349–2354 (1998).
[22] Tokunaga, M., Larrow, J.F., Kakiuc, F., and Jacobsen, E.N. *Science* **277**, 936–938 (1997).
[23] Zaks, A., and Dodds, D.R. *Drug Discovery Today* **2**, 513–531 (1997).
[24] Wong, C.-H., Halcomb, R.L., Ichikawa, Y., and Kajimoto, T. *Angew. Chem. Int. Ed. Engl.* **34**, 412–432 (1995).

[25] Hendrix, M., and Wong, C.-H. *Enantiomer* **1**, 305–310 (1996).
[26] Geysen, H.M., Meloen, R.H., and Barteling, S.J. *Proc. Natl Acad. Sci USA* **81**, 3998–4002 (1984).
[27] Gallop, M.A., Barret, R.W., Dower, W.J., Fodor, S.P.A., and Gordon, E.M. *J. Med. Chem.* **37**, 1233–1251 (1994).
[28] Gordon, E.M., Barret, R.W., Dower, W.J., Fodor, S.P.A., and Gallop, M.A. *J. Med. Chem.* **37**, 1385–1401 (1994).
[29] Vetter, D. *Drug Discovery Today* **3**, 404–408 (1998).
[30] Cowley, P.M., and Rees, D.C. *Current Med. Chem.* **4**, 211–227 (1997).
[31] Patel, D.V., and Gordon, E.M. *Drug Discovery Today* **4**, 134–144 (1996).
[32] Bunin, B.A., and Ellman, J.A. *J. Am. Chem. Soc.* **114,** 10997–10998 (1992).
[33] Choong, I.C., and Ellman, J.A. *Annu. Rep. Med. Chem.* **31**, 309–317 (1996).
[34] Drewry, D.H., Coe, D.M., and Poon, S. *Med. Res. Rev.* **19**, 97–148 (1999).
[35] Parlow, J.J., and Normansell, J.E. *Molecular Diversity* **1**, 266–269 (1995).
[36] Musser, J.H. *Annu. Rep. Med. Chem.* **27**, 301–310 (1992).
[37] Zamecnik, P.C., and Stephenson, M. L. *Proc. Natl Acad. Sci. USA* **75**, 280–284 (1978).
[38] Kiely, J.S. *Annu. Rep. Med. Chem.* **29**, 297–306 (1994).
[39] Thomas, S.M., and Burke, J.F. *Exp. Opin. Ther. Patents* **7**, 813–819 (1997).
[40] Olson, G.L., Bolin, D.R., Bonner, M.P., Bos, M., Cook, C.M., Fry, D.C., Graves, B.J., Hatada, M., Hill, D.E., Kahn, M., Madison, V.S., Rusieki, V.K., Sarabu, R., Sepinwall, J., Vincent, G.P., and Voss, M.E. *J. Med. Chem.* **36**, 3039–3049 (1993).
[41] Hruby, J., Al Obeidi, F., and Kazmierski, W.. *Biochem. J.* **268**, 249–262 (1990).
[42] Plucinska, K., Kataoka, T., Yodo, M., Cody, W L., He, J.X., Humblet, C., Lu, G.H., Lunney, E., Major, T.C., Panek, R.L., Schelkun, P., Skeean, R., and Marshall, G.R. *J. Med. Chem.* **36**, 1902–1913 (1993).
[43] Kazmierski, W., and Hruby, V.J., *Tetrahedron* **44**, 697–710 (1988).
[44] Hanessian, S., McNaughton-Smith, G., Lombart, H.-G., and Lubell W.D. *Tetrahedron* **53**, 12789–12854 (1997).
[45] Kubinyi, H. *Pharmazie* **50**, 647–662 (1995).
[46] Gubernator, K., and Bohm, H.-J. *Structure-Based Drug Design*. Wiley-VCH: Weinheim, 1998.
[47] Walters, W.P., Stahl, M.T., and Murcko, M.A. *Drug Discovery Today* **3**, 160–178 (1998).

2 Series Design in Synthetic Chemistry

Sergio Clementi, Gabriele Cruciani, Manuel Pastor and Torbjörn Lundstedt

Abbreviations

SAR	Structure–activity relationships
QSAR	Quantitative structure–activity relationships
PLS	Partial least squares
OLS	Ordinary least squares
MLR	Multiple linear regression
COST	Change one separate feature at a time
OVAT	One variable at a time
PCA	Principal components analysis
LV	Latent variable
PP	Principal property
FD	Factorial design
FFD	Fractional factorial design
CD	Composite design
DPP	Disjoint principal properties
CombC	Combinatorial chemistry
HTS	High-throughput screening
BB	Building block

Symbols

X	Data matrix
T	Score matrix
P	Loading matrix
E	Error matrix
D	D criterion
M	M criterion
n	Number of objects (rows) in X
p	Number of parameters (columns) in X
D_{ef}	D-efficiency criterion

2.1 Introduction

Most studies in medicinal chemistry, aimed at developing new or better drugs, involve synthesizing series of structurally related molecules. Faced with a table representing the structures and the biological activities, the scientist is always challenged to discover the hidden relationship between them. Strategies aimed to find such relationship rank between a wise observation, using only the chemical expertise of the scientists (structure–activity relationship, SAR) to the most sophisticated quantitative structure–activity relationships (QSAR) methodologies [1]. Ultimately, the objective is to obtain a formal model that allows the prediction of the activity of new compounds and leads to the discovery of new, more potent or more selective compounds. Even more, such a model would provide hints about the physico-chemical mechanism of the drug interactions that can in turn provide some insight into the mechanism involved in the activity.

Quite often, however, QSAR modeling fails to achieve these goals. In fact, two main points should be taken into account in order to derive sound models that allow reliable predictions of the activity of new structures. The first is the use of appropriate regression techniques, such as validated Partial Least Squares (PLS) models [2, 3], instead of the more widely used Ordinary Least Squares (OLS, also called Multiple Linear Regression, MLR) algorithms. The second, and probably the most important point, is the selection of an appropriate set of structures by means of design criteria, so that the database used is well-balanced and therefore provides very reliable models [4–6].

The importance of statistical design has not yet been fully recognized, although the concept of chemical diversity is becoming increasingly popular in designing new drugs. In order to find relationships between the structure and the activity, the scientist must start from a series in which the structure of the compounds shows relevant and ordered variation. Quite often, chemometricians were asked to find QSAR models from series of compounds obtained with no design criteria in mind. New structures are usually derived by changing one substituent at a time for each substitution site. Sets of molecules obtained in this way do not contain enough information for ranking the importance of individual features in affecting biological activity and for providing stable models to be used in predictions. Even worse, the size of these series is usually very large and required important investments in terms of money and resources.

The use of statistical design strategies to select test series in QSAR studies is particularly valuable, as it permits to rationalize both the advancement of knowledge and the investment of resources. In fact, by this approach it is possible to select only a few molecules to be synthesized and tested, but in such a way that they contain the widest information, and therefore they permit to derive reliable QSAR models while saving a lot of the resources presently required for a single study.

Although most of the design strategies have been reported in recent works [4–8], we wish to review some of the available techniques developed within the research teams in chemometrics [9–15]. The objective of this work is to help synthetic chemists to understand that a few molecules bearing simultaneous, well-balanced, structural variations contain much more information than many molecules where each variation is taken into account individually. Accordingly, although we are aware that it is easier synthesizing several molecules by similar laboratory procedures than few molecules involving individual synthetic pathways, we wish to point out the inefficiency of the widely used method and to provide design strategies oriented to avoid too much synthetic work.

2.2 The Design Approach

Optimization strategies in organic chemistry can be applied, besides process optimization, also to product optimization, where the goal is to find out the structural features that optimize the desired response in terms of biological properties.

In all such situations, optimization means keeping under control a number of factors that would possibly affect the response in order to find out the best combination of them, i.e. the values that each factor should simultaneously assume to produce the optimal response. Usually, the examination of the effect of each factor on the response is made in a very inefficient way, studying *O*ne *V*ariable *A*t a *T*ime (called OVAT, or also *C*hange *O*ne *S*eparate feature at a *T*ime, COST). The inefficiency of this approach is shown in Fig. 1: OVAT requires a large number of experiments to be performed, but the maximum response found depends exclusively upon the starting value examined, and the way in which the data are collected cannot provide any information whatsoever about the pair of factors that would have given the real optimum.

On the contrary, using the strategy suggested by statistical experimental design [16] permits us to gain information on how to reach the co-ordinates of the optimum, with the lowest possible number of experiments. The merit of this approach is twofold: not only does it permit us to save a lot of resources by dramatically diminishing the number of required experi-

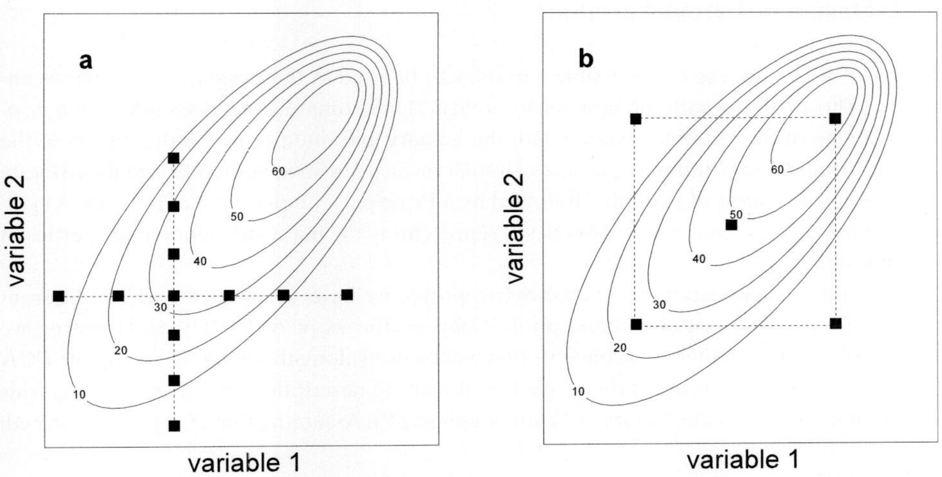

Figure 1. The experiments carried out according to an OVAT (COST) approach (a) fail to explore the response surface. Even if 12 experiments were performed, no one reaches a value higher than 40. A central factorial approach, using only 5 experiments covers a much wider region of the experimental space (b). With appropriate analysis tools, it would be possible to discover the approximate form of the surface.

ments, but it also furnishes us with greater information on the process, which in turn allows us to identify the combination of factors producing the optimal result.

When the relationship involves many factors, the design requires computer-assisted strategies able to span the operational space in the best possible way. In the particular case of the design of series of compounds, the operational space is discrete and contains all the compounds that potentially can be included in the series, described by the same variables used to build the relationships with the activity. Indeed, the design strategy should be applied in this space, in order to extract the most appropriate subset for the analysis. Different methods of design have been described, depending on the kind of variables used to describe the compounds and the criteria defining which is the best subset.

2.2.1 The Multivariate Approach

The preliminary idea of multivariate design in latent variables was developed by Wold and co-workers some 10 years ago [17]. In fact, experimental design provides a strategy for selecting the few most informative molecular structures in a series of homologs: it is possible to apply design criteria also with discrete systems, provided that these are multivariately characterized by a representative number of available data. Consequently, the chemometric strategy for QSAR problems relies on three major steps: (i) multivariate characterization of the discrete systems, giving the latent variables; (ii) a design in these latent variables; and (iii) modeling by PLS. Since the process is multivariate through all three steps, Wold et al. named the approach MV^3 [17].

2.2.2 Design in Latent Variables

If asked, "Which are the best possible variables to be used in the design?", the obvious answer is, "Those related with the biological activity". Unfortunately one does not have *a priori* knowledge of the variables related with the activity and often relies on the results of the analysis to validate the initial hypotheses. In such cases, a sensitive approach is to incorporate in the design an excess of variables, followed by a Principal Component Analysis (PCA) [18] which extracts a few Latent Variables (LVs) representing the main independent properties of the compounds.

PCA and its application to chemometric problems have been reviewed by different authors [18, 19], and a detailed description is beyond the scope of this article. However, we can introduce some of the basic concepts that will be used along this work. Basically, the PCA provides an approximations of the original multivariate description (the X matrix) in terms of two small matrices called scores (T) and loadings (P). Assuming that X is mean centered:

$$X = T \cdot P' + E$$

where E contains the part of the data not explained by the model.

The score matrix provides a simplified representation of the compounds in terms of a few variables (called Principal Components, Latent Variables or Principal Properties, depending on the context) that summarize the information in the original *X* matrix. The LVs have many interesting mathematical properties, but from the point of view of the design, the most important property is that LVs are orthogonal. This means that the properties described by each LV are completely independent of the properties represented by the other, or in other words, there is no correlation between them.

The second matrix obtained in the PCA is the loading matrix. This matrix describes the participation of the original *X* variables in the LVs and it is important to interpret their meaning. In many cases, the physico-chemical meaning of each LV can be recognized and they can be used as variables to summarize the information hidden in a large set of descriptors in a few (usually two or three) meaningful orthogonal variables. In this context the LVs are usually named Principal Properties (PPs).

PPs have been applied for describing amino acids in peptides [9, 17], aromatic substituents in general organic series [20–22] and heteroaromatic systems [23]. When the design affects many chemical elements simultaneously, i.e. amino acids in a peptide sequence, or substituents in a polysubstituted organic skeleton, each element can be described by a block of PPs. The blocks of PPs are then collected in a descriptor matrix. This is the matrix that will be analyzed by PLS in the last step in order to find out its relationship with the y-vector, or the Y-matrix, measuring the biological activity.

With respect to the second step, as the PPs are orthogonal to each other they are particularly suitable as design variables and the simplest design strategies, like the Factorial designs, can be applied thereafter on using PPs.

PPs can be used for the description of the compounds under study in two different ways: on one hand it is possible to use tabulated PP values from the literature [9, 21], describing substituents or fragments of the compounds under study. On the other hand, it is possible to run *ad hoc* PCA on a series of descriptors of candidate compounds. Obviously, the second alternative offers advantages in terms of giving a better description of the system under study, especially when the compounds are peculiar or there is some previous knowledge of which are the most relevant physico-chemical parameters. However, the first alternative is still a valid option for the design of the first series or when no PCA tools are available.

2.2.3 Factorial Designs

Factorial designs (FDs), fractional factorial designs (FFDs) and Box–Wilson or composite designs (CDs) are simple, straightforward, and therefore good in helping to understand the concept of design: spanning the variable space. The application of FDs and FFDs in continuous experimental spaces is straightforward, and its application in process optimization has been described in classical textbooks [16]. Basically, the FD consists of defining a number of levels (usually two or three) for each variable or PP involved and then to carry out an experiment for each possible combination of the levels defined. Two-level factorial designs are usually called 2^n, n being the number of variables involved. Complete factorial experience plans usually have the inconvenience of producing very large series, but fortunately this inconvenience is addressed by the use of fractional factorial plans, where only a fraction of the design (1/2, 1/4, etc. …) is selected.

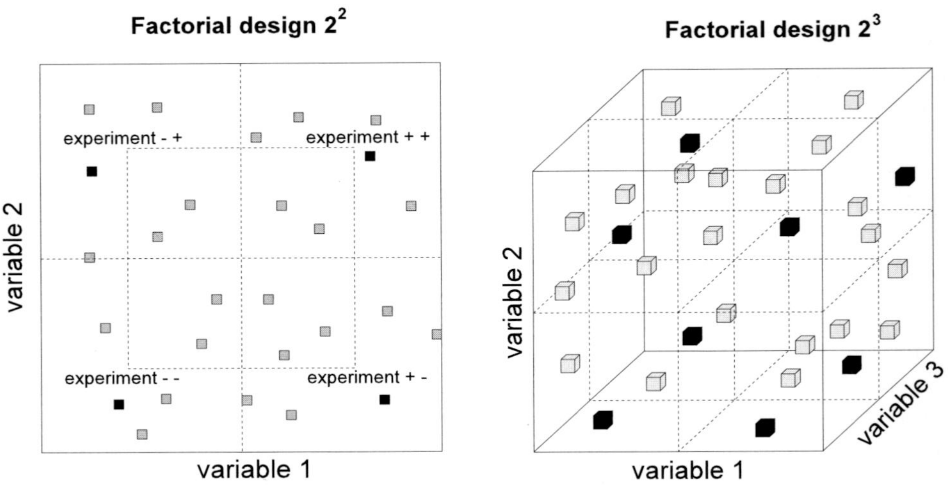

Figure 2. Examples of factorial designs using two (2^2) and three (2^3) variables. When the experimental space is discrete, the experiments can be chosen taking one object (substituent or compound) from each quadrant or octant. The objects chosen are represented as black squares.

In the problems under study, however, the space is discontinuous and often the experiments suggested by these methods do not fall on top of real compounds present in the parametric experimental space. Therefore, the design methodology should include some criteria or methods to extract from this PPs space a compound representing each one of the ideal experiments suggested.

Perhaps the simplest alternative can be to divide the space into quadrants or octants (Fig. 2), and then selecting one substituent representing each subspace (quadrant or octant), as labeled by the pair or triplet of signs corresponding to that subspace [17]. In a previous study [21], the substituents suggested as representative of each octant of the resulting three-dimensional PP space for organic substituents were NO_2, Br, Me, CO_2Et, COPh, OBu, Ph and OH. Other, more sophisticated methods of assigning substituents to the experiments suggested by the FDs have been suggested by different authors [13, 24–26], like the Nearest Neighbor Analysis, Cluster analysis or Euclidean Distances.

On taking into account polysubstitution we can always consider pairs or triplets of variables as representing a single substitution site [15]. For example, Fig. 3 represents a trisubstituted skeleton with substituents described by two PPs, as a six variables problem. The design matrix generated by a 2^6 FD would contain 64 compounds and is more convenient to use a saturated FFD 2^{6-3}, which contains only eight compounds.

However, within the FD strategy, the molecules included in the series should bear as many substituents as many sites one wishes to control. Therefore the FDs approach might not be easy to apply from the point of view of the synthetic feasibility of the compounds, simply because the compounds proposed contain too many substituents at the same time.

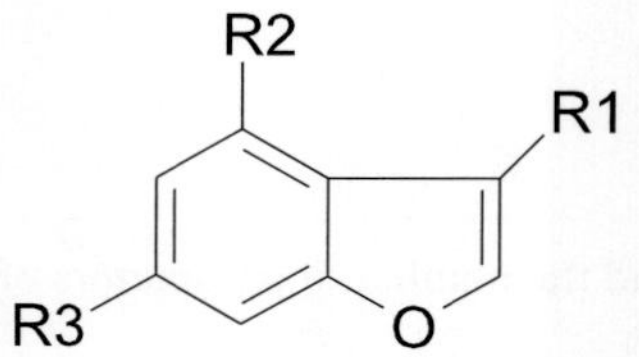

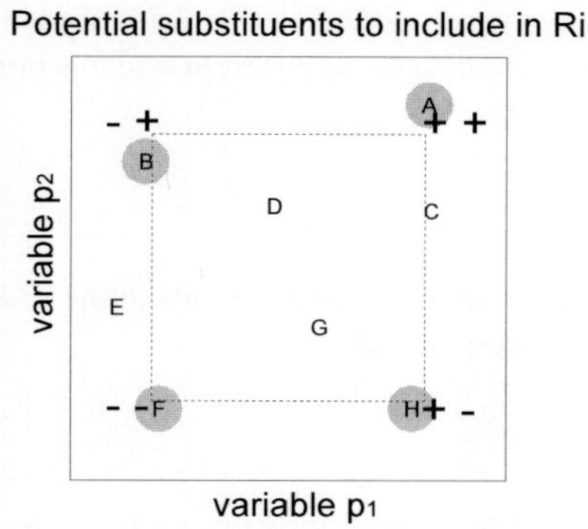

R1 p1	R1 p2	R2 p1	R2 p2	R3 p1	R3 p2
+	+	-	+	-	-
-	-	+	+	-	-
-	+	-	-	+	-
+	-	+	-	+	-
+	-	-	-	-	+
-	+	+	-	-	+
-	-	-	+	+	+
+	+	+	+	+	+

compound	R1	R2	R3
1	A	B	F
2	F	A	F
3	B	F	H
4	H	H	H
5	H	F	B
6	B	H	B
7	F	B	A
8	A	A	A

Figure 3. The substituents R1, R2 and R3 can be selected using a fractional factorial design (FFD). Firsts, the eight substituents candidates were plotted on a 2D space of PPs (p_1 and p_2) and four representatives (A, ++; B, −+; F, − −; H +−) were extracted. Then, these were combined according to the FFD scheme to obtain a series of eight compounds.

2.2.4 D-optimal Designs

FD, FFD and other related methods have in common that they suggest *ideal* experiments that, in a later stage, have to be associated with *real* molecules. A complete different approach starts from a set of *real* molecules, candidates to be included into the series. Provided that we can define an objective quality criterion for the series, it is possible to extract from the universe of candidates a few compounds yielding the best results, and therefore called "optimal" series. As a criterion of quality it has been proposed the value of

$$D = |(X'X)^{-1}|$$

usually referred to as "D criterion" [27]. It can be demonstrated that the series that have associated the lowest *D* value (D-optimal series) produce MLR models of optimum quality, in terms of the error associated with the estimations. From a geometric point of view, the members of D-optimal series are characterized by showing a large spread in the parametric space and a low correlation.

The addition of a new compound to the series always increases the value of D and therefore, in order to compare designs of different size it is preferable to use the M criterion or the D-efficiency (D_{ef}):

$$M = \frac{|X'X|}{N^p}$$

where N and p are the number of experiments in the series and the number of parameters of the postulated model, respectively.

$$D_{ef} = 100 \left[\frac{M_i}{M_2} \right]^{1/p}$$

FDs are D-optimal when each substitution site is controlled by a single parameter. In fact, the goodness of the designs can also be evaluated in terms of D_{ef}, which can be defined as the relative number of experiments required to obtain an equally good design with fully orthogonal variables: hence a full factorial design has a D_{ef} of 100%. On using two PPs the goodness of D-optimal and factorial designs is comparable in terms of D_{ef}. However, on increasing the number of PPs for describing each substitution site the efficiency of D-optimal designs becomes far better. For example, let us assume that we need to select the substituents to include in three different sites, defined by two PPs (six variables). The total number of possible molecules, allowing our eight selected substituents for each site, is $8^3 = 512$, of which the FFD 2^{6-3} design described above would select eight. In contrast, the D-optimality ap-

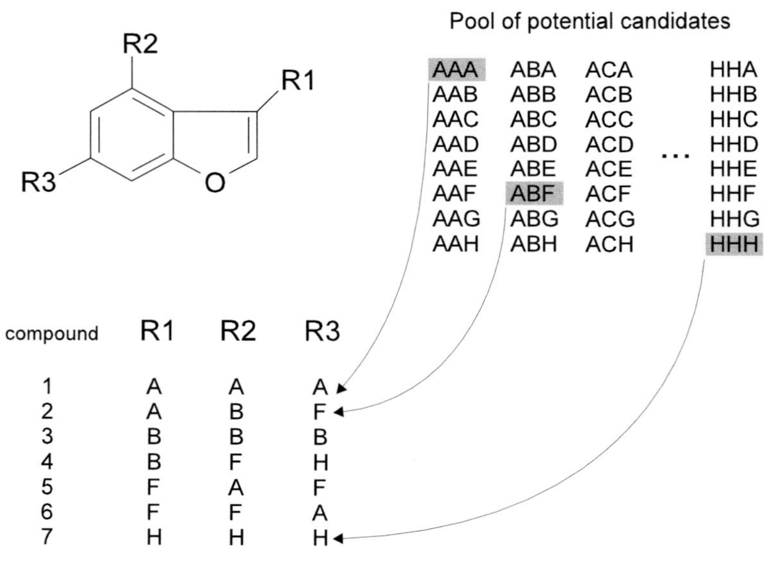

Figure 4. The substituents R1, R2 and R3 can be selected using D-optimal design. All the interesting compounds were collected in a pool of candidates. We can remove from this pool some compounds that are difficult to synthesize, toxic, etc. ... The D-optimal algorithm extracts from the pool the subset that optimizes the D criterion. Seven points are enough to test the six variables used in the design of the series.

proach works in a six-dimensional space with the actual values of the PPs. The D-optimality criterion then provides the selection of a minimum of seven points out of the 512 that meet the requirement of maximizing the quoted determinant, spanning in the best way the domain in the six-dimensional space (Fig. 4).

Independently of the goodness of the designs in terms of any statistical indices, the D-optimal design offers the possibility of incorporating different constraints into the designs. For example, we can remove from the pool of candidates highly polysubstituted compounds that are difficult to synthesize, as mentioned above. In the cited case, one should take into account also hydrogen as a selected substituent, so that the number of candidates become $9^3=729$, but the algorithm can select candidates excluding the 512 three-substituted compounds: the spanned domain will be smaller, but the strategy is preserved and the release of synthetic difficulties is highly appreciable.

Another good property of D-optimal designs is the possibility of selecting a number of molecules higher than the minimum, depending upon the resources that one is prepared to invest in synthesis, still keeping the "maximum span" criterion. However, it was recognized recently [28] that it is always very appropriate to add a central point (or a few points near the center of the domain) in order to explore better the systematic variations within the variable space.

The advantages of using D-optimal designs instead of fractional factorial designs in principal properties can be summarized as follows. It is possible to: (i) reduce the number of required structures; (ii) exclude molecules too difficult to synthesize; (iii) include molecules already available or tested; and (iv) reduce polysubstitution even controlling several sites.

The D-optimality algorithm is implemented in a number of computer programs (MODDE [28], NEMROD [29], EDISFAR [14], DESDOP [30], etc. ...). Among these the program DESDOP was developed in our group and is aimed at determining D-optimal designs in principal properties for QSAR studies [6].

2.2.5 Cluster Designs

As a consequence of their discrete nature, the organic compounds often appear to be clustered in the parametric space. Therefore, in order to span at best this parametric space one might select one or more representatives of each cluster, incorporating in the series all types of structures present in the space.

Clusters can be easily recognized by visual inspection of the plots of parameters or PPs, or can be highlighted by the chemometric methods of cluster analysis [31]. Cluster analysis was one of the first design methodologies used in QSAR studies [32] and has been also used in QSAR studies in the field of environmental chemistry [33], dealing with 66 pollutants categorized into five classes for which the soil sorption coefficient was available.

Since such sets of compounds usually contains several classes of substances it is needed to ensure that the selected training set comprises compounds representing all those chemical classes. Under these circumstances a single multivariate design may be suboptimal because of the risk of ignoring small classes with few members and only selecting training set compounds from the largest classes. The strategy proposed to overcome this problem is a training set selection recognizing clustering, by constructing local multivariate designs within each cluster. The chosen compounds arising from the local designs are finally united in the overall training set, which thus will contain members from all clusters.

In the general case, cluster design is applied to a number of candidate molecules, as in D-optimal design, but on a different criterion of representativeness. However, the same clustering approach can also be used for selecting the representative items of a multivariate characterization of discrete systems.

Clustering methodologies were used in our group when we derived the Principal Properties from molecular interaction fields computed with program GRID [34] for different heteroaromatic systems [23]. In this work, we recognized, by PCA, as many as 10 diverse clusters in a data set containing 40 items, thus implying that at least 10 systems were required to represent the whole space. Interestingly, the PCA results were in agreement with the cluster analysis results, but were able to explain further details with respect to the latter.

2.3 Optimization

Models obtained following the procedure described above can be used to predict the activity of new compounds. Strictly speaking, empirical models are valid only within the ranges of the parameters explored in the training series and can be used only to *interpolate*. Even if the models are used only to interpolate, it is still possible to find compounds with better properties than those already tested. However, it would be difficult to justify the investment on developing QSAR models only to obtain compounds that are not too different from those in the training series! The scientist should be aware that using the model to predict the activity of compounds very different from those in the training series is always risky, and they cannot blame the methodology when inaccurate predictions are made.

In order to check if the predictions are interpolations or extrapolations, the simplest test is to compare the values of the PPs on the predicted compounds with those in the training series, or simply inspect the position of the predicted compounds in a score plot together with the compounds in the training series. Other, more sophisticated methods have been suggested elsewhere [13].

It should be stressed that the problems of the extrapolation are strictly linked to the series design. A properly designed series should span the whole experimental space and the information extracted would be generally applicable. In contrast, a poorly designed series, apart from the problems related with the modeling, will produce results that may be relevant only in a small region of the experimental space. In general, no model is better than the series from which it has been extracted. This is another, important, reason to devote some effort to apply experimental design.

2.3.1 The CARSO Procedure

The final phase of an optimization problem is usually formulated as a response surface study. However, the coefficients of the polynomium describing the surface are usually computed by multiple linear regression, which is possible only if the data have been collected according to

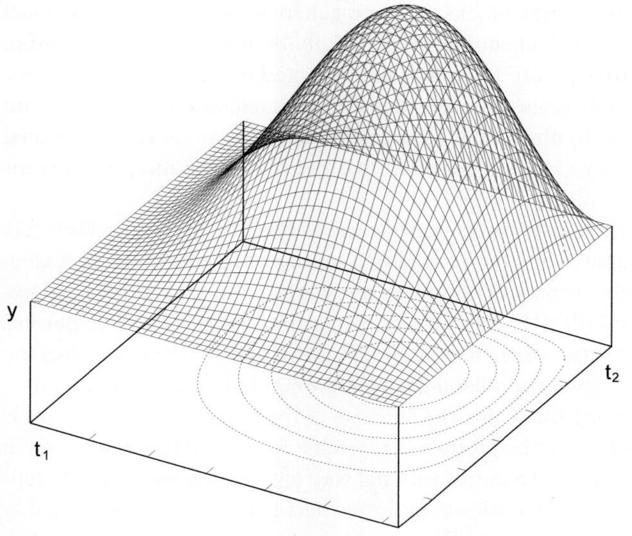

Figure 5. A schematic representation of a curved response surface. The CARSO method has been specifically developed to find the optimal solution in such situations.

a rotatable design. Unfortunately, this is not a common situation in QSAR, where often the series is not properly designed. The CARSO procedure [35] consists of modeling the response by PLS instead of by ordinary multiple regression. The PLS loadings are transformed thereafter into the coefficients of the polynomium, and the response surface is then studied by canonical analysis and by Lagrange analysis. Following these analyses, for the point of the surface with the highest Y value, it is possible to find out analytically the ranges allowed for reaching a certain response level. In our experience the CARSO procedure was found to be the most efficient chemometric tool for solving real QSAR problems (Fig. 5).

2.4 Disjoint Principal Properties

The first developed PP scales [20, 21] described statistically orthogonal (and hence chemically independent) effects. However, some of the PPs contain in a fairly mixed way the supposed "fundamental" effects (steric, electronic, lipophilic and H-bonding) that chemists are used to recognize. Even if this is not a drawback from the point of view of the design, it can be confusing and consequently we suggested [22] the use of Disjoint Principal Properties (DPPs) rather than PPs.

In this work, a set of 40 substituents was described by 86 variables. The PCA of this X matrix required five LVs, and their meaning, defined on the basis of their loadings, are not easily interpretable in terms of fundamental effects. In order to render more acceptable to synthetic chemists these statistical scales we could have rotated the PC score vectors to make them closer to some original descriptor, but in this case the orthogonality between the LVs would be lost. However, since orthogonality will be lost in any case, it seemed more appropriate to divide the original matrix into four blocks, collecting in each block only the descriptors "known" to measure each fundamental effect.

Since DPPs are derived from disjoint blocks of homogeneous variables, the chemical meaning of which is easily understood, chemists will hopefully be more confident in using scales where the effects of substituents are ranked in the expected order. Consequently, we suggested using four disjoint double scales, to represent the steric, electronic, lipophilic and hydrogen bonding capability of each substituent. These eight scales are no longer orthogonal to each other, but this is not strictly required provided that the regression method used for modeling is PLS.

Interestingly, the meaning of each scale is straightforward. In fact, the first steric DPP (s1) measures the size of the substituent, while the second (s2) measures its shape; the first electronic DPP (e1) describes the electronic effect on reactivity, while the second (e2) follows spectral data; the first lipophilic DPP (l1) follows size, while the second (l2) distinguishes between polar and non-polar substituents; finally the first H-bonding DPP (h1) measures the strength of the bond, while the second (h2) distinguishes between H-donors and acceptors.

Because of the co-linearity among the scales, the strategy of fractional factorial designs is clearly impractical with DPPs, although the groupings resemble fairly well those detected in the previous works [20, 21]. In order to develop a rational way of selecting some "really representative" substituents in this eight-dimensional space, we decided to use the D-optimality criterion, labeling each substituent by its signs. This approach furnished us a number of different sets of substituents, among which we preferred the twelve items containing, besides Br, Me, COPh, Ph, H and OH, also CN, $COCH_3$ or COOH, NMe_2 or NHMe, and SO_2Me.

Accordingly, we recommend selecting informative molecules for QSAR studies by using a D-optimal design in DPPs. The molecules are to be selected among those obtained allowing the 12 selected substituents for each substitution site, i.e., 12^n molecules where "n" is the number of sites to be controlled. The final PLS model to be derived thereafter will make use again of the DPPs in order to find out which "fundamental" effect of substituents for each substitution site is relevant to the structure–property relationship.

2.5 Combinatorial Chemistry

Combinatorial chemistry (CombC) techniques offer to the chemist the possibility of obtaining large series of compounds with a reasonable investment of time and resources. Initially, there was the belief that this approach would allow the synthesis of *all* the derivatives of a certain compound, and with the help of appropriate biological tests (high-throughput screening, HTS), it would lead to the direct discovery of the most active derivatives. Now, it is becoming clearer that it is not possible to obtain *all* the derivatives and, even if so, experimental errors and limitations on the HTS can miss the most active compounds. As a consequence, it has been suggested an alternative strategy [36], which, using the techniques of CombC, is aimed at producing highly informative series. These series can be used, together with thorough biological screening, for obtaining *models* of activity leading to obtaining more active compounds.

This strategy is based on a careful design of the combinatorial libraries; the building blocks (BBs) used must be selected in such a way that the information given by the products obtained is optimal. As illustrated in Fig. 6, this selection starts by a design of the BBs on the space of the relevant PPs. Once it has been carried out, the structural spaces of the BBs are

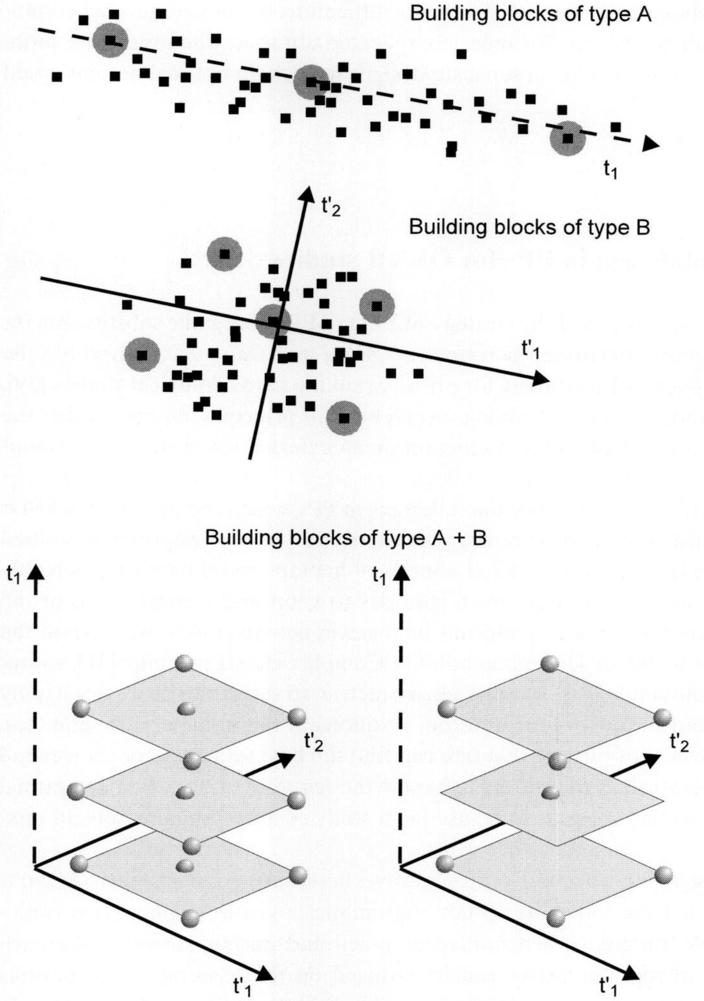

Figure 6. The building blocks of type A are described with one principal component (t_1), while the building blocks of type B are described with two principal components (t'_1 and t'_2). Three BBs of type A and five BBs of type B were selected. These were further combined to obtain 15 (5 × 3) possible compounds. This design can be further reduced, obtaining a reduced design of nine compounds that still contains a maximum amount of information about the investigated structural space.

well covered. By combining all of the selected BBs, the number of possible products in the designed library is given. The selected combinations are a good design to cover the structural space in the final library. However, if this procedure still results in too many combinations to be synthetically, practically or economically feasible, another selection/design must be performed. By using fractional factorial designs the number of compounds can be reduced, as in Fig. 6. In this selection, synthetic feasibility, BBs cost and availability can be taken into consideration, i.e. if one combination is suspected of being difficult to synthesize, another combination can to some extent be chosen. For more complicated libraries, when there are more than two building blocks which results in separate designs in each, D-optimal designs could be a better choice.

2.6 Examples

2.6.1 Applications of design in PPs for QSAR studies

Over the past few years, we have used the strategy of FFD in PPs of organic substituents for studying the toxicity of monosubstituted benzenes on *Algae* and *Daphniae* [38], within the framework of developing a general strategy for priority ranking in toxicological studies [39]. On using only four compounds as the training set we were in principle able to predict the toxicity of 100 compounds, and tested the prediction on an external set of three more compounds.

More recently, we preferred to use D-optimal designs in PPs, which are better than FFDs for handling polysubstitution [6], and therefore particularly suitable for peptides, described by the PPs for amino acids. At first we studied a series of heptapeptides behaving as highly selective NK-2 antagonists [37]: we used seven peptides to keep under control two of the positions in the sequence, obtaining a significant increase in activity. Later, we studied the binding of nonapeptides to Major Histocompatibility Complex class I proteins [41], stimulating the interest of immunologists for this chemometric strategy: again we used only nine peptides to keep under control four different positions in the nonapeptide sequence. Indeed, it is not by the wheel of fortune that one can find the best sequence for the desired response, but by a reliable strategy permitting to handle the required information. D-optimal designs were used for selecting solvents to be used in a study of water–organic solvent mixtures [40].

On using cluster design for selecting representative heteroaromatic systems [23], our recommended set included the following 10 heteroaromatics: pyrrole, tiophen, benzothiophen, pyridine, imidazole, quinoline, benzimidazole, uracil and purine. However, for each case study the number of representatives can be reduced on the basis of some previous knowledge. Accordingly, our second suggestion was to start with tiophen, pyridine, imidazole and quinoline, and then proceed depending on the characteristic of the data set:

a) If there is not a need for multiple interactions, then uracil and purine can be left out.
b) If there is no evidence of volume importance, benzothiophen, benzimidazole and purine can be neglected.
c) If there is no improvement with electron-rich heteroaromatics, pyrrole, indole, imidazole and benzimidazole can be excluded.

The importance of deriving a QSAR model on a well-balanced set of structures, even in the presence of a much larger number of available data for the same series, was shown in two studies regarding the antibacterial and antimycotic activity of benzofused heteroaromatic derivatives [42] as well as in the antibacterial activity of quinolones [43, 44], working in the PC space of a PCA performed *ad hoc* on the whole set of available or candidate structures.

In the first case, the structural description of over 100 thienyl- and furyl-benzimidazoles and benzoxazoles was multivariately characterized to identify three latent variables. A set of 16 informative molecules was derived thereafter on applying a central composite design criterion in these latent variables to all the available structures. The data were analyzed by a linear PLS model, which permitted the optimization of three structural features out of four. The fourth one, the substituent linked to the homocyclic ring of the bicyclic system was finally optimized by the CARSO procedure in terms of the substituents PPs, predicting two new compounds as possible optimal structures. Indeed, later analysis revealed the accuracy of these predictions.

In the second case, we took into account over 400 quinolones reported in the literature. Again, a chemometric approach based on multivariate characterization and design in the resulting latent variables permitted us to select a set of 32 molecules with a well-balanced structural variation on which to derive the QSAR models. Linear PLS modeling allowed ranking of the relative importance of individual structural features, and, by CARSO analysis, a new class of compounds was predicted, the lead of which was tested and shown to be as active as expected. This preliminary lead, after a proper modification, is presently being tested for further development.

It is worth noting that molecules selected according to a D-optimal design can be used both for linear modeling in preliminary screening, just as if selected by a FFD, and for response surface optimization using the CARSO procedure [35]. Usually, series obtained by proper D-optimal design exhibit enough variability in each PPs scale to permit a proper use of quadratic models. However, we would suggest that the QSAR study be performed in two phases: the first one for a preliminary screening, and a second one repeating the procedure with a finer control on substituents belonging to "good" subspaces, for the final response surface optimization.

2.6.2 Example on General Application of Series Design Techniques

In this section we wish to illustrate a typical example of the strategy to be used for selecting a series of molecules by design criteria. Let us consider a common skeleton with three substitution sites, where common organic substituents have to be placed. As suggested in reference [21], three PPs are required to describe in general organic substituents, labeling roughly their size, their electronic properties and their molecular shape, respectively.

In this particular example, we wish to reduce as much as possible the number of molecules to be synthesized and tested in the first series. Therefore, only the two first PPs, describing their steric and electronic properties, will be used in the design. Fig. 7 shows a large set of commonly used organic substituents represented in such 2D space. In this space it is possible to define four quadrants, which can be labeled as ++, +–, –+ and ––, representing the properties indicated in Table 1. According to our previous choice we have selected as the representative items of each quadrant CO_2Et, OBu, NO_2 and OH. In this choice, synthetic accessibility criteria should be applied and, if for example the n-Bu derivative is more accessible than the OBu, the first should be preferred. The values of PP1 and PP2 for these representatives have been also included in Table 1.

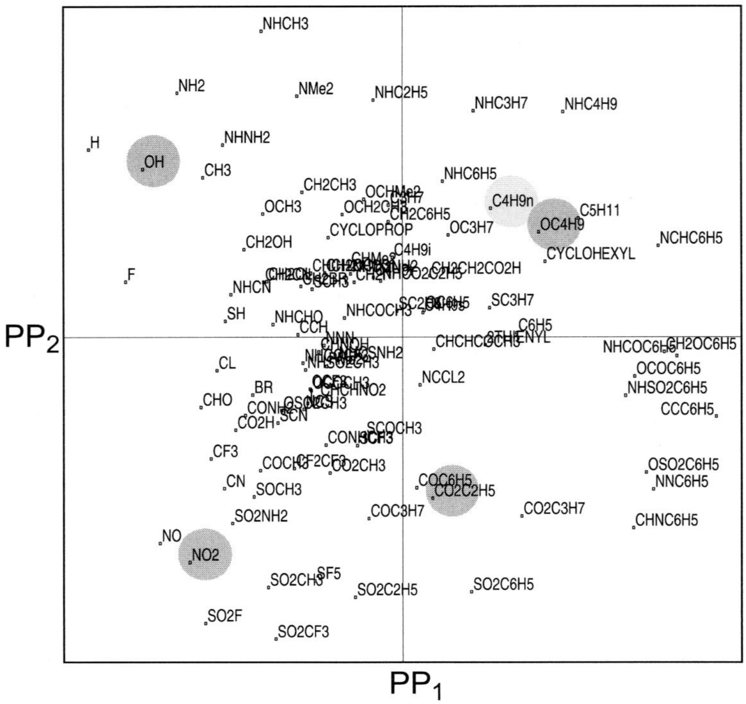

Figure 7. One hundred common organic substituents represented in the space of principal properties. PP_1 is mainly related with the size of the substituents, while PP_2 expresses their electronic properties. The four substituents highlighted with the dark gray circles are possible representatives of the four quadrants of this space. The n-Bu, highlighted in light gray is a possible alternative to OBu when these derivatives are not easily accessible.

Since the skeleton contains three substitution sites and the substituents are described by two PPs, the problem can be formulated as a six-variable design. For handling this problem we can build a FFD 2^{6-3} as the one represented in Table 2. In this matrix of signs, each couple of signs ++, +–, –+ or –– represents an organic substituent in the 2D PPs space, and if we accept the choices shown on Table 1 we obtain the series of eight compounds represented there.

Table 1. Four common organic substituents representative of the two-dimensional experimental space

	Properties		Quadrants			PP values	
	Size	Electronic	PP_1	PP_2	Candidate	PP_1	PP_2
1	Large	Acceptors	+	+	OBu	4.81	–0.63
2	Large	Donors	+	–	CO_2Et	3.67	–2.65
3	Small	Acceptors	–	+	OH	0.58	–0.15
4	Small	Donors	–	–	NO_2	1.08	–3.14

It is noteworthy to mention that all possible combinations of the suggested four substituents sitting in the three sites are 4^3. The FFD combination selected, from the eight possible fractions containing the same information, is the one that includes the line with all positive values.

If one wishes to keep under control a higher number of sites, the number of variables describing the problem increases: for four sites (eight variables) up to seven sites (14 variables) the FFD will require 16 molecules, while the D-optimal design is more convenient (9 to 15 molecules) in terms of a more economic design. Obviously, the same happens if one wishes to describe the substituents at each substitution site using all three PPs usually derived for substituents [21], amino acids [9], etc. ... In such cases three sites means the use of nine variables, five sites means 15 variables, etc. ... In the example reported above, the D-optimal method can be used to find a minimum of seven compounds out of the 64 possible candidates. Table 3 reports two possible solutions, both with D_{ef} values around 72%.

A special case is represented when compounds have equivalent substitution sites, as often happens for disubstituted compounds on a phenyl ring where the second substituent may sit at two equivalent ortho or meta positions. It is clear that under these circumstances the number of possible candidates decreases for taking into account equivalent structures. This problem is handled automatically by appropriate software, as, for instance, DESDOP [30], where equivalent positions are correctly defined in the problem formulation.

Table 2. Fractional factorial design 2^{6-3} used to fill three substitution sites (Rl, R2 and R3) with substituents described by two PPs (PP_1 and PP_2)

	R1		R2		R3		
	PP_1	PP_2	PP_1	PP_2	PP_1	PP_2	Compound[a]
1	–	–	–	+	+	+	NOB
2	+	–	–	–	–	+	CNO
3	–	+	–	–	+	–	ONC
4	+	+	–	+	–	–	BON
5	–	–	+	+	–	–	NBN
6	+	–	+	–	+	–	CCC
7	–	+	+	–	-	+	OCO
8	+	+	+	+	+	+	BBB

[a] N: NO_2, O: OH, B: OBu, C: CO_2Et, according to Table 1.

Table 3. Two possible series of compounds obtained using D-optimal design

	Compound[a]	Compound[a]
1	BBN	OOO
2	ONO	BBB
3	NCN	NOC
4	CNB	NCB
5	OOC	CBO
6	COO	OBN
7	OBB	BNN

[a] N: NO_2, O: OH, B: OBu, C: CO_2Et, according to Table 1.

2.6.3 Combinatorial Chemistry: A Design Example

A synthesis was planned between commercially available primary aliphatic amines and aromatic aldehydes as building blocks: 35 amines were characterized by 11 physico-chemical measured and calculated variables [45] and 44 aromatic aldehydes were characterized by 54 generated variables [46]. A PCA of the amines resulted in a one-component model that explained 72% of the variance in X. Three compounds were selected that covered the experimental space of the primary amines. Investigating the order of compounds from low to high t_1 value shows that the major reason for the separation of the amines is size. The compound with the lowest t_1 value is methylamine (C_1), the highest is dodecylamine (C_{12}), and in between there are amines of various sizes and branching of the carbon chain. Another PCA was performed on the aldehyde data set. This resulted in a two-components model which eplained 88% of the variance in X. A selection of five compounds from these principal components covers the experimental space of the aromatic aldehydes. SIMCA software [47] was used for the PCA, and a eigenvalue larger than two was used as the significant criterion for the extracted components.

All possible combinations between the original amine and aldehyde data sets would result in $35 \times 44 = 1540$ products. By using series design, a subset of eight structures has been selected (Fig. 8). Making all the combinations of the three amines with the five aldehydes would

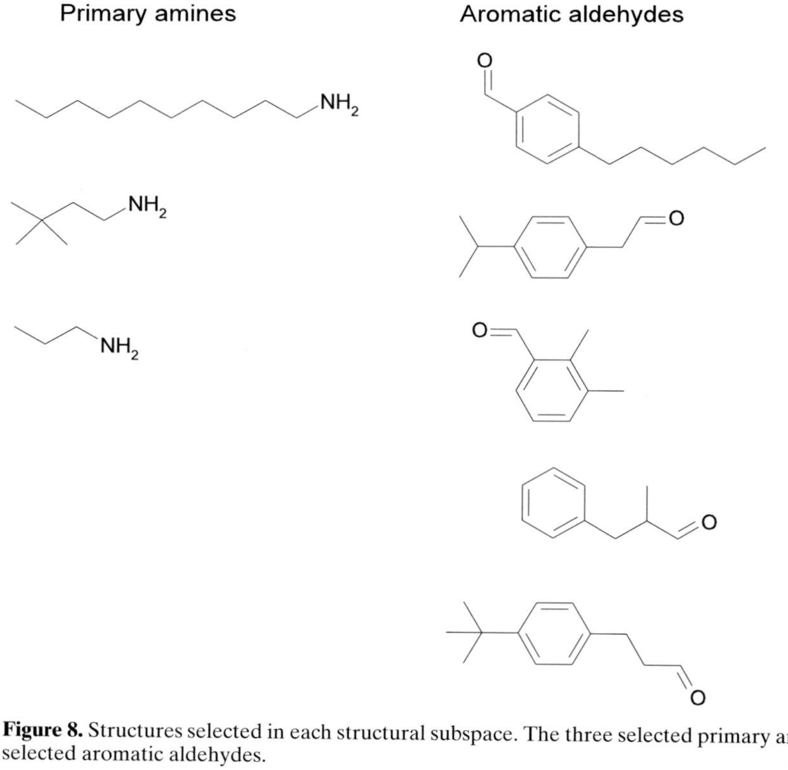

Figure 8. Structures selected in each structural subspace. The three selected primary amines and the five selected aromatic aldehydes.

result in $3 \times 5 = 15$ products. These 15 compounds are a good design for covering the investigated experimental space with only a small number of experiments. If there is a constraint (e.g. cost or time limit) it is possible to further reduce the design selecting only nine compounds, from this first design, as described above.

The same approach can be used for larger data sets and will be investigated further. It is important that clusters must be identified and separate designs made within each cluster. In this example, primary amines and aromatic aldehydes were used and no apparent groupings were observed within these classes. The first step in the chemical synthesis is to investigate the scope and limitations of the synthetic reactions, as different clusters of compounds, or even compounds within a cluster, may require different experimental settings. Even with automated solid-state syntheses using polymer beads as carriers of compounds, the synthetic steps need to be optimized so that reasonable yields are obtained for all compounds in the library. Otherwise unbalanced test data will result, with subsequent loss of information. Such optimization is easier accomplished with few compounds. Recently, robots that optimize organic syntheses on the basis of statistical modeling and experimental design have been developed. For a moderate number of compounds, say up to 1000, the robots can be programmed to perform all the synthesis, including the optimization of each individual step, in a reasonable time. These possibilities offer a further strong argument for limiting the number of compounds and just to prepare as few as are needed from the information point of view.

2.7 Conclusions

The results of a planned strategy, based on rational criteria are usually better than the results obtained from a random or COST way of working. This statement, on which most of us would agree, is also true with respect to the initial selection of the compounds used for the search of sound SARs.

In this article we have described practical methodologies for the design of series. They are based on the use of a few orthogonal PPs or DPPs to represent the information present in many correlated physico-chemical descriptors. With the use of these variables, both the applicability and the interpretation of the models is largely simplified. The criteria on which the series design are based are simple to understand and, to a large extent related with common sense. Simplest design methodologies, like FFD, are derived from methodologies used in process optimization, and only require some small adaptation. Other, more sophisticated methodologies, like D-optimal allow constraints to be added to the series obtained in terms of not including compounds that are too highly substituted, including previously synthesized compounds, and "repair" unfinished designs. Nowadays, with the help of specialized software the application of these techniques is simple and straightforward. Even so, series design is not a purely statistical subject. Chemical knowledge and expertise must be introduced at the design stage, and so we have mentioned the importance of recognizing clusters of compounds and incorporate compounds representing each different cluster.

In the field of combinatorial chemistry, series design techniques can also help to improve the information/cost ratio of the libraries and to work in a more rational way. Indeed, the joint application of QSAR and CombC techniques and methodologies can lead to new promising approaches.

The number of studies successfully carried out and illustrated in this work and in the cited references will hopefully help organic chemists to believe that these strategies can really help them professionally in their everyday work. Although it is hard to believe, optimization procedures do not yet constitute a significant aspect of the professional background of experimental scientists such as chemists, biologists, geologists and even engineers. We hope that things will shortly change, especially if academic lecturers will begin to think about optimization and will start teaching these topics.

Acknowledgments

The authors thank the Italian National Research Council (CNR) and MURST for sponsoring this research.

References

[1] Kubinyi, H. (ed.). *3D QSAR in Drug Design: Theory and Applications*. ESCOM: Leiden, 1993.
[2] Wold, S. *Quant. Struct.-Act. Relat.* **10**, 191–193 (1991).
[3] Baroni, M., Costantino, G., Cruciani, G., Riganelli, D., Valigi, R., and Clementi, S. *Quant. Struct.-Act. Relat.* **12**, 9–20 (1993).
[4] Cruciani, G., Clementi, S., and Baroni, M. Variable Selection in PLS Analysis. In: *3D QSAR in Drug Design: Theory and Applications*. ESCOM: Leiden, 1993.
[5] Sjöström, M., and Eriksson, L.,Application of Statistical Experimental Design and PLS Modeling in QSAR. In: *Chemometric Methods in Molecular Design*. VCH: Weinheim, 1995.
[6] Baroni, M., Clementi, S., Cruciani, G., Kettaneh-Wold, N., and Wold, S. *Quant. Struct.-Act. Relat.* **12**, 225–231 (1993).
[7] Wold, S., Berntsson, P., Eriksson, L., Geladi, P., Hellberg, S., Johansson, E., Jonsson, J., Kettaneh-Wold, N., Lindgren, F., Rånnar, S., Sandberg, M., and Sjöström, M. *Pharmacochem. Lib.* **16**, 15–24 (1991).
[8] Clementi, S., Cruciani, G., Baroni, M., and Skagerberg, B. *Pharmacochem. Lib.* **16**, 217–222 (1991).
[9] Hellberg, S., Sjöström, M., Skagerberg, B., and Wold, S. *J. Med. Chem.* **30**, 1127–1135 (1987).
[10] Carlson, R. *Design and Optimization in Organic Synthesis*. Elsevier: Amsterdam, 1992.
[11] Clementi, S., Cruciani, G., Riganelli D., and Valigi, R. Modelling and Chemometrics in Medicinal Chemistry. In: *New Perspectives in Drug Design*. Academic Press: London, 1994.
[12] Clementi, S., and Wold, S. How to Choose The Proper Statistical Method. In: *Chemometric Methods in Molecular Design*. VCH: Weinheim, 1995.
[13] Pastor, M., and Alvarez-Builla, J. *Quant. Struct.-Act. Relat.* **10**, 350–358 (1991).
[14] Pastor, M., and Alvarez-Builla, J. *J. Chem. Inf. Comp. Sci.* **34**, 570–575 (1994).
[15] Pastor, M., and Alvarez-Builla, J. *Quant. Struct.-Act. Relat.* **14**, 24–30 (1995).
[16] Box, G. E. P., Hunter, W. G., and Hunter, J. S., *Statistics for Experimenters*, J. Wiley & Sons: New York, 1971.
[17] Wold, S., Sjöström, M., Carlson, R., Lundstedt, T., Hellberg, S., Skagerberg, B., Wikström, C., and Ohman, *Anal. Chim. Acta* ***191***, 17–32 (1986).
[18] Wold, S., Esbensen, K., and Geladi, P. *Chemometrics and Intelligent Laboratory Systems* **2**, 37–52 (1987).
[19] Wold, S., Albano, C., Dunn, W.J., Edlund, U., Esbensen, K., Geladi, P., Hellberg, S., Johansson, E., Lindberg, W., and Sjöström, M. Multivariate Data Analysis in Chemistry. In: *Chemometrics*. Reidel: Dordrecht, 1984.
[20] van de Waterbeemd, H., and Testa, B. *Adv. Drug. Res.* **16**, 85–225 (1987).
[21] Skagerberg, B., Bonelli, D., Clementi, S., Cruciani, G., and Ebert, C. *Quant. Struct.-Act. Relat.*, **8**, 32–38 (1989).
[22] van de Waterbeemd, H., Costantino, G., Clementi, S., Cruciani, G., and Valigi, R. Disjoint Principal Properties of Organic Substituents. In: *Chemometric Methods in Molecular Design*. VCH: Weinheim, 1995.

[23] Clementi, S., Cruciani, G., Fifi, P., Riganelli, D., Valigi, R., and Musumarra, G. *Quant. Struct.-Act. Relat.*, **15**, 108–120 (1996).
[24] Austel, V. *Eur. J. Med. Chem.* **17**, 9–16 (1982).
[25] Austel, V, *Eur. J. Med. Chem.* **17**, 339–347 (1982).
[26] Austel, V. *Quant. Struct.-Act. Relat.* **2**, 59–65 (1983).
[27] Fedorov, V.V. *Theory of Optimal Experiments*. Academic Press: New York, 1972.
[28] MODDE, Umetri, Umeå. Sweden (1992).
[29] Phan-Tan-Luu, R. NEMROD, LPRAI, Universite d'Aix-Marseille III, Marseille, France (1981).
[30] DESDOP, Multivariate Infometric Analysis, Perugia, Italy (1993).
[31] Zupan, J. *Clustering of Large Data Sets*. Research Studies Press: Letchworth, 1982.
[32] Hansch, C., and Leo, A. *Substituent Constants for Correlation Analysis in Chemistry and Biology*. Wiley: New York, 1979.
[33] Eriksson, L., Johansson, E., Müller, M., and Wold, S. *Quant. Struct.-Act. Relat.* **16**, 383–390 (1997).
[34] Goodford, P.J. *J. Med. Chem.* **28**, 849–857 (1985).
[35] Clementi, S., Cruciani, G., Curti, G., and Skagerberg, B. *J. Chemometrics* **3**, 499–509 (1989).
[36] Lundstedt, T., Clementi, S., Cruciani, G., Pastor, M., Kettaneh, N. Andersson, P.M., Linusson, A. Sjöström, M., Wold, S., and Norden, B. Intelligent Combinatorial Libraries. In: *Computer-Assisted Lead Finding and Optimization*. Wiley-VCH: Weinheim, 1997.
[37] Clementi, S., Cruciani, G., Riganelli, D., Rovero, P., Pestellini, V., Maggi, C.A., and Baroni, M. *Tetrahedron Comput. Methodol.* **3**, 379–387 (1990).
[38] Tosato, M.L., Cesareo, D., Galassi, S., Viganò, L., Cruciani, G., Clementi, S., and Skagerberg, B. *Chimica Oggi* **6**, 41–45 (1988).
[39] Tosato, M.L., Viganò, L., Skagerberg, B., and Clementi S. *Environ. Sci. Technol.* **25**, 695–702 (1991).
[40] Cipiciani, A., Cruciani, G., and Primieri, S. *Gazz. Chim. Ital.* **120**, 757–763 (1990).
[41] Rovero, P., Riganelli, D., Viganò, S., Pegoraro, S., Revoltella, R., Fruci, D., Greco, G., Butler, R., Clementi, S., and Tanigaki, N., *Mol. Immunol.* **31**, 549–554 (1994).
[42] De Meo, G., Pedini, M., Ricci, A., Bastianini, L., Jacquignon, P.C., Bonelli, D., Clementi, S., and Cruciani, G. *Il Farmaco* **45**, 313–330 (1990).
[43] Bonelli, D., Cecchetti, V., Clementi, S. Cruciani, G., Fravolini, A., and Savino, A. *Quant. Struct.-Act. Relat.* **10**, 333–343 (1991).
[44] Cecchetti, V, Filipponi, E., Fravolini, A., Tabarrini, O., Bonelli, D., Clementi, M., Cruciani, G., and Clementi, S. *J. Med. Chem.* **40**, 1698–1706 (1997).
[45] Andersson, P.M., Linusson, A., Wold, S., Sjöström, M., Lundstedt, T., and Nordén, B. Design of Small Libraries for Lead Exploration. In: *Molecular Diversity in Drug Design*. Kluwer Academic Publishers, in press, 1999.
[46] Wold, S., Sjöström, M., Andersson, P.M., Linusson, A., Edman, M., Lundstedt, T., Nordén, B., Sandberg, M., and Uppgård, L. Multivariate Design and Modelling in QSAR, Combinatorial Chemistry, and Bioinformatics. In: *Molecular Modeling and Prediction of Bioactivity*. Plenum Press, in press, 1999.
[47] SIMCA, Umetri, Umeå. Sweden (1992).

3 Combinatorial Chemistry

Valery V. Antonenko, Reza Mortezaei and Nicolay V. Kulikov

The expanding needs of the biological field in general, recent advances in high-throughput screening, and the efforts of pharmaceutical companies to lower the cost of drug development, in particular, necessitate greater chemical diversity. New pharmacological targets constantly emerge. Traditionally, structural information on compounds that bind to a novel target was gathered from primary screening of collections of chemicals maintained by pharmaceutical companies. These libraries include substances synthesized in-house and acquired from external sources. The cost of maintaining such collections, the long-term stability of the compounds, and the occasional inability quickly to synthesize numerous analogs of an identified hit represent some of the problems associated with the traditional approach.

Combinatorial chemistry can be broadly defined as the simultaneous synthesis of a large number of chemical entities. Several approaches to chemical diversity have been developed. Depending on the specific technique, anywhere from dozens to hundreds of thousands compounds can be generated in a resource-efficient manner. This chapter will not cover the biological approaches to generating molecular diversity; it will focus only on combinatorial methods, which rely on organic synthesis. This subject was extensively reviewed [1–10].

3.1 Needs of the Drug Discovery Process

Different stages of the drug discovery process require different number, quality, and quantities of the synthesized materials. At the so-called lead discovery stage, the primary goal of the process is to identify chemical substances that bind to the desired target. This goal is usually achieved by developing a high-throughput screening (HTS) process to analyze in-house collection of chemicals. This process is normally automated to allow for the screening of a large number of samples. There is no preference to any particular chemical class in this search. Collections owned by large pharmaceutical companies contain anywhere from 100000 to over 500000 compounds. Being chemically diverse, the collections are somewhat biased towards chemical classes that have been historically developed and acquired by the company. Very small amounts of each sample are necessary for the primary screening; however, more and more samples become depleted. The expansion of chemical collections using conventional approaches is expensive. At the end of the lead discovery stage several compounds that bind to the target can be found, and these form the basis for a chemical program that will enable the development of a drug on the basis of the discovered initial leads. The ability quickly to synthesize numerous analogs of the identified leads is essential. However, not all leads and their analogs can be easily synthesized, and combinatorial chemistry offers a solution to most of these problems. Combinatorial chemistry allows for the rapid and inexpensive synthesis of hundreds of thousands of compounds, and large combinatorial libraries can significantly supplement the chemical diversity of the traditional collections, provided that sufficient amounts are produced for the primary screening. Analogs of the leads identified from chemical libraries can be easily synthesized.

After identifying the leads, the drug discovery process shifts to the lead optimization stage, at which synthesis of chemical analogs of the original leads aims at obtaining compounds that have the desired binding, selectivity, solubility, oral availability, metabolic stability, toxicity, duration of action, and other properties of a drug candidate. The synthetic program is focused around several chosen chemical structures. Synthesized compounds are tested in different assays, some of which require large quantities of material. In addition, most of the assays tolerate only very pure compounds. Anywhere from hundreds to a few thousands analogs are produced at the lead optimization stage. Traditionally, chemists have had to rely on sequential exploration of structure–activity relationships (SARs) of the synthesized analogs. Parallel synthesis makes this process more efficient and concise by enabling the simultaneous preparation of hundreds and even thousands of analogs in quantities sufficient for the lead optimization stage.

Combinatorial chemistry has already made a significant impact on drug discovery, with all pharmaceutical companies having established programs in this area. We continue to observe the emergence of numerous small companies that focus on different aspects of the combinatorial technologies. Further development of these technologies catalyzes progress in high-throughput chemical analysis and purification, initiates advances in automation, computation, information technologies, and other relevant areas.

3.2 Split/Pool Method for Creating Combinatorial Libraries

The "split/pool" method for the preparation of combinatorial libraries is outlined in Fig. 1.

In the given example, the library preparation is based on a three-step reaction. Three sets of building blocks, containing three reagents in each set, are used. The sets are $\{A_1, A_2, A_3\}$, $\{B_1, B_2, B_3\}$, and $\{C_1, C_2, C_3\}$. There should be enough resin beads or other solid support elements to synthesize all possible compounds from the given sets. The resin is subdivided into three equal portions, and the portions are placed into three reaction vessels. Each resin portion is reacted with its own building block from the first set. After removing the excess of the reagents and washing, all portions are recombined. The resulting batch contains a mixture of three compounds A_1, A_2, and A_3 attached to the resin. Importantly, each individual bead in the batch has one, and only one, building block attached to it. At the next step, the bead mixture is split again into three equal portions, and building blocks B_1, B_2, and B_3 are added individually to the reactors. As a result, three different compounds are formed in each of the three reactors, i.e. A_1B_1, A_2B_1, A_3B_1, A_1B_2, A_2B_2, A_3B_2, A_1B_3, A_2B_3, and A_3B_3. Each individual bead in all three portions contains only one compound. The resin is recombined again and split into three equal portions. Each portion contains the same mixture of nine different compounds. The addition of the last set of building blocks produces 27 different chemical entities listed in Fig. 1. In general, this strategy produces a x^y number of compounds from y sets of

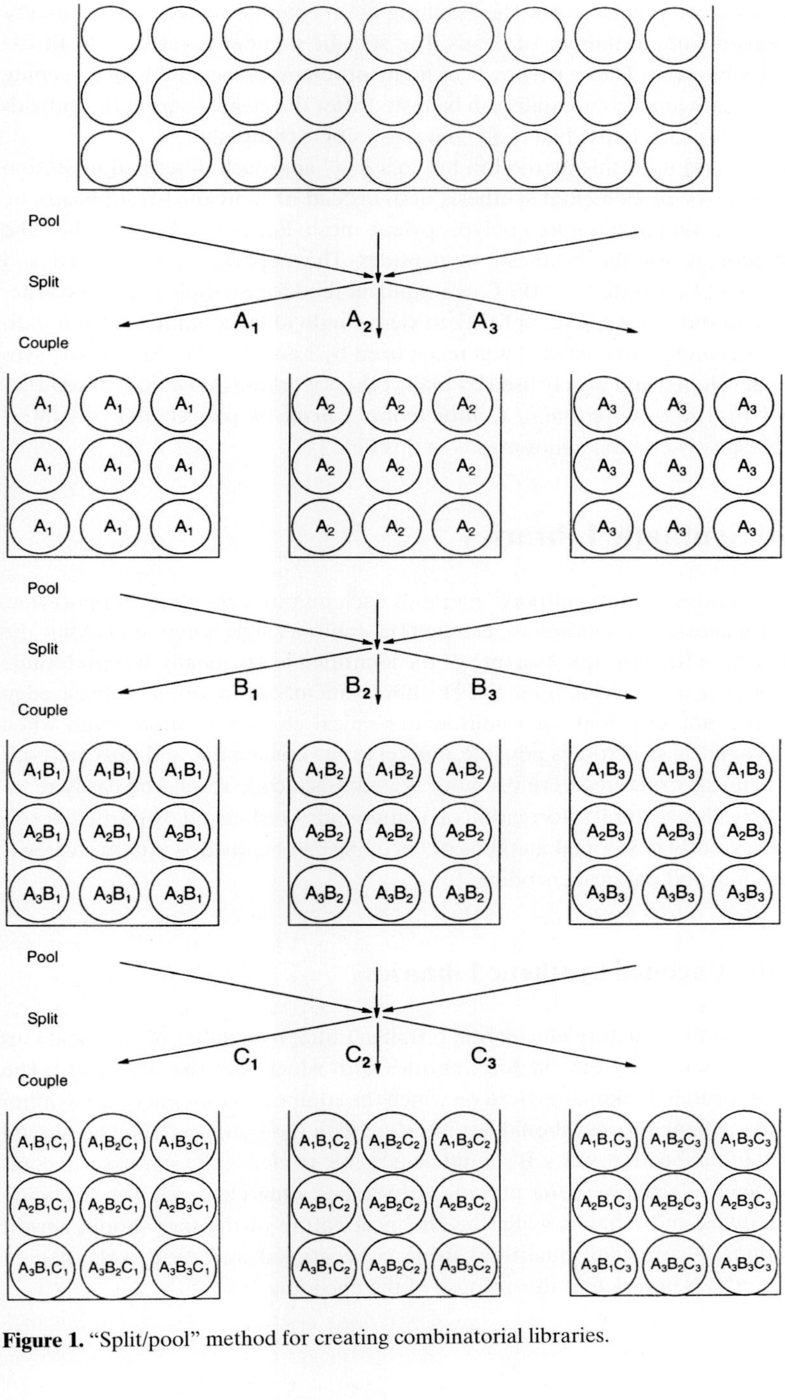

Figure 1. "Split/pool" method for creating combinatorial libraries.

chemicals containing *x* compounds each. The resulting library can be screened as a mixture of compounds cleaved from a number of beads. The specific number is selected to fit the needs of a particular bioassay. There are however many problems associated with screening mixtures. Preferably, individual compounds can be tested after the cleavage from the individually separated beads, as each individual bead contains a single compound.

In 1985, Houghten [11] used this method in his "tea bag" approach described in section 3.4.4.2 of this chapter. As an elemental synthesis unit, instead of individual resin beads, he used small batches of resin packed into a polypropylene mesh. Furka [12, 13] described the "split/pool" methodology for the synthesis of peptides. The peptides were cleaved and analyzed as mixtures and separated by HPLC as a rapid method for multiple peptide synthesis. In his work, Furka did not suggest explicitly to cleave individual compounds from individual beads. This seemingly obvious idea was recognized by Lam [14, 15]. "Split/pool" synthesis was further developed and widely used by many others working in the field of combinatorial chemistry [16]. The new *Journal of Combinatorial Chemistry* published a very interesting historical perspective on the major events in this field [17].

3.3 Encoded Synthetic Libraries

As was previously mentioned, in "split/pool" methods each unit of solid-phase support (i.e., polymeric bead, or macroscopic solid-phase element) contains a single compound. After the screening, it is important to know the structure of the identified leads; ideally, it is preferable to know the structures of all tested compounds. The library members are often synthesized in the amounts that are not sufficient for traditional chemical characterization. Even when elucidation of the chemical structure is possible, it is very time consuming, and thus methods for encoding of chemical structures have been developed. Each code should be easily readable and should carry all structural information on compounds synthesized on particular solid-phase carrier. Previously developed methods of encoding can be divided into two categories: chemical encoding and physical encoding.

3.3.1 Chemically Encoded Synthetic Libraries

One of the solutions to the structure elucidation problem utilizes a number of chemical tags that can unambiguously identify the chemical entities with which they are associated. The tags should be incorporated on the same bead on which the compound they encode is synthesized. Therefore, the tagging process should not interfere with the synthesis. The tags should not consume much of the bead capacity. It should be possible to cleave the synthesized compounds from the bead selectively in the presence of the encoding elements. The decoding process should be quick and reliable, while the chemical nature of the tags should permit their rapid determination in small quantities using conventional analytical technologies. Time-consuming methods may defeat the purpose of the encoding.

3.3.1.1 Oligonucleotide Tags

In 1992, Brenner and Lerner [18] were first to suggest in the literature the concept of chemical encoding, when they proposed a method for producing an oligonucleotide- encoded peptide library. Addition of each amino acid to the polymeric bead is followed by the attachment of two pre-selected oligonucleotides to a different "tag site" on the same bead. Each base pair encodes one, and only one, amino acid. In the "split/pool" strategy, the oligonucleotide chain grows in parallel with the peptide chain; thus, each unique peptide sequence is encoded by a unique oligonucleotide. The decoding process starts by amplifying the encoding oligonucleotide using the polymerase chain reaction (PCR). To reveal the primary structure, the encoding oligonucleotide is sequenced, unambiguously unveiling the amino acid sequence of the satellite peptide.

Oligonucleotide encoding was reported by Nielsen and colleagues [19], who used the construct presented in Fig. 2. α-Amino functionality of the serine residue was used as an attachment point for the growing peptide chain, while the β-hydroxyl of the side chain was used for the encoding oligonucleotide assembly. Therefore, the tag-to-peptide ratio used in this work was 1:1. Synthesis was carried out on controlled porous glass (CPG) beads.

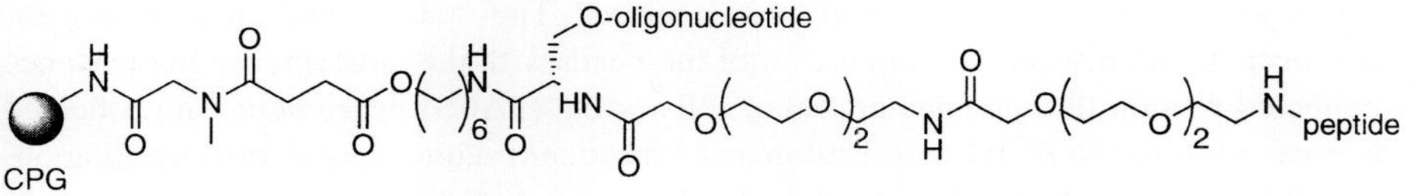

Figure 2. Nielsen construct for oligonucleotide encoding [19].

Scientists from the Affymax Research Institute reported [20] using the oligonucleotide tagging strategy for the encoding of a library containing 823 543 heptapeptides. The synthesis was carried out on 10 μm diameter monodisperse beads. The beads were used in sufficient number to synthesize 200 copies of each sequence in the library. The beads were made of polystyrene crosslinked with divinylbenzene and derivatized with a 1,12-diaminododecane linker. The capacity of the resin was 100 μmol g^{-1}, corresponding to 20 fmol per bead. The resin was further functionalized with two chemical linkers: a mixture of Fmoc-Thr(t-Bu)-OBt and succinimidyl activated ester of 4-O-(dimethoxytrityl)oxobutyrate. The former is introduced to support the growing peptide chain; the latter to support the synthesis of the encoding oligonucleotide. The resulting construct contained a 20:1 ratio of peptides to oligonucleotides on the bead. The structure of the construct is shown in Fig. 3.

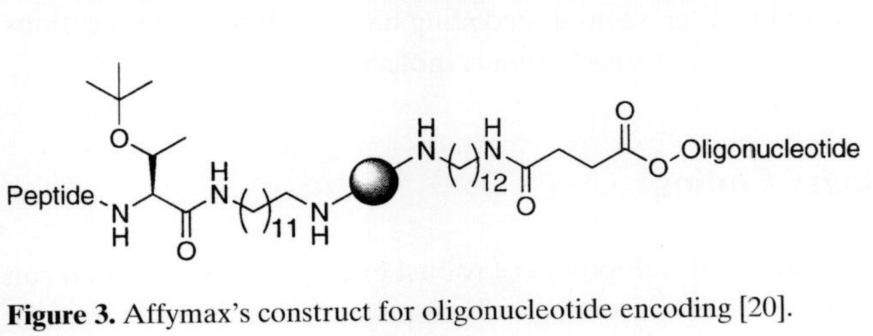

Figure 3. Affymax's construct for oligonucleotide encoding [20].

The peptides were synthesized using Fmoc-protecting groups for α-amino functionalities and *t*-butyl type protection for the side chains of amino acids. Oligonucleotides were assembled using dimethoxytrityl-protected 3′-O-methyl-N, N-diisopropyl phosphorimidates. To protect the encoded oligonucleotides from the depurination side reaction upon trifluoracetic acid (TFA)-mediated side-chain deprotection of *t*-butyl-type protecting groups, the susceptible 2′-deoxyguanosine (dG) unit was not used. 2′-Deoxyadenosine (dA) was substituted with 7-deaza-2′-deoxyadenosine (c7dA), which is stable to TFA treatment. To make the encoding compatible with the peptide synthesis, phosphorimidites carried 3′-O-methyl protection, which is, contrary to the β-cyanomethyl protective group, stable to piperidine used for Fmoc-deprotection.

The library was built to construct 7^7 analogs of the C-terminal heptapeptide RQFKVVT of opioid peptide dynorphin B. Amino acids Arg, Gln, Phe, Lys, Val, Thr and D-Val were used in a seven-step "split/pool" synthesis. Each amino acid was encoded by a unique dinucleotide. Thus, each heptapeptide was synthesized with the corresponding sequence of 14 nucleotide bases. At the end of the synthesis, the encoding sequences were framed by the degenerate DNA PCR primer sequence (55 bases). The library was screened against the fluorescently labeled anti-dynorphin B antibody, D32.39, with the fully deprotected peptides attached to the beads. A sample containing a sufficient number of beads to represent all synthesized sequences with high statistical certainty was analyzed. The beads, containing binding sequences became fluorescent. The top 0.17% of the beads with the most intense fluorescence was collected using a fluorescence-activated cell sorting (FACS) instrument. The collected beads were subjected to PCR amplification and subsequent sequencing of the encoding oligonucleotides. The sequences of the binding peptides were revealed from their primary structures.

Oligonucleotide tagging was developed and used successfully for the synthesis of peptides; many methodological issues, related mostly to the chemical compatibility of the approach, were solved. However, the compatibility of this encoding strategy with the synthesis of other classes of organic molecules remains limited.

3.3.1.2 Peptide Tags

Technology for automated Edman degradation of peptides to determine their sequence is well developed. Therefore, peptides can be used similarly to oligonucleotides as encoding molecules. Several groups reported using peptide encoding [21, 22]. Unfortunately, applications of the peptide encoding are very limited due to severe restrictions imposed on the scope of chemical methods that are compatible with this approach. With the development of alternative encoding strategies described further, peptide encoding has very limited applications in the modern approaches to the synthesis of small organic molecules.

3.3.1.3 Haloaromatic Binary Coding

Several laboratories have worked on the development of robust encoding strategies that can be used for combinatorial synthesis of pharmacologically appealing chemical libraries. In

1993, Still and his colleagues were first to report [23] using chromatographically resolvable haloaromatic reagents as tagging molecules. The tags are attached to the beads during library synthesis in a binary coding strategy to keep a molecular record of all chemical transformations to which the beads were subjected. The tags are incorporated via amide bond formation as part of a construct with a photolabile linker (Fig. 4) at the expense of the ligand synthesis sites. Each compound in the library is encoded by a limited set of tags. The presence, as well as the absence, of each member of the set carries information about the specific encoded structure. In a binary code, a set of n tags can encode $2^n - 1$ different structures. For example, 20 tags can encode $20^{20} - 1 = 1\,048\,575$ different library members. After being released from the beads by photolysis, the tags are detected by capillary gas chromatography using electron capture detection (ECGC), a detection method that is particularly sensitive to heavily chlorinated aromatics. The haloaromatic compounds were selected to ensure reliable and reproducible separation by ECGC, which is capable of detecting subpicomolar amounts of the tags. Consequently, the beads can be tagged at only 0.5–1% of the resin loading (0.5–1 pmol per bead) without detectable interference with the library synthesis.

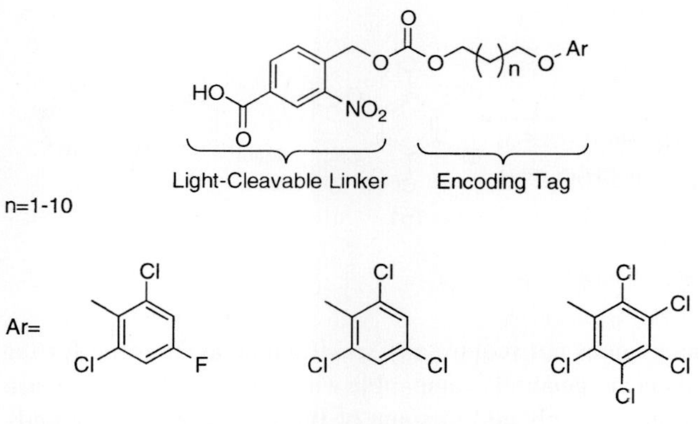

Figure 4. Haloaromatic tags on photolinker [23].

Later, Still and co-workers modified the original strategy by developing a new type of tagging reagent TnC [24], presented in Fig. 5. A derivative of vanillic acid (3-methoxy-4-hydroxybenzoic acid) was chosen as a linker. Synthesis of the tagging reagent TnC begins with a Mitsunobu reaction of a tag alcohol Tn with methyl vanillate, followed by LiOH hydrolysis of the methyl ester. This produces free acid TnA, which is converted to acid chloride TnB. Excess of diazomethane converts acid chloride TnB into the tagging diazoketone TnC. In the presence of rhodium reagents Rh2(OAc)4 or Rh2(O2CCF3)4 TnC forms an acylcarbene, which rapidly and cleanly reacts with benzene, forming a derivative of cycloheptatriene Tn1. Benzene was used as a soluble analog of polystyrene resin. Derivatives of TnC containing different numbers of methylene groups n have been prepared. The researchers found that diazoketones TnC are stable solids and can be stored at room temperature for months. They react easily with polystyrene resin, providing for binary encoding. At the decoding stage, the tags are oxidatively cleaved by ceric ammonium nitrate and analyzed by ECGC.

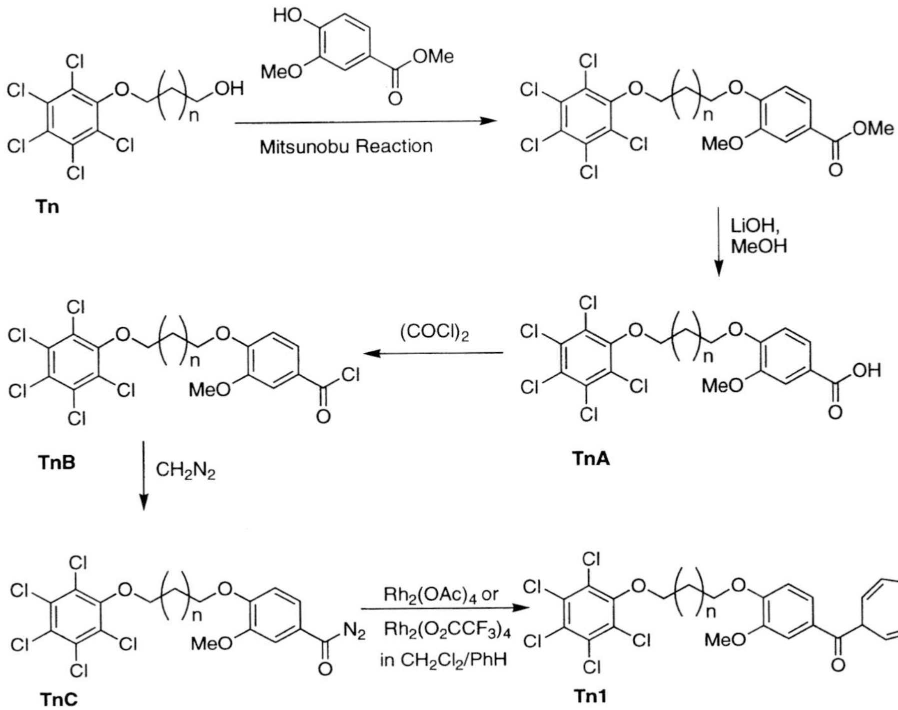

Figure 5. Acylcarbene-generating tagging reagents [24].

The acylcarbene tagging strategy does not require any specific functional group for tag attachment, and the tags and linkers are generally compatible with a wide range of chemical reactions. The acylcarbenes can unselectively add to some of the synthesized compounds. However, the bulk of the library is represented by the polymeric support, which accepts the major portion of the tag molecules. Because tags are added at molar levels corresponding to 1% or less of the library members, interference with the synthesis is minimal [25].

3.3.1.4 Secondary Amine Binary Coding

The Affymax Research Institute developed another robust encoding strategy [26, 27], based on secondary amine tags, which are incorporated into a polyamide backbone. The secondary amine binary coding scheme utilizes an amine-based polymeric resin that is differentially functionalized with sites for both ligand synthesis and tag addition (Fig. 6). The ligand synthesis site is derivatized with a N-Fmoc-protected photocleavable linker group [28]. This linker allows for the release of the ligand from the resin by exposure to ultra-violet (UV) light. The amino group of the tag site is protected with orthogonal to the Fmoc group functionalities (e.g., Boc or Alloc). The tag site occupies only 10% of the total number of amino groups on the resin. Each 130-μm diameter bead of the TentaGel S resin, which is recommended by the authors of this methodology, contains about 300 pmol of amino groups. Therefore, the theoretical yield of the ligand is 270 pmol per bead. Such quantities, being photoreleased in 100 µl of a solvent would provide adequate for the most biological assays concentration of the ligand.

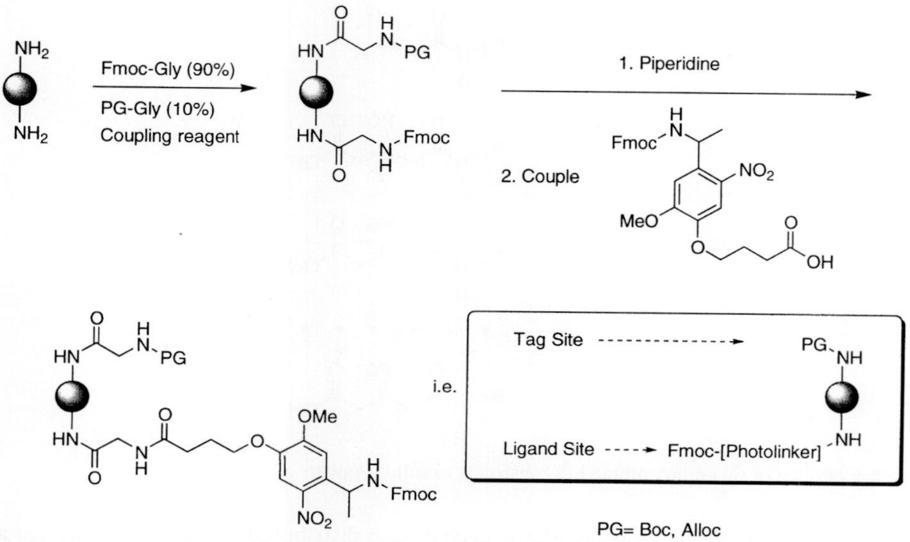

Figure 6. Resin construct for secondary amine binary coding [26, 27].

The tags are a set of relatively hydrophobic amines, such as HN(Et)(Bu), $HNMe(C_6H_{13})$, $HNBu_2$, $HNMe(C_7H_{15})$, $HN[CH_2CH(Et)C_4H_9]_2$. The set is selected to ensure reliable separation of the dansyl derivatives of the amines by reversed-phase HPLC. The tagging monomer units are synthesized by reaction of an N-protected iminodiacetic anhydride with a secondary amine from the set (Fig. 7).

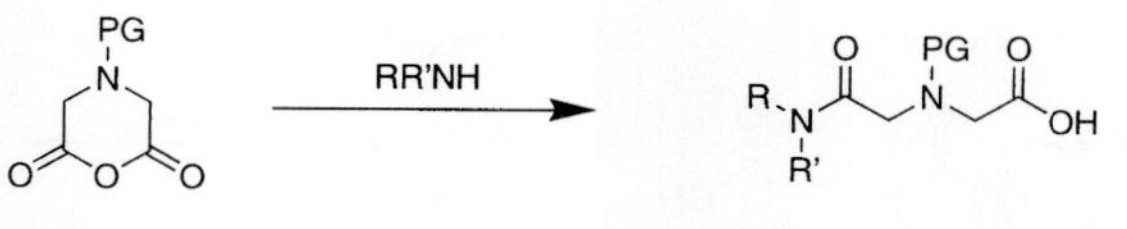

Figure 7. Preparation of tag monomers for secondary amine coding .

The resulting N-protected N-[(dialkylcarbamoyl)methyl]glycines are assembled into binary mixtures, which are incorporated into the tag sites of the resin beads by using HATU or other peptide-coupling reagents (Fig. 8). Addition of each new building block at the ligand synthesis site in the course of a "split/pool" combinatorial synthesis is accompanied by the incorporation of the preselected mixture of the monomer units at the tag addition site. Selection of the protecting groups allows for the addition of the tags either before or after the addition of the building block to the ligand. The ability to choose different protecting groups for the N-protected tag monomers helps resolve potential chemical compatibility issues.

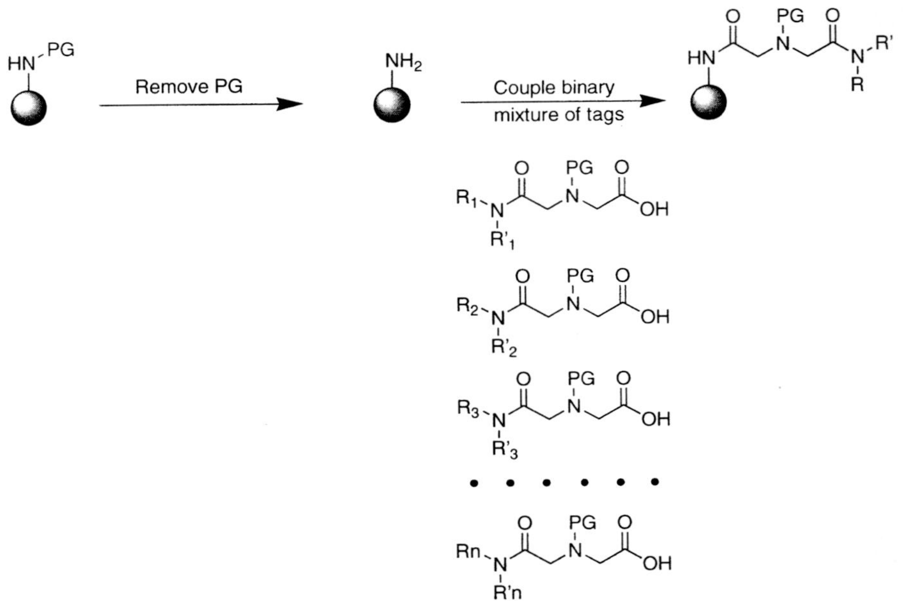

Figure 8. Coupling of secondary amine tag constructs to solid support.

Upon completion of the library synthesis, each bead is distributed into a separate well of a microtiter plate. The ligands are released from the beads into the assay medium by exposure to UV light at 365 nm. The tag residues remain covalently attached to the beads. After the screening, beads from the wells containing active compounds are collected for decoding. The decoding process is shown schematically in Fig. 9, and starts with the acid hydrolysis of the beads in 6 N hydrochloric acid. Under these conditions all amide bonds are hydrolyzed, releasing free secondary amine tags into the solution. After evaporation of the HCl, the amines

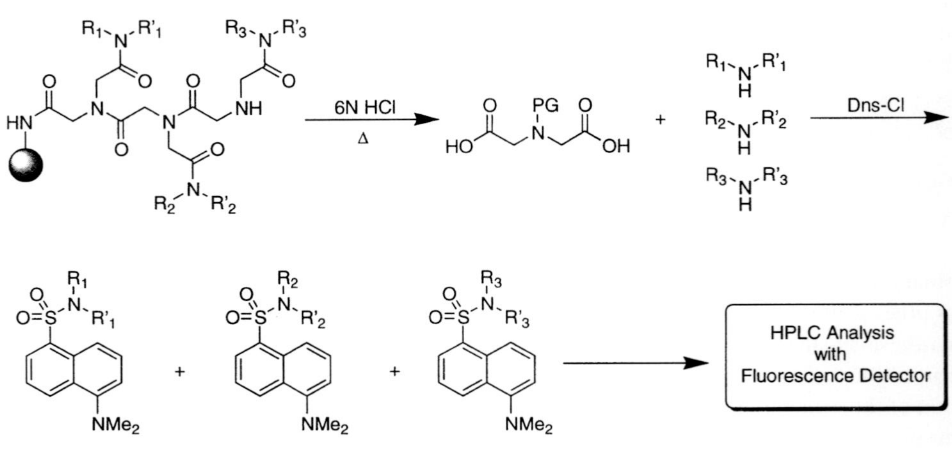

Figure 9. Decoding process for secondary amine binary coding.

are converted into the corresponding dansyl derivatives by treatment with dansyl chloride. Analysis of the obtained mixture of the dansylated tags is carried out by reversed-phase HPLC on a microbore column. Fluorescence detection allows for the reliable analysis of 20–30 fmol of a dansylated tag. Only 2–5% of the entire dansylated hydrolyzate from a single bead is sufficient for obtaining unambiguous results.

Secondary amine encoding method was used in the synthesis of a library of pyrrolidine-based inhibitors of angiotensin converting enzyme (ACE) via 2+3 cycloaddition [29].

3.3.1.5 Mass Encoding

As mentioned previously, conventional methods of characterization of chemical compounds cannot be applied to most combinatorial technologies due to the insufficient quantities of the analytes. One fortunate exception is mass spectroscopy. Matrix-assisted laser desorption and electrospray ionization mass spectroscopy (MS) have been used [30, 31] to identify compounds synthesized on a single bead. For beads with the diameters larger than 100 μm, a mass spectrum can be acquired in less than 1 minute. In 1996, Geysen and his colleagues from Glaxo Wellcome proposed a novel encoding method [32], which takes advantage of the modern MS techniques. In general, recording the chemical history of a compound synthesized on a bead mass encoding incorporates stable isotopes that give distinct isotopic patterns in mass spectra.

In one of the proposed strategies, the resin is derivatized with a linker (Fig. 10) to which an MS code is attached. The mass of the coding block is designed to appear in a convenient region of the mass spectrum. As an example, the code can be a dipeptide built from the combinations of the natural amino acids glycine (Gly) and alanine (Ala), and their ^{13}C-labeled derivatives: [^{13}C]Gly, [^{13}C]$_2$Gly, [^{13}C]Ala, and [^{13}C]$_4$Ala. As an example, the authors present mass spectra recorded in the 295–330 range for the following 10 mass codes: Gly-Gly, Gly-[^{13}C]Gly, [^{13}C]Gly -[^{13}C]Gly,Gly-[^{13}C]$_2$Gly, Gly-Ala, Gly-[^{13}C]Ala, [^{13}C]$_2$Gly-Ala, [^{13}C]$_2$Gly-[^{13}C]Ala, and [^{13}C]$_2$Gly-[^{13}C]$_4$Ala. After the code (Fig. 10), an additional linker is introduced; this should be orthogonal to the first one and is used for the release of the synthesized compound ABC. In a "split/pool" synthesis, the code defines the identity of the first building block A. The third building block C can be known from the final pool in which its addition was carried out. The identity of the building block B can be calculated from the molecular weight of the ligand and the molecular weight of monomers A and C.

Figure 10. Resin construct for mass encoding.

For unambiguous results, the set of building blocks used at the second step of the library synthesis should not contain any compounds with the same mass as building block B.

The code block can also consists of equimolar mixtures built from a set of isotopically labeled dipeptides: Gly-Gly, Gly-[^{13}C]Gly, Gly-[^{13}C]$_2$Gly, [^{13}C]Gly-[^{13}C]Gly, [^{13}C]Gly-[^{13}C]$_2$Gly, [^{13}C]$_2$Gly-[^{13}C]$_2$Gly. These dipeptides incorporate equally to their content in the mixture, because isotopic labeling does not affect chemical reactivity. The resulting mass spectra of these mixture codes serve as "bar codes".

In ratio encoding strategy, the encoded information is derived from just two peaks in the mass spectrum. A mixture of a reagent, common to all members of the library, is prepared from different ratios of the isotopic isomers of the reagent. The resulting mass spectrum will reveal two distinct peaks corresponding to each of the isomers. The relative heights of the peaks carry the encoded information.

The mass encoding strategy was used by Wagner and co-workers [33] for the synthesis of a peptoid library. A set of 20 $^{14}N:^{15}N$ ratio-encoded imidazoles was synthesized "to investigate pharmacokinetic applications of isotopic labeling".

3.3.2 Non-Chemical Encoding Methods

Chemical methods of encoding combinatorial libraries can be used for synthesis on very small elements of solid supports (i.e., 10 μm diameter beads). However, none of the procedures developed so far is totally compatible with all chemical transformations that might be necessary for the implementation of successful drug discovery programs. In principle, physical methods of encoding are fully compatible. Unfortunately, none of the conventionally used labeling techniques (bar codes, labels, etc.) is small enough to be used with resin beads. It is often necessary to obtain compounds via combichem approaches in substantially larger quantities than the single bead approach is capable of. As an alternative, macroscopic pieces of a solid support can be used, the larger physical dimensions allowing for the application of more space-demanding encoding strategies.

3.3.2.1 Radiofrequency Tags

IRORI, Inc. developed an automated solution for the synthesis of encoded chemical libraries. Their AccuTag™-100 system is based on radiofrequency (RF) tags. The RF tag is a small (8 × 1 mm) chip; this consists of a memory with a unique 40-bit alphanumeric code with 240 unique combinations, a circuit with which RF energy is converted into electrical energy, and an antenna that sends and receives RF signals. These signals can be initiated and read by a transceiver, which is connected to a computer. Any information about the structure or information about the sequence of chemical reactions required for the synthesis of a specific compound can be easily associated by the software with the unique signal of each tag [34, 35]. Each RF tag is hermetically enclosed into a glass shell and placed into a microreactor. IRORI offers several miniature reactors – MicroKans® and MicroTubes® (Fig. 11). MicroKan® is a small cylindrical container with mesh walls; the internal volume of the container is 330 μl. It addition to the RF tag, the container holds up to 30 mg of any commercially available resin. MicroTube® reactors are polypropylene or fluoropolymer-based tubes with a polystyrene-grafted surface [36], the RF tag being located within the tube. Microreactors can be combined in one reaction vessel for the addition of a common reaction block, and then redistributed with microreactors from other reactors for further chemical manipulations. By placing a microreactor above the reader, an operator allows the system to recognize the RF code and

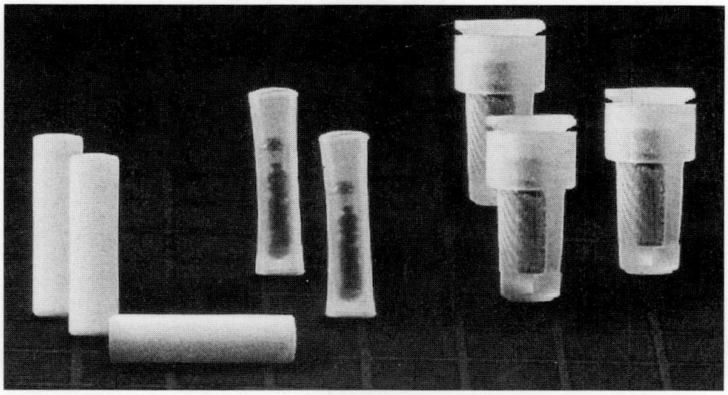

Figure 11. MicroKans® and MicroTubes®.

receives feedback with the physical location of the microreactor in the next chemical reaction. One of the major advantages of this technology is the ability to use conventional glassware that is still more versatile than the contemporary automated synthesizers.

Manual sorting of microreactors can become tedious when a large number (more than 1000–2000) of them is used. IRORI designed the AutoSort™-10K Microreactor Sorting System (Fig. 12) to solve this problem. The system accommodates up to 10 000 microreactors

Figure 12. IRORI AutoSort™-10K Microreactor Sorting System.

that can be distributed between chemical steps into 48 different containers. The distribution rate is 1000 microreactors per hour. The device is also very useful for sorting MicroKans® and MicroTubes® into microreactor carriers for further cleavage, which takes place on the Accu-Cleave-96 cleavage station. Twelve 8 × 12 microreactor carriers can be used simultaneously on the sorting system.

One of the first applications of the RF tagging technology was demonstrated by the synthesis of a 432-member library (18 × 8 × 3 array) of tyrphostins (Fig. 13) [37].

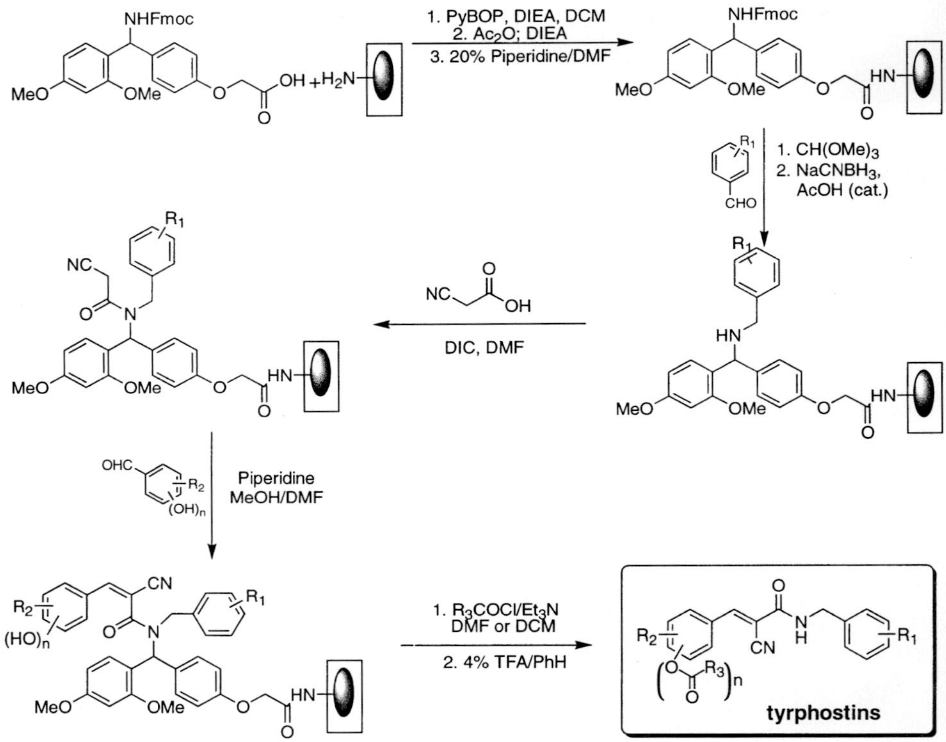

Figure 13. Synthesis of library of tyrphostins [37].

3.3.2.2 Laser Encoding

Another example of encoding that does not interfere with chemistry was suggested in 1997 by Xiao and co-workers [38]. The coding structure is an inert ceramic plate with a two-dimensional, laser-etched bar code. The encoded plate, which is a 3 mm × 3 mm square, is placed in the center of a laser optical synthesis chip (LOSC). The chip is a 1 cm × 1 cm square made of polypropylene grafted with polystyrene. The smallest possible size of the encoding ceramic plate is 0.5 mm × 0.5 mm. Unfortunately, the bar code cannot be modified during the course of the synthesis.

3.4 Parallel Synthesis and Positional Encoding

In contrast with the "split/pool" method, parallel synthesis does not require encoding, produces discrete compounds in measurable quantities, and allows for collection of a full SAR. It is becoming increasingly popular among other combinatorial methods.

Biological assays can be carried in a solid-phase format with the analyzed compounds immobilized on a solid support. Such assays usually require only very small amounts of compounds, though in HTS the cost of biological reagents is a very important determinant. All these factors call for the miniaturization of systems capable of simultaneous synthesis, and subsequent screening of chemical entities in large numbers. Compartmentalization of individual synthesis and screening sites in conventional three-dimensional systems does not allow for production of high-density arrays. These limitations were successfully overcome by synthesizing very dense arrays on laminar solid-phase supports.

3.4.1 Light-Directed Synthesis

In 1991, scientists from Affymax Research Institute combined photolithography, photochemistry, and solid-phase synthesis in a new technology [39–42], called Light-Directed, Spatially Addressable Parallel Synthesis. The principal points of the technology are illustrated in Fig. 14.

The synthesis occurs on a flat glass surface modified with an appropriate linker (e.g., 3-aminopropyl-triethoxysilane) to allow for the covalent attachment of protected amino acids. The entire synthesis area of the slide is derivatized with a photolabile protecting group (PG). At the first step of the synthesis, selected sites of the synthesis area (typically three squares per slide, 1.28 cm × 1.28 cm each) are exposed to UV light through photolithographic mask A. The variety of patterns available for photolithography is essentially unlimited. The exposure to light causes removal of the photolabile groups, thus elaborating amino functionality. At the next step, the synthesis area is treated with the reagents necessary for the elongation of the peptide chain. Only the sites that were previously photodeprotected will participate in a coupling reaction; the rest of the synthesis area remains protected and intact. Synthesis continues by illuminating another part of the surface through photolithographic mask B, followed by the next chain elongation reaction. By repeating the photodeprotection and coupling steps, highly dense arrays, each consisting of thousands of peptides, can be synthesized. Importantly, the primary structure of each peptide in the array is sufficiently defined by the sequence of coupling and photolysis steps, and by photolithography mask patterns. Therefore, the structure can be easily deduced from the (x, y) coordinates of the peptide on the slide. This eliminates the need for encoding–decoding procedures required by some other combinatorial technologies. After the completion of the synthesis, the synthesis area is exposed to reagents necessary for the elimination of side chain protecting groups. To assess the binding properties of all synthesized peptides, the entire array is incubated with a fluorescently labeled target molecule and scanned using a stage-scanning confocal fluorescence microscope [39]. Sites, containing peptides that bind to the target, become fluorescent. Affinity data on all peptides in the entire array are obtained in one step. An example of a fluorescent scan of 1024 undecapeptides synthesized on a 1.28 cm × 1.28 cm square is presented in Fig. 15.

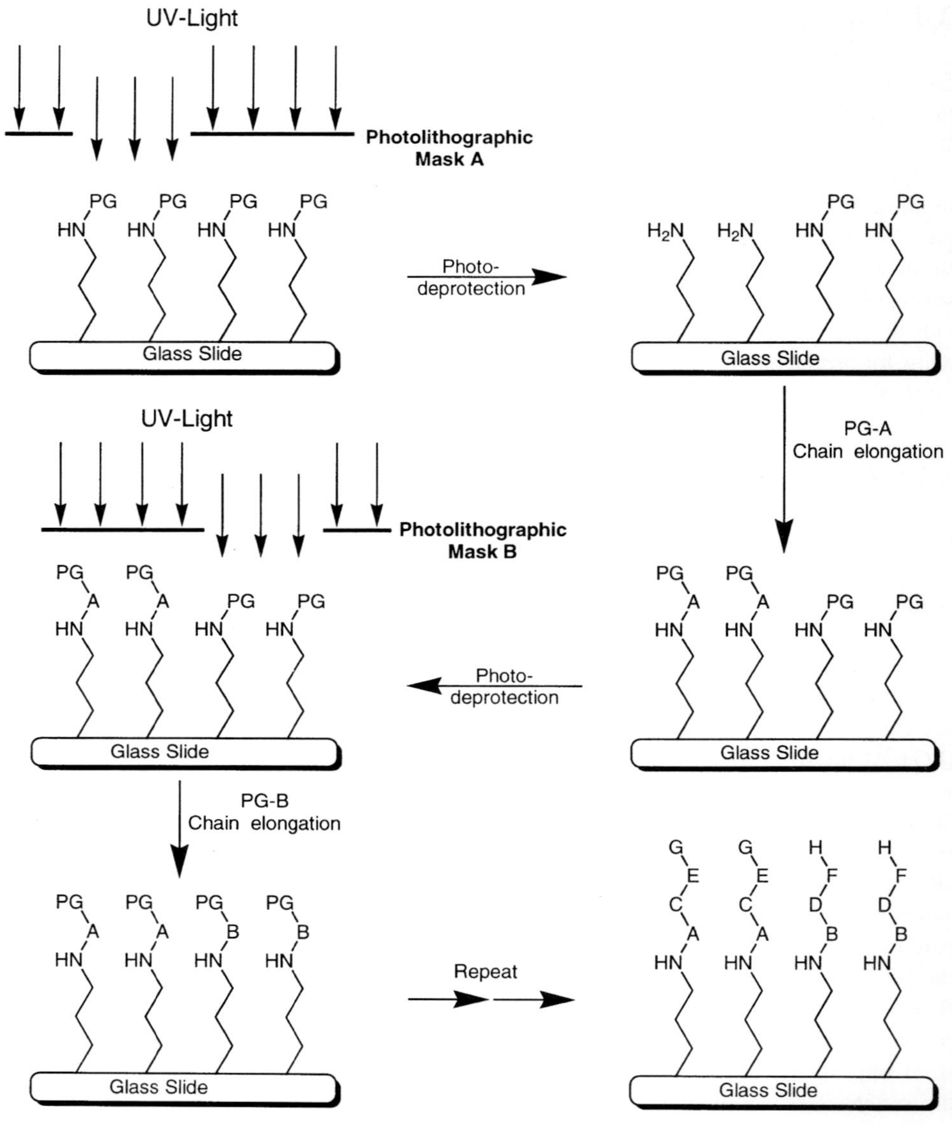

Figure 14. Principle of light-directed, spatially addressable synthesis.

The consumption of chemical reagents required for the synthesis of thousands of peptides comprising the array, together with the biological reagents necessary for bioassay, is very small, because the capacity of the flat glass surface is only $5 \div 20$ pmol cm^{-2}. Biological reagents used in this technology are recoverable and can be reused. Moreover, after performing an assay with one target molecule, the bound target can be easily dissociated from the array (for example, by treating it with 6 M guanidine hydrochloride), making the array avail-

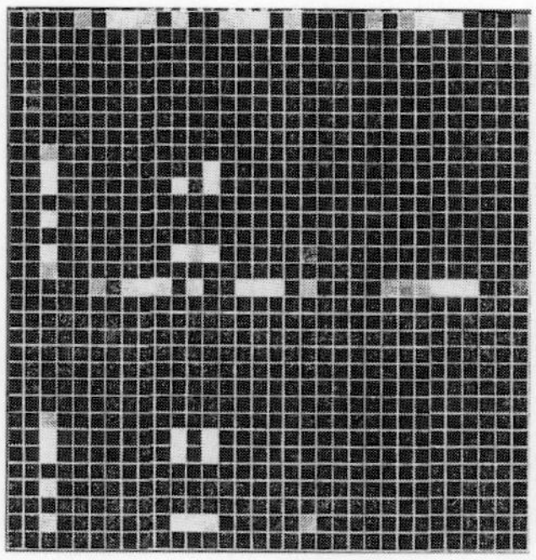

Figure 15. Fluorescence scan of 1024 peptides.

able for the subsequent screening with other targets. These arrays are reusable for at least six months.

Due to a very low capacity of the glass surface, traditional analytical tools cannot be used for adequate assessment of fidelity of the light-directed synthesis. A highly sensitive analytical method to characterize quality of peptides synthesized by this technology was developed [43]. This involves derivatization of the surface with an acid-cleavable Knorr linker (Fig. 16; **1**) followed by the attachment of a C-terminal chromophoric tag, Lys(Dabs) (Fig. 16; **2**).

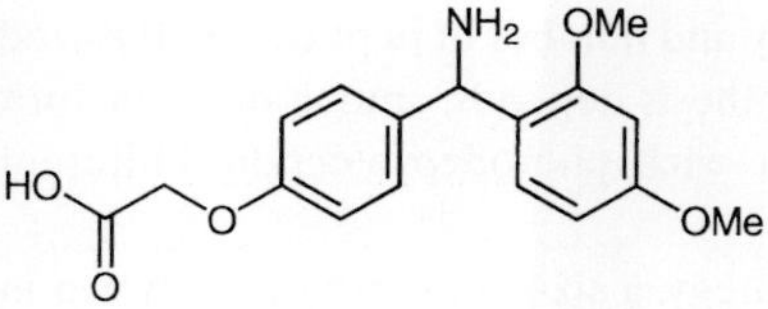

1

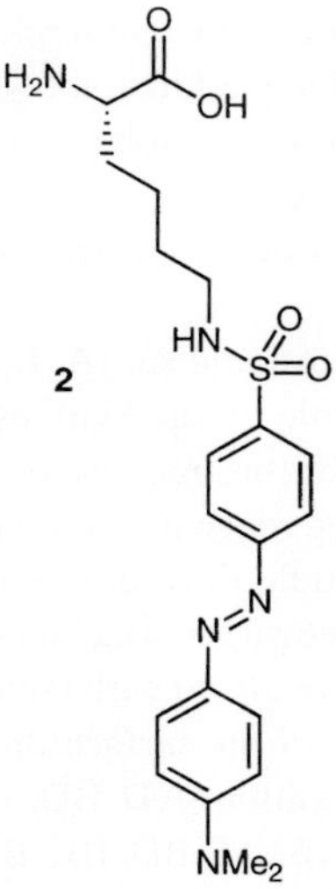

2

Figure 16. Structures of Knorr linker and Lys(Dabs) tag.

On completion of the synthesis, the synthesized peptide, containing C-terminal Lys(Dabs)NH_2, can be cleaved from the surface and characterized by HPLC. Its chromatographic behavior can then be compared with the retention time of the same compound synthesized using conventional methods and possessing a proven structure. This method was crucial for optimizing all methodological aspects of the light-directed synthesis. Originally, photoremovable nitroveratryloxycarbonyl group (Nvoc) (Fig. 17; **3**) introduced by Patchornik and co-workers [44] was utilized for the temporary protection of α-amino functionality of the growing peptide chain. However, photolysis under optimized conditions (12 min in 5 mM sulfuric acid/dioxane) sometimes causes side reactions and does not always proceed quantitatively.

Investigation of different α-amino photoprotective groups and conditions for their removal [43] resulted in selecting α-methyl-nitropiperonyloxycarbonyl (Menpoc) group [45] (Fig. 17; **4**). This can be removed quantitatively by photolysis under neutral conditions in 3–4 min.

3 **4**

Figure 17. Nvoc- and Menpoc-photolabile protecting groups.

The flat glass surface behaves noticeably different from the conventional supports for solid-phase synthesis [43]. Usually, higher concentrations of activated species (0.1–0.25 M) are necessary to achieve acceptable coupling efficiency. A number of coupling reagents have been examined and HATU [46] provided the most satisfactory results with short (5–15 min) coupling times [43].

With the same number of chemical steps, the diversity and number of peptides synthesized using light-directed, spatially addressable chemical synthesis depends, among other factors, on the patterns of photolithographic masks used for each photodeprotection. Different masking strategies can be used.

As an example of an orthogonal stripe masking strategy, a six-step synthesis is shown in Fig. 18.

The reactants are the ordered set {A, B, C, D, E, F}, and represent amino acids protected with a Nα-photoremovable group. Synthesis begins by exposing the left third of the entire synthesis area to UV-light through photolithographic mask A. Building block A is added in the presence of a coupling reagent, and reacts only at the previously illuminated region. UV-light exposure of the middle third of the synthesis area through photolithographic mask B activates it, allowing for covalent attachment of building block B. Similarly, the right third of the synthesis region is derivatized with building block C. Building block D is added after illuminating the upper third of the surface through mask D, thus creating in one coupling step three different chemical entities AD, BD, and CD. The process continues to generate the set of nine dipeptides, AD, AE, AF, BD, BE, BF, CD, CE, and CF, from six starting amino acids. In general, orthogonal masking strategy allows the generation of n^2 compounds from $2n$ starting materials in $2n$ chemical steps. For instance, a set of 400 dipeptides can be formed from 20 L- and 20 D-amino acids using only 40 coupling reactions. This orthogonal stripe synthesis strategy can continue to make longer peptides by subdividing previous stripes into a number of synthesis regions until the limit of photolithographic resolution is achieved (10–20 μm) [47]. With this resolution, 250000–1000000 compounds can be synthesized in 1 cm^2. Routinely, 50 μm resolution is practiced [39] and allows for the production of 40000 compounds in the same area.

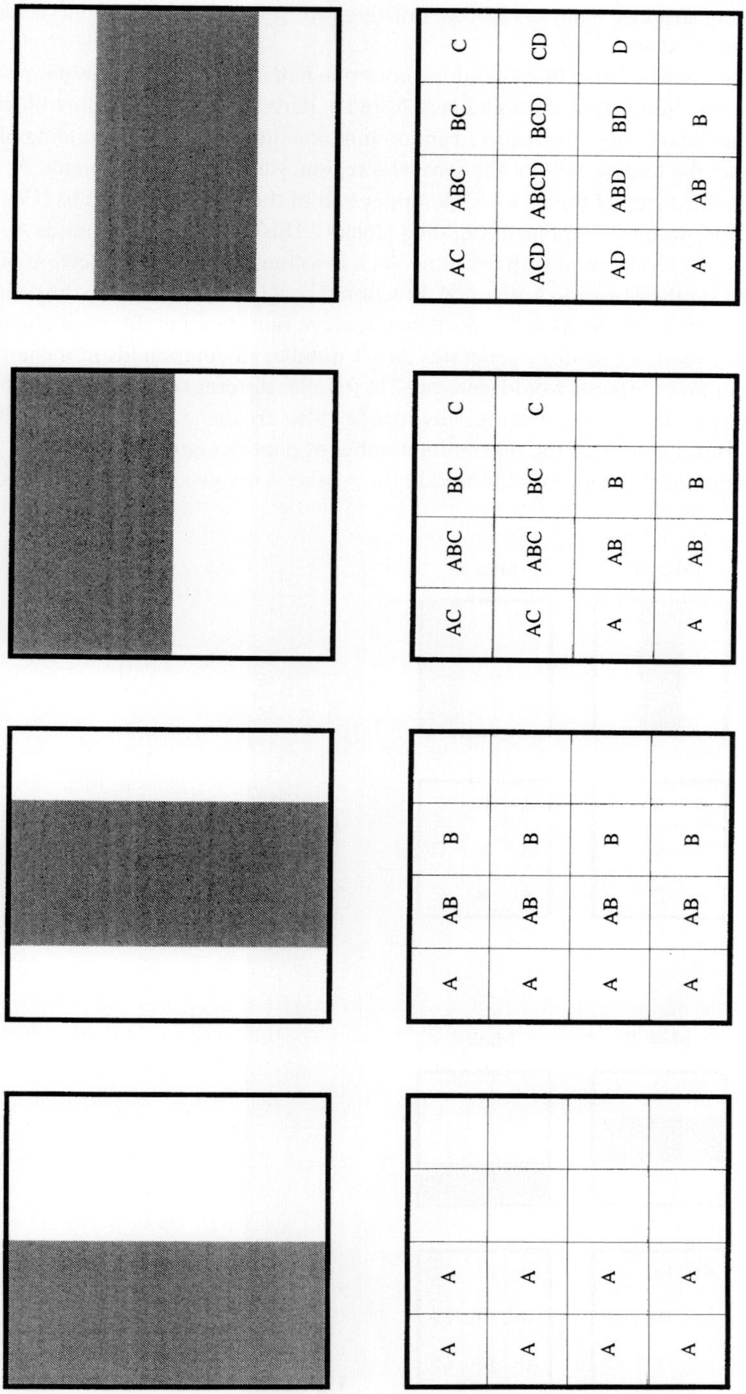

Figure 18. Orthogonal masking strategy.

Another masking strategy, binary synthesis, utilizing four photoprotected amino acids A, B, C, and D, is presented in Fig. 19.

In this approach, each building block is added after one half of the synthesis area is photolyzed. The left half is illuminated through mask A and is derivatized with building block B, while the other half bears only the protected amino functionality. Coupling of building block B after photolyzing the middle 50% of the synthesis region, yields three compounds, A, AB, and B. During the third step of the process, the upper half of the area is exposed to UV-light through mask C, followed by coupling of building block C. This produces compounds A, AB, ABC, B, BC, and C. Derivatization with building block D following photodeprotection of the middle 50% of the synthesis area in horizontal direction through mask D leads to the production of 16 unique regions of the glass support each (except one) bearing different chemical entities. The binary masking strategy generates $2n-1$ number of compounds in n chemical steps. A 16-step peptide synthesis would generate 216 (65 536) different areas, of which 65 535 would represent peptides or amino acids. This strategy also creates a distribution of chain lengths, ranging from 1 to n, with the maximum number of peptides being $n/2$ in length [39]. All possible permutations of amino acids used in the synthesis are generated, while preserv-

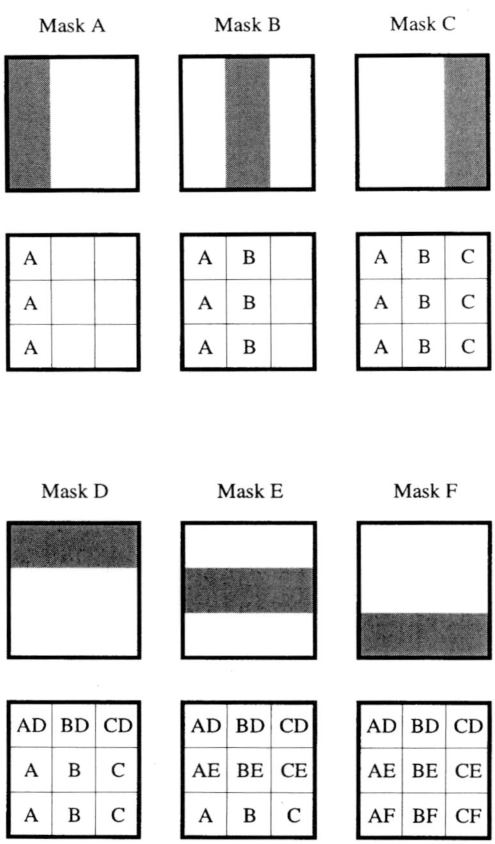

Figure 19. Binary masking strategy.

ing their order of introduction. Interestingly, all possible deletion sequences as well as all possible truncation sequences are obtained. This strategy was used in epitope mapping experiments with monoclonal antibody D32.39, which binds to the C-terminal region of dynorphin B. A binary synthesis of the decapeptide FLRRQFKVVT was carried out, generating 1023 different peptides. The results of this experiment identified heptapeptide RQFKVVT as the minimal binding sequence [48].

Light-directed, spatially addressable synthesis is a powerful technology for generating chemical diversity. Unfortunately, the technique is limited to peptides, oligonucleotides [39, 49] and other linear oligomeric structures [50].

3.4.2 Spot-Synthesis on Membranes

Frank reported [51–54] a method for parallel synthesis of peptides on cellulose membranes, derivatized with an appropriate linker to allow for synthesis of peptides. Reagents are delivered in drops to different sites on a flat surface. The volume of each drop, along with the other properties of the solid support, determines the diameter of the spot formed on the surface by the delivered reagents. This also determines the distance between the different spots, and therefore, the density of the array. Solvents of low volatility are used to prevent evaporation before a chemical step is completed. Reagents can be added to different sites on the array both manually and automatically, using liquid-handling equipment. In the course of the synthesis, common synthetic steps, such as deprotection of temporary protective groups and washing procedures, are carried out by treating the entire membrane with necessary reagents and solvents. At the end of the synthesis, all side-chain protecting groups are removed. The peptides comprising the array can be screened on solid support. Alternatively, if a cleavable linker is used (e.g. structures **5** [55], **6** [56], and **7** [57] presented in Fig. 20), individual synthe-

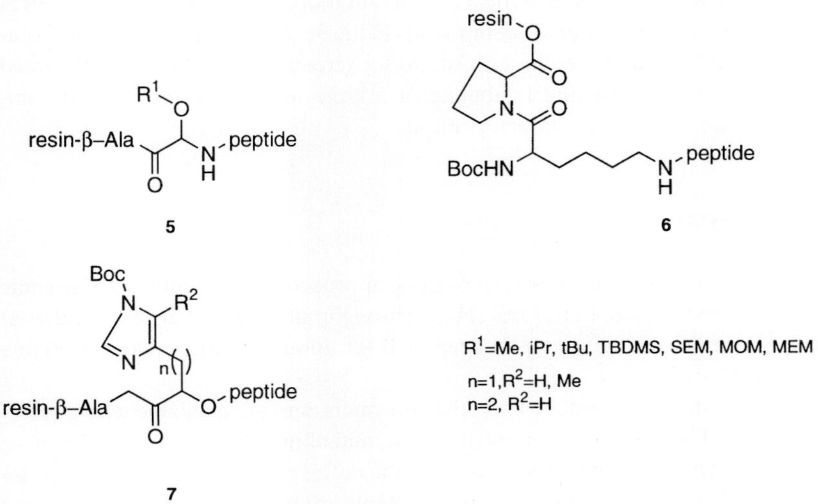

Figure 20. Cleavable linkers for spot synthesis.

sis sites can be physically separated and peptides can be released in solution by an appropriate reagent.

The amount of each peptide in the array is easy to control by the size of each spot. For instance, synthesis on Whatman 540 paper (capacity 0.4–0.6 μmol cm^{-2}) with the volume of reagents used for the coupling step 0.5–0.7 μL yields 25 nmol of a peptide on a spot, which is 7 mm in diameter [55]. These parameters are used in a standard format, an 8 × 12 array with 9-mm distance between the centers of the synthesis sites. Larger 40 × 50 arrays can be prepared on Whatman 50 paper with the help of automation. Drops 0.03–0.05 μL in volume form spots that are 1 mm in diameter, producing an array containing 2000 peptides, each of 1.0 nmol [55].

The spot-synthesis technique is simple and easily accessible for practical use. There are some limitations to the scope of chemistries that can be used with cellulose as a solid support. To eliminate some shortcomings of cellulose, Gao and Esnouf [58, 59] proposed using a different membrane, Immobilon AV-1, for the spot synthesis. Before synthesis of peptide arrays can be carried out, this polyvinylidene fluoride-based material is derivatized with ethylene diamine, followed by coupling of Fmoc-β-alanine as a spacer. In most other aspects, the peptide array synthesis is very similar to the spot synthesis proposed by Frank.

3.4.3 Manual Methods for Resin-Based Parallel Synthesis

The 96-well microtiter format has been a common platform for HTS for many years, and also offers many advantages as a platform for parallel synthesis. There are multiple liquid handling devices (multichannel pipettes, robotic systems, etc.), that significantly simplify and accelerate the process of the reagent delivery to the wells on the plate. Moreover, all 96 compounds can be handled as one synthesis entity. Their structures are spatially encoded by their location on the plate. Contemporary analytical instrumentation, such as mass spectrometers and HPLC systems, can also easily handle compounds in the 96-well microtiter format. Combined with the commonly used 96-well based systems for screening, this eliminates the need for time-consuming redistributing and relabeling of a large number of synthesized compounds for analytical and biological characterization.

3.4.3.1 HiTOPS System

Affymax Research Institute has developed an original approach for the synthesis of organic compounds in 96-deep-well plates. HiTOPS (High Throughput Organic Parallel Synthesis) System [60] (Fig. 21) utilizes a variety of 96-deep-well filtration microtiter plates available from Polyfiltronics/Whatman.

The plates are made of polypropylene and other polymers, and are available with a selection of different filters. The volume of each well is 2 mL and allows the use of up to 50 mg of resin. For larger-scale syntheses, several wells or even the entire row or the entire column can be used for the preparation of the same peptide. Reactants are retained in wells by the positive pressure of an inert gas (Fig. 22).

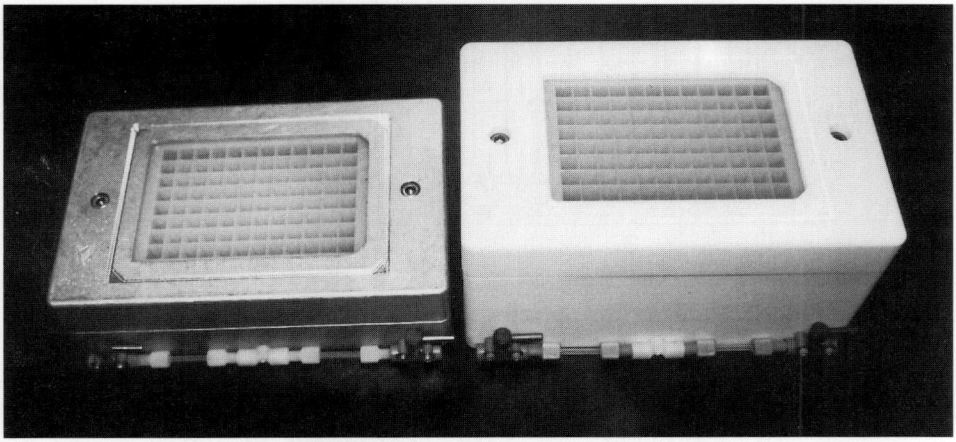

Figure 21. HiTOPS system. Synthesis Device (left) and Cleavage Device (right).

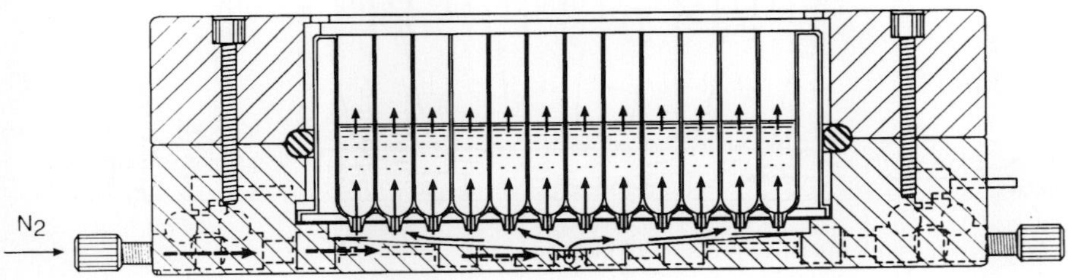

Figure 22. Positive pressure retains compounds in wells.

At the beginning of the synthesis, a filtration plate is placed into a Synthesis Device. The bottom portion of the device has an opening in the center connected to an internal channel, which leads to two valves trough a T-connector. One valve regulates the delivery of an inert gas to the bottom of the plate. The other valve links the bottom with a waste container connected to vacuum. When the inert gas valve is open, positive pressure is created at the bottom. Resins and reagents can be delivered to the plate and kept in place by the positive pressure for the duration of the synthesis. The pressure can be increased to allow the inert gas to bubble through the wells, providing for the mixing and creating an inert atmosphere. Mixing can also be achieved either by placing the device on an orbital shaker or by using magnetic stirring bars in each well. At the end of a chemical step, liquid reagents or washing solvents can be removed into the waste container by opening the vacuum valve (Fig. 23).

At all times the plate remains in the Synthesis Device. There is no need to disassemble the device in the course of the synthesis, making the synthetic process very ergonomic. One person can easily operate four devices simultaneously, synthesizing 384 compounds. For the final cleavage the plate is transferred to the Cleavage Device (Figs. 21 and 24), which is made of

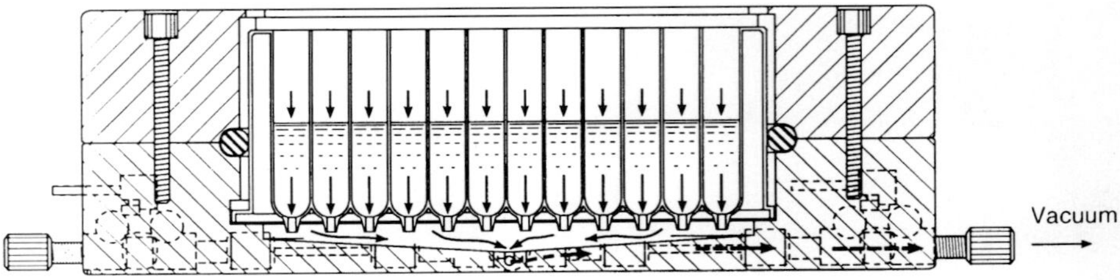

Figure 23. Reagent removal by vacuum.

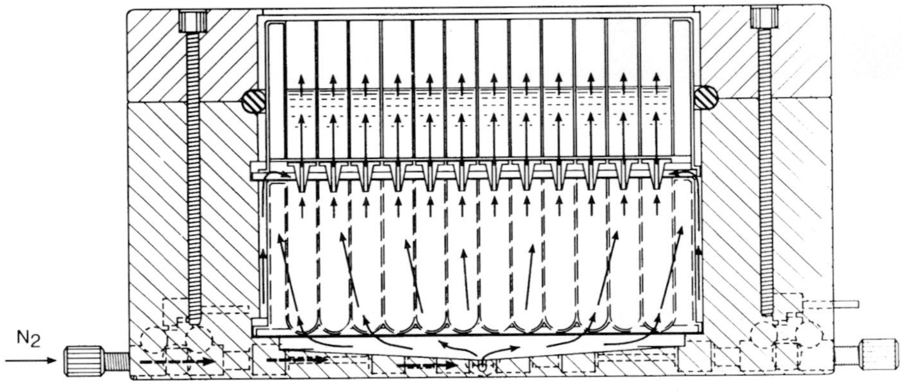

Figure 24. Cleavage under positive pressure of inert gas.

Teflon®. The Cleavage Device is deeper (in order to accommodate a collection plate) than the Synthesis Device, but otherwise has the same design.

After creating positive pressure at the bottom, a cleavage reagent is delivered to the top plate, which contains all resins with the synthesized compounds. On completion of the cleavage, a vacuum is applied and the filtrates are collected in the bottom 96-deep-well collection plate (Fig. 25). The collection plate is then placed in a vacuum centrifuge to eliminate the cleavage mixture.

The HiTOPS device was recently used in the synthesis of combinatorial libraries of diketopiperazines [61] for the development of highly selective inhibitors of collagenase-1. The HiTOPS system is easy to use, has an ergonomic design, and allows the use of up to 50 mg of a resin per well. The other advantage of this method is the commercial availability of the described devices. Affymax has licensed its technology to Whatman/Polyfiltronics, which manufactures the HiTOPS System.

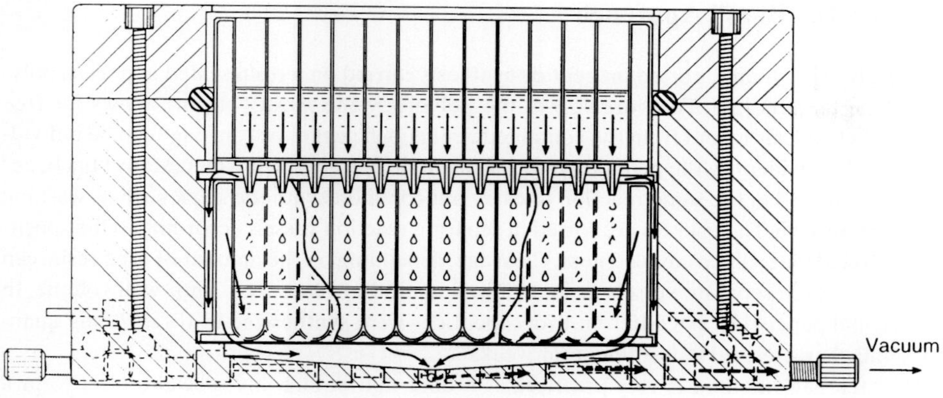

Figure 25. Collection of cleaved compounds in HiTOPS Cleavage Device.

3.4.4 Synthesis on Macroscopic Polymer Supports

Traditionally, solid-phase synthesis of peptides is carried out on polymeric supports comprised of small beads of a polymer. In parallel synthesis approaches that are not spatially addressable, there is a need for easy manipulation with all members of the array in the course of the synthesis. It is often easier to do that when each member of the array is synthesized on a single macroscopic piece of solid support or on a small batch of resin enclosed in an appropriate shell. Such macroscopic supports are also easy to label.

3.4.4.1 Synthesis on Multipins

In 1984, Geysen and co-workers [62] introduced an alternative to the synthesis on polymeric beads. As a solid support, they proposed to use "pins"; these are reusable polyethylene rods, which have a diameter of 4 mm diameter and a length of 40 mm, and are grafted with polyacrylic acid. The pins are attached to a supporting block arranged in an 8×12 array to fit into a 96-well microtiter plate. The 96-well plate is used as 96 separate reaction vessels. Each well can be used for the attachment of a different building block. Washing and deprotection steps can be carried out in a separate common reaction vessel. After the synthesis is completed and side-chain protecting groups are removed, peptides can be assayed while attached to the pins [63, 64]. If a cleavable linker was used, individual peptides are separated from the solid support in a 96-well microtiter plate containing appropriate reagents [56, 65], or by exposing the array to vapor-phase ammonolysis [66, 67]. Originally, the capacity of each pin was 10–100 nmoles. Later, a much greater variety of multipins of different capacities with a wide selection of linkers became commercially available from Chiron Technologies Pty. Ltd. In the latest development, the pins can be attached to the radiofrequency tags to allow for encoding in the "split/pool" mode.

3.4.4.2 "Tea Bags" Approach

In 1985, R. Houghten reported on peptide synthesis carried on a resin sealed in porous polypropylene packets [11]. The pore size of the polypropylene mesh (74 μm) allows for free access of chemicals to and from the contained resin. Each packet or "tea bag" can be individually labeled to identify the peptide synthesized on the entrapped polymer. Many "tea bags" can be combined in the same reaction to carry out common synthetic steps, such as washing and deprotection. The packets are sorted in separate reaction vessels according to the specific amino acid that will be coupled next. Cleavage, depending on the amount of used resin, can be carried out in separate vessels or in a 96-well microtiter plate with 2-mL well volume. In the original paper [11], 248 different tridecapeptides were synthesized in 10- to 20-mg quantities and characterized in less than four weeks.

The "tea bag" method is very practical, as it does not require any special tools, except a sealing device to make the "tea bags". A number of mesh materials of different porosity are available from Spectrum Medical Industries, Inc. Almost all commercially available resins can be used, as long as the size of the beads is larger than the mesh size; in addition, the mesh should be stable to reaction conditions. The scale of the synthesis is easy to control by the size of the packet chosen to contain the desired amount of a polymer support. Multiple synthesis can be carried out manually or on a synthesizer. The "tea bags" can be mechanically labeled, or RF tags can be used for this purpose.

3.4.4.3 Synthesis on Cellulose and Other Laminar Supports

Similarly to the "tea bag" approach, pieces of any solid-phase support suitable for solid-phase synthesis can be used for parallel peptide preparation. One of the first materials used in this manner was paper. Frank used paper disks [68, 69] (Whatman 3MM, 1.5 cm diameter) packed into columns of a multicolumn continuous flow synthesizer. Prior to the synthesis, the paper was derivatized with p-alkoxybenzyl linkage to allow for the cleavage with trifluoroacetic acid. The disks can be easily labeled with a pencil, sorted, and combined depending on the common amino acid to be coupled at the next step.

Cotton has mechanical properties superior to those of paper, and was used in a similar way by Lebl and colleagues [70, 71]. Researchers from the Rockefeller University used polystyrene-grafted polyethylene film [72, 73], while scientists from Pfizer used material prepared from two sheets of polypropylene mesh with resin sandwiched in between [74]. The "sandwich" was fused together with a polymer with a low melting point.

3.5 Tools for Increasing Productivity in Combinatorial Chemistry

Medicinal chemistry, being largely an experimental discipline, is often a tedious and labor-intensive process, which requires attention of highly skilled specialists to seemingly simple synthetic procedures. Therefore, it is always desirable to automate the synthesis process, allowing the scientists to concentrate on more intellectually demanding activities. Combinatorial chemistry deals with much larger numbers of compounds, than traditional medicinal chemistry, making the automation a necessity. The field of combinatorial chemistry is still very young, and the market for automation often crystallizes from the widely different needs of various organizations. Fortunately, many major instrumentation companies have begun paying attention to this new area. Historically, the same companies that pioneered the combinatorial approaches to drug discovery carried out the original work in automation. To avoid the problems associated with building reliable instruments, some major pharmaceutical companies have formed alliances with instrumentation firms, and presently, a number of options are available for automating specific synthetic needs.

The scope of the methods practiced in modern combinatorial chemistry is very large. Compounds are synthesized in solution and on solid phase, in a wide range of quantities; large and small, encoded and unencoded libraries fit different needs of the drug discovery process. Chemistry optimization is more demanding in the case of combinatorial chemistry. Optimal conditions for the reliable and high-yield synthesis of all library members should be discovered through experimentation.

Not all stages of combinatorial chemistry can easily be automated. Collecting building blocks necessary for the library synthesis, weighing them out, and preparing the appropriate solutions is often better left to chemists. Post-synthesis operations, such as cleavage, evaporation, distribution for analytical and biological characterization can be easily more time-consuming than the synthesis itself.

The following section will describe some manual and automated devices that can improve the productivity of different types of combinatorial chemistry approaches.

3.5.1 Instruments for Solution-Phase Chemistry

3.5.1.1 MultiReactor™ (RoboSynthon, Inc.)

The MultiReactor is a manual synthesizer produced by RoboSynthon, Inc. [75] (Fig. 26). It has a relatively simple design and consists of a reaction block (dimensions 582 mm × 160 mm × 132 mm) holding up to 24 reaction vessels, glassware, and a control unit. The software helps manage reaction conditions in the reaction vessels. The volume of each reaction vessel is 20 mL. The reaction block maintains identical temperatures in all reaction vessels (−60 to 200 °C). Each reaction vessel is supplied with an air condenser, though water-cooled condensers can be ordered as an option. To achieve temperatures below ambient, the reaction block should be used with an external liquid cooling system. For effective stirring with the stirring bars in each reaction vessel, stirring is mediated by an individual motor, located beneath each of the reaction vessel chambers. The MultiReactor control unit permits setting up the desired temperatures and mixing values directly on the instrument. The Windows™-based program allows for the more sophisticated control of the temperature and mixing profiles. This is helpful for carrying out reactions with viscous reagents or for reactions on solid phase. Using solution-phase chemistry, the MultiReactor can be used for production of more than 200 compounds per month in 100- to 200-mg quantities [76]. RoboSynthon also manufactures the MaxiReactor designed for large-scale synthesis with six reaction vessels (500 ml each).

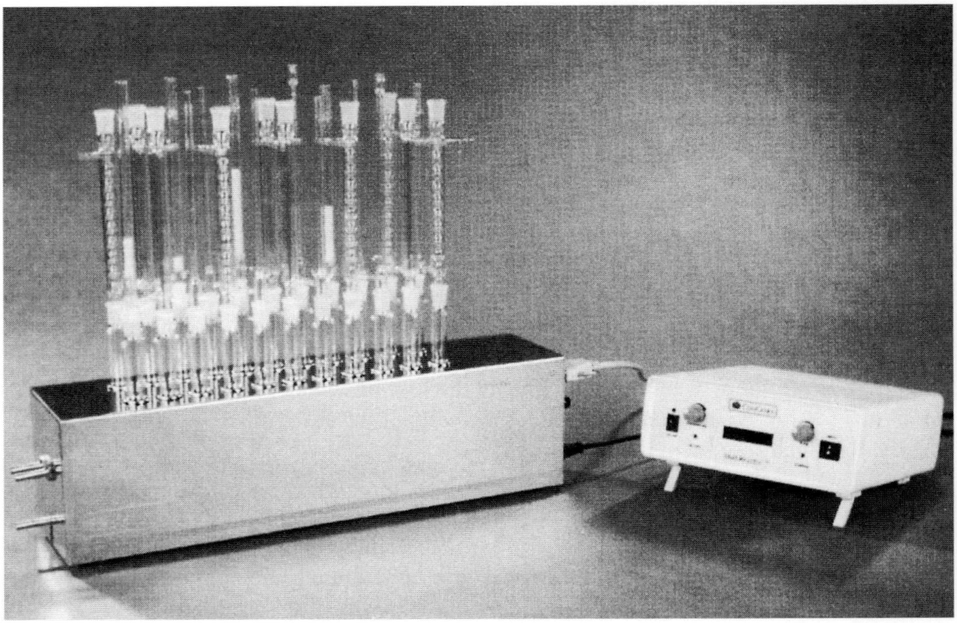

Figure 26. MultiReactor™ (RoboSynthon, Inc.).

3.5.1.2 STEM Reacto-Station™ (STEM Corporation)

STEM corporation [77] has developed a line of Reaction Stations (Fig. 27), which could be used as stand-alone units or in combination with other automated systems, such as autosamplers and robots. Reaction Stations can accommodate, depending on the specific model, 10, 25 or 50 reaction vessels. All vessels can be heated or cooled simultaneously, with a maximum of $\pm 0.5\,°C$ temperature differences between the different positions. The manufacturer designed the stirring system to ensure effective mixing of viscous solutions with a magnetic bar. An interesting feature of the stirring system is a "soft start". This allows for a slow acceleration of the mixing speed until it reaches the value specified by the operator (400 to 2000 rpm). To improve the efficiency of mixing, flat-bottomed reactors can be obtained from the company. The desired temperature (up to 150 °C) can be easily set on the instrument's keypad. To accommodate refluxing, reflux modules can be used; these are available in forced air and water cooling options. An example of the air-cooled module attached to the top of the heated reaction station is shown in Fig. 28. The cooling modules allow for isolating the content of the reaction vessels from the atmosphere. They have a removable lid with a port for supplying an inert gas. The lid also has gaskets that can be pierced to introduce or withdraw chemicals under the inert conditions. STEM Corporation also provides a unique chilling and stirring system (RS2480) which uses a CFC-free refrigerated gas chiller. The chiller is designed to control the temperature automatically through the keypad.

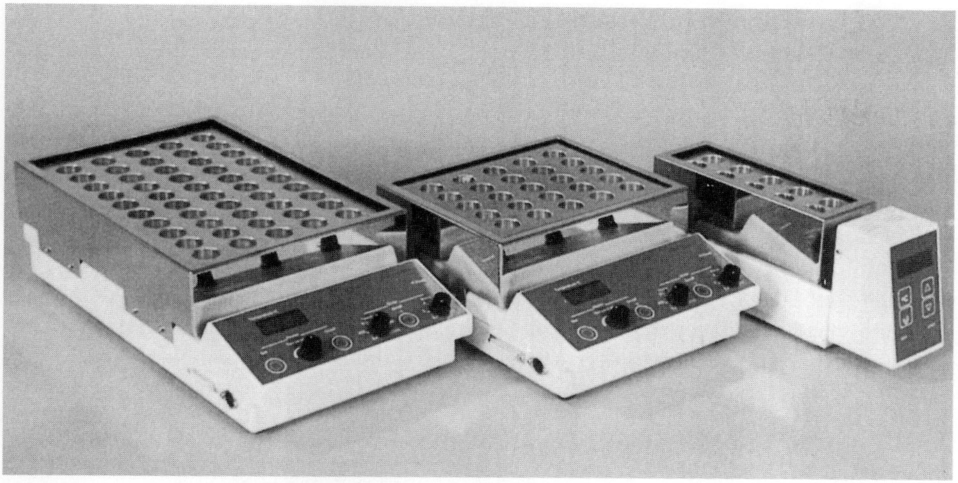

Figure 27. STEM Reacto-Station™ (STEM Corporation).

Figure 28. RS 1000 Reacto-Station with air-cooled reflux module (STEM Corporation).

3.5.2 Manual and Semiautomated Instruments for Solution-Phase Chemistry

Traditional medicinal chemistry laboratories are beginning to practice parallel synthesis and combinatorial chemistry more often. The development of manual and semiautomated devices, which improve the productivity of the synthesis process, is a reflection of this recent trend. Fully automated systems are expensive, and it is also not always easy to find adequate space in the laboratory to locate such systems. The synthesizers may have counterintuitive software and require a lot of maintenance; therefore, it is often necessary to have dedicated users to operate them. Ad hoc usage by everyone in need is difficult due to the complexity of the software and scheduling issues, but the market is beginning to respond by offering smaller, substantially less expensive devices that are easy to use without learning complex software.

3.5.2.1 Quest 210 (Argonaut Technologies)

The Quest 210 [78] is a semi-automated instrument with 20 reaction vessels located on both sides of the synthesizer. The vessels are cylindrical in shape and are made of a transparent fluoropolymer. They are available in two sizes: 5 and 10 mL. At the bottom of each reaction vessel there is a frit, which retains the resin inside the vessel during the solid-phase syntheses. The Quest 210 is designed for both solution- and solid-phase reactions. With a 45-cm (18-in) footprint, the entire system fits into a regular chemical hood and is mounted on a rotating table (Fig. 29) to provide easy access to either side of the synthesizer. Reagents are added by a user through the top of each reactor. The mixing is achieved by up-and-down movements of a magnetic bar placed in each cylindrical vessel. Using a controller, the user can set the

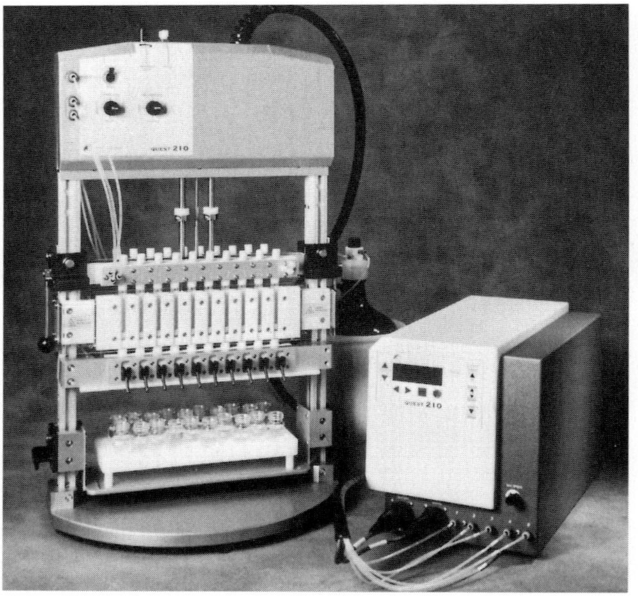

Figure 29. Quest 210 synthesizer (Argonaut Technologies, Inc.).

agitation speed reaction, while the temperature of each of the two groups of reaction vessels can be set between –40 and 150 °C. An optional Solvent Wash Module is also available from the Argonaut Technologies. This automates multistep washing protocols with up to four different solvents. The liquid content of each vessel can be collected individually, with all vessels also capable of being emptied simultaneously to the waste container. A collection rack simplifies the collection of the final products after cleavage from the solid support or after the liquid–liquid extractions into the 20- or 40-mL scintillation vials. An example of a solution-phase synthesis of ten β-ketoesters carried out at −40 °C on the Quest 210 is shown in Fig. 30 [78].

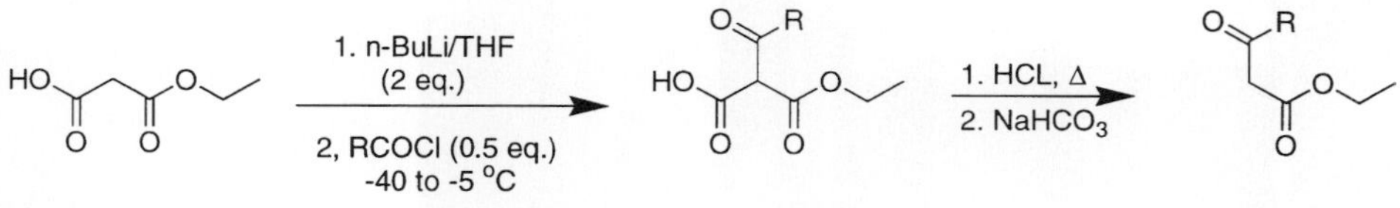

Figure 30. Synthesis of β-ketoesters on Quest 210 synthesizer [78].

3.5.2.2 APOS 1200 System (Rapp Polymere GmbH)

Rapp Polymere GmbH [79] has designed the APOS 1200 synthesizer (Fig. 31) for solid-phase parallel synthesis. This has 12 cylindrical glass reaction vessels, each of which has a

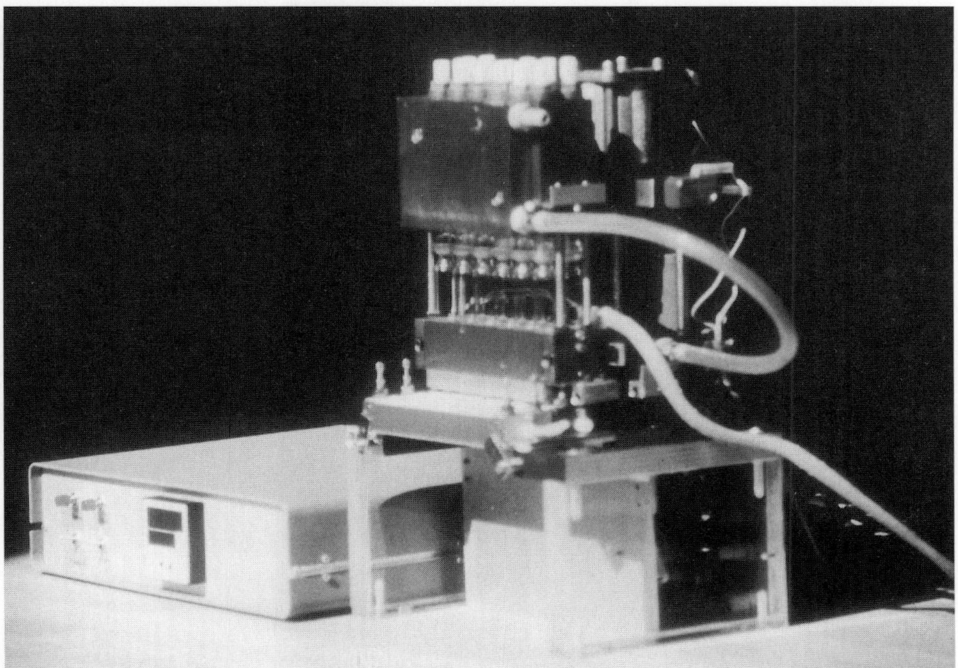

Figure 31. APOS 1200 synthesizer (Rapp Polymere GmbH).

Figure 32. Reaction block of APOS 122 synthesizer.

glass frit at the bottom and can hold up to 200 mg of resin and 3 mL of solvents. The design of the reaction block is shown in Fig. 32, the reagent vessels forming two rows of six. The temperature can be controlled within the −60 to 150 °C range. The reagents are delivered manually through the top of each reactor, or automatically using a robotic system. All reactors can be connected to 12 condensers organized in a cooling block. All reactions can be carried out under inert atmosphere, the agitation being accomplished by aspirating an inert gas.

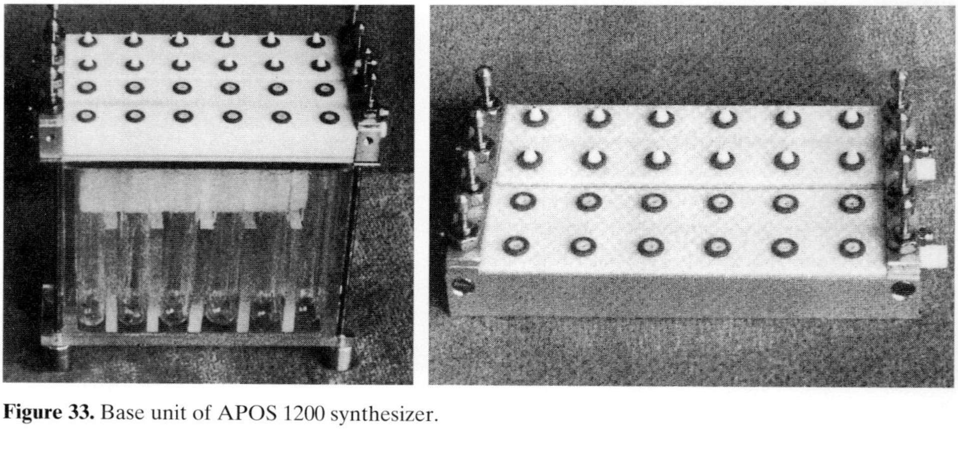

Figure 33. Base unit of APOS 1200 synthesizer.

APOS design includes a unique base unit design (Fig. 33) with two positions, which align automatically with the reaction block. One position provides gas flow through the reactors during the synthesis, and the other position is used for filtering and washing. The controller directs the alteration between the two positions. There are two types of base units; one is used to carry out reactions as explained before, and the second is for collecting compounds after cleavage. The collector base accommodates 12 tubes or vials with volumes of 10 to 15 mL.

3.5.2.3 ReacTech Synthesizer (Advanced Chemtech)

The ReacTech synthesizer [80] accommodates up to 40 separate reactors with the volume of 8 mL in a Teflon reaction block. The synthesizer can be used for both solution- and solid-phase chemistry. The reactors are organized into groups of ten. This allows for independent control over four different sets of reactions in each group. The reactors are sealed with chemically resistant Teflon-coated septa. The reagents can be delivered manually with a syringe under inert atmosphere. It is possible to deliver simultaneously a choice of up to three common reagents or solvents to all reactors within one reaction group. The precision of the reagent delivery is controlled by a 50-mL measuring vessel with a photocell. The software controls the temperature of the syntheses (within a -70 to 150 °C range) and other parameters including variable-speed vortex mixing.

3.5.2.4 SAS (MultiSynTech GmbH)

MutiSynTech GmbH [81] manufactures a semiautomatic synthesizer for solid-phase synthesis SAS. The synthesizer works with three different reaction blocks for different scale chemistries. One of the blocks can accommodate 40 removable glass or polypropylene reaction vessels with a choice of a glass or PTFE frit at the bottom. The other two options accommodate 60 vessels (5 mL each) or 96 reactors (2 mL each). The synthesizer controls the reaction temperature within the -30 to 150 °C range. Mixing is accomplished by a magnetic bar, which levitates above the bottom of the reactor; this helps prevent damage to the resin beads. Each reaction vessel can be easily removed from the reaction block in the course of the synthesis, which can be carried out under the inert atmosphere. Washing can be done automatically with up to six different solvents. The cleavage from the resin is also automated.

3.5.2.5 Systems Based on 96-Well Microtiter Format

Meyers, together with colleagues from Sphinx Pharmaceuticals, were first to report [82, 83] a system based on a 96-well microtiter format for parallel synthesis. Each well of the reaction block, which is a polypropylene plate, has a polyethylene frit and a small hole at the bottom. The clamping device seals all holes by forcing the bottoms of each well against a gasket. After loading the resin and the reagents, the wells are sealed with eight-strip caps. Mixing can be carried out manually or by placing the plate on a shaker. At the end of each chemical step, the clamping device and the cap strips are removed, and the resin is washed. After the cleavage, all products are collected in tubes or into another microtiter plate.

A very similar approach was undertaken and advanced into a line of commercially available products by Charybdis Technologies [84]. Their Calypso System™ can be used with the disposable deep-well filtration plates or with the reusable Multi-Well Reaction Arrays. The clamping mechanism is designed for easy sealing either both sides of the plates simultaneously or each of them individually. The reagents can be delivered to the plate when the bottom is sealed. The top cover can then be connected to the clamping frame, sealing the plate. The top cover has holes, which allow reagent delivery into the sealed wells by piercing the gasket. In fully assembled form, the Calypso System™ can be heated, cooled, and agitated. Charybdis Technologies offers specialized Multi-Well Reaction Arrays (PTFE and Multi-Temp) for solution- and solid-phase synthesis applications. The PTFE Arrays are available in three different formats: 96-, 48-, and 24-well. They have either round-bottomed wells for carrying out solution-phase reactions, or bottom-filtration wells with frits for solid-phase applications. They can be used at the ambient temperature up to 150 °C. The Multi-Temp Arrays are also available in the same three formats as the PTFE Arrays. The all-glass reaction wells are assembled within a glass-filled PTFE shell, which has an internal cavity through which a fluid can be circulated for temperature control. The Multi-Temp Arrays can be used to carry out reactions at temperatures from -80 to 150 °C. To provide an inert atmosphere, the Gas Manifold Systems should be used. The Vacuum Manifold System permits draining of the Calypso Reaction Block™ Assembly. The content of the wells can be either sent to waste or collected.

Robbins Scientific offers the FlexChem™ System, which is their version of the "clamping" approach [85, 86]. It is based on the company's own 96-well plate which has channels molded between the wells to prevent cross-contamination. The volume of each well is 2 mL. Three different materials, i.e. rubber, Viton™, and ChemTuf™ are used as the sealing gaskets. ChemTuf™, designed by Robbins Scientific, is a version of Viton that is laminated with a thin layer of fluoropolymer. The company has studied the sealing efficiency and solvent resistance of each gasket at different temperatures [87]. It claims that complete sealing was achieved at low or elevated temperatures (up to 100 °C) even with low boiling solvents, such as dichloromethane. The FlexChem rotating oven accommodates up to four reaction blocks and provides efficient mixing for solid-phase reactions. The approaches to washing and cleaving the resin are similar to those developed by Charybdis Technologies.

The most ergonomic design of a 96-well based device for parallel synthesis of organic compounds was introduced by Affymax Research Institute; their HiTOPS system is described in section 3.4.3.1 of this chapter.

3.5.3 Automatic Systems for Solid-Phase Synthesis

3.5.3.1 Automated RAM™ Synthesizer (Bohdan Automation, Inc.)

The RAM synthesizer (Fig. 34), which has a modular design, is a fully automated workstation designed to execute the entire process of solution- and solid-phase synthesis [88, 89]. The Automatic Weigh Station allows the necessary reagents to be placed into preweighed vials.

Figure 34. RAM™ synthesizer (Bohdan Automation, Inc.).

The station records the weight of the reagents and prepares solutions of specified concentrations. The Vortex Mixer module is used to solubilize dry reagents and for liquid–liquid extractions, the reagent vials being moved by a robotic arm. The reaction block contains 48 glass reaction vessels in a 6 × 8 array. Reactions requiring temperature control (−40 to 150 °C), refluxing, elevated pressures of up to 2 atm, and inert atmosphere can be carried out on the synthesizer in a fully automated mode. Reagents are added to the reaction block through a septum by a cannula, which consists of three concentric tubes. The inner tube pierces the septum and delivers and withdraws liquids to and from the reactors, while the inert gas is delivered through the middle tube. The outer tube is used for venting the vessels to allow for displacement of atmospheric gases.

Using this instrument it is possible to produce over 50 compounds per week on solid support, or 100–250 compounds in solution.

3.5.3.2 Nautilus™ 2400 Synthesizer (Argonaut Technologies, Inc.)

The Nautilus 2400 synthesizer (Fig. 35) is a highly sophisticated instrument designed to carry out parallel synthesis of 24 compounds [89, 90]. It is also very useful for chemistry optimization. The instrument contains six modules: an autosampler, the Reagent Solvent Enclosure, the Reaction Vessel Enclosure, a gas chiller, and a computer.

The Reagent Solvent Enclosure is used to store solvents and common reagents. The Reaction Vessel Enclosure has three banks of reaction vessels. Each bank accommodates eight reactors, which are available in three sizes: 8, 15, and 23 mL. The temperature of the reaction vessels can be controlled within the −40 to 150 °C temperature range. As it is often important to have visual control of some chemical reactions, the content of each reactor is easily visible at all times during the synthesis. Reagents are delivered to the reaction vessels under the inert atmosphere through Teflon tubing that reaches the bottoms of the reactors. At the

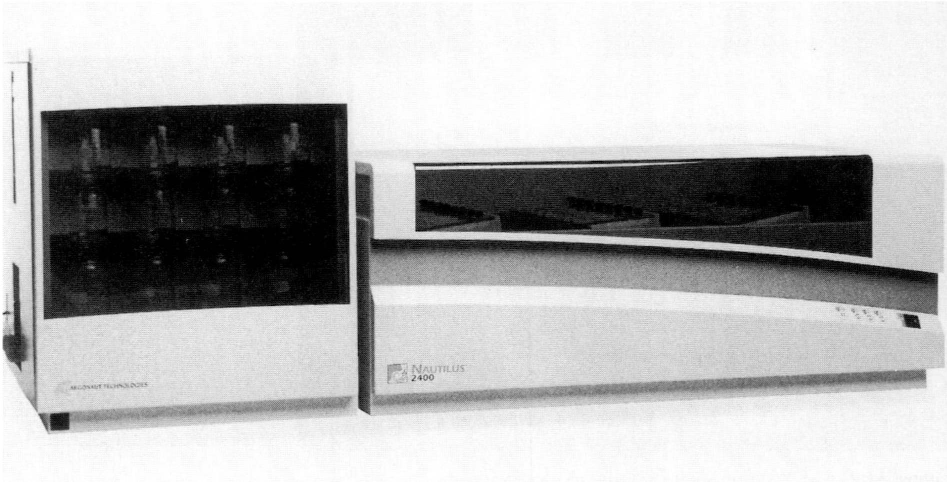

Figure 35. Nautilus synthesizer (Argonaut Technologies, Inc.).

bottom of each reactor there is a disposable 30-μm frit. The smallest amount that can be delivered automatically is 200 μL. Mixing is achieved by inverting the reactors by 217°, using a rocking agitator. After the synthesis, compounds can be cleaved from the resin automatically, and delivered to any location on the fraction collector. Porco and co-workers reviewed [91] some chemistries developed with the help of the Nautilus 2400 synthesizer.

3.5.3.3 Trident™ Library Synthesizer (Argonaut Technologies, Inc.)

Argonaut Technologies has developed the Trident synthesizer for parallel synthesis of up to 192 compounds. The volume of each reaction vessel is 4 mL, and the reactors are organized into modular cassettes. The synthesizer can work with four cassettes simultaneously; each of the cassettes having 48 reaction vessels organized in a 6×8 array. The reagents are delivered serially to eight reaction vessels using a delivery manifold. The minimum deliverable volume is 50 μL, and the volume can be adjusted by 10 μL increments. The temperature range is the same as for the Nautilus, -40 to 150 °C. Mixing is achieved by variable-speed oscillation. The cassettes can be easily removed from the synthesizer and placed into a separate agitation–thermal unit, making the synthesizer available for the next synthesis. The cassettes are placed on the synthesizer for reagent delivery and for carrying out washing cycles. If necessary, manual reagent delivery can be carried out under an inert atmosphere. Both solution- and solid-phase reactions can be carried out on the Trident. After the automated final cleavage from solid support, compounds can be distributed to the appropriate vessels by the fraction collector.

3.5.3.4 Syro II Synthesizer (MultiSynTech GmbH)

The Syro II synthesizer is manufactured by the MultiSynTech GmbH, a German-based company. The synthesizer evolves from the previous model of a peptide synthesizer [92], and is based on a robotic liquid dispenser with two moving arms. All solvents and reagents are stored under an inert atmosphere. The design of the reaction block, which is available as a part of the semiautomated system is described in section 3.5.2.4.

3.5.3.5 Advanced ChemTech Synthesizers

Advanced ChemTech was the first company to release a combinatorial library synthesizer. The company offers several instruments for combinatorial chemistry. All fully automated models are based on a robotic system.

The Model 496 Multiple Organic Synthesizer has two robotic arms for delivering regents and solvents to the Teflon reaction block. Standard monomer racks are designed to home 10, 36 or 128 vessels with the reagents under an inert atmosphere. The robotic arms pierce the septum at the top of the rack to access a reagent for delivery to the reaction block. In addition, six 100-mL reservoirs are available to store common reagents. The content of the reservoirs is protected from the atmosphere by a specially designed double septum flushed with an inert gas. The needles of the robotic arms are washed on the inside and on the outside in a needle wash station located on the instrument. The tip of the needle is designed to spray liquids evenly and efficiently in all directions. This allows for efficient washing of the walls of each reaction well on the reaction block. The block has 96 reaction vessels organized in a 8×12 array. Each reactor has a Teflon frit at the bottom, the content of the reactors being separated from the atmosphere by a gasket, located at the top of the block. Reaction mixtures in the block are agitated by a vortex mixer under an inert gas. Groups of 24 reaction vessels can be filtered simultaneously; the filtration process is assisted by positive pressure of an inert gas. The published working temperature range for the synthesizer is -70 to $150\,°C$. Cleavage from the solid support can be carried out automatically, after which the cleavage solutions are transferred into individual beakers in a collection tray placed under the reaction block

The Model 440 Multiple Organic Synthesizer is smaller than the 496 MOS (60 cm W $\times$ 64 cm D $\times$ 90 cm H; $24 \times 25 \times 36$ in), but otherwise is very similar to that instrument. The major difference is the reaction block, which has only 40 larger reaction vessels, the volume of each being 8 mL.

The 384 High Throughput Synthesizer is essentially a large version of the 496 model. It is designed for parallel synthesis of 384 compounds, and has four reaction blocks with 96 reaction vessels in each. Most of the other parameters are identical to those of the 496 MOS.

3.5.4 HP 7686 Solution-Phase Synthesizer (Hewlett Packard)

Scientists from Neurocrine Biosciences were first who reported on synthesis of 350 individual analogs of triazine 8 (Fig. 36) [93] using the HP 7686 Solution-Phase Synthesizer (Fig. 37).

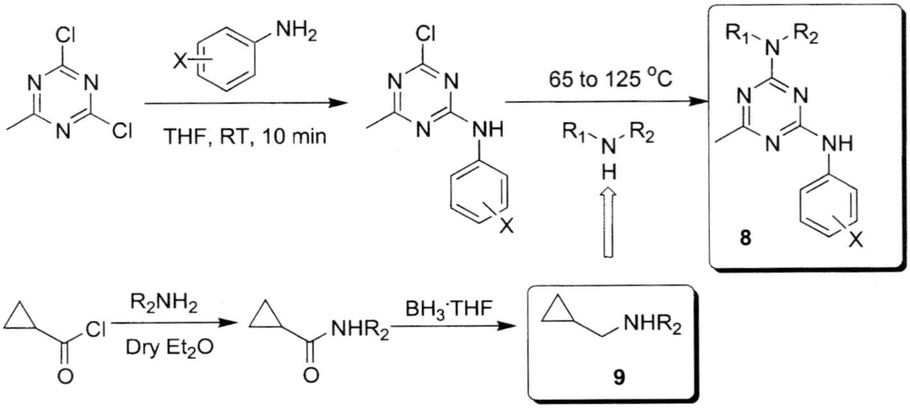

Figure 36. Synthesis of 350 triazine analogs on Hewlett-Packard HP 7686 Solution-Phase Synthesizer [93].

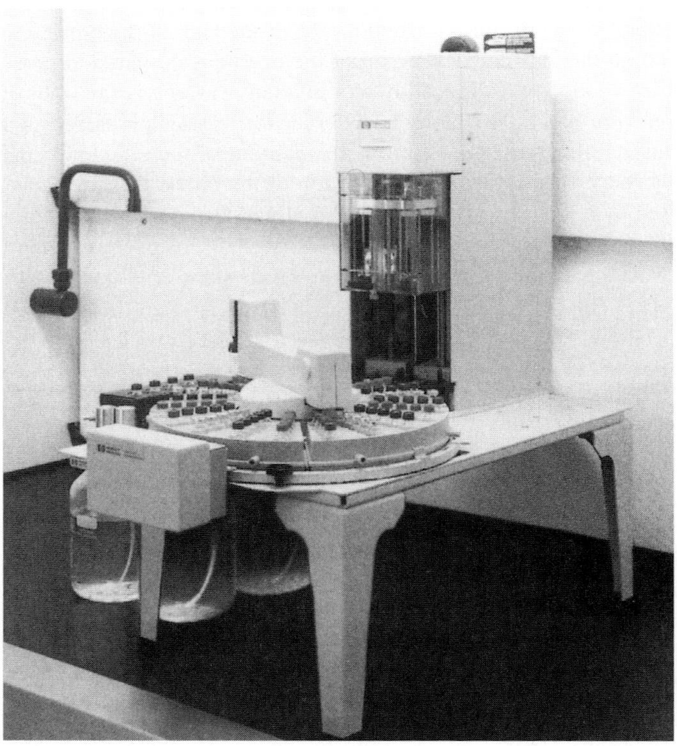

Figure 37. Hewlett-Packard HP 7686 Solution-Phase Synthesizer.

Almost 70% of the synthesized analogs of triazine **8** were isolated in 70–95 + % purity and in milligram quantities. Preparation of amines **9** in an additional reaction sequence on the HP 7686 synthesizer demonstrates its application for automating chemical reactions requiring inert atmosphere and absolute solvents.

The synthesizer easily fits into a chemical hood. Its carousel hosts up to 100 vials that can hold reagents, reaction intermediates, and/or HP solid-phase extraction cartridges (100 or 300 mg size) for purification of synthesized compounds. It has a heating station capable of maintaining temperatures in the range of 30 to 120 °C (in 1 °C increments) for a time specified by the operator. The content of the vials can be easily transferred in wide volume increments (0.001 to 1.6 mL) to any location on the synthesizer, and with high precision.

Reagent transfers are carried out by syringes (0.5, 1.0, 2.5, 10.0, or 25.0 mL, 2.5 mL standard; reagent carryover on the instrument is minimized by an internal/external needle wash system) with no tubing lines to fill out, and therefore, with no materials to waste. This feature is especially useful when expensive and/or rare reagents are used in the synthesis. Many other commercially available synthesizers are almost useless in a fully automated mode, when it comes to reagent consumption issues, and require manual addition of valuable building blocks. The HP 7686 Solution-Phase Synthesizer is capable of carrying out both liquid–liquid and solid-phase extractions. It is equipped with a bar code reader/mixer, which offers three pre-set speeds for automatically mixing vial contents. The HP 7686 offers automatic evaporation of solvents and solvent exchange. Evaporation is achieved by blowing a stream of an inert gas into a vial while the needle tracks the surface of the liquid as it descends. Evaporation can be carried out with or without heating.

References

[1] Gallop, M.A., Barrett, R.W., Dower, W.J., Fodor, S.P.A., and Gordon, E.M. *J. Med. Chem.* **37**, 1233–1251 (1994).
[2] Gordon, E.M., Barrett, R.W., Dower, W.J., Fodor, S.P.A., and Gallop, M.A. *J. Med. Chem.* **37**, 1385–1401 (1994).
[3] Janda, K.D. *Proc. Natl. Acad. Sci. USA* **91**, 10779–10785 (1994).
[4] DeWitt, S.H. *Pharmaceutical News* **1**, 11–14 (1994).
[5] Special Issue: Combinatorial Chemistry, Foote, Ch.S., and Valentine, J.S. (eds.) *Acc. Chem. Res.* **29**, 112–170 (1995).
[6] Terrett, N.K., Gardner, M., Gordon, D.W., Kobylecki, R.J., and Steele, J. *Tetrahedron* **51**, 8135–8173 (1995).
[7] *Combinatorial Peptide and Nonpeptide Libraries: A Handbook*. Jung, G. (ed.). VCH: Weinheim; New York; Basel, 1996.
[8] *Molecular Diversity and Combinatorial Chemistry: Libraries and Drug Discovery*. Chaiken, I.M., and Janda, K.D. (eds.) American Chemical Society: Washington, DC, 1996.
[9] *Combinatorial Chemistry and Molecular Diversity in Drug Discovery*. Gordon, E.M., and Kerwin, J.F., Jr. (eds.) Willey-Liss, Inc.: New York, 1998.
[10] Terrett, N.K. *Combinatorial Chemistry*. Oxford University Press: New York, 1998.
[11] Houghten, R.A. *Proc. Natl. Acad. Sci. USA* **82**, 5131–5135 (1985).
[12] Furka, A., Sebestyen, F., Asgedom, M., and Dibo, G. *Cornucopia of peptides by synthesis*. In Highlights of Modern Biochemistry, Proceedings of the 14th International Congress of Biochemistry, Prague, Czechoslovakia, 1988; VSP: Utrecht, The Netherlands, 1988; vol. 13, pp. 47.
[13] Furka, A., Sebestyen, F., Asgedom, M., and Dibo, G. *Int. J. Pept. Protein Res.* **37**, 487–493 (1991).
[14] Lam, K.S., Salmon, S.E., Hersh, E.M., Hruby, V.J., Al-Obeidi, F., Kazmierski, W.M., and Knapp, R.J. In: *Peptides: Chemistry and Biology: Proceedings of the Twelfth American Peptide Symposium, June 16–21, 1991, Cambridge, Massachusetts, U.S.A.* Smith, J., and Rivier, J.E. (eds.) ESCOM, Leiden; 492–495 (1992).

[15] Lam, K.S., Salmon, S.E., Hersh, E.M., Hruby, V.J., Kazmierski, W.M., and Knapp, R.J. *Nature* **354**, 82– 84 (1991).
[16] Lam, K.S., Lebl, M., and Krchnak, V. *Chem. Rev.* **97**, 411–448 (1997).
[17] Lebl, M. *J. Comb. Chem.* **1**, 3–24 (1999).
[18] Brenner, S., and Lerner, R.A. *Proc. Natl. Acad. Sci. USA* **89**, 5381–5383 (1992).
[19] Nielsen, J., Brenner, S., and Janda, K.D. *J. Am. Chem. Soc.* **115**, 9812–9813 (1993).
[20] Needels, M.C., Jones, D.G., Tate, E.H., Heinkel, G.L., Kochersperger, L.M., Dower, W.J., Barrett, R.W., and Gallop, M.A. *Proc. Natl. Acad. Sci. USA* **90**, 10700–10704 (1993).
[21] Nikolaev, V., Stierandova, A., Krchnak, V., Seligmann, B., Lam, K.S., Salmon, S.E., and Lebl, M. *Pept. Res.* **6**, 161–170 (1993).
[22] Kerr, J.M., Banville, S.C., and Zuckermann, R.N. *J. Am. Chem. Soc.* **115**, 2529–2531 (1993).
[23] Ohlmeyer, M.H.J., Swanson, R.N., Dillard, L.W., Reader, J.C., Asouline, G., Kobayashi, R., Wigler, M., and Still, W.C. *Proc. Natl. Acad. Sci. USA* **90**, 10922–10926 (1993).
[24] Nestler, H.P., Bartlett, P.A., and Still, W.C. *J. Org. Chem.* **59**, 4723–4724 (1994).
[25] Still, W.C. *Acc. Chem. Res.* **29**, 155–163 (1996).
[26] Ni, Z-J., Maclean, D., Holmes, C.P., Murphy, M.M., Ruhland, B., Jacobs, J.W., Gordon, E.M., and Gallop, M.M. *J. Med. Chem.* **39**, 1601–1608 (1996).
[27] Ni, Z.-J., Maclean, D., Holmes, C.P., and Gallop, M.M. Encoded Combinatorial Chemistry: Binary Coding Using Chemically Robust Secondary Amine Tags. In: *Methods in Enzymology*, vol. 267, *Combinatorial Chemistry*. Abelson, J.N. (ed.). Academic Press: San Diego; pp. 261–272 (1996).
[28] Holmes, C.P., and Jones, D.G. *J. Org. Chem.* **60**, 2318–2319 (1995).
[29] Maclean, D., Schullek, J.R., Murphy, M.M., Ni, Z.-J., Gordon, E.M., and Gallop, M.A. *Proc. Natl. Acad. Sci. USA* **94**, 2805–2810 (1997).
[30] Stankova, M., Issakova, O., Sepetov, N.F., Krchnak, V., Lam, K.S., and Lebl, M. *Drug. Dev. Res.* **33**, 146–156 (1994).
[31] Zambias, R.A., Boulton, D.A., and Griffin, P.R. *Tetrahedron Lett.* **35**, 4283–4286 (1994).
[32] Geysen, H.M., Wagner, C.D., Bodnar, W.M., Markworth, C.J., Parke, G.J., Schoenen, F.J., Wagner, D.S., and Kinder, D.S. *Chem. Biol.* **3**, 679–688 (1996).
[33] Wagner, D.S., Markworth, C.J., Wagner, C.D., Schoenen, F.J., Rewerts, C.E., Kay, B.K, and Geysen, H.M. *Comb. Chem. High Throughput Screening* **1**, 143–153 (1998).
[34] Nicolau, K.C., Xiao X.Y., Parandoosh, Z., Senyei, A., and Nova, M. *Angew. Chem. Int. Ed.* **34**, 2289–2291 (1995).
[35] Moran, E.., Sarshar, S., Cargill, J.F., Shahbaz, M.M., Lio, A., Mjalli, A.M.M., and Amstrong, R.W. *J. Am. Chem. Soc.* **117**, 10787–10788 (1995).
[36] Zhao, C., Shi, S., Mir, D., Hurst, D., Li, R., Xiao, X., Lillig, J., and Czarnik, A.W. *J. Comb. Chem.* **1**, 91–95 (1999).
[37] Czarnik, T., and Nova, M. *Chemistry in Britain* 39–41, (1997).
[38] Xiao, X.Y., Zhao, C., Potash, H., and Nova, M.P. *Angew. Chem. Int. Ed. Engl.* **36**, 780–782 (1997).
[39] Fodor, S.P.A., Read, L.J., Pirrung, M.C., Stryer, L., Lu, A.T., and Solas, D. *Science* **251**, 767–773 (1991).
[40] Pirrung, M.C., Read, J.L., Fodor, S.P.A., and Stryer, L. U.S. Patent 5,143,854, (1992).
[41] Jacobs, J.W., and Fodor, S.P.A. *Trends. Biotechnol.* **12**, 19–26 (1994).
[42] Pirrung, M.C. *Chem. Rev.* **97**, 473–488 (1997).
[43] Antonenko, V.V., Kulikov, N.V., Tumelty, D., Sundaram, A., and Jacobs, J.W. In: *Peptides 1994*, Maia H.L.S. (ed.). ESCOM: Leiden, pp. 451–452 (1995).
[44] Patchornik, A., Amit, B., and Woodward, R.B. *J. Am. Chem. Soc.* **92**, 6333–6335 (1970).
[45] Holmes, C.P., and Kiangsoontra, B. In *Peptides: Chemistry, Structure and Biology*, Hodges, R.S., and Smith, J.A. (eds.) ESCOM: Leiden, pp. 110–112 (1994).
[46] Carpino, L.A. *J. Am. Chem. Soc.* **115**, 4397–4398 (1993).
[47] Lipshutz, R.J., Morris, D., Chee, M., Hubbel, E., Kozal, M.J., Shah, N., Shen, N., Yang, R., and Fodor, S.P.A. *BioTechniques* **19**, 442–447 (1995).
[48] Holmes, C.H., Adams, C.L., Kochersperger, L.M., and Mortensen, R.B. *Biopolymers* **37**, 199–211 (1995).
[49] Pease A.C., Solas, D., Sullivan, E.J., Chronin, M.T., Holmes, C.P., and Fodor, S.P.A. *Proc. Natl. Acad. Sci. USA* **91**, 5022–5026 (1994).
[50] Cho, C.Y., Moran, E.J., Cherry, S.R., Stephans, J.C., Fodor, S.P.A., Adams, C.L., Sundaram, A., Jacobs, J.W., and Schultz, P.G. *Science* **261**, 1303–1305 (1993).
[51] Frank, R., Guler, S., Krause, S., and Lindermaier, W. In: *Peptides 1990*, Giralt, E., and Andreu, D. (eds.) ESCOM: Leiden, pp. 151–152 (1991).

[52] Frank, R. *Tetrahedron* **48**, 9217–9232 (1992).
[53] Frank, R. *Bioorg. Med. Chem. Lett.* **3**, 425–430 (1993).
[54] Frank, R., Kieß, M., Lahmann, H., Behn, Ch., and Gausepohl, H. In: *Peptides 1994*, Maia H.L.S. (ed.) ESCOM: Leiden, pp. 479–480 (1995).
[55] Frank, R., In: *Immunology Methods Manual*, Lefkovits, I. (ed.) Academic Press, vol. 2, pp. 777–795 (1997).
[56] Bray, A.M., Maeji, N.J., and Geysen, H.M. *Tetrahedron Lett.* **31**, 5811–5814 (1990).
[57] Hoffman, S., and Frank, R. *Tetrahedron Lett.* **35**, 7763–7766 (1994).
[58] Gao, B., and Esnouf, M.P. *J. Biol. Chem.* **271**, 24634–24638 (1996).
[59] Gao, B., and Esnouf, M.P. *J. Immunol.* **157**, 183–188 (1996).
[60] Antonenko, V.V. In: *Combinatorial Chemistry and Combinatorial Technologies Methods and Applications*. Miertus, S., and Fassina, G. (eds.) Marcel Dekker, pp. 205–232 (1998).
[61] Szardenings, A.K., Antonenko, V., Campbell, D.A., DeFrancisco, N., Ida, S., Shi, L., Sharkov, N., Tien, D., Wang, Y., and Navre, M. *J. Med. Chem.* **42**, 1348–1357 (1999).
[62] Geysen, H.M., Meloen, R.H., and Barteling, S. J. *Proc. Natl. Acad. Sci. USA* **81**, 3998–4002 (1984).
[63] Geysen, H.M., Rodda, S.J., and Mason, T.J. *Mol. Immunol.* **23**, 709–715 (1986).
[64] Geysen, H.M., Rodda, S.J., Mason, T.J., Tribbick, G., and Scoofs, P.G. *J. Immunol. Methods* **102**, 259–274 (1987).
[65] Bray, A.M., Maeji, N.J., Valerio, R.M., Campbell, R.A., and Geysen, H.M. *J. Org. Chem.* **56**, 6659–6666 (1991).
[66] Bray, A.M., Maeji, N.J., Jhingran, A.G., and Valerio, R.M. *Tetrahedron Lett.* **32**, 6163–6166 (1991).
[67] Bray, A.M., Jhingran, A.G., Valerio, R.M., and Maeji, N.J. *J. Org. Chem.* **59**, 2197–2203 (1994).
[68] Frank, R., and Doring, R. *Tetrahedron* **44**, 6031–6040 (1988).
[69] Blankenmayer-Menge, B., and Frank, R. *Tetrahedron Lett.* **29**, 5871–5874 (1988).
[70] Lebl, M., Gut, V., and Eichler, J. In: *Peptides 1990*. Giralt, E., and Andreu, D. (eds.) ESCOM: Leiden, pp. 1059–1060 (1991).
[71] Lebl, M., and Eichler, J. *Peptide Res.* **2**, 297–300 (1989).
[72] Berg, R.H., Almdal, K., Watsberg Pedersen, W., Holm, A., Tam, J.P., and Merrifield, R.B. *J. Am. Chem. Soc.* **111**, 8024–8026 (1989).
[73] Berg, R.H., Almdal, K., Watsberg Pedersen, W., Holm, A., Tam, J.P., and Merrifield, R.B. In: *Innovation and Perspectives in Solid Phase Synthesis: Peptides, Polypeptides and Oligonucleotides, Macro-Organic Reagents and Catalysts-1990*. Epton, R. (ed.) SPCC: Birmingham, pp. 453–459 (1990).
[74] Terrett, N.K., Gardner, M., Gordon, D.W., Kobylecki, R.J., and Steele, J. *Chem.-Eur. J.* **3**, 1917–1920 (1997).
[75] Internet: http://robosynthon.com.
[76] Powers, J.L., and Scott, W. *Genet. Eng. News*. **18**, 14–18 (1998).
[77] Internet: http://www.stemcorp.com.
[78] Internet: http://www.argotech.com.
[79] Internet: http://www.rapp-polymere.com.
[80] Internet: http://www.peptide.com.
[81] Internet: http://www.multisyntech.com.
[82] Meyers, H.V., Dilley, G.J., Durin, T.L., Powers, T.S., Winssinger, N.A., Zhu, H., and Pavia, M.R. *Mol. Diversity*, **1**, 13–20 (1995).
[83] Meyers, H.V, Dilley, G.J., Durgin, T.L., Powers, T.S., Winssinger, N.A., and Pavia, M.R. *Methods Mol. Cell. Biol.* **6**, 1–7 (1996).
[84] sBaiga T. J. Integrated Instrumentation for High-throughput Organic Synthesis. Presented at Cambridge Healthtech Institute's Third Annual Symposium "High-Throughput Organic Synthesis", March 5–6, Coronado, California (1998).
[85] Internet: http://www.robosci.com.
[86] Stanchfield, J. A flexible Modular System for Performing High-Throughput Synthesis of Small Molecules. Presented at Cambridge Healthtech Institute's Third Annual Symposium "High Throughput Organic Synthesis", March 5–6 Coronado, California (1998).
[87] Stanchfield, J. FlexChem™: A Modular System for High Throughput Synthesis of Small Molecules. *Robbins Innovations*, **5**, No.3, 1–6 (1997).
[88] Harness, J.R. Automation of High-Throughput Synthesis. Automated Laboratory Workstations Designed to Perform and Support Combinatorial Chemistry. In *Molecular Diversity and Combinatorial Chemistry: Library and Drug Discovery*. ACS Conference Proceedings Series, Chaiken, I.M., and Janda, K.D. (eds.) American Chemical Society, Washington, DC, pp. 188–198 (1996).

[89] Rivero, R.A., Greco, M.N., Marynoff, B.E. Equipment for the High-Throughput Organic Synthesis of Chemical Libraries *In A Practical Guide to Combinatorial Chemistry*, Czarnik, A.W., Dewitt, S.H. (eds) American Chemical Society, Washington, DC, pp. 281–307 (1997).

[90] Gooding, O., Hoeprich, P.D., Labadie, J.W., Porco, J.A., van Eikeren, P., and Wright, P. Boosting the Productivity of Medicinal Chemistry Through Automation Tools. Novel Technological Developments Enable a Wide Range of Automated Synthetic Procedures, In *Molecular Diversity and Combinatorial Chemistry: Libraries and Drug Discovery*. ACS Conference Proceedings Series. Chaiken, I.M., and Janda, K.D. (eds.) American Chemical Society, Washington, DC, pp. 199–206 (1996).

[91] Porco, J.A., Deegan, T.L., Devonport, W., Gooding, O.W., MacDonald, A.A., Newcomb, W.S., and van Eikern, P. Automated chemical synthesis: chemistry development on the Nautilus 2400™. *Drugs of the Future* **23**, 71–78 (1998).

[92] Floyd, Ch., Lewis, Ch., Patel, S., and Whittaker, M. The Automated Synthesis of Organic Compounds – Some Newcomers Have Some Success. In *Proceedings of the International Symposium on Laboratory Automation and Robotics 1996*. Zymark Corporation, Zymark Center, Hopkinton, MA, pp. 51–76 (1997).

[93] Whitten, J.P., Yun Feng Xie, Erickson, P.E., Webb, De Souza, E.B., Grigoriadis, D.E., and McCarthy, J.R. *J. Med. Chem.* **39**, 4354–4357 (1996).

4 Tubes and Cubes, Chips and Tips: Tools for Solid-Phase Organic Synthesis

Anthony W. Czarnik

4.1 Introduction

Starting in the early 1990s, a new ability to "pan" rapidly for biological activity created new challenges – and therefore opportunities – for the chemist to contribute to what is now a clearly successful stratagem for discovery [1–6]. These contributions come at two rather different levels: (i) In the creation of vast numbers of diverse, cataloged chemical entities to screen, termed *lead generation*; and (ii) in increasing the efficiency of *lead optimization*, the process by which the structure–activity relationship (SAR) for an identified structure of interest is determined. Lead generation activities have been reported for nearly a decade, producing new compounds in numbers ranging from tens (low-molecular weight organics) to billions (polynucleotides, polypeptides). When creating very large numbers of samples, it is by necessity not possible to evaluate analytically the success of each reaction, nor the homogeneity of even most of the resulting products. This synthesis philosophy demands a reliance on a "numbers game" approach to discovery – not unlike that required by natural products screening.

The creation of large numbers of new samples is facilitated by the use of synthesis on solid supports, in large part because the isolation of intermediate products is so easy. In this chapter, I would like to recount three distinct projects with which I have been involved that relate to solid-phase synthesis.

4.2 Tubes: The Diversomer Project

At many pharmaceutical companies, the generation of leads is typically less of a bottleneck than is the optimization of leads. This is not to say that enhancement of a compound library by any means is undesirable. Rather, unlike most smaller companies that commonly have no compound library of their own, screens of an internal compound collection often yield more interesting "hits" than one can rapidly generate SAR data for. Obtaining such quantitative data on the products of multistep syntheses requires a level of sample characterization that is not possible with large libraries. Thus, methods for enhancing the efficiency of lead optimization will by necessity: (i) permit the use of multistep reaction schemes and most synthetic methods; and (ii) afford smaller libraries (i.e., hundreds of compounds), with each member present in sufficient quantity for characterization and multiple biological testing (i.e., 1 mg). This article describes work carried out in the BioOrganic Chemistry group at Parke-Davis between 1991 and 1996, and the development of the Diversomer method for achieving these goals [7–12].

It is worth recapitulating the strategy behind solid-phase organic synthesis [13, 14]. The rapid synthesis of hundreds of compounds requires simultaneous reactions to be run. As any chemistry graduate student knows, it certainly is possible to have many reactions running at any given time – *n* reactions simply requires *n* reaction vessels. However, multistep reactions conducted in parallel become a problem, as isolation of the desired product requires a purification step. Anyone can set up 40 solution-phase reactions in parallel, but if each reaction is a three-step synthesis the product isolation steps quickly become unmanageable. Merrifield addressed an analogous problem in the context of peptide synthesis some 35 years ago by the employment of solid-phase-supported synthesis. In brief, an insoluble bead with reactive chemical groups on its surface is charged with the first amino acid of a peptide. The addition of more amino acids can be accomplished using large excesses of reagents because their removal (and that of reaction byproducts) requires only copious washing of the bead. This ease of intermediate isolation lowers the intrinsic technical barrier to the application of automation to long syntheses. As a result, the modern-day synthesis of a polypeptide is a largely automated process.

Parallel synthesis is a different problem from long sequential synthesis, but each falls to the solid-phase approach. Organic synthesis on an insoluble matrix has been studied intermittently by various laboratories over several decades. Even so, the reader will note that solid-phase organic synthesis bears some disadvantages that are largely self-evident: (i) While the isolation of reaction intermediates is greatly simplified, their purification becomes impossible (by conventional means); (ii) the monitoring of reaction progress on solid support is less straightforward than the same reaction when run in solution; (iii) the scale of reactions run on solid supports is typically less than that of reactions run in solution; and (iv) insoluble reagents and catalysts are no longer useful when starting materials are located on the surface of an insoluble matrix. In the polypeptide and polynucleotide areas, these problems have been largely ameliorated by extensive research to identify soluble reagents and very high-yielding conditions that limit the need to purify intermediates. Bioanalytical methods permit the characterization of minute amounts of these compounds, making reaction scale less of an issue. While it seems likely that similar extensive research as applied to the more vast realm of organic synthesis will have similar beneficial effects, such research has not yet been completed and remains an important area for study.

Our group chose to utilize solid-phase methods despite these temporal limitations. The monitoring of solid-phase reactions is achievable using gel-phase ^{13}C NMR spectroscopy (among other methods); our group used this method successfully from the onset of our research. A great many organic reaction conditions do not require the use of insoluble reagents and catalysts, and for those that do alternatives frequently exist. The amount of product required for convenient characterization by NMR and mass spectrometry (ca. 1 mg) is readily accessible via solid-phase synthesis, and is likewise vastly more than is required for SAR studies on a single target. Finally, while intermediate purification is an important issue, mitigating circumstances and solutions exist: (i) Ready removal of excess reagents facilitates their use to drive reactions to completion; (ii) reactions products need not be homogeneous for useful evaluations, as contaminants are highly unlikely to demonstrate equipotent ligand properties; and (iii) cyclative cleavage of reaction products from the resin does afford a measure of purification.

This last concept warrants comment. Binding assays of small molecules are best accomplished in solution, which requires the ultimate cleavage of an organic compound from the solid support. We view such cleavage as occurring in one of three ways, as shown in Fig. 1. This figure shows in a schematic way how building blocks may be added to a polymeric support, such as the Merrifield resin. Each "BB" represents a different chemical entity added in the form of a reagent. For example, "BB_1" might represent the class of organic compounds known as carboxylic acids with attachment to a hydroxyl-bearing resin by way of an ester linkage. There are thousands of commercially available carboxylic acids, and thus thousands of potential Resin-*-BB_1 entities. Subsequent additions of different reagents (BB_2, BB_3, etc.) will occur in a sequential fashion, although not necessarily linear as shown in the figure. The first method of cleavage of product from the resin, which we call *cyclative cleavage*, has the added benefit of retaining on the resin those truncated reaction products that are unable to cyclize. In fact, the NMR spectra of many compounds we prepared in this way demonstrate few impurities. The amounts of impurities we do observe can be determined accurately from the NMR spectrum. In most cases, such impurities do not interfere with the biological testing of products, but in cases where high purity is required, automated purification schemes are being developed.

We are working under the premise that syntheses of even several hundred compounds will benefit by the use of automation. While our group was hardly the first to apply robotics to organic synthesis, our goal of applying robotics to parallel synthesis using solid-phase methods

1

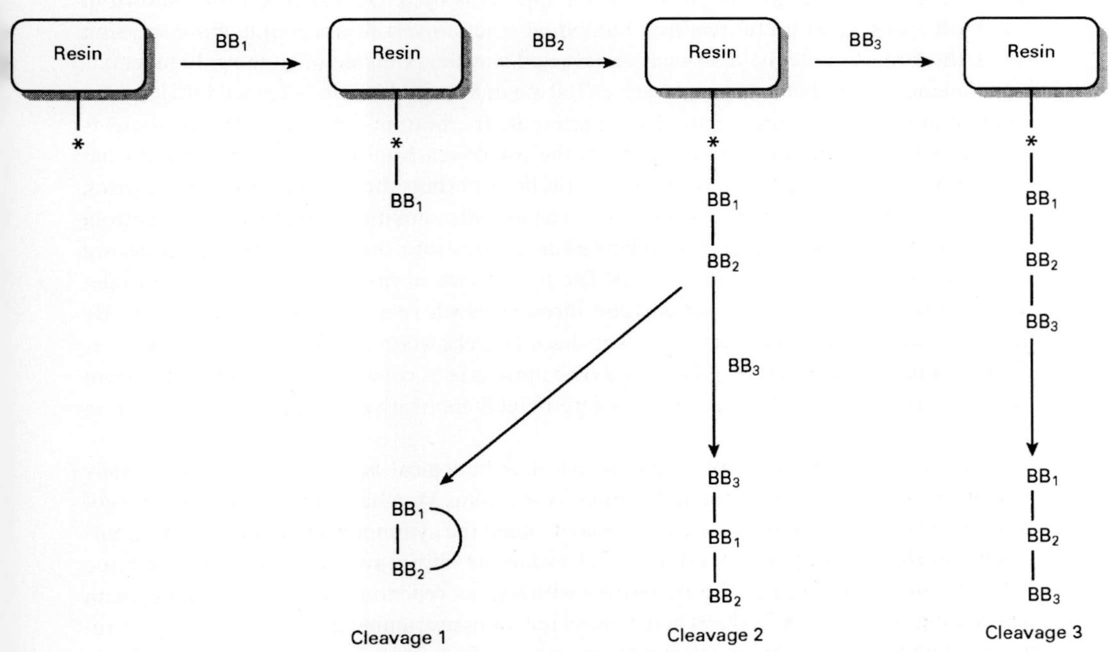

Figure 1. Three conceptually different ways of cleaving a product from a solid synthesis support.

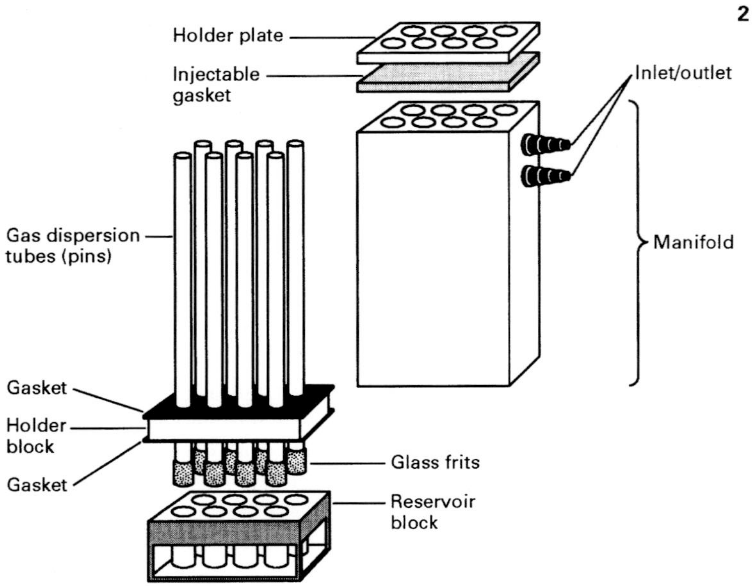

Figure 2. Schematic of the Diversomer parallel synthesis apparatus.

is novel, and required the design of a novel apparatus by S. DeWitt. The result, shown in Fig. 2, offered several useful features. Individual reaction vessels are simple gas dispersion tubes; the largest model held 40 such vessels, and was thus capable of running 40 reactions simultaneously. A polymer support, such as 100 mg of Merrifield resin, is loaded into the tube and contained in the bottom fritted compartment. The bottom portion of the apparatus is used to hold reaction solvents, and permits the use of reaction temperatures ranging from −78 to 200 °C. The top portion of the apparatus body permits the introduction of inert gases, allowing reactions that require anhydrous or anaerobic conditions. By heating the bottom portion of the apparatus and introducing chilled gases into the top, reflux conditions are readily achieved in all 40 reaction tubes. The top surface of the apparatus consists of inlet holes centered above each reaction tube through which reagents can be introduced. By applying inert and air-tight sealing gaskets here and elsewhere on the apparatus, strict exclusion of atmosphere can be maintained. The apparatus is covered in an issued U.S. patent (US Patent No. 5,324,483), and was considered highly innovative when it was introduced in 1993.

When a core molecular structure demonstrates biological activity, its SAR is typically established by synthesizing structural variants for testing. Members of the hydantoin class of compounds (e.g., phenytoin) are useful in controlling the symptoms of epilepsy, making targeted libraries of hydantoins of interest. Our hydantoin library synthesis is shown in Scheme 1. Any parallel synthesis benefits by starting with a set of reactions that work efficiently with a wide range of reagents. In this scheme, those reactions are: amine deprotection (Step 1), reaction of an amine with an isocyanate to give a urea (Step 2), and acid-catalyzed cyclization of the urea to give a hydantoin (Step 3). Each of these reactions is known to be general and

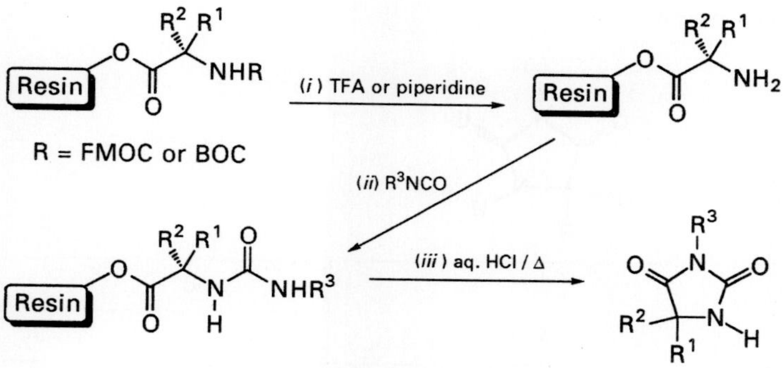

Scheme 1. Route to the synthesis of a hydantoin library.

typically high-yielding in solution, although one cannot predict how such reactions will translate to chemistry on a solid support. Thus, one or two reactions are carried out on a solid-supported starting material before running a set of 40 compounds, to ensure that appropriate conditions can be found.

The actual synthesis of hydantoins was carried out as follows. Eight resins containing different protected amino acids were each placed into five gas dispersion tubes, affording a total of 40 reaction vessels. Deprotection as shown in Step 1 afforded the corresponding amino acid-charged resins, now possessing free amine ($-NH_2$) groups. Each resin-bound amino acid was then reacted with five different isocyanates (Step 2), to afford a total of 40 different ureas. In the final step, treatment of all reaction tubes with 6 M HCl and heat resulted in cyclative cleavage and release of the product hydantoins from the resin, which were dissolved individually in methanol, concentrated, and analyzed. Products and weight yields are shown in Table 1; it should be noted that we now determine mole yields instead using ^{1}H NMR and an internal standard. The reader will note that one reaction (#32) yielded no product. One accepts this deletion as the cost of making large numbers of compounds; if the trend indicates that it should be made, it will be synthesized individually. Each of the 39 product hydantoins in this library was then available for biological testing. Using this same procedure, our group additionally completed the syntheses of benzodiazepine, benzisothiazolone, and other libraries.

4.3 Cubes: Matrix Chemistry

If drug discovery is facilitated by using machines to assist in the synthesis of large collections of new compounds, then should chemists perform such syntheses? Clearly, the answer is "yes." Of course, even with the aid of automation, chemical synthesis can only be done with extensive training. The real question is not whether chemists should or should not engage in the activity, but rather "Are there any tools that really empower the chemist to 'play their game' better than without them?" One such tool is actually an intellectual one. It is the embrace of an approach to synthesis called "combinatorial chemistry", which has been defined as, "a new subfield of chemistry with the goal of synthesizing very large numbers of chemical

Table 1. Hydantoins generated in array

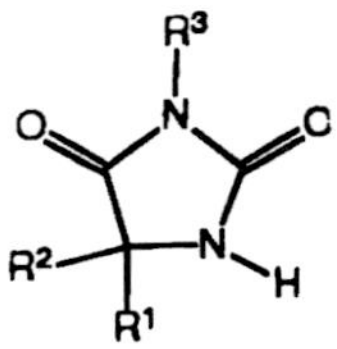

no.	R^1	$R^{2\,a}$	$R^{3\,a}$	yield[b] mg	yield[b] %
1	H	Me	H	4.1	67
2	H	Bn	H	2.5	38
3	H	H	H	3.3	65
4	H	*sec*-Bu	H	3.1	42
5	H	*i*-Bu	H	4.9	61
6	H	*i*-Pr	H	4.9	58
7	H	2-MeInd	H	5.0	35
8	Ph	Ph	H	1.4	5
9	H	Me	Bu	1.6	17
10	H	Bn	Bu	3.9	47
11	H	H	Bu	1.0	13
12	H	*sec*-Bu	Bu	5.3	48
13	H	*i*-Bu	Bu	0.7	7
14	H	*i*-Pr	Bu	0.9	8
15	H	2-MeInd	Bu	0.9	5
16	Ph	Ph	Bu	1.6	5
17	H	Me	allyl	0.3	4
18	H	Bn	allyl	2.4	29
19	H	H	allyl	3.7	48
20	H	*sec*-Bu	allyl	3.6	36
21	H	*i*-Bu	allyl	5.0	64
22	H	*i*-Pr	allyl	1.6	14
23	H	2-MeInd	allyl	1.9	11
24	Ph	Ph	allyl	2.1	7
25	H	Me	$2\text{-}CF_3C_6H_4$	2.6	23
26	H	Bn	$2\text{-}CF_3C_6H_4$	2.2	23
27	H	H	$2\text{-}CF_3C_6H_4$	2.9	28
28	H	*sec*-Bu	$2\text{-}CF_3C_6H_4$	5.7	46
29	H	*i*-Bu	$2\text{-}CF_3C_6H_4$	4.7	37
30	H	*i*-Pr	$2\text{-}CF_3C_6H_4$	4.9	33
31	H	2-MeInd	$2\text{-}CF_3C_6H_4$	3.0	15
32	Ph	Ph	$2\text{-}CF_3C_6H_4$	0.0	0
33	H	Me	$4\text{-}MeOC_6H_4$	3.1	22
34	H	Bn	$4\text{-}MeOC_6H_4$	3.5	32
35	H	H	$4\text{-}MeOC_6H_4$	5.6	46
36	H	*sec*-Bu	$4\text{-}MeOC_6H_4$	11.5	81
37	H	*i*-Bu	$4\text{-}MeOC6H_4$	3.2	21
38	H	*i*-Pr	$4\text{-}MeOC_6H_4$	4.1	24
39	H	2-MeInd	$4\text{-}MeOC_6H_4$	4.9	22
40	Ph	Ph	$4\text{-}MeOC_6H_4$	3.0	7

[a] Benzyl (Bn), 2-methylindolyl (2-MeInd), 2-trifluorotolyl ($2\text{-}CF_3C_6H_4$), 4-methoxyphenyl ($4\text{-}MeOC_6H_4$).

[b] Yields based upon reported loading of commerically available functionalized resins (0.34–1.04 mequiv g).

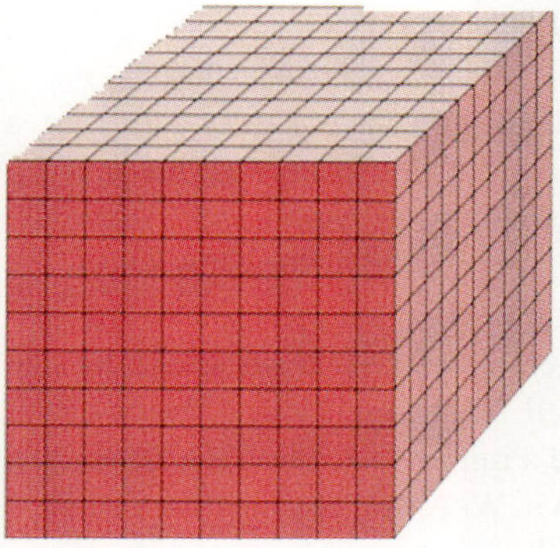

Figure 3. A $10 \times 10 \times 10$ matrix of products can be made via $10+10+10$ directed sorting synthesis steps.

entities by condensing a small number of reagents together in all combinations defined by a given reaction sequence".

In an effort requiring both chemists and engineers, electronic memory tags have now been developed as tools available to all chemists engaged in "split/pool" combinatorial syntheses. The term "split/pool" refers to a procedure described first by A. Furka, in which microreactors (large or small pieces of chemically reactive, but insoluble polymer) are repooled for each step in a reaction sequence such that $a+b+c$ reactions (i.e., flasks) result in the production of $a \cdot b \cdot c$ reaction products (i.e., compounds). As an example, one starting material can be converted to 10 related products by using one type of reaction and 10 types of reagents. Using two sequential reactions and 10 types of each required reagent (20 total), 100 related products can be made. Three reactions incorporating ten reagents each (30 total) can yield 1000 related products. The resulting $10 \times 10 \times 10$ cubic matrix of products (Fig. 3) can be produced either as one mixture, several less complex mixtures, or as 1000 individual discrete compounds, depending on the method employed. The "split/pool" method essentially views the cube in Fig. 3 as a set of planes, each of which results in one step from one reaction using one reactant type. Thus, the efficiency of synthesis can be increased by making every compound in that plane in a single synthesis vessel (flask, beaker, etc.). The total number of reaction steps, and therefore vessels, required is simply the total number of planes that exist.

As each matrix element resides in a unique set of three planes, making every compound in the cube requires that individual reactors (matrix elements) find their way into the correct vessel at each step of the synthesis. If one could look at a chemical intermediate and know its structure, this process of sorting reactors between synthesis steps would be very simple. As we cannot, reactors must be encoded in a way that can be easily read [15–17]. Many types of encoding are known, including: spatial, graphical, chemical, spectrometric, electronic, and physical. Encoding gives each reactor a name with respect to a relational database; if that name can be read quickly and easily, then automation may be used at each sorting step.

4.4 Chips: Radiofrequency-Tagged Microreactors

The tools that we have developed utilize electronic identification devices. Figure 4 illustrates one such device, the radiofrequency (RF) ID tag. Similar devices have been used for years to tag laboratory mice subcutaneously, and more recently these have been used to provide security for automobile ignition keys, to secure building entrance identification, and for a variety of other functions. The application of RF tags to combinatorial syntheses was reported in 1995 by two groups working at IRORI and at Ontogen. The RF ID tag is about the size of a "flea" stirbar and is encased in a thick-walled glass shell. As shown in Fig. 4, the essential components are: an antenna (the largest component), and a microelectronic chip. Each chip has a unique, non-volatile 40-bit ID code laser-etched into it. With 40 bits available, a total of 2^{40} (over 1 trillion) unique ID codes is possible. This virtually inexhaustible range allows one to guarantee the uniqueness of all present and future RF tag ID codes. Additional bits are used to perform extensive digital error detection, which prevents incorrect reporting. (Bar codes are a more common graphical embodiment of encoding, but of course one that is substantially less information-rich).

A transceiver controlled by a computer is used to interrogate and receive the ID code of each RF tag. The transceiver antenna transmits a specially modulated, 125 kHz electromagnetic field. This field is of very low energy and not harmful. When an RF tag is held within about 1 cm from the transceiver's antenna, energy is picked up by the RF tag's antenna. A rectifier in the chip converts this energy to microwatt levels of DC power, which is enough to power-up the logic circuitry on the chip. In a very real sense, the RF tag is similar to a crystal radio (which does not require an external power source), except that the device serves as both receiver and transmitter. It is "self-contained", in that the chip uses no internal batteries and has no external metallic connections.

A synchronization signal modulated onto the transceiver's signal allows the chip to respond with its ID code (a serial sequence of ID bits) and error-checking bits. The time elapsed between placing the chip on the transceiver and seeing the ID code on the interfaced computer screen is about 0.5 s.

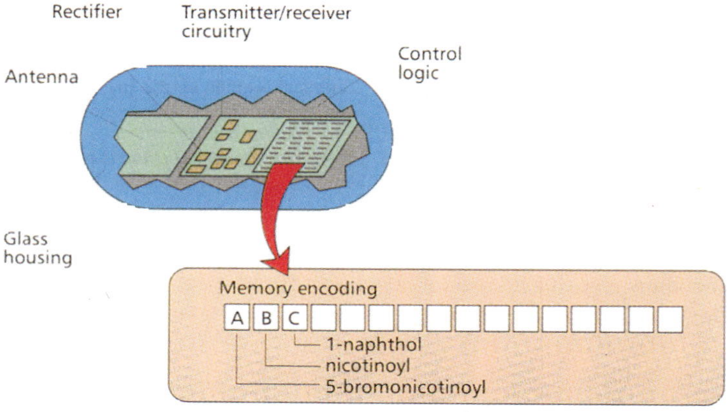

Figure 4. Schematic of an Rf memory chip useful for encoded synthesis.

If one could physically associate this chip with a compound undergoing synthetic transformations, then the ID code on the chip would permit one to pick up a sample at any point of the combinatorial synthesis and know which reaction(s) it had already been through. Of course, just knowing the reaction history of a sample is not equivalent to knowing the chemical content of a sample, but it is much better than nothing. If each reactor ("compound-carrying unit") must be present in a series of reaction flasks containing lots of other reactors, then having them tagged makes it possible to put them into the right flask for each step of the synthesis. Moreover, if that tag can be read easily, then that movement of reactors can be automated. Electronic tagging permits all of these benefits.

How this works in practice is detailed as follows. After a compound has been identified for which several hundred to several thousand derivatives would be of value, a synthetic route is chosen that: (i) permits linkage to a solid-phase support; (ii) utilizes reaction steps that appear possible to optimize to ≥90% yield; and (iii) affords reagents in each step for which desirable variants can be purchased (or, less optimally, can be made trivially). In the synthesis itself, one of the significant advantages of the "microreactor" approach becomes evident: one can use standard laboratory glassware and equipment to accomplish the library synthesis. There is no need for the automation of liquid-handling steps, and indeed no need for automation at all until rather large libraries are desired (*vide supra*).

Our Chemistry group at IRORI (led on this project by Dr. Shuhao Shi) synthesized an 18×8×3 array (432 members) of tyrphostins (**1**), the parent of which is a known tyrosine kinase inhibitor [18]. We started with a set of 432 reactors called MicroTubes, which are polystyrene-grafted polypropylene tubes (about 2.5 cm (1 in) long), each containing one RF tag. The polystyrene was chemically functionalized with aminomethyl groups to afford 48 micromoles of aminomethyl group per MicroTube (Scheme 2). At this level of loading, a product of MW 400 could be obtained in as much as 19 mg. (The molecular weights for the expected products in this 432 member library ranged from 278 to 594 Daltons.)

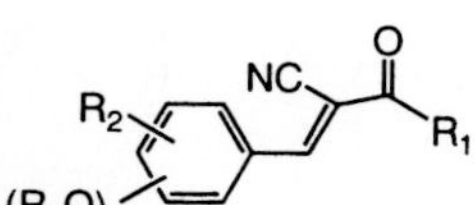

1

For the first reaction step, 24 (8×3) MicroTubes were placed into each of 18 flasks, and a different variant of reagent 1 was then added to each flask; the reaction then occurred. We had previously spent time optimizing the conditions for each reaction step so that we knew for how long the 18 reactions should be allowed to proceed. After completion of all variants of the first step, all 432 MicroTubes were placed into a single large beaker and washed three times sequentially with DMF, methanol, and dichloromethane; a final wash with ether facilitated drying. It is extremely important to note that the washing of resin is not a rinse, just as the washing of dirty socks is not simply a rinse—absorbed impurities require both agitation and time in order to be desorbed from the support.

For the second reaction step, every MicroTube needed to be reacted with an identical reagent, so one reaction was carried out with all 432 MicroTubes, after which they were washed thoroughly. For reaction Step 3, 54 (18×3) MicroTubes were placed into each of eight flasks; computer software permitted us to read the RF tag of each MicroTube and know the flask

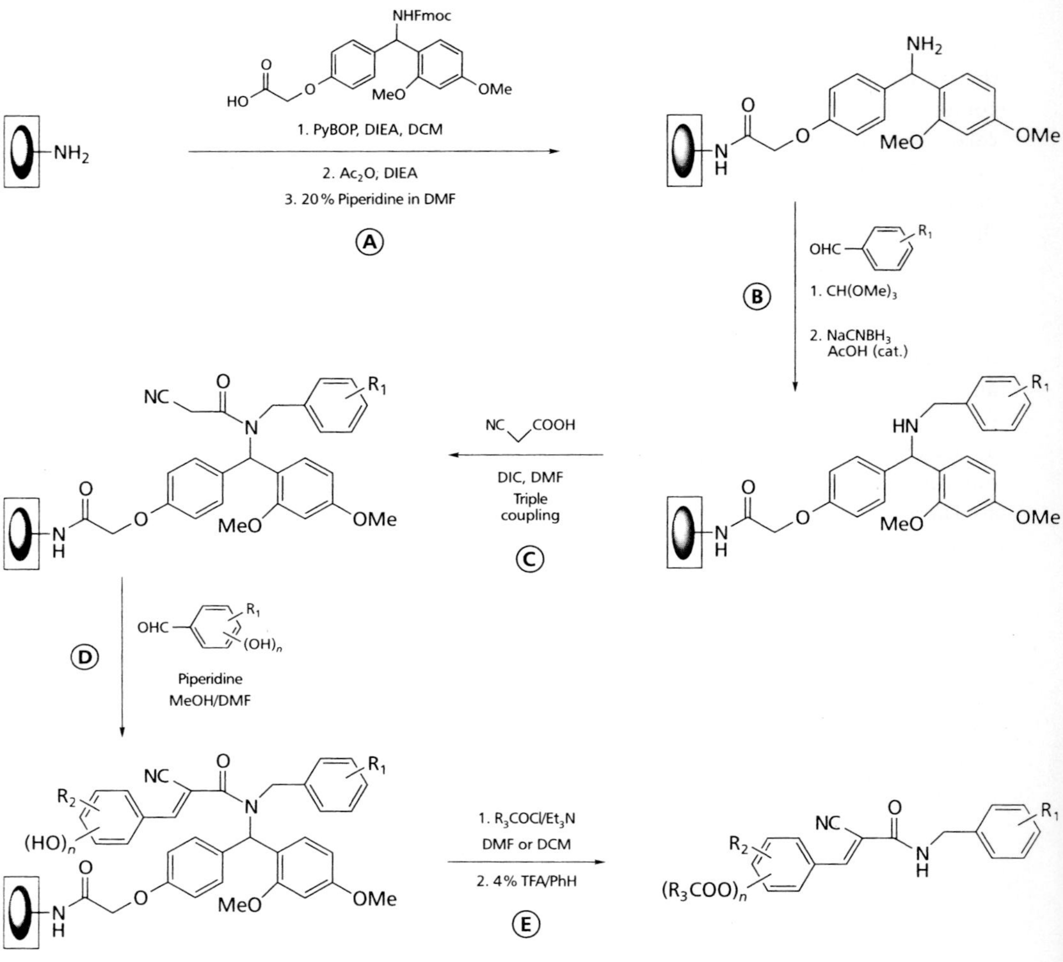

Scheme 2. A combinatorial synthesis of 432 tyrphostins (prepared using 29 reaction vessels).

into which it would be placed. Again, a different variant of reagent 3 was added to each of the eight flasks, the reactions were carried out, and the collected 432 MicroTubes washed. For the fourth synthetic step, 144 (18 × 8) MicroTubes were placed into each of three flasks. At this stage, reading the ID codes is the only convenient way to know which MicroTube goes into which flask, but with software the task is trivial. This number of MicroTubes (144) can fit conveniently into a 500-mL round-bottomed flask, with adequate room available for solvents and reagents. After the reaction, all MicroTubes are pooled and washed extensively.

With the synthetic steps complete, and if all the reactions have worked as planned, one has 432 MicroTubes each with a different tyrphostin derivative covalently attached to its surface. For most applications, the final step is to cleave each compound from the MicroTube into a separate vial. One MicroTube is placed into each of 432 vials, after which a "cleavage" solution of 4% trifluoroacetic acid in benzene is added. After about 2 h, almost all the compound has been cleaved from the surface of the MicroTube and has passed into the solution in the vial. Before removing the MicroTubes, each ID code is read for a final time and each vial is labeled. Because benzene was used in the cleavage solution, each vial may be frozen and the cleavage solution removed conveniently by lyophilization.

Each labeled vial contains the product from a single MicroTube. This sample is just that – a sample – to be distinguished from a compound, which is a pure sample. To establish the content of each sample, as well as to characterize the compound(s) in it, we evaluated each of the 432 samples. It is simply not practical today to obtain 432 NMR spectra, even though this approach would yield among the best information. Instead, we first carried out a thin-layer chromatographic (TLC) analysis of all samples, which indicated in every case one clear "spot" on the plate. TLC is an inexpensive and rapid analysis method that in most cases can be conducted on every member of a compound library. Subsequently, 24 samples chosen at random were characterized using proton NMR, MS, and HPLC methods. All of the NMR spectra obtained revealed the desired compound as the major component, and low ($\leq 10\%$) levels of organic impurities present. All mass spectra provided the correct molecular weights. As determined by HPLC (254 nm detection), the homogeneity of the 24 samples ranged from 56% to 100%. We are now confident in using all of these library samples for biological

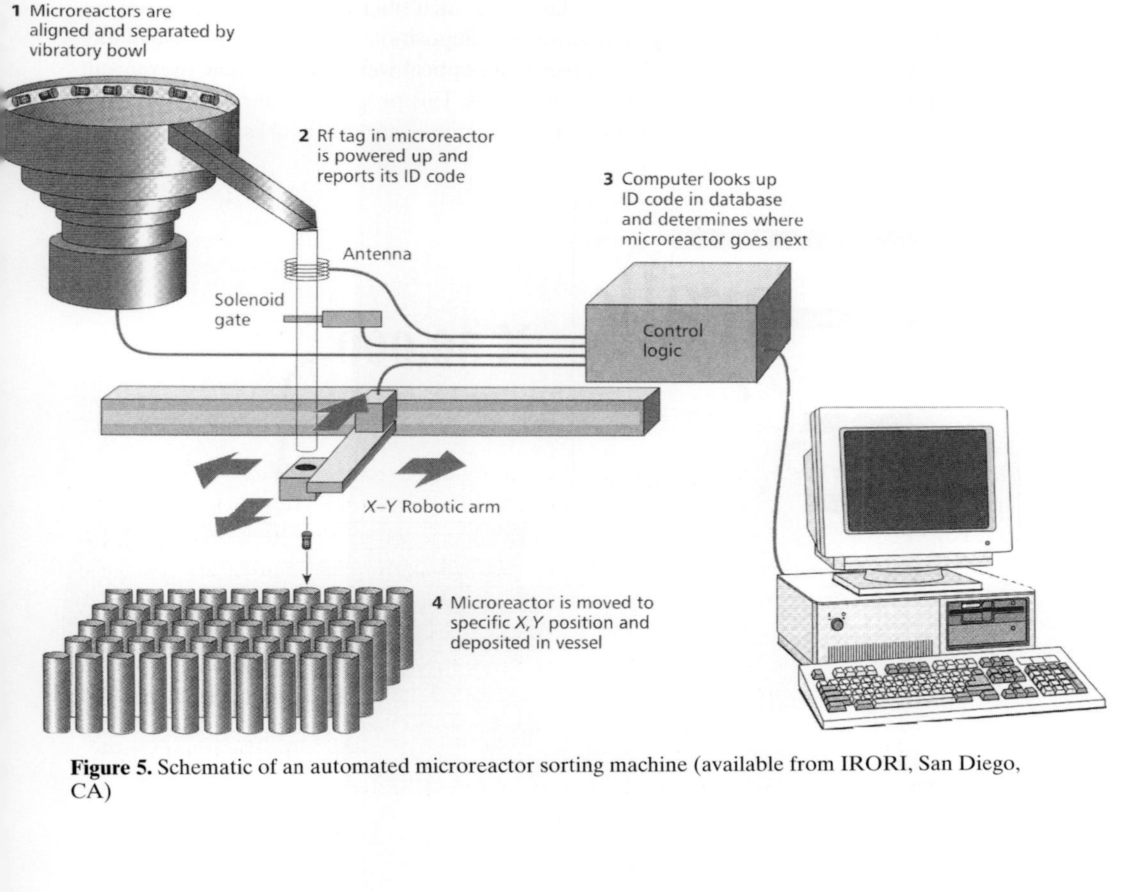

Figure 5. Schematic of an automated microreactor sorting machine (available from IRORI, San Diego, CA)

screening purposes, because activity will be readily apparent, and the most active compounds will be resynthesized, purified, and retested.

For scale-up to a 10000-member tyrphostin library using this same approach, the hand-reading of MicroTube ID codes after each reaction step becomes cumbersome. To facilitate this process, IRORI developed a machine capable of doing the MicroTube reading and movement steps in an automated manner. As shown in the schematic in Fig. 5, up to 10000 MicroTubes (or resin-bearing containers called MicroKans) can be placed into the vibrating bowl. When switched on, the bowl feeds the reactors one at a time through an on-board transceiver. The code is read and sent to the computer, which recognizes it from its database and instructs the X–Y conveyor arm where to deposit the MicroTube. The system does not involve any liquid handling, and is designed for overnight operation, with the sorting of 10000 reactors in about 10 h.

4.5 Tips: Self-Assembled, Randomly Ordered Microarrays

My present company, Illumina, is developing the fiber-optic bundle sensor platform devised by Professor David Walt and used for genotyping and other applications. This approach begins with a commercially available bundle containing hundreds to millions of individual optical fibers (Fig. 6) [19–21]. As shown in the figure, each fiber consists of a glass core surrounded by a glass cladding of somewhat different composition. Because the refractive indices of core and cladding differ, the fiber serves as an optical waveguide capable of transmitting light over long distances, with little intensity loss. This property is, in part, what makes optical fibers so useful in the construction of telecommunication networks. The Walt group

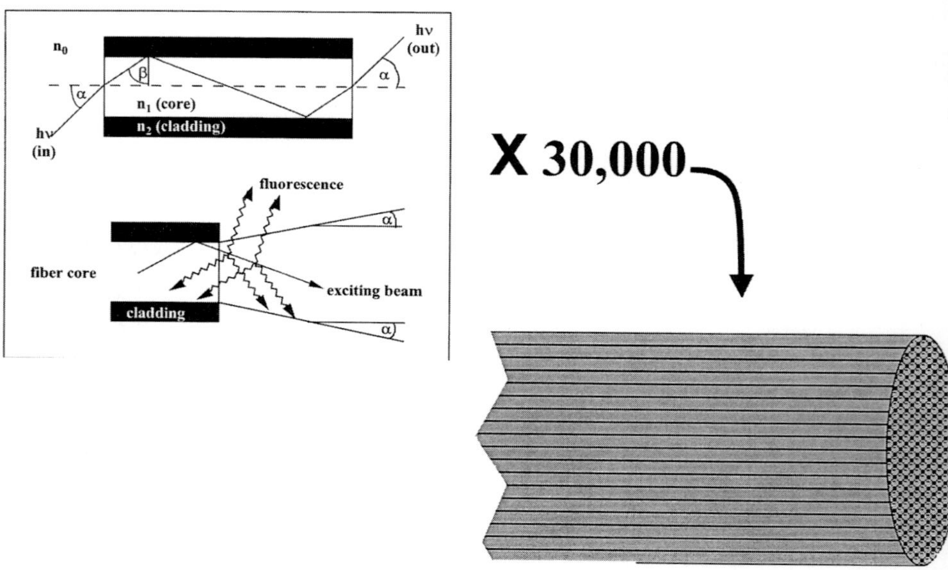

Figure 6. Fiber-optic imaging bundle.

showed that when the distal end of a fiber is derivatized with a fluorescent chemosensing molecule, excitation from the proximal end led to fluorescence at the distal end, with some of that light returning down the fiber and back to the proximal end. The result is a remote sensor.

Use of an appropriate cladding material will permit a fiber's core to etch faster than its clad. When such an etch is carried out on a fiber-optic bundle, the result is a bundle tip with one "well" per fiber. Remarkably, when the tip is brought into contact with a suspension of well-sized beads, self-assembly occurs to yield a bead-filled array. While this bulk self-assembly process is not really akin to the molecular assembly processes of current academic interest, the result is perhaps equally striking. The atomic force micrographs (AMF) shown in Fig. 7 reveal an etched bundle tip with 4-μm wells being filled by 3.5-μm polystyrene beads. Experience reveals that many types of bead material will self-assemble into etched bundle tips.

By functionalizing the beads with receptors, the result is a dense array of those receptors; densities as high as 250000 elements in a 1-mm^2 area have been achieved. One of the inviting aspects of this approach is its operational simplicity. Any chemical library, combinatorial or otherwise, that can be formatted onto beads can be densely arrayed with no precision equipment whatsoever. All array-forming processes are bulk processes.

Unlike array synthesis methods in which the synthesis history (and, therefore, often chemical identity) of each element is known, array assembly methods result in the random ordering of functional elements. In a very real sense, the Illumina approach exchanges the challenging problem of array synthesis for the interesting problem of post-assembly registration. The analogy of this situation to that of parallel versus split/pool combinatorial synthesis is an excellent one, and the same approach that enables element identification in split/pool synthesis also enables post-assembly registration: element encoding.

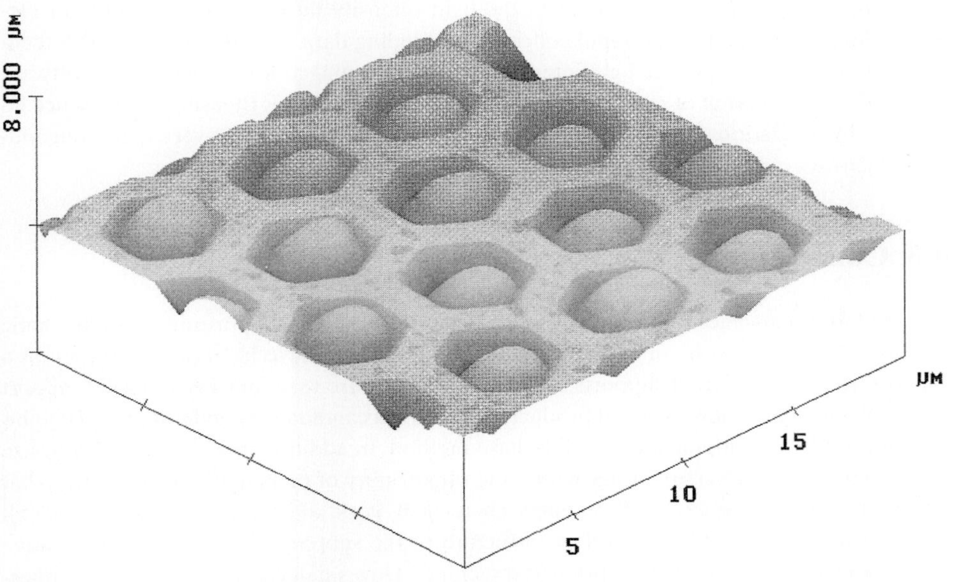

Figure 7. Atomic Force Micrograph view of an assembled bead array.

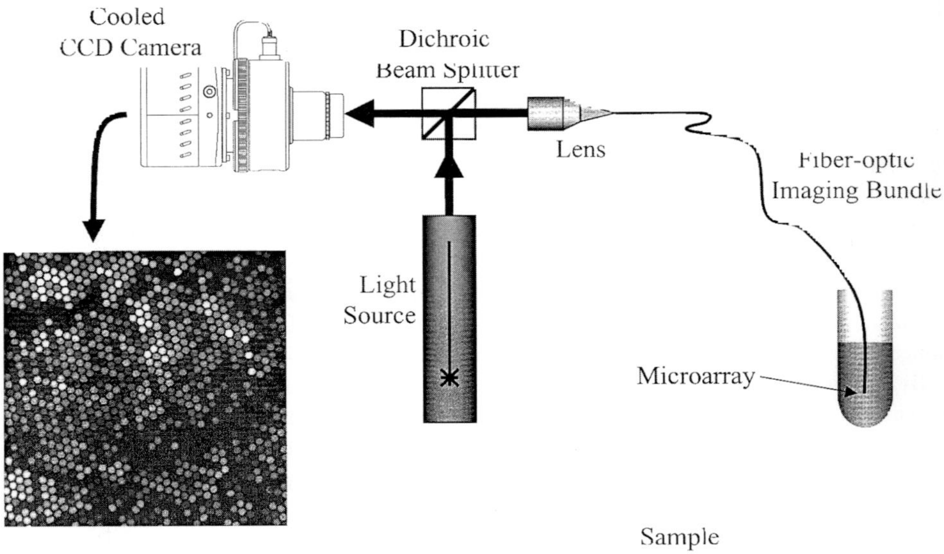

Figure 8. Fiber-optic sensor imaging system.

Light signals can be generated using a wide assortment of assay methods, all of which transduce a binding event into a luminescence event. When interfaced with near-stock hardware and software for fiber-optic sensing platforms, useful array-based sensing experiments may be accomplished (Fig. 8). The use of a commercial charge-coupled device (CCD) for data collection permits measurement of the light intensity information from all fiber elements simultaneously. For the rapid collection of binding data from dense arrays, this technology is highly enabling. The light intensity data can be imaged using commercial software to afford a constellation of results such as that shown in the figure. By using fluorescence labels and hybridization directly analogous to those already in use, these very dense oligonucleotide arrays can be used for parallel analysis of DNA-containing solutions.

4.6 Conclusion

The modern renaissance of solid-phase synthesis offers vast new opportunities for synthetic chemists. Perhaps 10% of all known solution reaction types have been demonstrated on a support. How will different supports affect the rates of these reactions, i.e., is a solid support like a solvent? Any known natural product synthesis is a candidate for at least partial combitorialization by developing a new solid-phase method. In addition, the ability to synthesize our own molecular diversity brings with it the opportunity of making libraries in a way that makes their screening easier. How many assays will work when a library member is still attached to the support? How will the structure of the support affect this? What creative ways exist to release compounds prior to screening? How small can we get? All these questions and more await our discovery.

References

[1] DeWitt, S.H., and Czarnik, A.W. *A Practical Guide to Combinatorial Chemistry.* ACS Books: Washington, 1997.

[2] Wilson, S.H., and Czarnik, A.W. *Combinatorial Chemistry: Synthesis and Application.* Wiley & Sons: New York, 1997.

[3] Czarnik, A.W. *Analyt. Chem.* **70**, 378A–386A (1998).

[4] Czarnik, A.W., and Winograd, N. Combinatorial Chemistry. In: *McGraw-Hill Yearbook of Science & Technology*, vol. 66. 1998.

[5] Mei, H.-Y., and Czarnik, A.W. Oligonucleotide Libraries as a Source of Molecular Diversity". In: *Combinatorial Chemistry and Molecular Diversity in Drug Discovery.* Gordon, E.M., and Kerwin, J.F. (eds.) Wiley: New York, 1997.

[6] Czarnik, A.W. *ChemTracts-Organic Chemistry* **8**, 13 (1995).

[7] DeWitt, S.H., Kiely, J.S., Stankovic, C.J., Schroeder, M.C., Cody, D.M., Pavia M.R. *Proc. Natl. Acad.Sci. USA* **90**, 6404–6413, (1993)

[8] Czarnik, A.W., and DeWitt, S.H. *Chem. Brit.* **32**, 43 (1996).

[9] DeWitt, S.H., and Czarnik, A.W. *Curr. Opin. Biotechnol.* **6**, 640 (1995).

[10] DeWitt, S.H., Schroeder, M.C., Stankovic, C.J., Strode, J.E., and Czarnik, A.W. *Drug Dev. Res.* **33**, 116 (1994).

[11] Czarnik, A.W., DeWitt, S.H., Schroeder, M.C., Stankovic, C.J., and Strode, J.E. *Polym. Prep.* **35**(2), 985 (1994).

[12] DeWitt, S.H., and Czarnik, A.W. *Acc. Chem. Res.* **29**, 114 (1996).

[13] Czarnik, A.W. *Biotechnol. Bioeng. (Comb. Chem.)* **61**, 77–79 (1998).

[14] MacDonald, A.A., Dewitt, S.H., Ghosh, S., Hogan, E.M., Kieras, L., Czarnik, A.W., and Ramage, R. *Mol. Diversity* **1**, 183 (1996).

[15] Czarnik, A.W. *Proc. Natl Acad. Sci. USA* **94**, 12738 (1997).

[16] Czarnik, A.W., and Nova, M.P. *Chem. Brit.* **33**, 39 (1997).

[17] Czarnik, A.W. *Curr. Opin. Chem. Biol.* **1**, 60 (1997).

[18] Shi, S., Xiao, X.-Y., and Czarnik, A.W. *Biotechnol. Bioeng. (Comb. Chem.)* **61**, 7–12 (1998).

[19] Zhao, C., Shi, S., Mir, D., Hurst, D., Li, R., Xiao, X.-Y., Lillig, J., and Czarnik, A.W. *J. Comb. Chem.* **1**, 91–95 (1999).

[20] Czarnik, A.W. *Modern Drug Discovery* **1**, 49–55 (1998).

[21] Li, R., Xiao, X.-Y., and Czarnik, A.W. *Tetrahedron Lett.* **39**, 8581–8584 (1998).

5 Stereoselective Synthesis of Enantiomerically Pure Drugs*

Maurizio Benaglia, Mauro Cinquini, and Franco Cozzi

Abbreviations

acac	acetylacetonate
BINAP	2,2′-bis(diphenylphosphino)-1,1′-binaphthyl
Boc	t-butyloxycarbonyl
COD	cyclooctadiene
DBU	diazabicycloundecene
d.e.	diastereoisomeric excess
DEAD	diethylazadicarboxylate
DIBAL-H	diisobutylaluminum hydride
DIPEA	diisopropylethylamine
DIPT	diisopropyltartrate
DMF	dimethylformamide
d.r.	diastereoisomeric ratio
e.e.	enantiomeric excess
LA	Lewis acid
LDA	lithium diisopropylamide
LHMDS	lithium hexamethyldisylazane
MEM	methoxyethoxymethyl
Ms	CH_3-SO_2
PMB	4-MeO-C_6H_4-CH_2
Py	2-pyridyl
Red-Al	sodium bis(2-methoxyethoxy)aluminum hydride
TADDOL	tartaric acid derived diol
TBS	t-butyldimethylsilyl
TEA	triethylamine
THF	tetrahydrofuran
Tf	CF_3-SO_2
TFAA	trifluoroacetic anhydride
TMS	trimethylsilyl
Tol	4-Me-C_6H_4
TPS	t-butyldiphenysilyl
TS	transition state
Ts	4-Me-C_6H_4-SO_2

* Dedicated to Professor Fernando Montanari on the occasion of his 75th birthday.

5.1 Introduction

The intrinsically chiral and non-racemic nature of the living world often results in its different interactions with the enantiomers of a given substance. If this substance is a drug, it might well be that only one of the two isomers is capable of exerting the desired therapeutic effect, while the other can be inert (but still must be metabolized by the organism), harmful (causing even dramatic biological damages), or responsible for difficult-to-predict (and possibly undesirable) side effects.

Well aware of this fact and stimulated by the new policy statements issued by the regulating agencies [1–3], the pharmaceutical industry has, over the past decade or so, systematically begun to develop chiral drugs in enantiomerically enriched and possibly pure form. Currently [3], 68% of the 1200 drugs under development world-wide are chiral, and 75% of these (or an amazing 51% of the total number) are developed as single isomers.

This new trend has caused a tremendous change in the industrial small and large-scale production approach to enantiomerically pure drugs, leading to the revisiting and updating of old technologies (as the resolution of racemic mixtures), and to the development of new methodologies for their large-scale preparation (as the use of stereoselective syntheses and of biocatalyzed reactions).

All of these methods have their own advantages and drawbacks. The aim of this chapter is to review the recent trends in the *stereoselective synthesis* of chiral drugs by presenting a restricted selection of relevant applications and results. Some of these have been chosen because they serve to illustrate the pro and cons of a certain approach rather than because of their own practical value.

The production of enantiomerically pure drugs by resolution and biocatalyzed processes are the topic of other chapters of this book.

5.2 Classification of the Methods for the Synthesis of Enantiomerically Pure Compounds

5.2.1 Definition of Stereoselective Synthesis

"Stereoselective process" is a comprehensive expression that refers to a vast variety of chemical transformations capable of generating one stereoisomer of a substance in excess over the other(s) [4].[1]

The stereoselective process is not necessarily a synthesis, i.e., a reaction in which bonds are formed; indeed, stereoselectivity can also be observed when some bonds of a molecule are broken to produce two or more species.

More specifically, the stereoselective synthesis of an enantiomerically enriched compound requires that the two enantiomers of the product (that therefore must be a molecule contain-

1 Since Mislow's original proposal was made [4], the definitions of *stereoselective synthesis* and of *stereoselectivity* have remained virtually unchanged and are largely agreed upon: see for instance the definition of these terms reported in ref. [5].

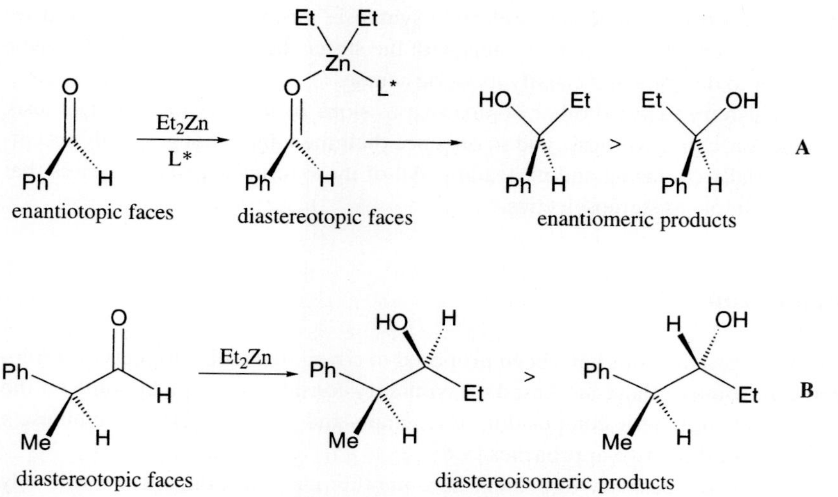

Figure 1. Stereoselective addition of diethylzinc to different aldehydes (L* = chiral non-racemic ligand).

ing at least one stereogenic element as a center or an axis or a plane of stereogenicity) are produced in unequal amounts, or, ideally, free from each other.[2]

The efficiency of a stereoselective process, determined by the ratio between the concentrations of the obtained stereoisomers, is defined as stereoselectivity or stereoselection. This ultimately depends on the difference in the free energy of activation ($\Delta\Delta G^{\ddagger}$) of the reaction pathways leading to the stereoisomeric products: obviously, the higher this difference, the more stereoselective the process.

It is common practice to speak of *diastereoselective synthesis* when the products are diastereoisomers, and of *enantioselective synthesis* when they are enantiomers. In both cases, however, the stereoisomers are obtained in unequal amounts because of the diastereoisomeric relationships between the competitive transition states (TS) leading to different stereoisomers. Consider for instance the reactions reported in Fig. 1.

Both are stereoselective syntheses resulting from diastereoisomerically related attacks of the organometallic species on diastereotopic carbonyl faces. Reaction A generates enantiomers and results in the enantioselective production of one enantiomer over the other. Reaction B generates diastereoisomers and results in the diastereoselective production of one diastereoisomer over the other. The argument that the enantiotopic faces of benzaldehyde in A become diastereotopic only upon reagent complexation, whereas the 2-phenylpropanal faces in B are diastereotopic '*per se*' clearly cannot be used to classify these processes in a different way. Indeed, both reactions can be considered of *type 2a* according to the Mislow's classification of stereoselective processes [4].

---

2 It is not the aim of this chapter to deal either with the definitions or with the methods of evaluation of the stereoisomeric composition of a mixture of stereoisomers. For an authoritative recent overview of these subjects, see ref. [5] and Chapter 6.

In conclusion, the expression "enantioselective synthesis" seems inappropriate, and we will only use the term enantioselectivity to deal with the stereochemical outcome of a given process rather than to describe and classify its steric course.

Finally, we will also try to avoid other popular expressions such as asymmetric synthesis, asymmetric induction, chiral synthesis, and so on, since their meaning is not generally accepted and thus potentially confusing and misleading. All of these terms classify processes that are better defined simply as stereoselective.

5.2.2 Classification

Several exhaustive classifications have been proposed to cover all possible aspects of stereoselective synthesis. Some of these are based on symmetry considerations [4, 6], some on the stereogenicity and enantiomeric composition of reactants and products [7], and some use a sort of combination of these two approaches [5, 8].

Therefore, it seems unnecessary to present here another classification, and we will only report some examples of reactions illustrating stereoselective synthetic approaches relevant to the topic of this chapter, i.e., the preparation of *enantiomerically enriched* compounds. Stereoselective reactions leading to unequal amounts of *racemic* diastereoisomers will not be considered.

Similarly, those reactions carried out on an enantiomerically pure compound in which no new stereogenic element is created (e.g.: the $NaBH_4$ reduction of a chiral non-racemic aldehyde, the S_N2 type conversion of a chiral non-racemic tosylate into the corresponding bromide, the conversion of a stereogenic element in a new one) will also be disregarded.

5.2.2.1 Stoichiometric Reactions

5.2.2.1.1 Chiral Non-Racemic Substrates

Perhaps the simplest way to control the absolute stereochemistry at a newly formed stereogenic element involves the use of an enantiomerically pure substrate.[3] In this case, the stereochemical information is transferred from the already present stereogenic element(s) (most likely, one (or more) stereocenter(s)) to the new one(s), in a process often, but somehow misleadingly, referred to as "asymmetric induction".[4]

The (*S*)-configurated aldehyde of reaction A possesses two diastereotopic carbonyl faces. Thus, the two possible attacks of the Grignard reagent (on the *Re* or on the *Si* face) are diastereotopically related. The reaction leads to the formation of an excess of the *syn* isomer over the *anti* one [9], as predicted by the Cram–Felkin–Anh rule [10].

---

3 The substrate is a reactant whose structural features are almost entirely maintained throughout a reaction sequence. A reagent contributes to the target molecule only a part of its atomic components.

4 On the basis of simple symmetry considerations it is obvious that in this type of process the asymmetry of the substrate is maintained throughout the reaction to the product independently of the creation of a new stereocenter.

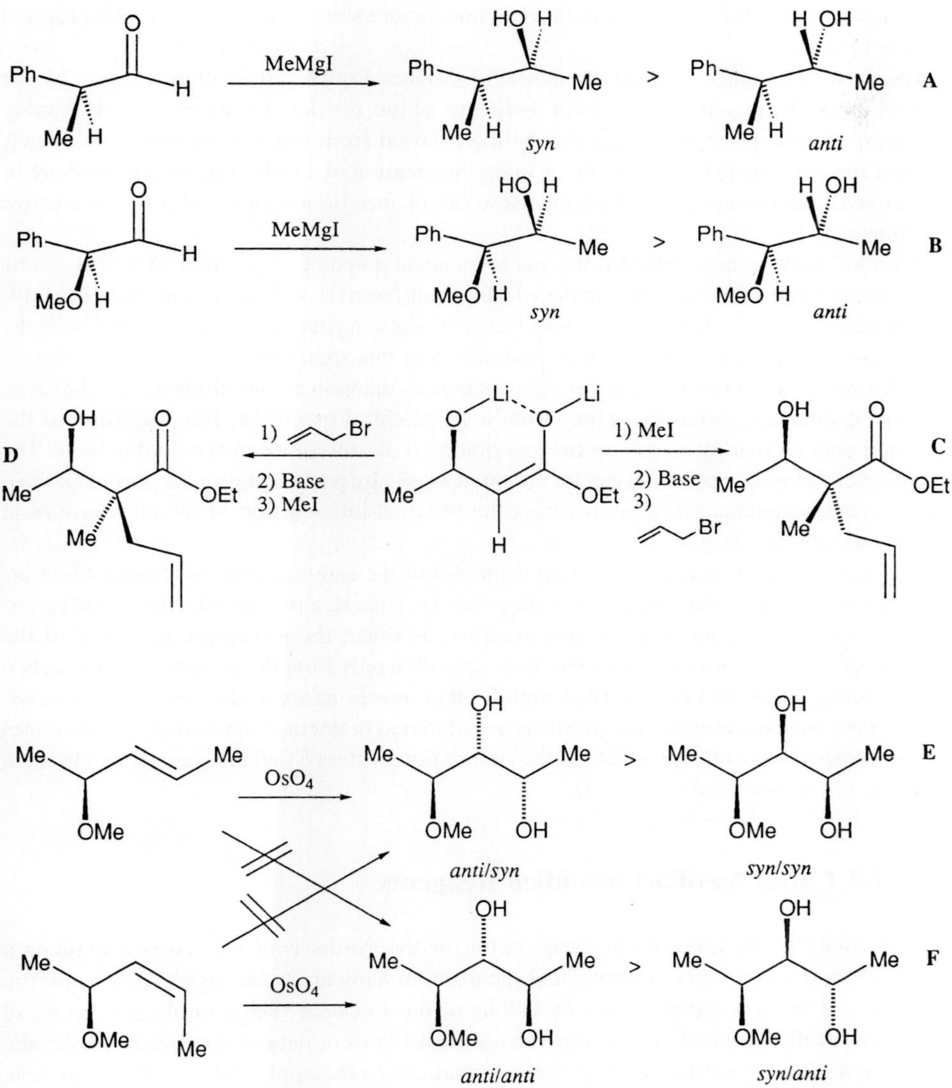

Figure 2. Examples of stereoselective formation of new stereocenters by reaction of chiral non-racemic substrates with achiral reagents.

This model, as are many other rationalizations of the steric course of a stereoselective reaction, is based on a combination of stereochemical factors (such as the preferential conformations around the bond(s) connecting the stereocenter and the reacting carbon, and the relative steric requirements of the substituents at the stereocenter) and stereoelectronic effects (such as the substrate/reagent orbital interactions and the direction of attack of the nucleophile on the carbonyl).

Reaction B differs from A only in that it is much more stereoselective in the generation of the *syn* isomer.

This result originates from co-ordination of the metal of the nucleophilic reagent by the two oxygens of the substrate to form a chelate whose restricted conformational mobility maximizes the transfer of stereochemical information from the pre-existing to the newly formed stereocenter [11]. As we will see later, the creation of a well-ordered intermediate or TS is one of the most popular ways to ensure or enhance the efficiency of a stereoselective reaction.

It is also worth mentioning that the stereochemical outcome of reactions A and B can be reversed by forcing the chiral aldehydes to adopt a different reacting conformation. Thus, different stereoisomers arising from stereodivergent reaction courses can be obtained from the same substrate, greatly expanding the applicability of this approach.

Reactions C and D in Fig. 2 are strictly related to reaction B and illustrate another possibility of obtaining stereodivergency from a single chiral precursor. The alkylation of the chelated enolate readily occurs on the less-hindered diastereoface of the double bond. The two diastereoisomeric products can be obtained simply by performing an alkylation/enolization/alkylation sequence and by inverting the order of introduction of the alkyl groups at C-2 of the substrate [12].

Reactions E and F in Fig. 2 represent examples of the simultaneous creation of two stereocenters under the influence of an existing one. In general, a process like this could generate four diastereoisomers [13]. In this example, however, the stereospecificity [5][5] of the osmylation reaction makes the two *syn* diols accessible only from the (*E*)-alkene and the two *anti* diols only from the (*Z*)-one. Thus, both E and F reactions are at the same time stereoselective (as two diastereoisomeric products are obtained in unequal amounts from each one) and stereospecific (as the products of the osmylation of the (*E*)-alkene cannot be obtained from the (*Z*)-isomer, and vice versa).

5.2.2.1.2 Chiral Auxiliary Modified Reagents

The reactions of Fig. 2 are particularly useful for the synthesis of molecules containing a stereocenter that is preserved throughout the reaction sequence to the target compound. The substrates of these reactions generally belong to the so-called "chiral pool", a large set of enantiomerically enriched or pure derivatives available from natural sources. However, the very origin of these substrates poses some limitations to the applicability of this approach, both in terms of substitution patterns at the stereocenters and of their stereochemical configuration.

A possible way to circumvent this problem is provided by the "chiral auxiliary" approach. A chiral auxiliary is an enantiomerically pure residue that is covalently attached to a (generally) achiral substrate to allow its stereocontrolled transformation, and is then removed (and possibly recycled) to release the desired product.

---

5 A reaction is said to be stereospecific when it proceeds through a mechanism that leads to the formation of a single stereoisomeric product from a single stereoisomeric substrate.

The main features of an efficient chiral auxiliary are: (i) Commercial availability at low cost in both enantiomeric forms; (ii) the facility of insertion on the substrate and removal from the product in high yields and non-racemizing conditions; and, obviously, (iii) powerful and predictable stereodirecting ability.

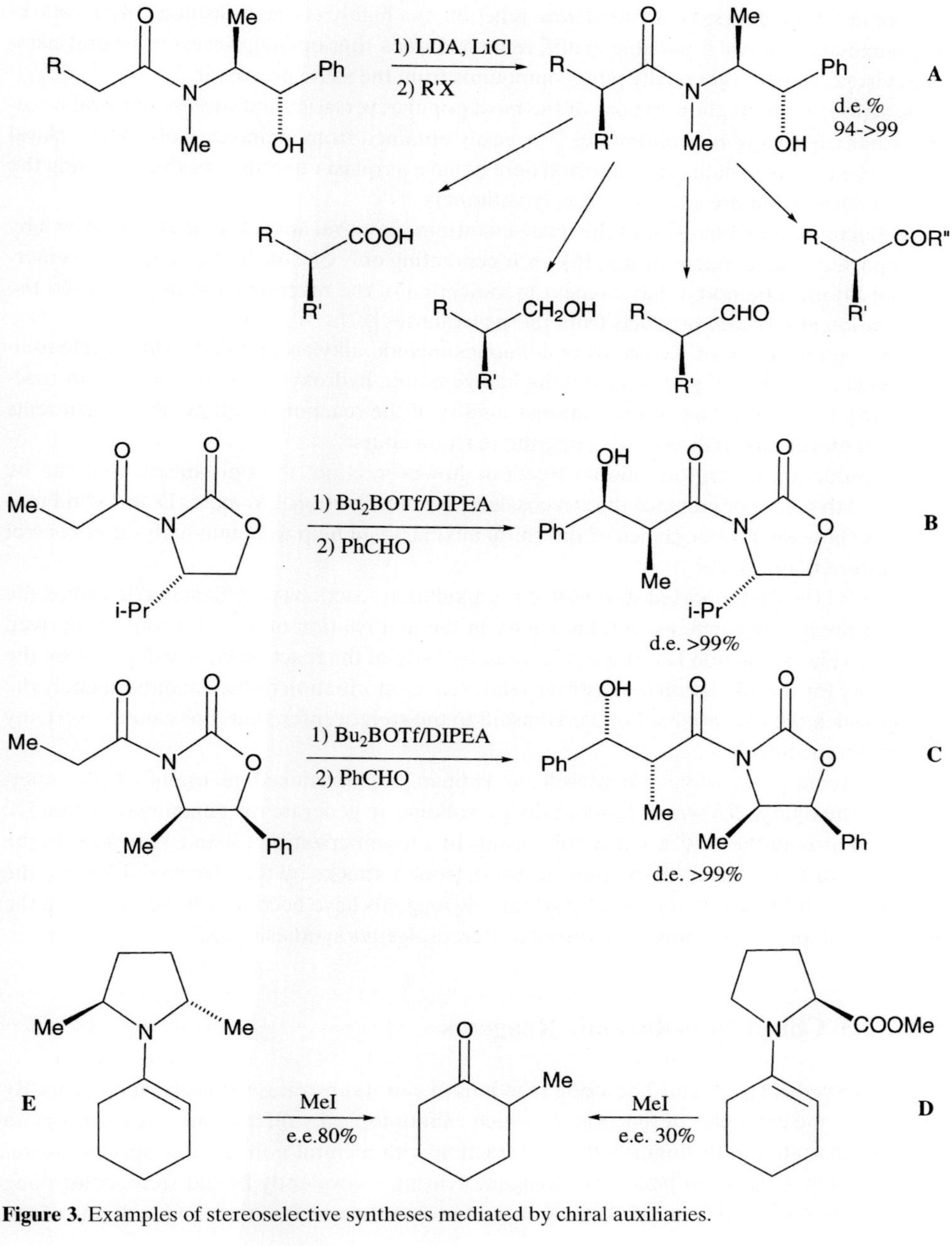

Figure 3. Examples of stereoselective syntheses mediated by chiral auxiliaries.

Many chiral auxiliaries that fulfill all of these requirements are currently available from the "chiral pool" and from synthetic and semisynthetic sources. A very limited collection of this type of stereoselective syntheses is presented in Fig. 3 to illustrate the fundamental principles of the chiral auxiliary strategy.

In reaction A of Fig. 3, a very cheap chiral auxiliary such as pseudoephedrine is able to control the steric course of an alkylation reaction in a highly efficient fashion [14]. Remarkably, auxiliary removal is possible in different conditions, thus opening access to several classes of virtually enantiomerically pure compounds from the same precursor.

Reactions B and C show the use of the most popular, versatile, and successful chiral auxiliaries, namely Evans' oxazolidinones [15], easily obtained from aminoalcohols of the "chiral pool". The two oxazolidinones reported here behave as quasi-enantiomers, thus avoiding the use of their much more expensive real enantiomers.

The enantiomeric behavior of the quasi-enantiomeric chiral auxiliaries is clearly shown by the reported aldol condensations [16], each generating only one of the possible four isomeric aldols. It must be noted that a recent modification of the reaction conditions allowed the preparation also of *anti* products from these auxiliaries [17].

Other applications of Evans' oxazolidinones include alkylation, Diels–Alder cycloaddition, Michael addition, and electrophilic halogenation, hydroxylation, and amination reactions [15]. In all cases, the conformational rigidity of the reaction TS allows the substituents at the stereocenters to direct efficiently the reaction course.

The reduction of conformational freedom, however, is not the only means that can be exploited to secure or enhance the stereoselectivity of a reaction. Examples D and E in Fig. 3 illustrate how the proper choice of the chiral auxiliary can help to obtain high stereocontrol by a different approach.

In one of the first reported stereoselective alkylation reactions [18], Yamada described the use of a methyl prolinate as chiral auxiliary in the methylation of a cyclohexanone-derived enamine (Fig. 3; reaction D). The low stereoselectivity of the reaction clearly depends on the possibility for the alkylation to occur on a different conformation of the enamine, namely the one in which the C-C double bond is transoid to the stereocenter, that thus cannot exert any useful stereocontrol.

To overcome this problem, Whitesell and Felman [19] introduced the use of a C_2-symmetric chiral auxiliary, (*S,S*)-*trans* 2,5-dimethylpyrrolidine, to generate the enamine (reaction E). Rotation around the N–C bond simply results in a topomerization [5] and, therefore, in the reduction of the number of the non-stereocontrolled attacks by the electrophile. Since the appearance of this seminal paper, C_2-symmetric reagents have become a standard tool in the hands of the organic chemists interested in stereoselective synthesis [20].

5.2.2.1.3 Chiral Non-Racemic Reagents

The examples of Figs. 2 and 3 describe reactions of chiral substrates with achiral reagents. By contrast, some examples of reactions in which enantiotopically related faces or groups of an achiral substrate are distinguished by interaction with a chiral non-racemic species are reported in Fig. 4. This can be either a reagent bearing a covalently bound stereocontrolling group (as in A, B, and C in Fig. 4), or a promoter co-ordinatively interacting with one of the

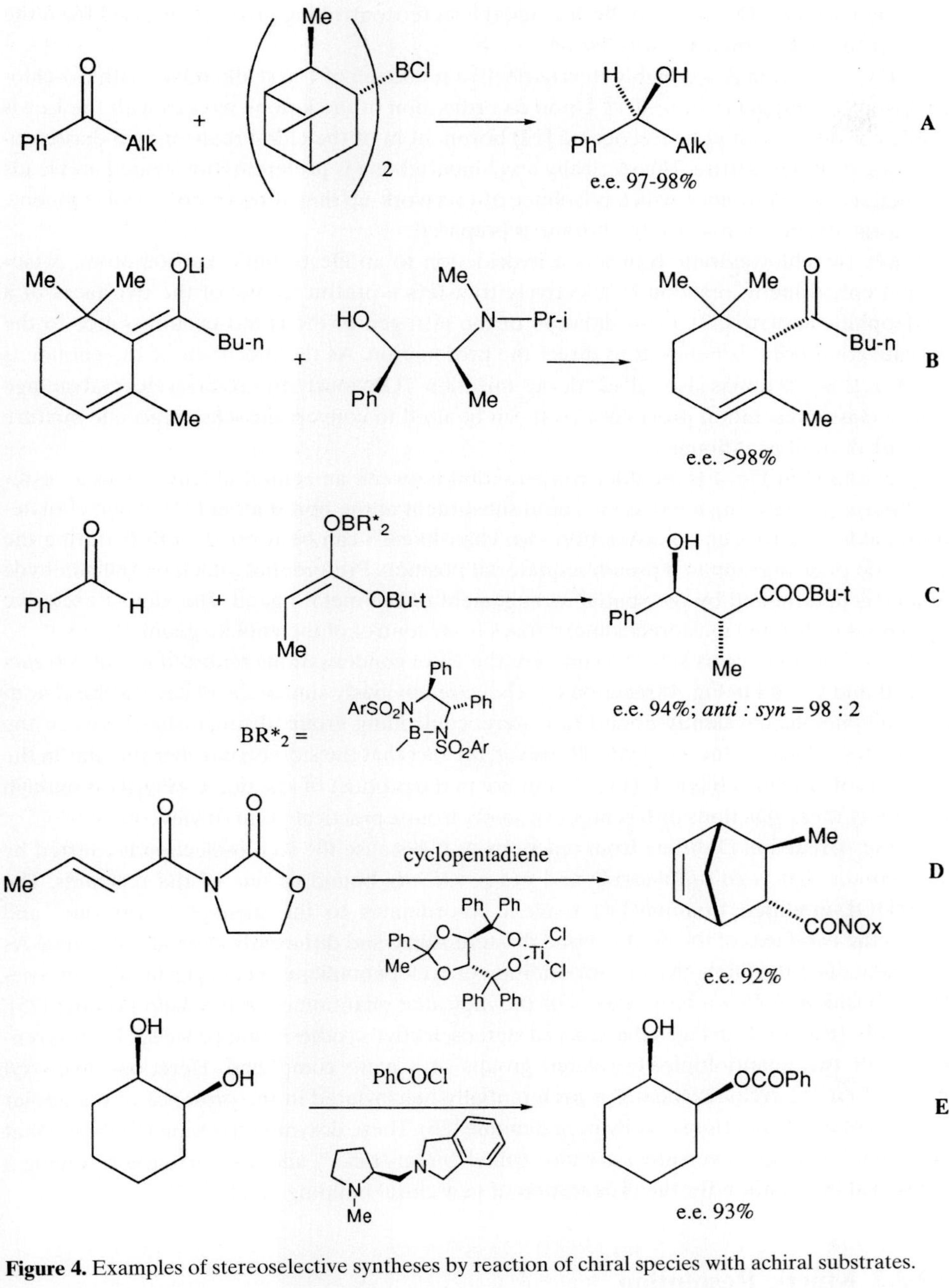

Figure 4. Examples of stereoselective syntheses by reaction of chiral species with achiral substrates.

reaction partners (D and E). In both cases, the stereocontrolling group is released from the product upon its formation or isolation.

In Fig. 4, reaction A is a highly stereoselective reduction of 1-aryl alkanones with (–)-chloro diisopinocampheylborane [21]. Upon co-ordination of the ketone oxygen with the Lewis acidic chirotopic and non-stereogenic [22] boron atom of the chiral reagent, two diastereoisomeric complexes arise. The sterically less hindered one is preferentially formed and leads the major (*S*)-enantiomer, which is isolated after a work-up that allows recovery of α-pinene, the chiral alkene from which the borane is prepared.

While the chloroborane transfers a hydride ion to an electrophilic carbon atom, *N*-isopropyl ephedrine in reaction B selectively transfers a proton to one of the two faces of a nucleophilic enolate [23]. Co-ordination of the nitrogen of the chiral proton source to the enolate counterion is believed to direct the protonation. As the precursor of the enolate is racemic, this reaction is also called "deracemization". Obviously, this offers a clear advantage over a classic resolution procedure, as it can be used to convert the whole racemic mixture into the desired enantiomer.

Reaction C in Fig. 4 is an aldol condensation between an achiral aldehyde and an ester enol borinate featuring a bidentate chiral substituent at the boron atom [24]. Upon enolate-boron/aldehyde-oxygen co-ordination, two chair-like TS can be formed, both featuring the aldehyde phenyl group in a pseudo-equatorial position. Preferential attack on the aldehyde *Si* face is determined by the spatial arrangement of the metal ligand. The almost exclusive formation of the *anti* diastereoisomers arises from control of the enolate geometry.

It is interesting at this stage to compare the aldol condensations reported in Fig. 3 (reactions B and C) and in Fig. 4 (reaction C). They are obviously similar: in all cases a chiral non-racemic enolate, covalently bound to a stereocontrolling group, distinguishes between the enantiotopic faces of the aldehyde. However, the fact that the stereocontroller remains in the products of reactions B and C (Fig. 3), but not in the product of reaction C (Fig. 4), is enough to classify these reactions differently, especially from a practical point of view.

In Fig. 4, reaction D differs from reactions A–C because the stereoselection is exerted by a promoter that is *co-ordinatively* and not *covalently* bound to one of the reactants. The TADDOL-modified titanium(IV) reagent co-ordinates to the dienophile carbonyl and makes the two faces of the double bond diastereotopic and differently sterically shielded. As a consequence, the Diels–Alder cycloaddition of cyclopentadiene occurs in a highly stereoselective fashion to give a large excess of the indicated enantiomer of the *endo* product [25].

Finally, reaction E in Fig. 4 illustrates a stereoselective synthesis that proceeds by differentiation of two enantiotopically related groups of a *meso* compound. Here, one hydroxyl group of *cis*-1,2-cyclohexanediol is preferentially benzoylated in the presence of one molar equivalent of an enantiomerically pure diamine [26]. These desymmetrization reactions (that have many biological versions) are also called "*meso*-tricks", and are currently receiving a great deal of attention for the preparation of new chiral building blocks [27].

5.2.2.2 Kinetic Resolution

In all the examples of stereoselective synthesis seen so far, at least one of the reactants (either the substrate or the reagent) is achiral. When two or more chiral reactants are exposed to each other, various phenomena can occur. Their classification essentially depends on both the enantiomeric composition and absolute configuration of the interacting species.

In the simplest case a racemic substrate is allowed to react with an enantiomerically pure compound, e.g., of (*R*)-configuration. The TS of the reactions leading to the (*R*,*R*)- and (*R*,*S*)-isomers are diastereoisomerically related and differ in energy. Thus, different reaction rates can be expected for the two processes, and either by stopping the reaction before completion or by using a deficiency of the enantiomerically pure reagent, enrichment of the racemic mixture in the slower-reacting enantiomer can be obtained.

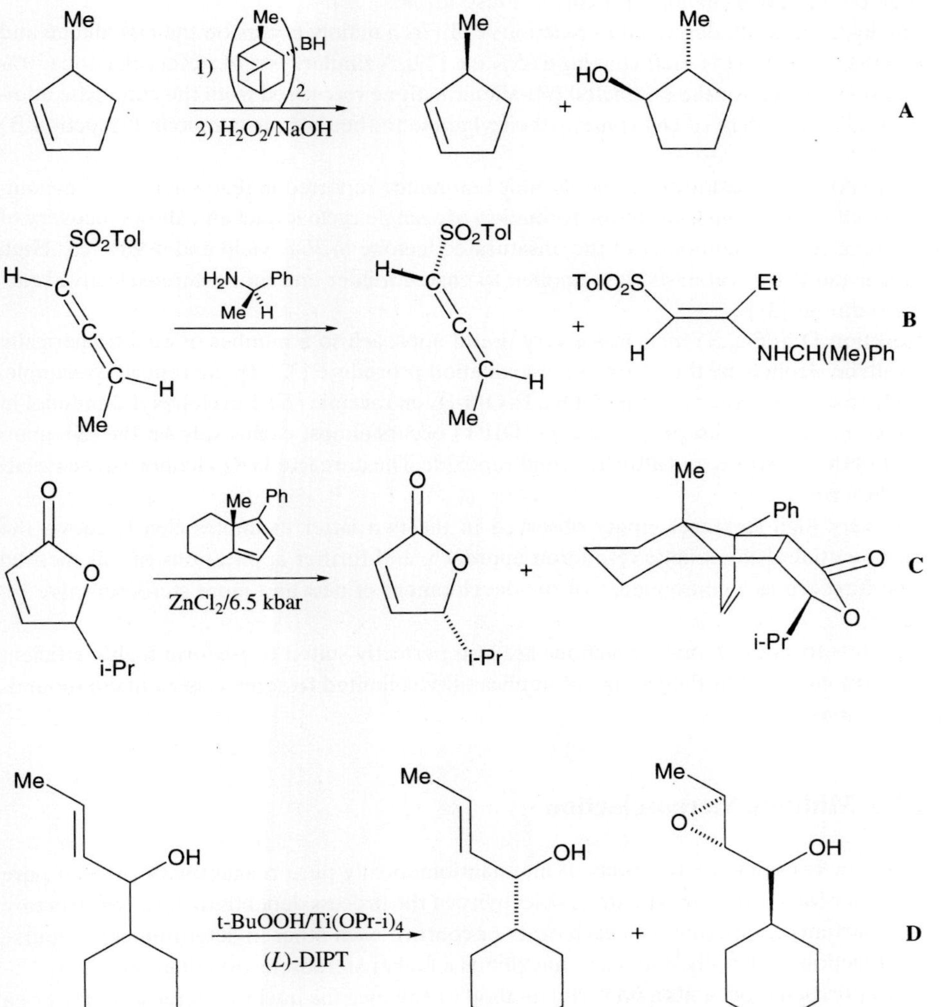

Figure 5. Examples of stereoselective syntheses of enantiomerically enriched compounds by kinetic resolution.

This process, that "resolves" enantiomers on the basis of their different reaction rates, is called "kinetic resolution" [28]. Conceptually it is similar to the reactions of Fig. 4, as the enantiomerically pure reagent (in this context, the kinetic resolving agent) differentiates between the *two* enantiomerically related *molecules* of the racemic substrate. Obviously, in the reactions of Fig. 4 it is *one molecule* that features the enantiomerically related entities (e.g., the two carbonyl faces of benzaldehyde) which are differentiated by the chiral non-racemic reagent.

A few examples of kinetic resolutions are reported in Fig. 5. These have been selected among those reactions leading to the preparation of enantiomerically enriched compounds that do not feature a "handle" for classical resolutions.

The hydroboration described in reaction A of Fig. 5 mainly occurs on the (*R*)-alkene and leaves the unreacted (*S*)-olefin having a 65% e.e. [29]. A similar level of stereoselection (70% e.e.) was estimated for the unreacted (*R*)-allenic sulfone recovered from the conjugate addition of half equivalent of (*R*)-α-methylbenzylamine to the racemic compound (reaction B) [30].

Diels–Alder cycloaddition of the racemic butenolide reported in reaction C with an enantiomerically pure diene leads to the formation of a single cycloadduct and allows recovery of the unreacted (*S*)-enantiomer of the unsaturated lactone in 36% yield and >98% e.e. High pressure and $ZnCl_2$ catalysis were applied to ensure milder and more stereoselective reaction conditions [31].

Reaction D in Fig. 5 represents a very useful approach to a number of enantiomerically pure allylic alcohols by the Sharpless' epoxidation procedure [32]. In the reported example, the reaction carried out with t-BuOOH/Ti(OPr-i)$_4$ on racemic (*E*)-1-cyclohexyl-2-butenol in the presence of (+)-diisopropyltartrate (DIPT) occurs almost exclusively on the (*S*)-enantiomer of the substrate, and affords a single epoxide. The unreacted (*R*)-alcohol can be isolated in 96% e.e.

The very high stereoselectivity observed in the two latter examples clearly shows the great potential of the kinetic resolution approach, and further applications of this method can be foreseen as a consequence of the development of new and more stereoselective reagents.

Needless to say, enzymatic reactions are also perfectly suited to perform highly efficient kinetic resolutions, but their range of applicability is limited to some classes of compounds and reactions.

5.2.2.3 Multiple Stereoselection

Let us now consider the reactions of an enantiomerically pure reagent with the two pure enantiomers of a substrate. The stereoselectivity of the process depends on how the structure of the reactants co-operate with each other or contrast each other in determining the course of the reaction. Generally, one reaction exhibits a higher stereoselection than the other.

This approach represents a powerful method to enhance the intrinsic stereoselectivity of a given reactant and is called "double" or, more generally, "multiple stereoselection". The reasons for the increased stereoselectivity of these processes are evident from the examples collected in Fig. 6 [33].

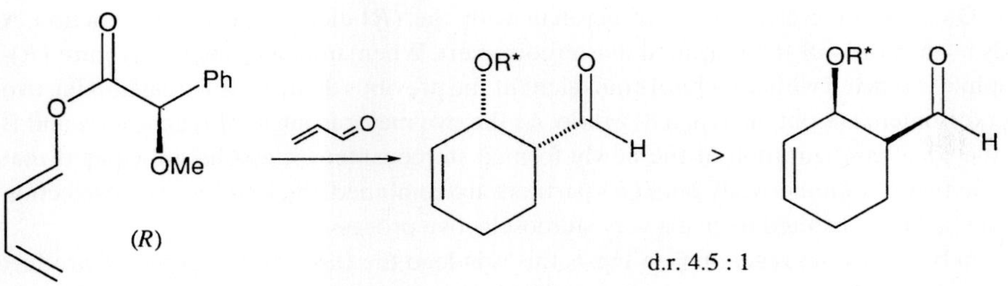

A

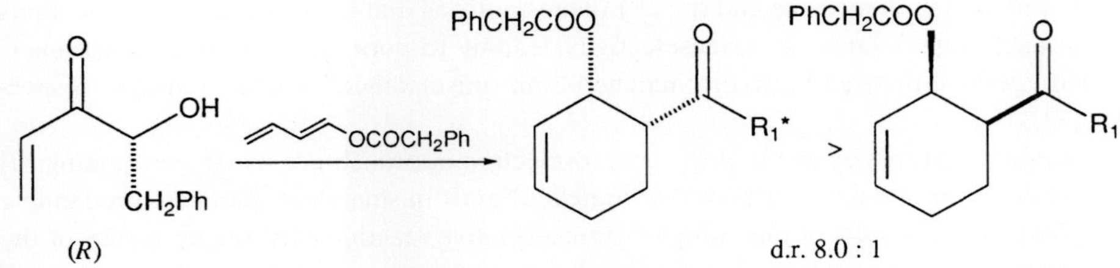

B

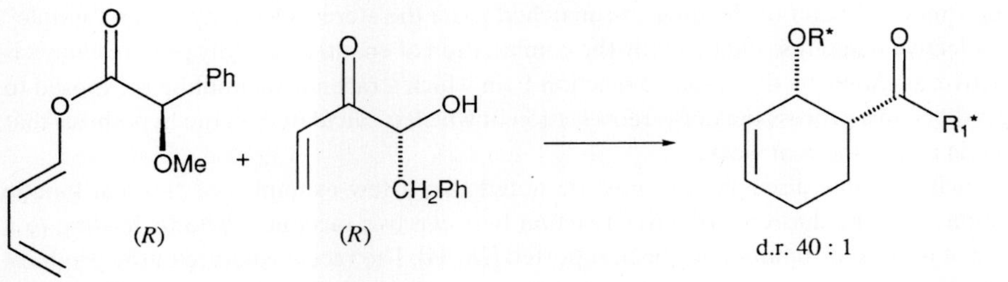

C

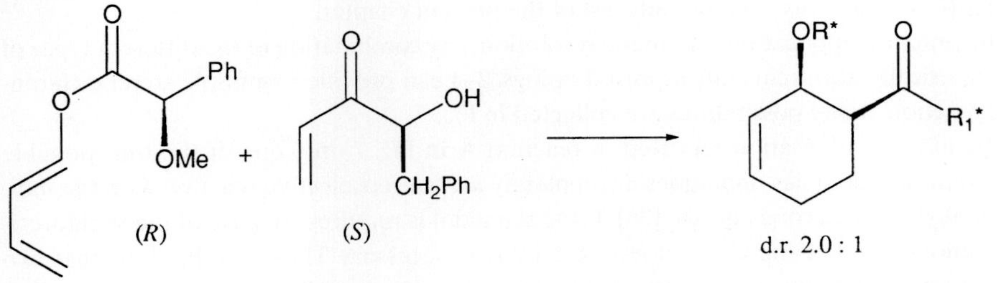

D

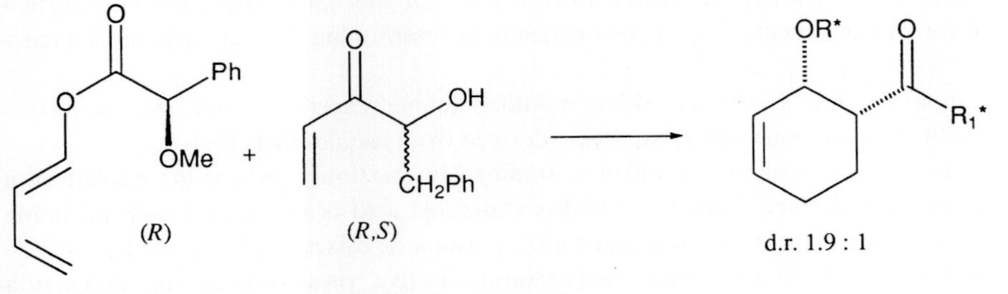

E

Figure 6. Stereoselective synthesis by mulitiple steroselection.

The Diels–Alder cycloaddition of acrolein with the (*R*)-diene reported in reaction A affords a 4.5:1 ratio of the indicated diastereoisomers. When an enantiomerically pure (*R*)-dienophile is reacted with an achiral equivalent of the previous diene (Fig. 6; reaction B), two diastereoisomers are obtained in a 8:1 ratio. As the two major isomers of reactions A and B have the same configuration at the newly formed stereocenters, it can be anticipated that when the two enantiomerically pure (*R*)-partners are combined, the intrinsic stereoselection of each reactant is exalted to give a very stereoselective process.

As can be seen from reaction C in Fig. 6, this is indeed the case and the expected product is obtained virtually as a single isomer (d.r. 40:1). The control experiment (reaction D) carried out with an (*S*)-dienophile and the (*R*)-diene confirms that in this case the two reactants contrast each other's intrinsic stereoselectivity, leading to poor stereocontrol. Reactions C and D are defined to occur between "matched" and "mismatched" reaction partners, respectively [33].

Three main features of this multiple stereoselection approach are worth mentioning: (i) The configuration of the products of the "matched" and "mismatched" pairs are predictable on the basis of the results of the "simple" stereoselective reactions; (ii) the efficiency of the stereocontrol seems also predictable (although not in a very precise manner) by simply multiplying (matched pair) or dividing (mismatched pair) the stereoselectivity of the "simple" stereoselective reactions; and (iii) only the combination of enantiomerically pure reactants is productive, as shown by the result of reaction E, in which a racemic dienophile is exposed to the (*R*)-diene in a process the diastereoselection of which is calculated on the hypothesis that no kinetic resolution is at work.

In concluding this discussion, it must be noted that a few examples of "mutual kinetic resolution", i.e., the diastereoselective reaction between two racemic reactants leading to a mixture of diastereoisomers, have been reported [28, 34]. The racemic nature of the products excludes these reactions from the interest of the present chapter.

In the practical applications of kinetic resolution, any combination of the different types of enantiomerically pure reactants reported in Figs. 2–4 can provide a multiple stereodifferentiating reaction. Some possibilities are collected in Fig. 7.

In the aldol condensation reported in reaction A in Fig. 7 only one of the four possible stereoisomeric aldehydes undergoes a completely *anti* stereoselective reaction with the indicated dialkylboron ketone enolate [35]. If the *syn* aldol is required, the use of the trichlorotitanium enolate and of the C-3 epimeric aldehyde is necessary. These results show that two stereocenters of the same reagent can constitute a sort of internally matched or mismatched pair, and can operate synergically or antithetically in determining the steric course of a reaction.

In example B in Fig. 7 a chiral auxiliary modified ketene reacts in a totally stereoselective fashion with an enantiomerically pure imine derived from lactaldehyde [36].

The same high level of stereocontrol is obtained in reaction C, where the enolate of a 2-pyridylthioester derived from (*R*)-3-hydroxybutanoic acid is condensed with an imine bearing a chiral auxiliary to produce a single stereoisomeric β-lactam [37].

Extension of this methodology to a "triple stereoselective" process is reported in reaction D. This completely stereoselective reaction takes advantage of the matching combination between the enolate and the imine chiral auxiliary, which, in its turn is also internally matched with the other imine stereocenter [37].

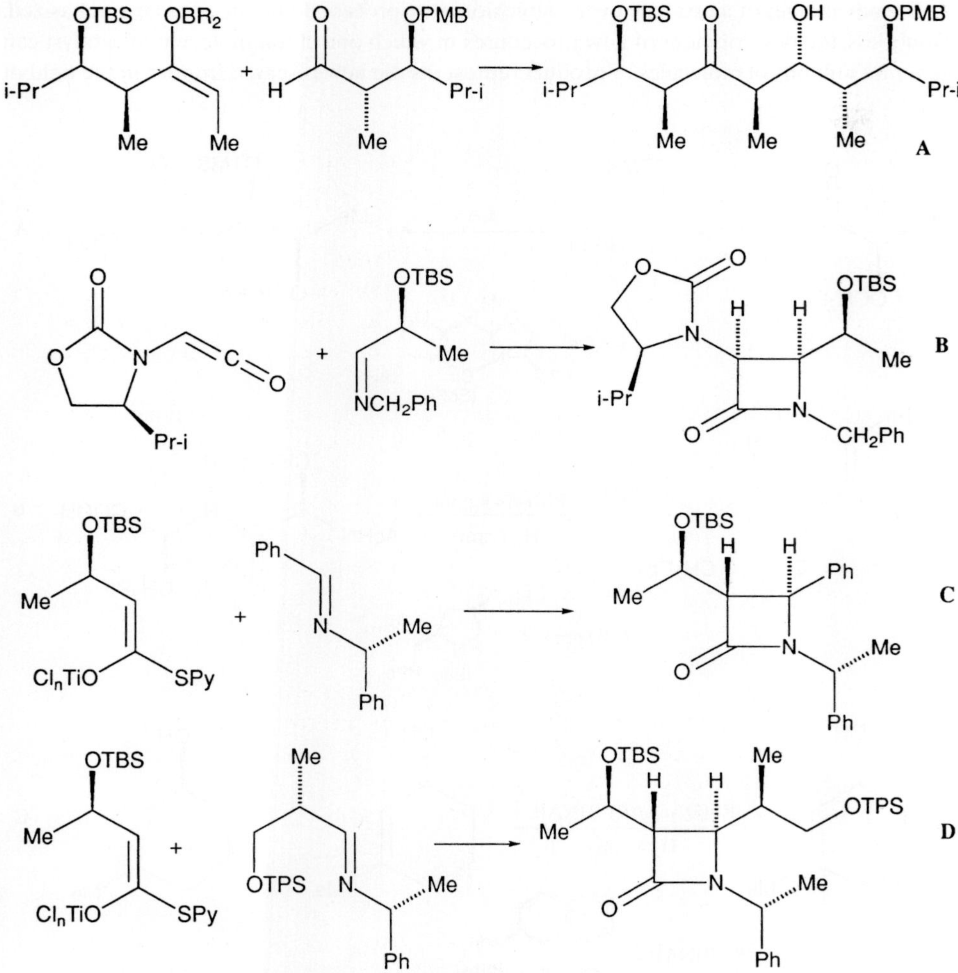

Figure 7. Completely stereoselective syntheses by multile stereoselective reactions.

5.2.2.4 Catalytic Methods

In section 2.2.1.3 the stereoselective reactions of an achiral substrate with a stoichiometric amount of a chiral non-racemic species were described.

As pointed out previously, reactions of this type exist in which the component responsible for the stereoselection (e.g., in examples D and E of Fig. 4 the chiral LA and the diamine, respectively) does not form a covalent bond with either the substrate or the reagent or the product, and is spontaneously released upon product formation. If this species is also able to lower the activation energy of the reaction, all the conditions required to establish a catalytic stereoselective synthesis are fulfilled [38].

The advantages of a catalytic over a stoichiometric process do not need to be emphasized. Doubtless, the development of new procedures in which one chiral molecule of catalyst can generate millions of molecules of product represents the new research frontier in the field of

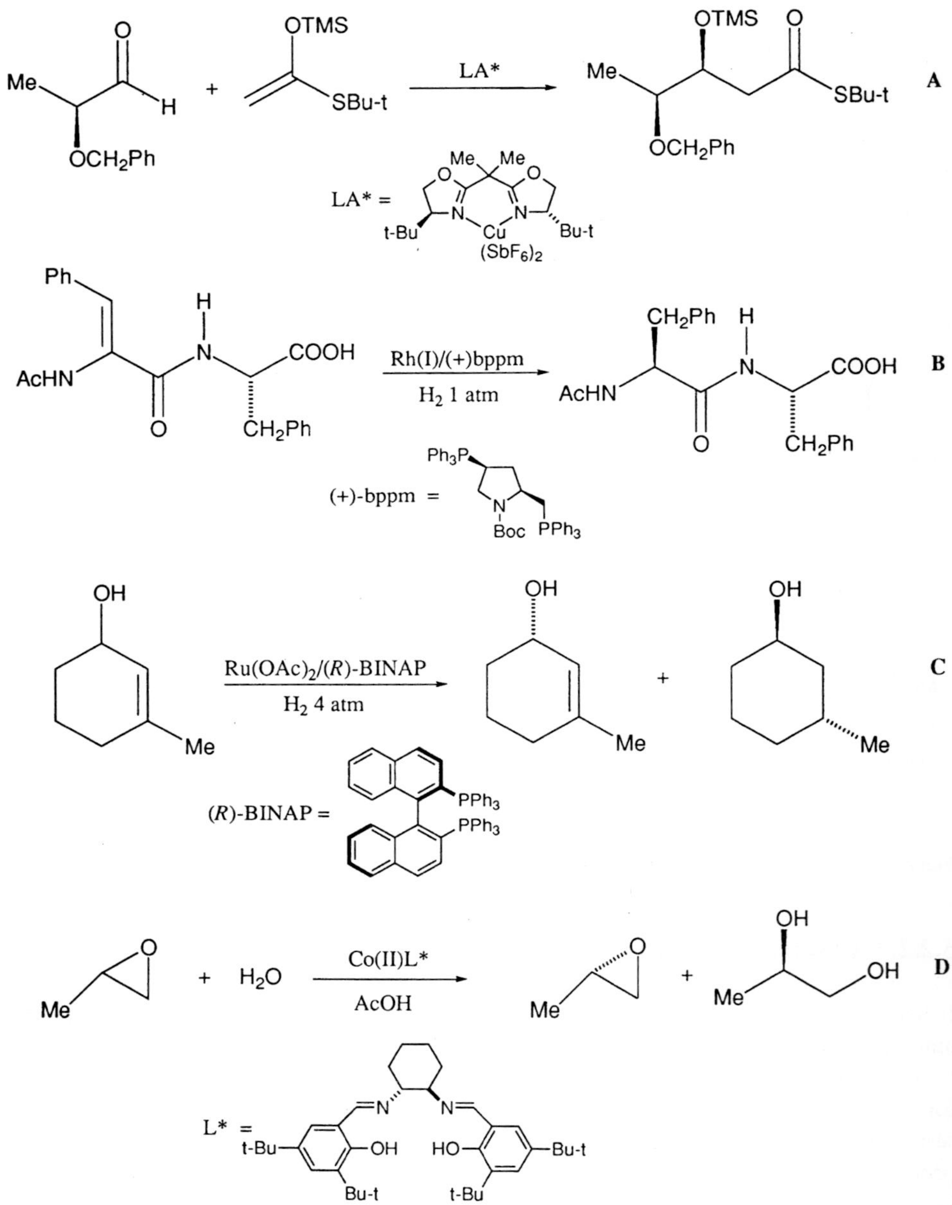

Figure 8. Multiple steroselective syntheses and kinetic resolutions using chiral catalysts.

stereoselective synthesis. Many spectacular results have already been achieved [38], and it can be safely stated that at least one stereoselective catalytic version of all the fundamental reactions of organic chemistry is now available. Remarkably, some of them have found relevant industrial applications [38–40].

A compilation of different catalytic stereoselective processes seems unnecessary for the purpose of a classification of the stereoselective syntheses. However, a few examples of some less frequently encountered catalytic kinetic resolutions and chiral catalyst mediated multiple stereodifferentiating reactions will be presented in Fig. 8.

Reaction A in Fig. 8 describes the Mukaiyama-type aldol condensation between (*S*)-benzyloxypropanal and a silyl ketenethioacetal catalyzed by a (*S,S*)-bisoxazoline/Cu(II) complex. A 98.5:1.5 ratio of the (*3S*):(*3R*) aldols was obtained from this matching combination, while the reaction carried out on the (*R*)-aldehyde was stereorandom [41].

Rh(I)/(+)bppm-catalyzed hydrogenation of the (*S*)-dehydro-dipeptide of reaction B is extremely stereoselective (d.r. = 161 : 1) affording (*S,S*)-N-acetylPhePheOH virtually as the sole product. The use of the enantiomeric ligand leads to the formation of the (*S,R*)-isomer with good selectivity (d.r. 25 : 1), thus showing that the chiral catalyst largely overrides the substrate in directing the course of the reaction [42].

In example C in Fig. 8 the $Rh(OAc)_2$/(*R*)-BINAP complex preferentially catalyzes the hydrogenation of the (*R*)-enantiomer of racemic 3-methyl-2-cyclohexenol to give (*1R,3R*)-3-methylcyclohexanol in 95% e.e. and 49% yield, and leaving the unreacted (*S*)-alcohol, that is recovered in 99% e.e. and 46% yield [43].

Finally, a spectacular example of epoxide hydrolysis catalyzed by a chiral Co(II)/salen complex is reported in reaction D in Fig. 8. From this reaction, (*S*)-propylenoxide is recovered in 98.6% e.e. and 44% yield, while (*R*)-1,2-propandiol is obtained in 98% e.e. and 50% yield [44].

5.3 Examples of Stereoselective Synthesis of Enantiomerically Pure Drugs

5.3.1 Synthesis of Sanfetrinem

The different stereoselective routes explored toward the synthesis of the tricyclic β-lactam sanfetrinem cilexetil (GV 118819) [45] perfectly illustrate advantages and drawbacks of the approach to enantiomerically pure drugs based on the use of commercially available chiral non-racemic substrates.

This highly potent, broad-spectrum antibiotic contains five stereocenters. Three of these are already present in the available substrate selected as starting material, namely azetidin-2-one 1 (Fig. 9) [46].

Early synthetic efforts involved TMSOTf-catalyzed addition of the cyclohexanone-derived TMS enolether to the *N*-TMS protected **1**, followed by *N*-deprotection to give ketone **2**. While the stereocontrol at C-4 of the β-lactam ring was complete and only *trans* products were obtained, the stereoselection at the newly formed stereocenter was rather poor since the reaction yields a 70:30 mixture of epimers in favor of the indicated one.

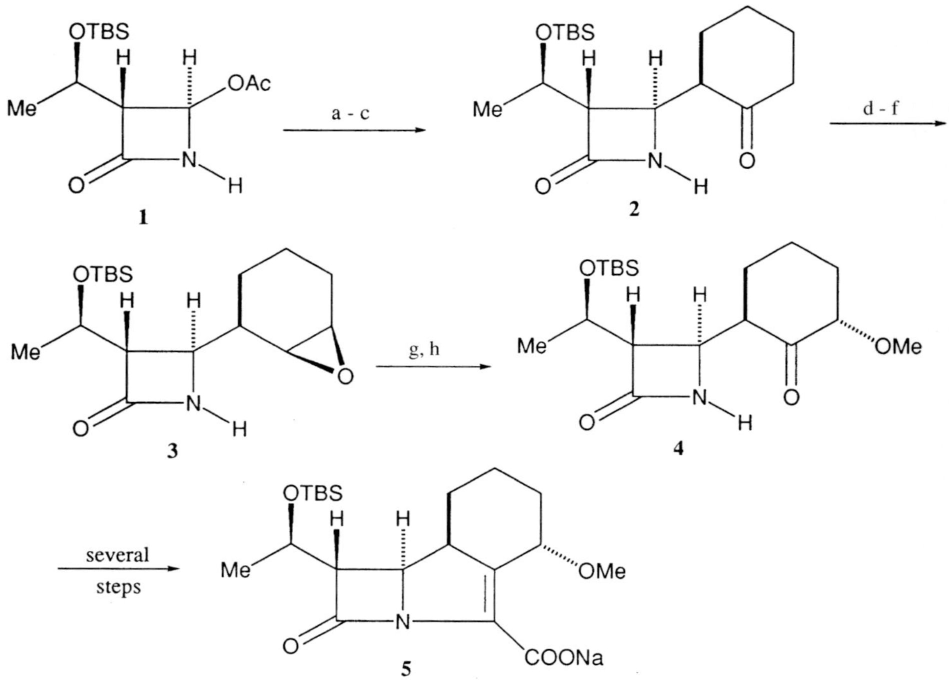

Figure 9. Stereoselective synthesis of Sanfetrinem. Reagents: a, TMSCl, TEA; b, trimethylsilyloxycyclohex-1-ene, TMSOTf; c, KF; d: $TsNHNH_2$; e: LDA; f: Mg monoperoxyphthalate; g: MeOH, H^+; h: pyridine – SO_3

Manipulation of the carbonyl function to epoxide **3** was accomplished by conversion to the tosylhydrazone, Shapiro olefination, and epoxidation. The latter step occurred with high stereocontrol, as did the epoxide ring opening with acidic methanol. Oxidation to ketone **4** was then performed, and from this product the desired derivative **5** was obtained by known procedures.

Although this route allowed the preparation of multikilogram quantities of the target molecule, it clearly suffered from some drawbacks, the major ones being: (i) The poor stereocontrol of the reaction leading to ketone **2**; (ii) the low overall chemical yield; and (iii) the high number of steps involved in the reaction sequence.

An intramolecular approach to an advanced intermediate was then designed to overcome some of these problems (Fig. 10).

In Fig. 10, reaction of **1** with chlorosilane **6** yielded the *N*-silyl derivative **7** that was not isolated but directly transformed into alkene **8** (the precursor of epoxide **3** of Fig. 9) under Sakurai's conditions. Both the yield (50%) and the stereoselectivity (80:20 for the indicated isomer) were not very high, but the road was paved toward a better process.

This was eventually obtained by replacing the silicon reagent with a boron one, as described in Fig. 10 [47]. Here, treatment of **1** with a mild LA gave the unstable imine **9** that underwent reaction with a variety of differently boron-disubstituted cyclohexen-2-yl

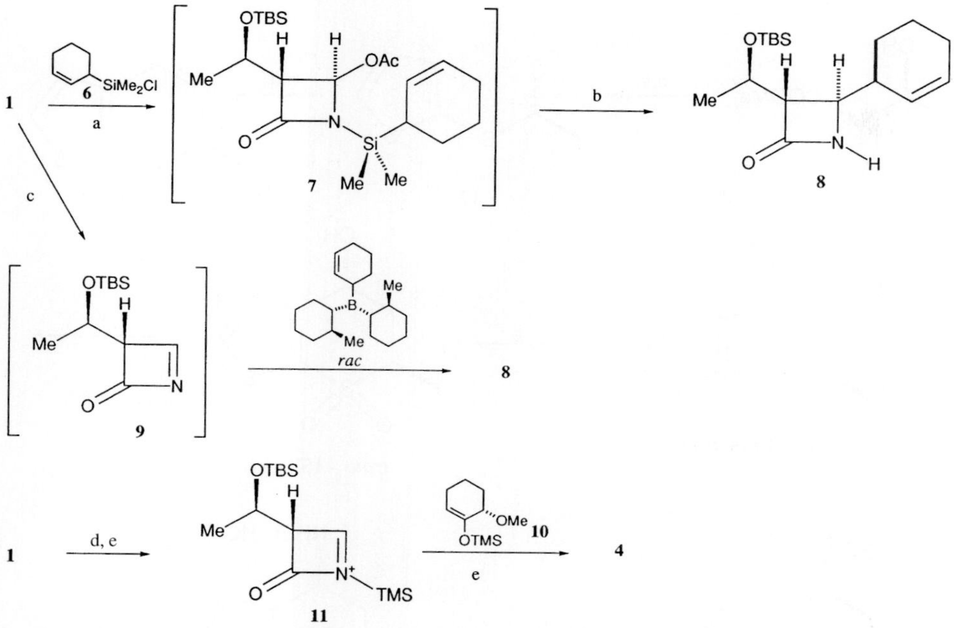

Figure 10. Alternative stereoselective syntheses of Sanfetrinem. Reagents: a, TEA; b, TMSOTf; c, Et_2Zn; d, TMSCl, TEA; e, $SnCl_4$.

boranes. By proper choice of the LA (diethylzinc was found to be the best), of the reaction conditions (from 0° to 23 °C in a hexane/THF mixture as solvent), and of the boron ligands (2-methylcyclohexane), the alkene **8** could be obtained in excellent yield (96%) and stereoselection (d.r. 95:5).

Further improvement of the process could be achieved only upon drastic reduction of the number of steps required to produce ketone **4**. This result was envisaged in the convergent approach described in Fig. 10 [48], that allowed the generation of **4** directly from **1**.

The new approach took advantage of the availability in enantiomerically pure form of (*S*)-2-methoxycyclohexanone, the precursor of enolether **10**. This was reacted in the presence of an excess of tin tetrachloride with the iminium ion **11**, obtained *in situ* from **1** by *N*-silylation and exposure to the LA, to afford directly ketone **4** in high yield (up to 79%) and excellent stereoselectivity (up to 96:4 d.r.). This high level of stereocontrol appears to be the result of a matching combination of enantiomerically pure reactants in a multiple stereoselective process, and this is currently exploited for the large-scale preparation of sanfetrinem.

5.3.2 Synthesis of MK-0507

In Fig. 10, while most of the structural features of substrate **1** are maintained throughout the reaction sequence to **4**, it is not so straightforward to trace back the molecular framework of MK-0507 **17** (Fig. 11) to its synthetic precursor.

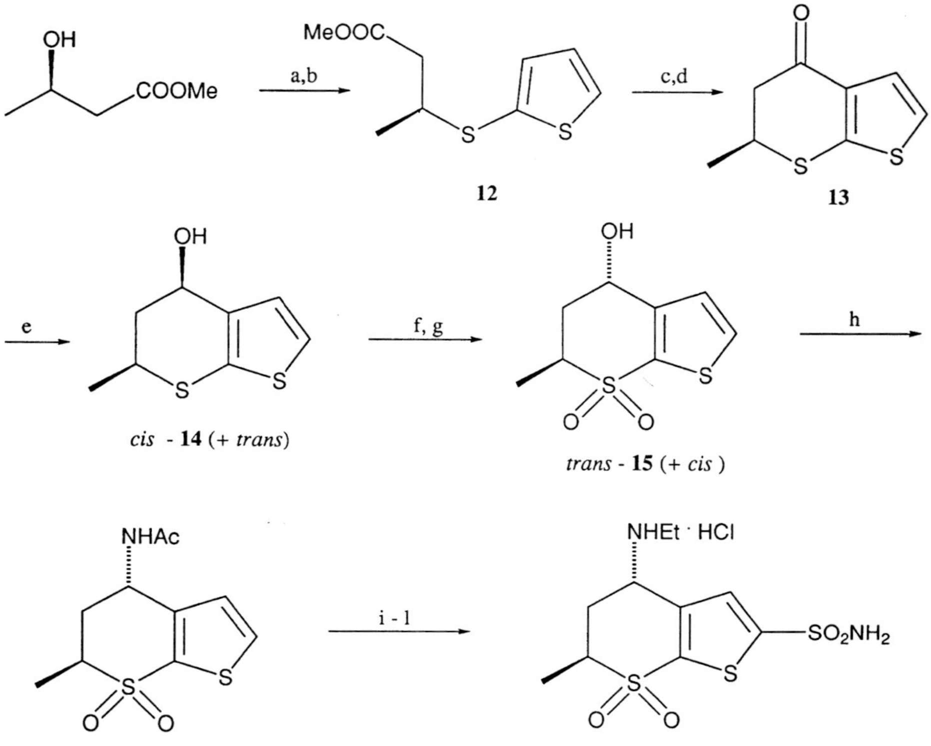

Figure 11. Stereoselective synthesis of MK-0507. Reagents: a, TsCl, pyridine; b, 2-(lithiomercapto)-thiophene, formamide/THF; c, 12N HCl; d, TFAA, toluene; e, $LiAlH_4$, toluene; f, H_2O_2, Na_2WO_4; g, 1N H_2SO_4; h, MeCn, H_2SO_4; i, HSO_3Cl, $SOCl_2$; j, NH_4OH, THF; k, $BH_3 \cdot Me_2S$, THF; l, HCl, EtOAc.

This topically active carbonic anhydrase inhibitor is a powerful controller of the elevated intraocular pressure associated with glaucoma, and has been prepared by Merck in >32% yield in a multistep synthesis starting from (*R*)-3-hydroxybutyrate [49]. Although the key ketone reduction step has been recently improved by researchers at Zeneca by replacing a chemical process with a biological one [50], the Merck synthesis well illustrates the potentialities of the approach to enantiomerically pure drugs based on the use of chiral non-racemic substrates.

Readily available (*R*)-3-hydroxybutyrate was first transformed into its tosylate and then treated with 2-(lithiomercapto)thiophene in 50:50 formamide:THF solvent to give ester **12** in >97% e.e. Ester hydrolysis in acidic medium followed by treatment with trifluoroacetic anhydride in toluene at room temperature gave ketone **13**, the reduction of which in different conditions gave almost exclusively the *cis*-alcohol **14**.

A Ritter reaction to be carried out on the sulfone **15** was then selected to establish the *N*-substituent at C-4. Surprisingly, however, this reaction did not occur exclusively in the expected inversion mode, but *largely* with inversion on the *cis*-isomer and *mainly* with retention on the *trans*-one! To take advantage of this dichotomy of behavior, the typically obtained 95:5 *cis:trans* mixture of isomers of compound **15** was transformed in the presence of dilute acids into a 76:24 *trans:cis* mixture. Thus, the Ritter reaction resulted in the isolation of a mixture of *trans* and *cis* isomers of acetamide **16**, in which the desired *trans* isomer was satisfactorily predominant (d.r. = 89:11).

Sulfonimidation at C-2, and borane reduction of the acetamide gave the final product **17** after crystallization of its hydrochloride. As mentioned above, a biotransformation is currently used to replace the chemical reduction in the plant-scale production of the drug [50].

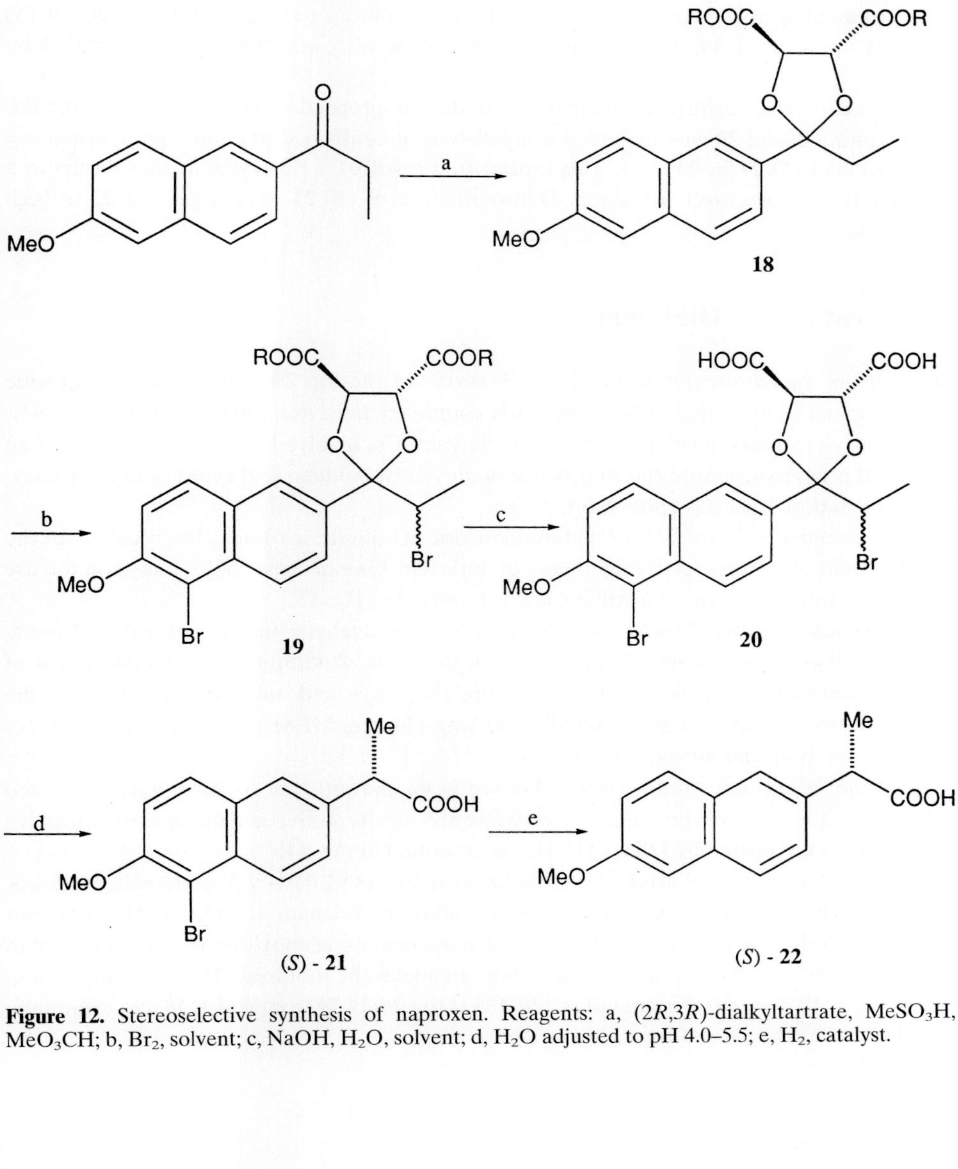

Figure 12. Stereoselective synthesis of naproxen. Reagents: a, (2*R*,3*R*)-dialkyltartrate, $MeSO_3H$, MeO_3CH; b, Br_2, solvent; c, NaOH, H_2O, solvent; d, H_2O adjusted to pH 4.0–5.5; e, H_2, catalyst.

5.3.3 Synthesis of Naproxen (Chiral Auxiliary Approach)

The Zambon synthesis of the non-steroidal anti-inflammatory agent (*S*)-2-(6-methoxy-2-naphthyl)propanoic acid (naproxen) is a landmark-setting application of the chiral auxiliary approach in the industrial stereoselective synthesis of an enantiomerically pure drug [51]. The chiral auxiliary employed, a (2*R*,3*R*)-dialkyltartrate, is a paradigmatic representative of this class of stereocontroller, being cheap, readily available, easily introduced on the substrate and removed from the product, and eventually recycled (although as its parent acid).

The process is relatively short and requires five steps from 1-(6-methoxy-2-naphthyl)-propan-1-one (Fig. 12). The ketone was converted in high yield into ketal **18**, that was then brominated at the carbon α to the ketal carbon atom to give a 94:6 mixture of (*S*) and (*R*) epimers of **19**. Unfortunately, bromination also occurred in the naphthalene nucleus.

Ester hydrolysis, performed under basic conditions, proceeded without affecting the d.r. The resulting diacid **20** was subjected to hydrolysis at controlled pH and high temperature to afford acid (*S*)-**21** in 99% e.e. This crucial step involves a 1,2-aryl shift that occurs in a completely stereocontrolled fashion. Debromination of (*S*)-**21** gave naproxen **22** in high overall yield.

5.3.4 Synthesis of Diltiazem

The calcium antagonist diltiazem **26**, which is one of the top 20 selling drugs world-wide according to a 1990 estimate [39], is currently commercialized as a single (*S*,*S*)-stereoisomer. While early synthesis developed by Tanabe Seiyaku Co. involved classical resolution of an advanced precursor, a more recent process employed by Andeno [39] exploits an early enzymatic resolution of an α,β-epoxyester.

The unavoidable loss of 50% of material intrinsic in both these routes, clearly called for the development of stereoselective syntheses of diltiazem. One of these [52] is based on the use of an oxazolidinone as chiral auxiliary, and is reported in Fig. 13.

N-Acyloxazolidinone **23** was condensed with p-anisaldehyde to give, after dehydration, the Michael acceptor **24**, mainly in the shown (*Z*) form. Addition of the *S*-lithium salt of 2-amino-thiophenol afforded an 82:18 mixture of 2,3-*syn* and -*anti* adducts **25**. From the major isomer, diltiazem was obtained upon ring closure, MEM group removal, hydroxy group acetylation, and nitrogen alkylation.

Far from being industrially viable, this synthesis showed that a chiral auxiliary-based approach to the drug was possible, and very recently an alternative and much more attractive route has been reported by DSM [53]. This is described in Fig. 14.

The α-chloroacetic acid ester of the readily available (1*S*,2*R*)-1-*N*,*N*-dimethylaminoindanol **27** was reacted with potassium t-butoxide and p-anisaldehyde to give the Darzens condensation product **28** as an 89:11 mixture of diastereoisomeric epoxides. The one reported in the figure is the major component of the mixture and has the required (2*R*,3*S*) configuration. Treatment with 2-amino-thiophenol at 120 °C led to adduct **29** from which **26** was eventually obtained by known procedures.

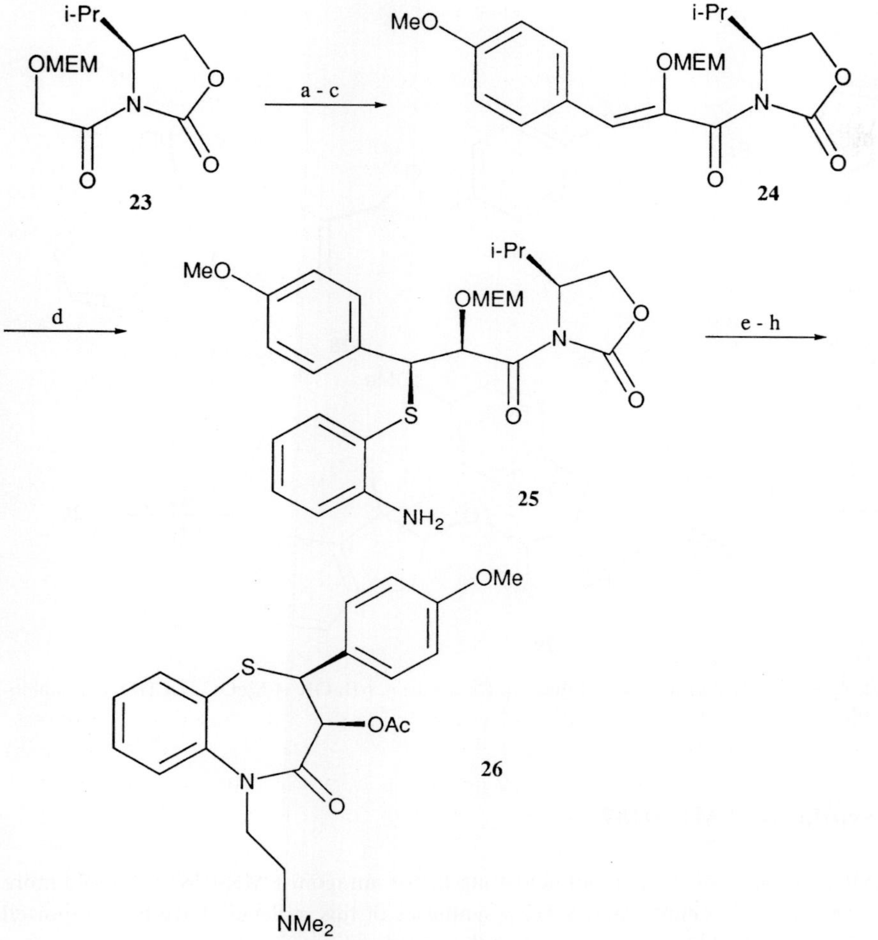

Figure 13. Stereoselective synthesis of diltiazem. Reagents: a, LDA, 4-$MeOC_6H_4CHO$; b, MsCl, TEA; c, DBU; d, 2-(lithiomercapto)aniline; e, Me_3Al; f, $TiCl_4$; g, Ac_2O; h, alkylation.

The stereoselective synthesis of epoxide **28** appears to be the major advantage of this route, since it generates an intermediate already involved in an industrial synthesis of diltiazem. The recent development of basic research methods for the catalytic enantioselective epoxidation of electron poor alkenes [54] can provide a new entry to **26** that is more practical than the approaches of both DSM and others [55].

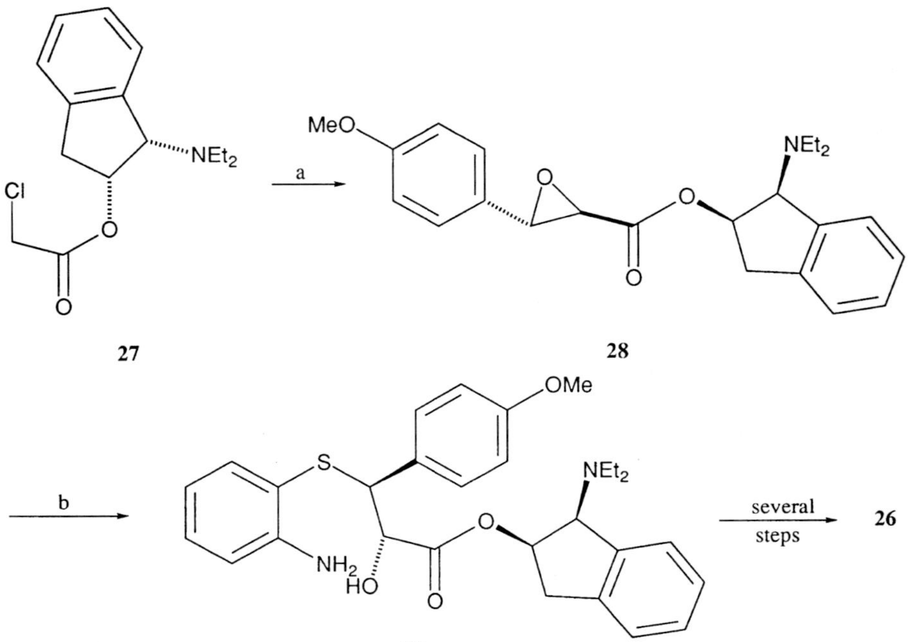

Figure 14. Stereoselective synthesis of diltiazem. Reagents: a, t-BuOK, 4-MeOC_6H_4CHO; b, 2-aminothiophenol, 120 °C.

5.3.5 Synthesis of MK-0287

The (*S*,*S*)-enantiomer of the platelet activating factor antagonist MK-0287 is 20-fold more potent than the (*R*,*R*)-enantiomer. Several syntheses of this molecule have been reported [56]. The one described in Fig. 15 is based on the use of a chiral non-racemic reagent.

γ-Ketoester **30** in Fig. 15 was reduced with high enantioselectivity (up to 90% e.e.) to alcohol **31** by the use of *in situ*-generated *B*-chlorodiisopinocampheylborane. Lactonization to **32** occurred in standard conditions. After introduction of a protected β-hydroxy-substituted ethylsulfonyl side chain on the aryl ring of **32**, as in **33**, the stereocenter at C-5 of the furanone ring was exploited to stereoselectively create the C-2 one.

To this end, the lactone carbonyl was reduced, and the resulting hydroxyl group protected as its silyl ether and reacted with 3,4,5-trimethoxyphenyl magnesium bromide to give the 2,5-*trans*-configurated diarylfuran in large excess over the *cis*-one. Desilylation of the primary hydroxyl group in the side chain led to MK-0287 **34** in 30% overall yield from the precursor of **30** [56].

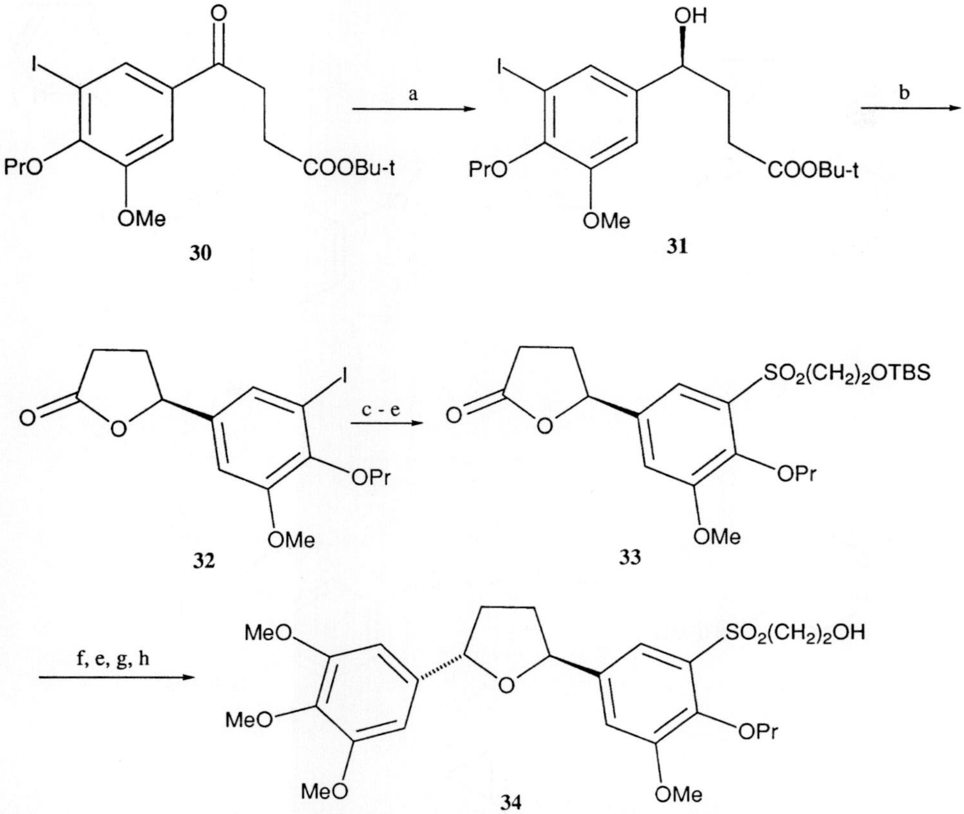

Figure 15. Stereoselective synthesis of MK-0287. Reagents: a, *B*-chlorodiisopinocampheylborane, THF; b, NaOH, MeOH, then pyridinium tosylate, toluene; c, $(HOCH_2CH_2S)_2$, Cu, DMF; d, Mg monoperoxyphthalate, MeCN, H_2O; e, TBSCl, imidazole, DMF; f, DIBAL-H, toluene; g, 3,4,5-trimethoxyphenyl magnesium bromide, THF; h, $(COOH)_2$, MeOH, H_2O.

5.3.6 Synthesis of L-738,372 and L-743,726

The possibility of using a non-covalently bound chiral modifier to direct the steric course of a reaction represents a very appealing method for the stereoselective synthesis of enantiomerically pure drugs. The preparation of the single stereocenter-containing L-738,372 and L-743,726 illustrate this approach (Fig. 16) [57].

The non-nucleosidal, reverse-transcriptase inhibitor L-738,372 **36** could be obtained in up to 99.5% e.e. by addition of lithium 2-pyridylacetylide to substrate **35** in the presence of quinine (Fig. 16), followed by *N*-1 protecting group removal in acidic conditions.

The structurally related product L-743,726 **39** was similarly prepared by addition of lithium cyclopropylacetylide to ketone **37** in the presence of a *N*,*N*-disubstituted norephedrine derivative. This occurred in 98% e.e. at 0 °C to give adduct **38** that was subsequently cyclized to afford **39**.

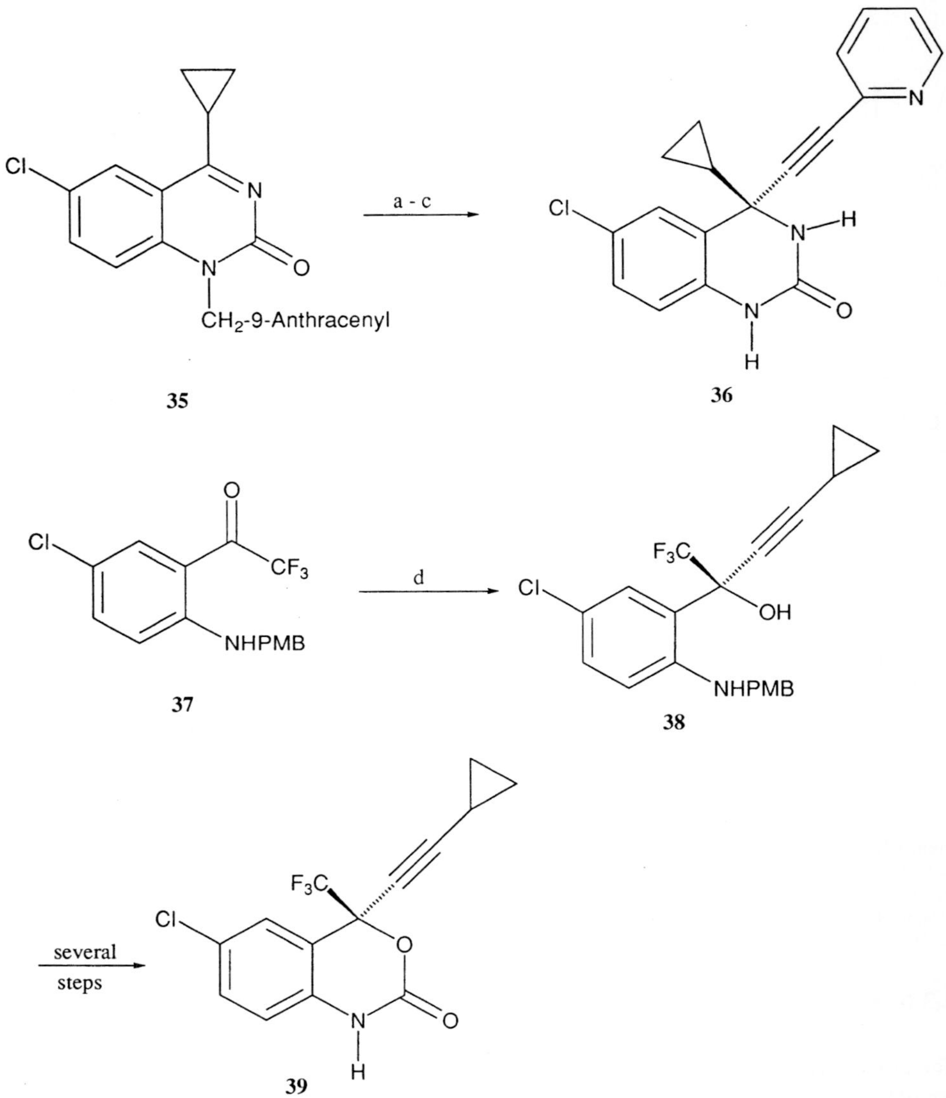

Figure 16. Stereoselective synthesis of L-738,372 and L-743,726. Reagents: a, quinine, BuLi, then lithium 2-pyridylacetylide; b, camphorsulfonic acid; c, trifluoroacetic acid; d, lithium cyclopropylacetylide, *N*-pyrrolidinyl norephedrine.

5.3.7 Synthesis of (*S*)-Propranolol

The kinetic resolution of racemic tertiary β-hydroxyamines by enantioselective *N*-oxide formation was developed by the Sharpless' group as a corollary of the allylic alcohol epoxida-

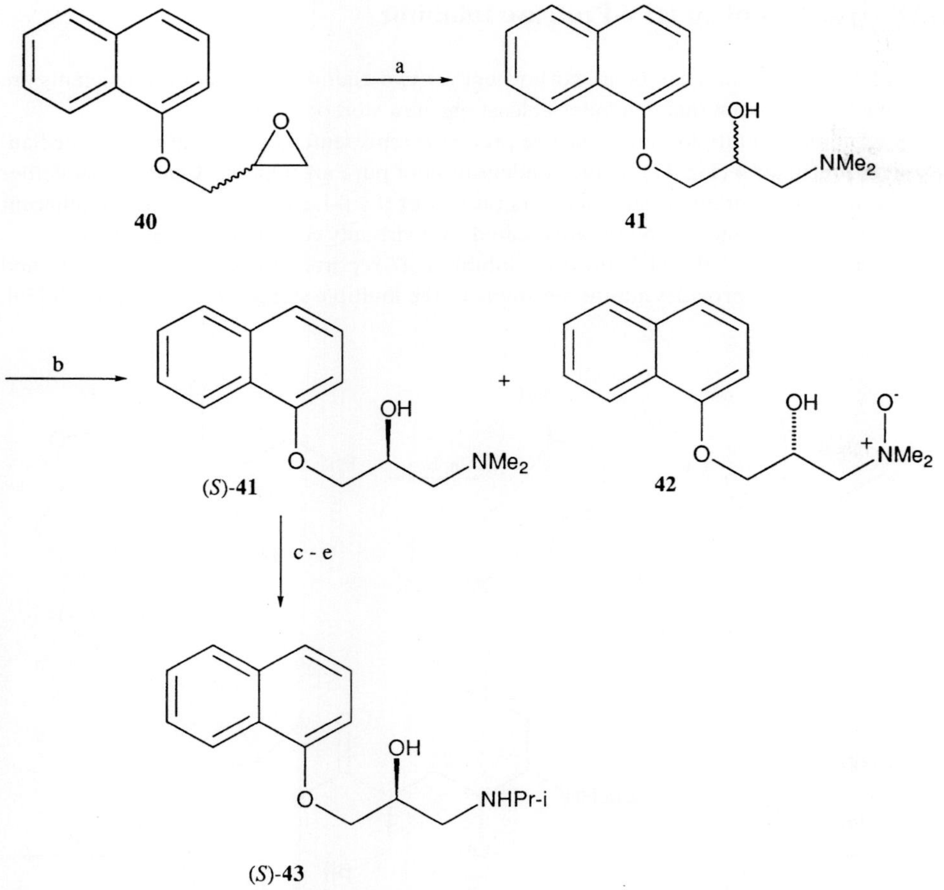

Figure 17. Steroselective synthesis of (*S*)-propranolol. Reagents: a, dimethylamine, H_2O; b, t-BuOOH, (+)-DIPT, $Ti(OPr\text{-}i)_4$, CH_2Cl_2; c, MeI, DMF; d, NaH, DMF; e, i-propylamine.

tion procedure. The recognized importance of β-hydroxyamines as β-blocker drugs provided an obvious application for the method [58].

For instance, (*S*)-propranolol can be obtained in five steps with an 11% overall yield (including the crucial resolution process) as described in Fig. 17.

Racemic epoxide **40** was first transformed into β-dimethylamino alcohol **41** by nucleophilic epoxide ring opening. Treatment of this racemic mixture (1.0 mol equiv.) with (+)-DIPT (1.2 mol equiv.), $Ti(OPr\text{-}i)_4$ (2.0 mol equiv.), and t-BuOOH (0.6 mol equiv.) allowed to obtain *N*-oxide **42** and to recover the unreacted (*S*)-enantiomer of **41** in 56% yield and 92% e.e. Standard manipulations of this material including *N*-methylation, epoxide formation, and addition of isopropylamine, gave (*S*)-**43**.

5.3.8 Synthesis of an HIV-Protease Inhibitor

Multiple stereoselection can be at work whenever two enantiomerically pure reactants are combined in a reaction that generates at least one new stereocenter.

An example of multiple stereoselective process is represented by the synthesis of the Sanfetrinem precursor **4** (Fig. 11) by the condensation of pure azetidinone **1** and silylenolether **10**, in which two contiguous stereocenters (at C-4 of the β-lactam ring and at the adjacent C-4' in the cyclohexanone moiety) are created with virtually complete stereocontrol.

The preparation of the HIV-protease inhibitor **47** reported by Merck researchers and described in Fig. 18 provides another example of the multiple stereoselective approach [59].

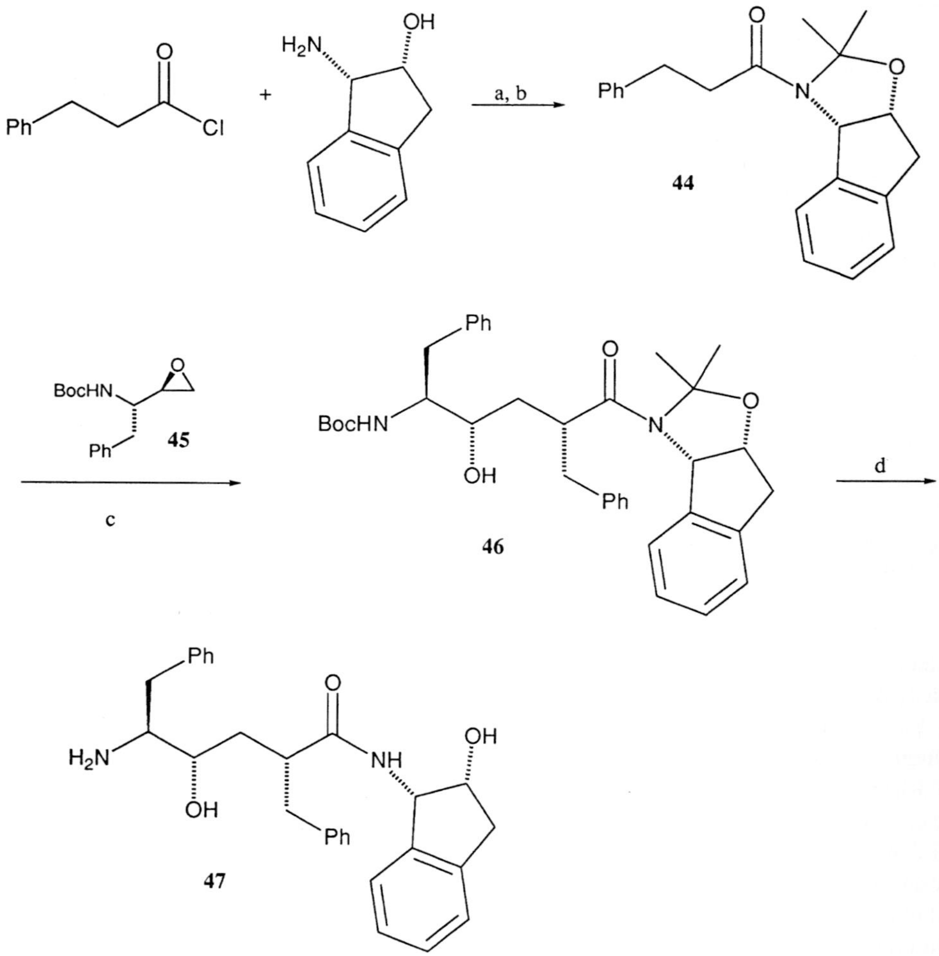

Figure 18. Stereoselective synthesis of an HIV-protease inhibitor. Reagents: a, NaOH, AcOPr-i; b, 2-methoxypropene, MsOH ; c, BuLi; d, acid hydrolysis.

(1*S*,2*R*)-1-Aminoindanol was transformed into amide **44** by acylation with 3-phenylpropionyl chloride followed by acetonide formation. Treatment of a mixture of this amide and epoxide **45** (obtained from the corresponding alkene by epoxidation and diastereoisomer separation) with 2 mol equiv. of BuLi at −78 °C gave adduct **46** as a single (2*R*,4*S*,5*S*)-isomer in more than 90% yield. From this compound the target molecule **47** was obtained by protecting groups removal in acidic conditions.

The remarkably high stereoselection observed in the formation of the C-2 stereocenter in **46** seems to be mainly due to the intrinsic stereodirecting ability of **44**. Indeed, alkylation of this amide with both alkyl [59] and allyl [60] halides led to a 97 : 3 mixture of C-2 epimers. One can speculate therefore, that the stereocenters of epoxide **45** co-operate only marginally with those of **44** in determining the stereochemical outcome of the reaction and inducing the formation of a single product.

5.3.9 Synthesis of Mibefradil

Mibefradil is the first calcium antagonist known to selectively block T-type calcium channels. It is highly effective against hypertension and angina pectoris, and significantly improves the benefit–risk balance in comparison with drugs which have similar targets.

Chemically, mibefradil is a relatively simple molecule that contains two stereocenters, and was originally obtained as a single isomer by a highly efficient resolution process coupled with the racemization and recycling of the undesired enantiomer. More recently, however, a more convenient procedure toward enantiomerically pure mibefradil was developed.

Central to this route was an asymmetric hydrogenation of an early intermediate catalyzed by a Ru complex containing a chiral non-racemic diphosphine ligand [61, 62]. The reaction sequence is described in Fig. 19.

4-Fluorophenylacetic acid was transformed into the unsaturated acid **48** by reaction with 2 mol equiv. of i-PrMgCl, followed by acetone addition, dehydration, and crystallization. The tetra-substituted double bond was then hydrogenated under high pressure in an *ad hoc* designed continuous-stirred tank reactor system and in the presence of the Ru complex **49** (substrate/catalyst ratio = 1000) to afford (*S*)-acid **50** in 93.5% e.e. Crystallization of its sodium salt upgraded the e.e. to 98%.

The thus established stereocenter served in the following steps. Conversion of **50** to **51** was achieved by acyl chloride formation followed by Friedel–Crafts reaction with ethylene and $AlCl_3$. Addition of the lithium enolate of t-butylacetate occurred from the less-hindered face of the cyclic ketone to give, after reduction and tosylation, adduct **52**. Tosylate displacement with amine **53**, esterification with methoxyacetylchloride, and addition of HCl gave mibefradil **54**.

5.3.10 Synthesis of Naproxen and Ibuprofen (by C–H Bond Formation)

Among the many stereoselective routes currently employed to prepare enantiomerically pure NSAID α-arylalkanoic acids [63], a pre-eminent position is held by the enantioselective catalytic hydrogenation of α-arylsubstituted acrylic acids.

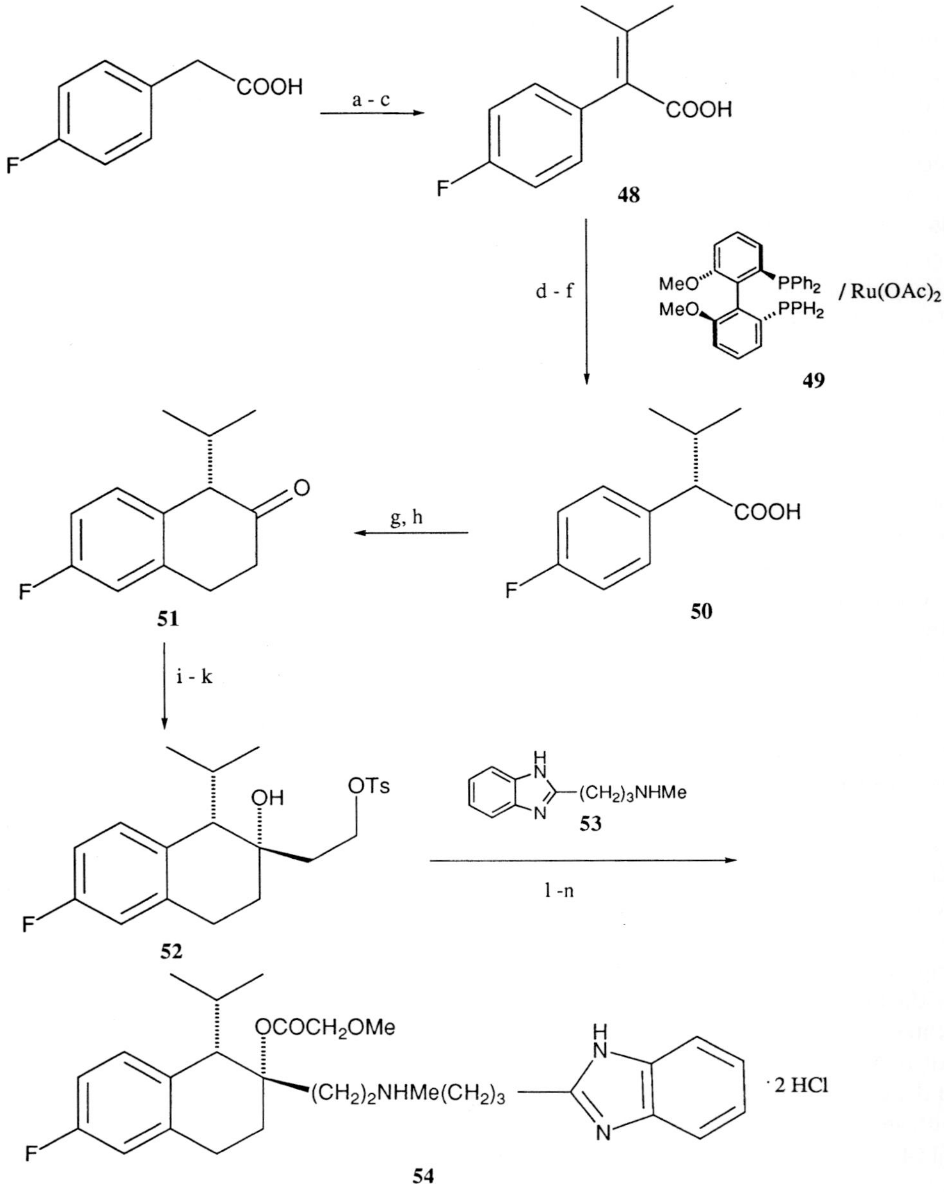

Figure 19. Stereoselective synthesis of mibefradil. Reagents: a, i-PrMgCl, THF; b, acetone; c, H_2SO_4; d, H_2, 270 bar, MeOH/TEA; e, MeONa; f, HCl; g, $SOCl_2$; h, ethylene, $AlCl_3$; i, $LiCH_2COOBu$-t; j, RedAl; k, TCl, TEA; l, TEA; m, $MeOCH_2COCl$, TEA; n, HCl.

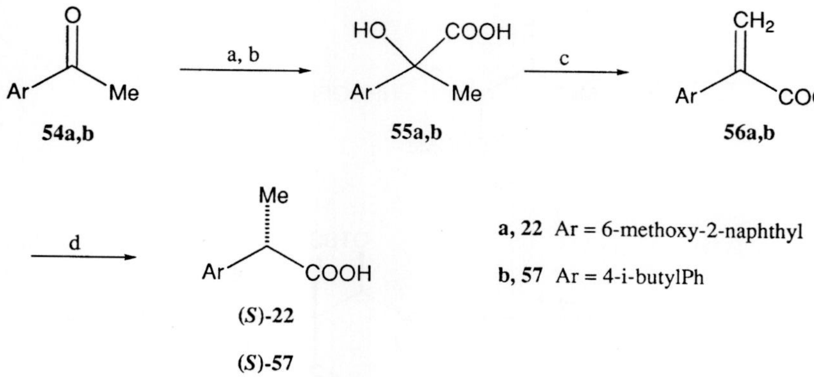

Figure 20. Stereoselective synthesis of naproxen and ibuprofen by hydrogenation. Reagents: a, CO_2, electrolysis; b, H_3O^+; c, H^+, $-H_2O$; d, H_2, $Ru(acac)_2$/BINAP.

By this route, Monsanto researchers were able to obtain both (*S*)-naproxen **22** and (*S*)-ibuprofen **57** by exploiting the Ru/BINAP-catalyzed hydrogenation originally developed by Noyori and his group. In the synthesis of both drugs the required acrylic acid was obtained by the relevant procedure outlined in Fig. 20.

Starting from the readily available parent arenes, ketones **54a,b** were prepared by standard Friedel–Crafts chemistry. Environment-friendly electrochemical carboxylation of these gave α-hydroxy acids **55a,b**, from which acrylic acids **56a,b** were easily obtained. Their reduction, catalyzed by the modified Noyori complex $Ru(acac)_2$/BINAP, afforded in high yields the (*S*)-enantiomers of acids **22** and **57** having 98% and 96% e.e., respectively. Improvements of both processes have been reported [64, 65].

5.3.11 Synthesis of a Carbapenem and Penem Precursor

As mentioned above, in describing the synthesis of sanfetrinem (see section 3.1.1.1), enantiomerically pure 4-acetoxy-substituted azetidin-2-one **1** is the key starting material for the preparation of a variety of β-lactam antibiotics of the carbapenem and penem families. About 30 different syntheses of this compound have been reported in the literature [46, 66], but only a few are industrially viable.

Among these is the Takasago process outlined in Fig. 21, which exploits an enantioselective catalytic reduction of a C–O double bond [67].

Starting from benzamide, ketoester **58** was obtained by standard chemistry. Reduction with hydrogen in the presence of the Noyori's catalyst Ru/(*R*)-BINAP gave *anti*-β-hydroxyester **59** in high e.e. (97%). The amide and ester groups were hydrolyzed in acidic and basic conditions, respectively, to give the β-amino acid. This was ring-closed by the Mukaiyama method to afford azetidinone **60** after protection of the hydroxy group. Finally, Ru-catalyzed oxidation at C-4 afforded the target product **1** in 99.2% e.e. and 58% overall yield from **58**.

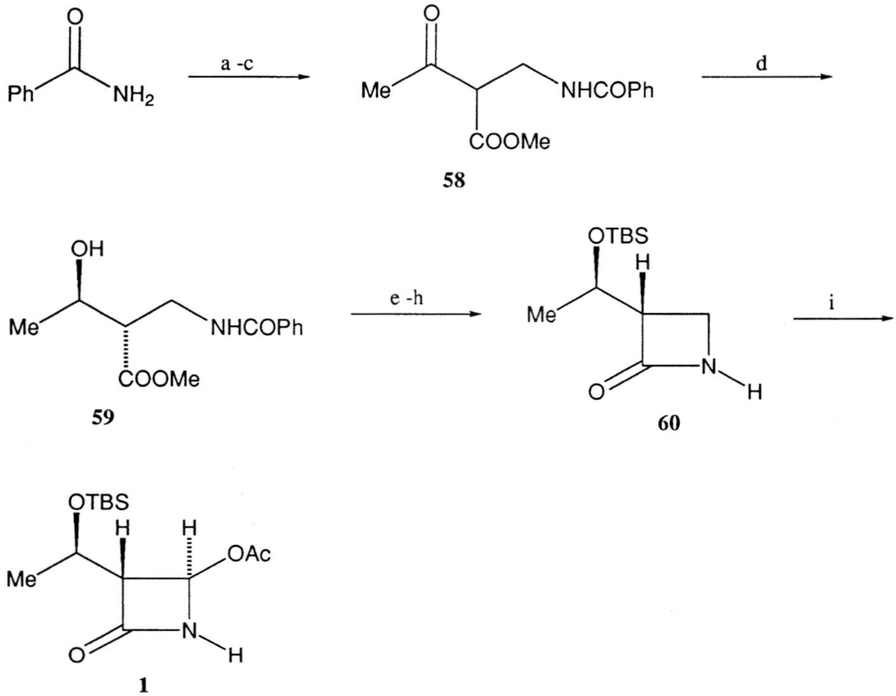

Figure 21. Stereoselective synthesis of 4-acetoxyazetidin-2-one **1.** Reagents: a, HCHO, Na_2CO_3, $CHCl_3$; b, $SOCl_2$, i-Pr_2O; c, MeCOCHNaCOOMe, THF; d, H_2, Ru/BINAP; e, HCl; f, TEA, MeCN-MeOH; g, PySSPy, PPh_3, MeCN; h, TBSCl, imidazole, DMF; i, Ru, $MeCO_3H$, CH_2Cl_2, AcOH.

5.3.12 Synthesis of Sertraline

Sertraline is an inhibitor of the synaptosomal serotonin uptake, and has been developed by Pfizer as a single isomer. After early attempts that were based on a classical resolution procedure [68], a catalytic stereoselective entry to the enantiomerically pure drug was established [69].

In this process (Fig. 22), ketone **61** underwent reduction with an over-stoichiometric amount of borane in THF in the presence of 5 mol % of the proline-derived Corey's oxazaborolidine **62** to afford alcohol **63** in 100% yield and 92% e.e. S_N2 displacement carried out on the alcohol mesylate with a higher-order cuprate gave adduct **64**, that was transformed first into tetralone **65** by a triflic acid-promoted Friedel–Crafts-type reaction, and then into sertraline **66** (86% e.e.) by imine formation and borohydride reduction.

It is worth mentioning that when *B*-chlorodiisopinocampheylborane was used as a stoichiometric chiral reducing agent to convert **61** into **63**, the latter was obtained in a slightly lower e.e. (88%) than that observed in the catalytic reaction [69].

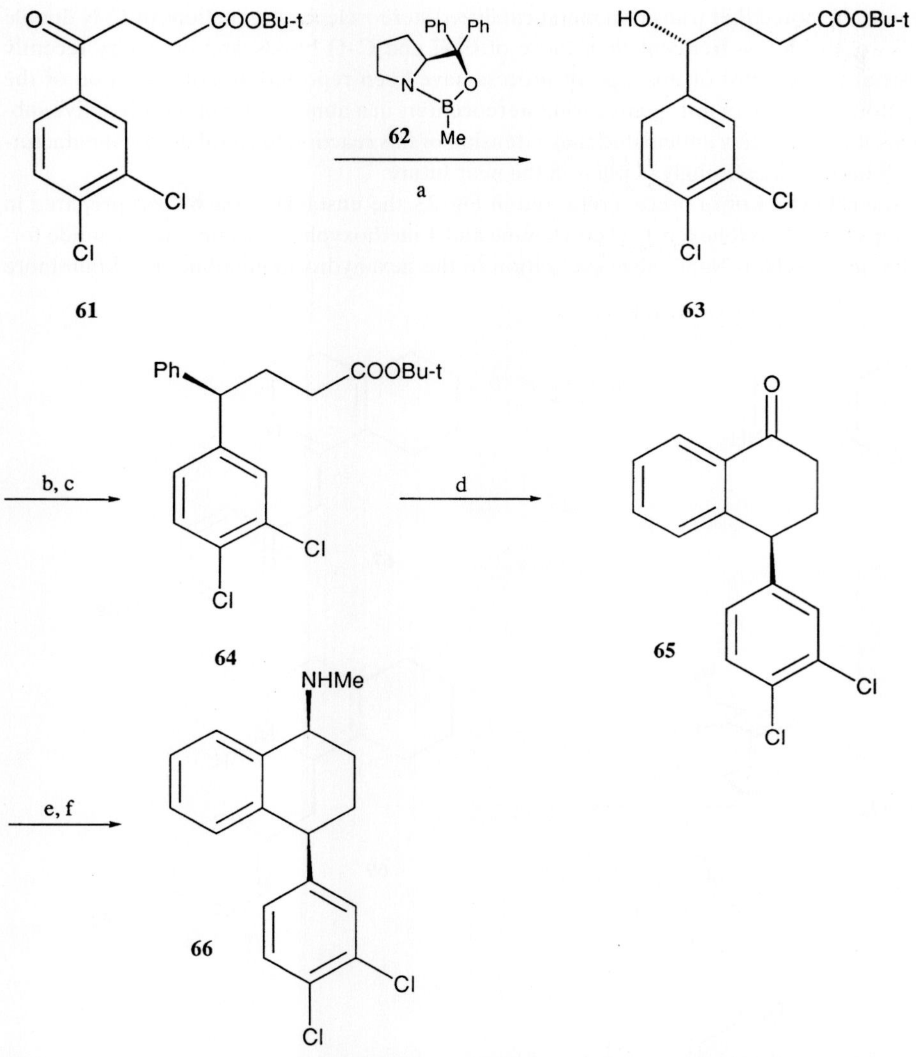

Figure 22. Stereoselective synthesis of sertraline. Reagents: a, $BH_3 \cdot THF$; b, MsCl, TEA, CH_2Cl_2; c, $Ph_2CuCNLi_2$, Et_2O; d, TfOH, benzene; e, $MeNH_2$, $TiCl_4$, toluene; f, $NaBH_4$, MeOH.

5.3.13 Synthesis of Dextromethorphan

The alkaloid dextromethorphan is an antitussive drug manufactured by Lonza in enantiomerically pure form. While early synthesis involved the tedious resolution of an octahydroisoquinoline intermediate with mandelic acid, a more recent process takes advantage of the catalytic reduction of a C–N double bond promoted by a chiral Ir/ferrocenylphosphine complex [70].

It must be noted that transition metal-catalyzed stereoselective reductions of C–N double bonds are much less frequent than those of C–C and C–O bonds, and only very recently industrial applications of this type of process have been reported. In consideration of the ubiquitous presence of nitrogen-bearing stereocenters in a number of biologically active substances, it can be safely anticipated that extension of this reaction to chiral drugs manufacturing will become increasingly popular in the near future.

In the catalytic Lonza process reported in Fig. 23, the unstable imine **67** was prepared in two steps from 2-cyclohexen-1′-yl ethylamine and 4-methoxyphenylacetic acid *via* amide formation and Bischler–Napieralski cyclization to the hexahydroisoquinoline. To obtain more

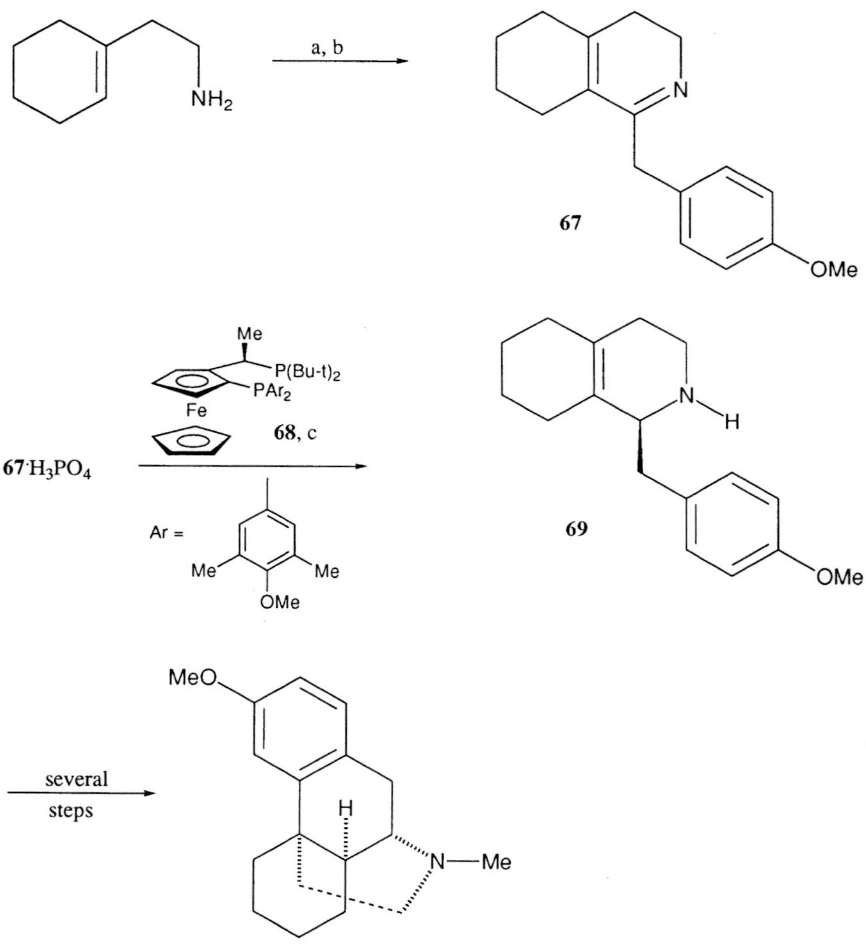

Figure 23. Stereoselective synthesis of dextromethorphan. Reagents: a, 4-methoxyphenylacetyl chloride; b, $POCl_3$; c, $Ir(COD)BF_4$, H_2, toluene, H_2O, NaOH, Bu_4NX.

reproducible results in the hydrogenation, **67** was transformed into its phosphate salt and then reduced in the presence of the catalyst [Ir(COD)(**68**)]BF_4 featuring an enantiomerically pure ferrocenylphosphine ligand [70].

Working with a substrate/catalyst ratio = 1500, amine **69** was obtained in 89% e.e. From this, an established reaction sequence led to the desired compound **70**.

5.3.14 Synthesis of (1*S*,2*R*)-1-Aminoindanol

The synthesis of enantiomerically pure (1*S*,2*R*)-1-aminoindanol became a subject of frontline industrial research activity after the Merck group discovered a series of HIV-protease inhibitors that contained this moiety. Merck's discovery eventually led to the development of indinavir sulfate as one of the leading drugs for the treatment of AIDS, and, as a consequence, to an industrially feasible, large-scale preparation of (1*S*,2*R*)-1-aminoindanol in enantiomeric pure form [71, 72].

Merck's researchers immediately realized that Jacobsen's epoxidation method catalyzed by chiral Mn/salen complexes provided a very straightforward entry to this compound (Fig. 24), since this procedure is known to epoxidize with very high e.e. *cis*-aryl/alkyl substituted alkenes. In fact, when a chlorobenzene solution of indene was exposed to a 1.5 M aqueous solution of NaOCl, 3 mol % of 4-(3-phenylpropyl)pyridine-*N*-oxide, and 0.75 mol % of catalyst (*S*,*S*)-**71** [73], (1*S*,2*R*)-indene oxide **72** was obtained in >90% yield and 88% e.e.

Conversion of **72** to the desired compound **73** was accomplished *via* a Ritter process with oleum in acetonitrile followed by addition of water and treatment with tartaric acid [71, 72].

Alternatively, Sepracor scientists employed the enantiomeric catalyst (*R*,*R*)-**71** to epoxidize indene with NaOCl in dichloromethane to give (1*R*,2*S*)-**72**. This was ring-opened with ammonia and transformed into oxazoline **74** by reaction with benzoylchloride in Schotten–Baumann conditions followed by addition of 80% sulfuric acid. Hydrolysis of **74** afforded **73**.

(1*S*,2*R*)-1-Aminoindanol was then used for the synthesis of indinavir sulfate **80** (Fig. 24) [72]. Amide **44** (see Fig. 18) was allylated to give **75** with high stereocontrol (d.r. >97:3). The double bond was epoxidized by a three-steps procedure involving addition of *in situ*-generated *N*-iodosuccinimide, iodohydrin formation with $NaHCO_3$ in water, and MeONa-promoted ring closure to afford **76**. Addition of the enantiomerically pure monoprotected piperazine **77** gave **78** after removal of the protecting groups.

N-Alkylation with 3-picolyl chloride led to **79** and eventually to the target compound **80** by formation of the sulfate salt. Thus, the two stereocenters established by the epoxidation procedure served to control the formation of two other stereocenters at the benzyl and hydroxyl bearing carbon atoms. This stresses the centrality of the stereoselective preparation of **73** to the synthesis of indinavir sulfate.

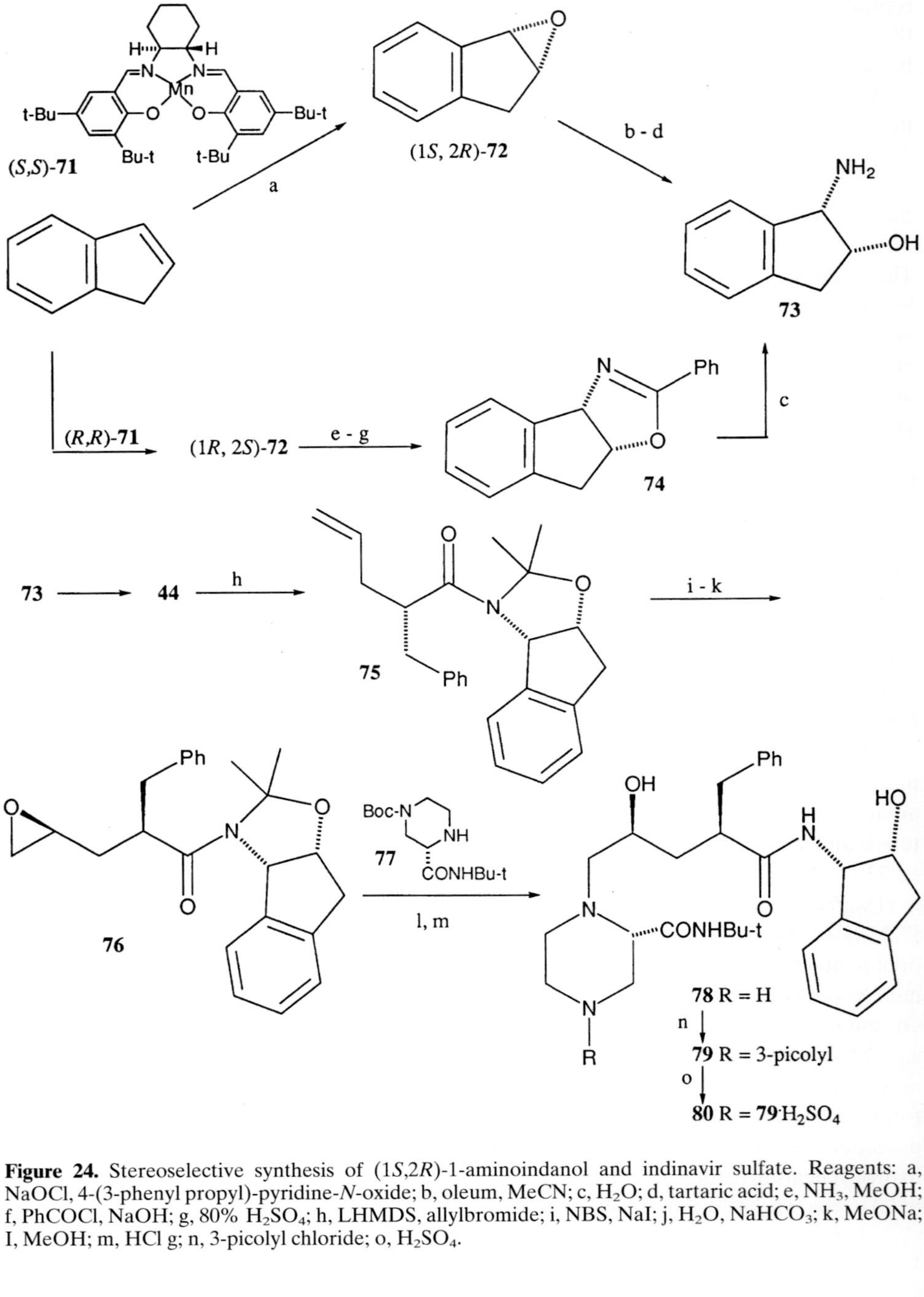

Figure 24. Stereoselective synthesis of (1*S*,2*R*)-1-aminoindanol and indinavir sulfate. Reagents: a, NaOCl, 4-(3-phenyl propyl)-pyridine-*N*-oxide; b, oleum, MeCN; c, H_2O; d, tartaric acid; e, NH_3, MeOH; f, PhCOCl, NaOH; g, 80% H_2SO_4; h, LHMDS, allylbromide; i, NBS, NaI; j, H_2O, $NaHCO_3$; k, MeONa; I, MeOH; m, HCl g; n, 3-picolyl chloride; o, H_2SO_4.

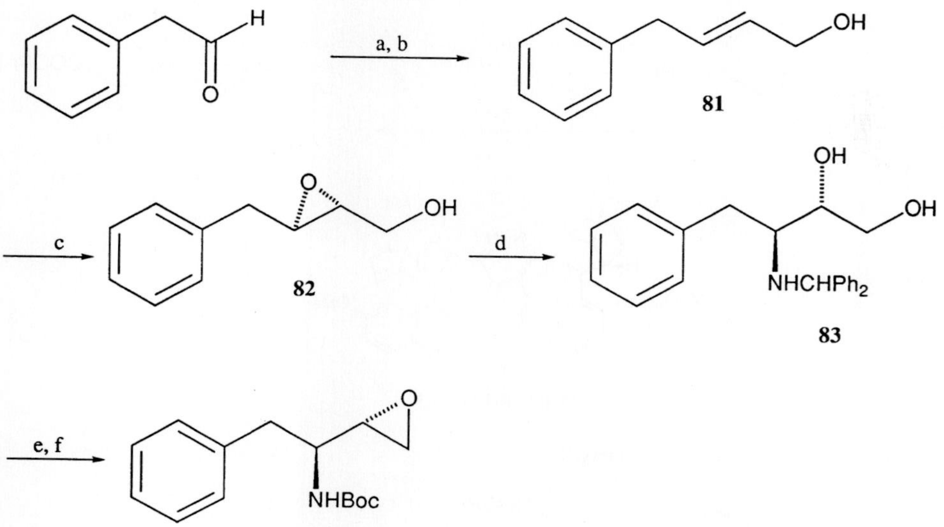

Figure 25. Stereoselective synthesis of a fragment of the HIV-protease inhibitor Ro 31-8959. Reagents: a, $Ph_3PCHCOOEt$, CH_2Cl_2; b, DIBALH, Et_2O; e, t-BuOOH, $Ti(OPr\text{-}i)_4$, (–)-DIPT; d, Ph_2CHNH_2, $Ti(OPr\text{-}i)_4$, $ClCH_2CH_2Cl$; e, $(Boc)_2O$, H_2; $Pd(OH)_2/C$, AcOEt; f, PPh_3, DEAD,. $CHCl_3$.

5.3.15 Synthesis of a Fragment of the HIV-Protease Inhibitor Ro 31-8959

In the course of the synthesis of the potent HIV-protease inhibitor Ro 31-8959 [74], a simple entry to epoxide (*S*,*S*)-**84** (Fig. 25) was needed. Of the many attempted routes, the one based on Sharpless' catalytic epoxidation, reported in Fig. 25, proved to be very successful [75].

Phenylacetaldehyde was readily converted into allylic alcohol **81** by a standard olefination and reduction protocol. Epoxidation with t-BuOOH and catalytic $Ti(OPr\text{-}i)_4$ in the presence of catalytic (–)-DIPT gave (2*R*,3*R*)-3-benzylglycidol **82** in >99% e.e. Addition of diphenylmethanamine and $Ti(OPr\text{-}i)_4$ in refluxing 1,2-dichloroethane led to aminodiol **83**, from which aminoepoxide (*S*,*S*)-**84** was obtained by hydrogenolysis and *N*-protection followed by an intramolecular Mitsunobu reaction [76].

5.3.16 Synthesis of (2*R*,3*S*)-Phenylisoserine Precursor of the Taxol and Taxotere Side Chains

The anticancer drugs Taxol and Taxotere feature a (2*R*,3*S*)-*N*-benzoyl phenylisoserine as side chain. A number of stereoselective syntheses of this moiety have been reported. Among them, the preparation based on the catalytic asymmetric aminohydroxylation protocol recently developed by Sharpless and reported in Fig. 26 seems particularly attractive [77].

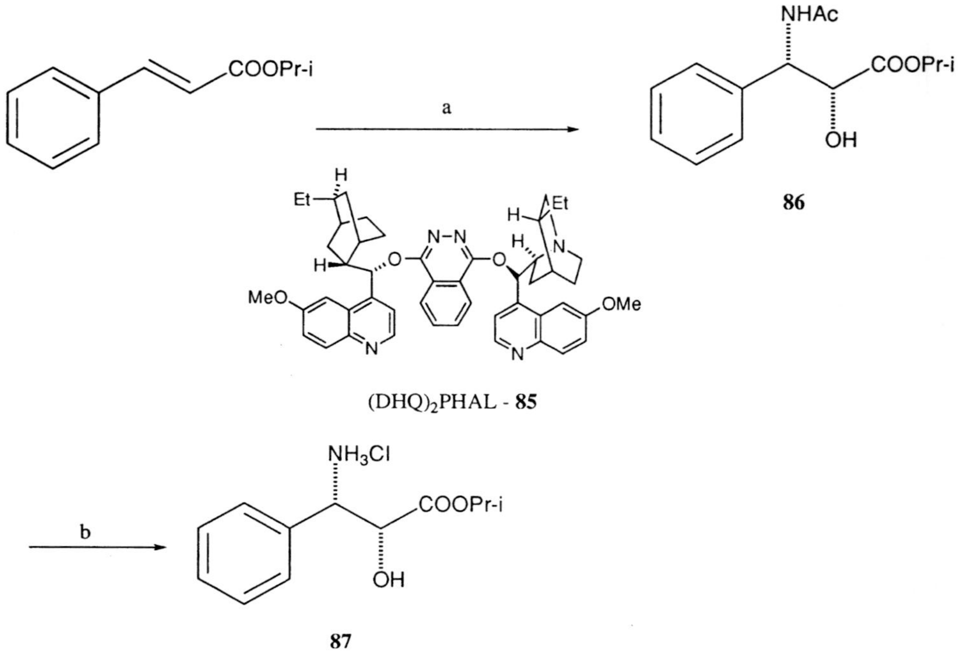

Figure 26. Stereoselective synthesis of Taxol and Taxotere side chain. Reagents: a, AcNHBr, LiOH, $K_2OsO_2(OH)_4$, **85**, t-BuOH, H_2O; b, 10% HCl, 100 °C.

Readily available i-propylcinnamate was treated with 1.10 mol equiv. of *N*-bromoacetamide, 1.07 mol equiv. of LiOH, and catalytic amounts of both Os(VIII) (0.015 mol equiv.) and the dihydroquinine-derived ligand **85** (0.01 mol equiv.) in a t-BuOH/water mixture as solvent. From this reaction, carried out on a 630 mmol scale, *N*-acetyl α-hydroxy-β-aminoester **86** was isolated by crystallization in 71% yield and 99% e.e. When the reaction was run on a smaller scale, chromatographic purification gave the product in 81% yield. Acid hydrolysis of **86** afforded (2*R*,3*S*)-phenylisoserine **87** as its hydrochloride in 68% overall yield from the cinnamate.

5.3.17 Synthesis of Naproxen (by C–C Bond Formation)

While catalytic stereoselective versions of many reactions forming C–H, C–O, and C–N bonds are nowadays well established, the discovery of equally efficient processes to form C–C bonds that can find industrial application has been less successful [38, 78].

One of the few available examples is represented by the synthesis of cilastatine by a chiral Cu complex promoted cyclopropanation reaction developed by Sumitomo Chemical Co. [78]. Another is the catalytic asymmetric hydrocyanation of vinylarenes developed at DuPont [79]. In this process (Fig. 27) sugar-derived phosphinites are used in combination with a Ni catalyst to prepare enantiomerically enriched precursors of the NSAID naproxen.

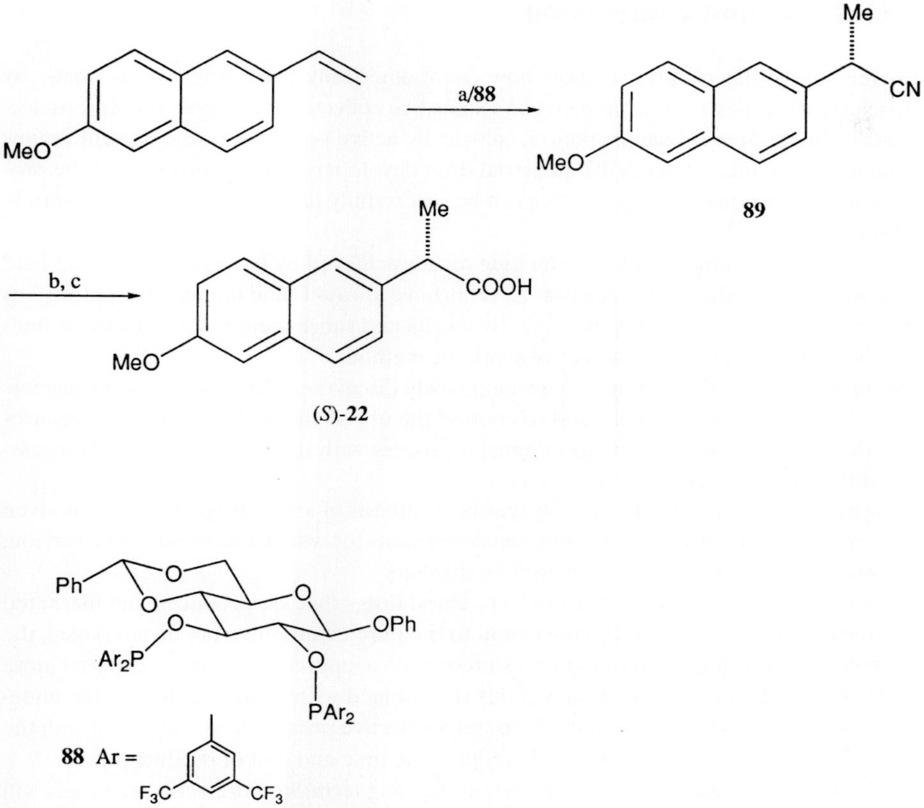

Figure 27. Stereoselective synthesis of naproxen. Reagents: a, HCN, Ni(COD)/**88**; b, crystallization; c, acid hydrolysis.

The exocyclic double bond of 6-methoxy-2-vinyl naphthalene underwent a totally regioselective, Markovnikov-type addition of HCN in the presence of less than 1 mol% of (*D*)-glucose-derived phosphinite ligand **88** and of $Ni(COD)_2$ in toluene to afford nitrile **89** in more than 85% yield and up to 85% e.e. The e.e. was improved to >99% by two crystallizations. Acid hydrolysis of 89 gave (*S*)-22 in high yield and unaltered e.e.

The complete regioselectivity, the absence of byproducts, the ready availability at low cost of the ligand employed, and the chemical and stereochemical efficiency of this reaction make it a good candidate to become the first catalytic, stereoselective industrial manufacturing of naproxen, even if the use of HCN can be a drawback. In this context, it must be remembered that Union Carbide Co. has recently patented a route to (*S*)-ibuprofen based on a $Rh(CO)_2$(acac)/chiral phosphite-catalyzed hydroformylation of 4-isoutylstyrene that occurs in 82% e.e. [63, 80].

5.4 Outlook and Conclusion

It was the aim of this chapter to show how enantiomerically pure drugs can be made by stereoselective synthesis. Far from being an exhaustive collection of all possible stereoselective entries to chiral non-racemic pharmacologically active compounds, the present review was meant to convince (if needed!) industrial drug developers that the problem of the production of an enantiomerically pure drug can be successfully tackled and solved by a variety of approaches.

The unbelievable progress and outstanding results achieved by basic research in the field of stereoselective synthesis over the past 25 years have always found immediate applications in the pharmaceutical industry, which eagerly awaits and subsequently develops these findings in the search for new stereoselective synthetic methods.

Nowadays, it is common practice to develop newly discovered chiral drugs not as racemic mixtures but as pure enantiomers, and to confine the use of classical resolution procedures only to the first stages of the drug development process, with the sole aim of rapidly assessing the different activities of stereoisomers.

In contrast, in designing an industrially feasible synthesis of a new drug, preference is given to those reaction sequences that include transformations for which a stereoselective version, and possibly a catalytic one, is already well established.

As for the so-called "racemic switches", i.e., chiral drugs that are produced and marketed as racemic mixtures and for which conversion to the pure enantiomer can be envisaged, the scenario is slightly different. In this case, a stereoselective approach can be considered more useful than a resolution step (especially if this is combined with a racemization of the undesired isomer and by its recycling), only if the stereoselective process "connects" well with the existing synthetic procedure or effectively reduces the time and cost of production.

However, it can be easily anticipated that, in the long term, stereoselective synthesis will dominate the stage of chiral drug production. Indeed, it is the authors' belief that stereoselective synthesis will provide a great opportunity for new and exciting chemistry in both the academic and the industrial worlds.

References

[1] Stinson, S.C. *Chem. Eng. News*, September 19, 38 (1994).
[2] Stinson, S.C. *Chem. Eng. News*, October 9, 44 (1995).
[3] Stinson, S.C. *Chem. Eng. News*, October 20, 38 (1997).
[4] Mislow, K. *Introduction to Stereochemistry*, Benjamin, New York (1965).
[5] Eliel, E.L., Wilen, S.H., and Mander, L.N. *Stereochemistry of Organic Compounds*. Wiley Interscience: New York, 1994.
[6] Izumi, Y., and Tai, A. *Stereo-differentiating Reactions*. Academic Press: New York, 1977.
[7] Atkins, R.S. *Stereoselective Synthesis*. Wiley Interscience: New York, 1995.
[8] Helmchen, G. In: *Houben-Weyl vol. E-21 a: Stereoselective Synthesis*. Thieme Verlag: Stuttgart, 1995.
[9] Masamune, S., Ali, S.A., Snitman, D.L., and Garvey, D.S. *Angew. Chem. Int. Ed. Engl.* **19**, 557 (1980).
[10] Anh, N.T. *Top. Curr. Chem.* **68**, 145 (1980).
[11] Reetz, M.T. *Angew. Chem. Int. Ed. Engl.* **23**, 556 (1984).
[12] Frater, G. *Helv. Chim. Acta* **62**, 2825, 2829 (1979); ibid. **63**, 1783 (1980).

[13] Poli, G., and Scolastico,C. In: *Houben-Weyl vol. E-21 e: Stereoselective Synthesis.* Thieme Verlag: Stuttgart, 1995.
[14] Myers, A.G., Yang, B.H., Chen, H., McInstry, L., Kopecky, D.J., and Gleason, J.L. *J. Am. Chem. Soc.* **119**, 6496 (1997).
[15] Ager, D.J., Prakash, I., and Schaad, D.R. *Aldrichimica Acta* **30**, 3 (1997).
[16] Evans, D.A., Bartroli, J., and Shih, T.L. *J. Am. Chem. Soc.* **103**, 2127 (1981).
[17] Walker, M.A., and Heathcock, C.H. *J. Org. Chem.* **56**, 5747 (1991).
[18] Yamada, S., Hiroi, K., and Achiwa, K. *Tetrahedron Lett.* **11**, 4233, (1969).
[19] Whitesell, J.K., and Felman, S.W. *J. Org. Chem.* **42**, 1663 (1977).
[20] Whitesell, J.K. *Chem. Rev.* **89**, 1581 (1989).
[21] Dhar, R.K. *Aldrichimica Acta* **27**, 43 (1994).
[22] Mislow, K., and Siegel, J. *J. Am. Chem. Soc.* **106**, 3319 (1984).
[23] Fehr, C. *Angew. Chem. Int. Ed. Engl.* **35**, 2567 (1996).
[24] Corey, E.J., and Kim, S.S. *J. Am. Chem. Soc.* **112**, 4976 (1990).
[25] Narasaka, K., Inoue, M., and Okada, N. *Chemistry Lett.* 1109 (1986).
[26] Oriyama, T., Imai, K., Hosoya, T., and Sano, T. *Tetrahedron Lett.* **39**, 397 (1997).
[27] Poss, C.S., and Schreiber, S.L. *Acc. Chem. Res.* **27**, 9 (1994).
[28] Kagan, H.B., and Fiaud, J.C. *Top. Stereochem.* **18**, 249 (1988).
[29] Brown, H.C., Ayyangar, N.R., and Zweifel, G. *J. Am. Chem. Soc.* **86**, 397 (1964).
[30] Cinquini, M., Colonna, S., and Cozzi, F. *J. Chem. Soc., Perkin Trans.* **1**, 247 (1978).
[31] Wegener, B., Hansen, M., and Winterfeldt, E. *Tetrahedron: Asymmetry* **4**, 345 (1993).
[32] Martin, V.S., Woodard, S.S., Katsuki, T., Yamada, Y., Ikeda, M., and Sharpless, K.B. *J. Am. Chem. Soc.* **103**, 6237 (1981).
[33] Masamune, S., Choy, W., Petersen, J.S., and Sita, L.R. *Angew. Chem. Int. Ed. Engl.* **24**, 1 (1985).
[34] Heathcock, C.H., Pirrung, M.C., Lampe, J., Buse, C.T., and Young, S.D. *J. Org. Chem.* **46**, 2290 (1981).
[35] Evans, D.A., Dart, M.J., Duffy, J.L., and Rieger, D.L. *J. Am. Chem. Soc.* **117**, 9073 (1995).
[36] Palomo, C., Aizpurua, J.M., Mielgo, A., and Linden, A. *J. Org. Chem.* **61**, 9186 (1996).
[37] Annunziata, R., Benaglia, M., Chiovato, A., Cinquini, M., and Cozzi, F. *Tetrahedron* **51**, 10025 (1995).
[38] *Catalytic Asymmetric Synthesis.* Ojima, I. (ed.) VCH: New York, 1993.
[39] Sheldon, R.A. *Chirotechnology.* Marcel Dekker: New York, 1993.
[40] Kotha, S. *Tetrahedron* **50**, 3639 (1994).
[41] Evans, D.A., Murry, J.A., and Kozlowski, M.C. *J. Am. Chem. Soc.* **118**, 5814 (1996).
[42] Ojima, I., Kogura, T., Yoda, N., Suzuki, T., Yatabe, M., and Tanaka, T. *J. Org. Chem.* **47**, 1329 (1982).
[43] Kitamura, M., Kasahara, I., Manabe, K., Noyori, R., and Takaya, H. *J. Org. Chem.* **53**, 708 (1988).
[44] Tokunaga, M., Larrow, J.F., Kakiuchi, F., and Jacobsen, E.N. *Science* **277**, 936 (1997).
[45] (a) *Drugs of the Future* **21**, 1238 (1996). (b) Ghiron, C., and Rossi, T. In: *Targets in Heterocyclic Systems.* Attanasi, O.A., Spinelli, D. (eds.) Italian Society of Chemistry, vol. 1, pp. 161–186 (1997). (c) For a new synthesis see: Murayama, T., Mitsuhashi, S., and Miura, T. Eur. Patent Application 97400458-2, 1998, to Takasago Int. Corp.
[46] Berks, A.H. *Tetrahedron* **52**, 331 (1996).
[47] Rossi, T., Biondi, S., Contini, S., Thomas, R.J., and Marchioro, C. *J. Am. Chem. Soc.* **117**, 9604 (1995).
[48] Rossi, T., Marchioro, C., Paio, A., Thomas, R.J., and Zarantonello, P. *J. Org. Chem.* **62**, 1653 (1997).
[49] Blacklock, T.J., Sohar, P., Butcher, J.W., Lamanec, T., and Grabowski, E.J.J. *J. Org. Chem.* **58**, 1672 (1993).
[50] Blacker, H.J., and Holt, R.A. In: *Chirality in Industry II.* Collins, A.N., Sheldrake, G.N., and Crosby, J. (eds) J. Wiley & Sons: New York, pp. 245–261, 1997.
[51] Giordano, C., Villa, M., and Panossian, S. In: *Chirality in Industry.* Collins, A.N., Sheldrake, G.N., and Crosby, J. (eds) J. Wiley & Sons: New York, pp. 303–312, 1992; and references cited therein.
[52] Miyata, O., Shimada, T., Ninomiya, I., and Naito, T. *Tetrahedron Lett.* **32**, 3519 (1991).
[53] Kaptain, B., and Verzijl, G.K.M. Eur. Patent Application 98200484, 1998, to DSM.
[54] Chang, S., Galvin, J.M., and Jacobsen, E.N. *J. Am. Chem. Soc.* **116**, 6937 (1994). See also: Lygo, B., and Wainwright, P.G. *Tetrahedron Lett.* **39**, 1599 (1998).
[55] (a) Palmer, J.T. Eur. Patent Application 03042904, 1989, to Marion Laboratories; (b) Coffen, D.L., Medan, P.B., and Schwatz, A. Eur. Patent Application 0343474, 1988, to Hoffman La Roche.

[56] Thompson, A.S., Tschaen, D.M., Simpson, P., McSwine, D.J., Reamer, R.A., Verhoeven, T.R., and Shinkai, I. *J. Org. Chem.* **57**, 7044 (1992).
[57] Shinkai, I. *Pure Appl. Chem.* **69**, 453 (1997).
[58] Miyano, S., Lu, L.D.L., Viti, S.M., and Sharpless, K.B. *J. Org. Chem.* **50**, 4350 (1985).
[59] Askin, D., Wallace, M.A., Vacca, J.P., Reamer, R.A., Volante, R.P., and Shinkai, I. *J. Org. Chem.* **57**, 2771 (1992).
[60] Maligres, P.E., Weissman, S.A., Upadhyay, V., Cianciosi, S.J., Reamer, R.A., Purick, R.M., Sager, J., Rossen, K., Eng, K.K., Askin, D., Volante, R.P., and Reider, P.J. *Tetrahedron* **52**, 3327 (1996).
[61] Crameri, Y., Foricher, J., Hengartner, U., Jenny, C.-J., Kienzle, F., Ramuz, H., Scalone, M., Schlageter, M., Schmid, R., and Wang, S. *Chimia* **51**, 303 (1997).
[62] Schmid, R., Foricher, J., Cereghetti, M., and Schönholzer, P. *Helv. Chim. Acta* **74**, 370 (1991).
[63] Stahly, G.P., and Starrett, R.M. In: *Chirality in Industry II.* Collins, A.N., Sheldrake, G.N., and Crosby, J. (eds) John Wiley & Sons: New York, pp. 19–40, 1997.
[64] Wan, K.T., and Davis, M.E. *Nature* (*London*) **370**, 449 (1994).
[65] Manimaran, T., Wu, T.-C., Klobucar, W.D., Kolich, C.H., Stahly, G.P., Fronczek, F.R., and Watkins, S.E. *Organometallics* **12**, 1467 (1993).
[66] Cozzi, F., Annunziata, R., Cinquini, M., Poletti, L., Perboni, A., and Tamburini, B. *Chirality* **10**, 91 (1998).
[67] Akutagawa, S. In: *Chirality in Industry I.* Collins, A.N., Sheldrake, G.N., and Crosby, J. (eds) John Wiley & Sons: New York, pp. 325–339, 1992.
[68] Welch, W.M., Kraska, A.R., Sarges, R., and Koe, K.B. *J. Med. Chem.* **27**, 1508 (1984).
[69] Quallich, G.J., and Woodall, T.M. *Tetrahedron* **48**, 10239 (1992).
[70] Imwinkelried, R. *Chimia* **51**, 300 (1997).
[71] Senanayake, C.H. *Aldrichimica Acta* **31**, 3 (1998); and references therein.
[72] Reider, P. J. *Chimia* **51**, 306 (1997).
[73] Hughes, D.L., Smith, G.B., Liu, J., Dezeny, G.C., Senanayake, C.H., Larsen, R.D., Verhoeven, T.R., and Reider, P.J. *J. Org. Chem.* **62**, 2222 (1997).
[74] Parkes, K.E.B., Bushnell, D.J., Crackett, P.H., Dundsdon, S.J., Freeman, A.C., Gunn, M.P., Hopkins, R.A., Lambert, R.W., Martin, J.A., Merrett, J.H., Redshaw, S., Spurden, W.C., and Thomas, G.J. *J. Org. Chem.* **59**, 3656 (1994).
[75] Shum, W.P., and Cannarsa, M.J. *Chirality in Industry II.* Collins, A.N., Sheldrake, G.N., and Crosby, J. (eds) John Wiley & Sons: New York, pp. 363–380, 1997.
[76] Castejon, P., Pasto', M., Moyano, A., Pericas, M.A., and Riera, A. *Tetrahedron Lett.* **36**, 3019 (1995).
[77] Bruncko, M., Schlingloff, G., and Sharpless, K.B. *Angew. Chem. Int. Ed. Engl.* **36**, 1483 (1997).
[78] Scott, J.W. *Top. Stereochem.* **19**, 209 (1989).
[79] Casalnuovo, A.L., and Rajanbabu, T.V. *Chirality in Industry II.* Collins, A.N., Sheldrake, G.N., and Crosby, J. (eds) John Wiley & Sons: New York, pp. 309–333, 1997.
[80] Babin, J.E., and Whiteker, G.T. PCT Int. Pat. Appl. WO 93/03839 (1993) to Union Carbide.

6 Resolution of Enantiomers of Chiral Drugs

Gottfried Blaschke and Bezhan Chankvetadze

Abbreviations

CAMP	chiral additive to the mobile phase
CD	cyclodextrin
CDR	chiral derivatizing reagent
CE	capillary electrophoresis
CEC	capillary electrochromatography
CEKC	capillary electrokinetic chromatography
CLRC	closed-loop recycling chromatography
CSP	chiral stationary phase
CZE	capillary zone electrophoresis
DNBG	dinitrobenzoylphenylglycine
DNBTyr	dinitrobenzoyltyrosine
FFF	field-flow fractionation
GC	gas chromatography
HPLC	high-performance liquid chromatography
IB	ibuprofen
I.S.	internal standard
LC	liquid chromatography
SFC	supercritical fluid chromatography
SMB	simulated moving bed
TLC	thin-layer chromatography

6.1 Introduction

Enantioseparations are intensively used as an access to intermediates and chiral pools of many synthetic methodologies for obtaining enantiomerically pure compounds. In addition, analytical enantioseparation is the most reliable instrumental technique that allows the estimation of the success of any enantioselective synthesis. Therefore, a chapter on this non-synthetic topic hopefully can be of some use also for the readers of this volume, which comprises entirely "synthetic" topics.

The property by which a molecule may exist as two mirror forms, which are able to rotate a beam of polarized light in opposite directions, is known as *chirality* [1]. The chiral molecules contain the elements (center, axis or plane) of chirality. The chiral molecule which contains only one element of chirality may exist in two possible configurations which are not superimposable on each other. These compounds with different configurations are called enantiomers.

After its discovery in 1848 by the French scientist Louis Pasteur [2], chirality as a phenomenon has served and continues to serve as a source of inspiration of generations of chemists. There is no doubt that this is an exciting field from the academic point of view. However, a pure academic interest is insufficient for a flourished development of the field of chirality. In addition, even more intensive impulses for a rapid development of this field are provided by industrial interests. This relates basically to pharmaceutical, food, chemical, and agrochemical fields.

The importance of chirality in pharmaceutics stems from the observation that drug action in general requires a steric molecular interaction with receptor sites. The receptor sites are constructed from L-amino acids and/or D-sugars and, consequently, are chiral. Therefore, the biological activity of many chiral drugs at their receptor sites are markedly different. In addition, stereoselectivity may associate any step of drug residence in a living body. Drug absorption, metabolism, pharmacokinetics, transport to the receptor sites and excretion from the body all may be stereoselective processes for a chiral compound. Ample examples of stereoselective (even stereospecific) actions of chiral drugs are known [3–7].

Continuously increasing interest in academia and a growing number of examples of dramatic, sometimes even fatal differences between the biological activity of drug enantiomers led to the situation where governmental and regulatory acts were released in the USA, European Community and Japan which consider one enantiomer as the impurity of the other enantiomer. This requires from the drug companies an independent investigation of both enantiomers of a chiral drug [8].

Chiral drugs represent a significant part of the global pharmaceutical market. Of the top 100 drugs world-wide, 50 are marketed as pure enantiomers. Their sales were to the value of $ 42.8 billion in 1997, or 51% of the total of $ 85.2 billion for these top 100 [9].

The final decision whether a given chiral drug will be marketed in an enantiomerically pure form, or as a racemic or non-racemic mixture of both enantiomers, will be made weighting all medical, financial and social proficiencies of one or another form. Independent from a final decision the kinetic, pharmacological and toxicological properties of individual enantiomers need to be clearly characterized. Therefore, pure enantiomers of biologically active chiral compounds may be required at the very early stage of a drug development process. Enantioseparation may be the method of choice because it is, in general, both cheaper and more rapid to develop compared with an enantioselective synthesis.

Enantiomers possess equal free energy in an achiral environment and cannot be resolved without covalent or non-covalent interactions with a chiral reagent. The reagent can be an enantiomerically enriched or pure chiral resolving agent, a chiral stationary phase or additive to the mobile phase, as well as a stereoselective catalyst. Depending on the function of the chiral reagent the techniques of enantioseparations can be classified in different groups as given in the outline of this chapter.

Based on the general synthetic scope of this book, the subject of enantioseparations is treated in this chapter mainly from the viewpoint of micropreparative- and preparative-scale production of enantiomerically pure compounds. Analytical aspects are treated to a very limited extent and a quantification of enantiomers in biological fluids as well as stereoselective aspects of drug metabolism, pharmacokinetics and pharmacodynamics are almost completely omitted.

6.2 Crystallization

6.2.1 Spontaneous Crystallization

This is the method which allowed Pasteur to discover molecular chirality 150 years ago [2]. The prerequisite for working of this technique is the occurrence of a chiral compound as a crystalline conglomerate (racemic mixture) rather than a racemic compound. In bulk, such a conglomerate is optically inactive. However, individual crystals contain only one enantiomer. This is not the case for a racemic compound which does not form a conglomerate. In the latter case crystals contain equal amounts of both enantiomers.

The theory and experiment of direct crystallization of enantiomers is quite well understood at present [10]. There are a number of variables which affect the resolution by direct crystallization in practice. Several technological schemes based on this principle are realized on the commercial scale. These are, for example, the Merck process used for the production of antihypertensive drug methyldopa [11], a process developed by Harman and Reimer for (–)-menthol, which is separated as an ester [12], the process patented by Industria Chimica Profarmaco for the resolution of naproxen enantiomers as the ethylamine salt [13], the production of L-glutamic acid by the Japanese company Ajinomoto on a scale in excess of 10 000 tons annually as early as the 1960s [14], etc. In general, it seems that spontaneous crystallization is a very useful technique for the enantioseparation of the naturally occurring α-amino acids. All of them may be resolved either directly or as derivatives [10].

The important advantage of a direct crystallization is that a chiral compound does not necessarily possess functional groups which are required for diastereomeric salt formation. The disadvantages include its limited applicability because less than 10% of all crystalline racemates occur as a conglomerates, poor predictability and very sensitive dependence from the experimental conditions. Thus, as mentioned by some authors, even Louis Pasteur would not have succeeded with his very first enantioseparation had he performed the experiment with the mixed cesium-rubidium salt of tartaric acid at temperatures higher than 27 °C [4].

6.2.2 Diastereomeric Crystallization

Although considered to be too empirical and based on a trial-and-error approach, diastereomeric crystallizations remain a method of choice for obtaining enantiomerically pure synthetic drugs in relatively small amounts.

The principle of this technique relies on the following: the enantiomers possess the same free energy and all physico-chemical properties except chiro-optical ones. However, the free energy and, consequently, physico-chemical properties of the diastereomers are different. Thus, to be resolved using this technique, a mixture of enantiomers must be transformed in a mixture of diastereomeric compounds by stereoselective interaction with an optically pure reagent.

The resulting covalent diastereomeric compounds possessing different free energies may be resolved based on their different physico-chemical properties and then be transformed back to the optically pure starting material. The technique of covalent diastereomer formations is very laborious and time-consuming, and sometimes needs to be performed under conditions where a racemization may occur. Therefore, for preparative purposes this technique represents just a theoretical interest. However, it is still used in analytical-scale enantioseparations where the enantiomers must not necessarily be collected after the separation.

Not only covalent diastereomeric compounds but also diastereomeric non-covalent complexes differ in terms of their free energy. Non-covalent diastereomeric complexation is a more suitable technique for the preparation of pure enantiomers because the formation and destruction of non-covalent complexes are commonly more gentle and less laborious.

The diastereomeric crystallization relies on a different solubility of diastereomeric salts.

The first enantioseparation based on a diastereomeric salt formation was performed by Pasteur in 1853 [2, 10]. In this example, racemic tartaric acid was resolved as diastereomeric salts with (+)-cinchotoxine or (+)-quinotoxine. Diastereomeric complexes may also be of charge-transfer or inclusion type.

Thus, in contrary to spontaneous crystallization, a resolving agent is required in diastereomeric crystallizations. The resolving agent should meet certain requirements:

- Availability in high chemical and enantiomeric purity.
- Stability in use and storage.
- Availability in both enantiomeric forms.
- Ease of recovery and reuse.
- Low toxicity.
- Low price and ease of preparation.
- Reasonable solubility.
- Low molecular weight.

Some commonly used resolving agents are summarized in Table 1 [15–48]. The formation of non-covalent diastereomeric salts is driven by ionic interactions. Therefore, suitable functional groups (acidic or basic) are required to be present in both counterparts. This makes impossible a direct application of the diastereomeric crystallization technique to several classes of chiral compounds such as alcohols, aldehydes, ketones, diols, thiols, dithiols, and phenols. This is a critical disadvantage of this technique. The compounds of the above-mentioned groups may be transformed to their more polar derivatives and resolved as such. However, this requires an additional reaction step, and reagents, and the recovery of the starting material after the resolution may not always be easy.

An alternative technique for the resolution of enantiomers which lack polar functional groups is a diastereomeric complex formation via non-ionic mechanisms. These complexes may be of the external or the inclusion type. Chiral amino acids, metal complexes or host compounds may serve as useful complexing agents (Table 2) [49–64].

Diastereomeric crystallization is intensively used for the preparation of enantiomerically pure chiral drugs on a relatively small scale. However, applications of this technique on the production scale are also described (Table 3) [5].

Table 1. Commonly used resolving agents for polar chiral compounds [15–48]

Chiral compounds	Resolving agents	Reference
Acids and lactones	Alkaloids (brucine, strychnine, quinidine, quinine, chinchonine, chinchonidine, dehydroabietylamine,	[15–19]
	quinotoxine, cinchotoxine), terpene derivatives (endo-bornylamine, 1R-3-endo-aminoborneol,	[20–23]
	pinelylamines, (+)-3-aminomethylpinane) ephedrine and ephedrine analogs, synthetic amines (d-methylbenzylamine and its derivatives, d-1-naphthylethylamine, 2-amino-1,2-diphenylethanol), amino acids	[24, 25–27]
	and bases derived thereof	[28–31]
Bases	Tartaric acid and derivatives (dibenzoyl- and di-p-toluyltartaric acids)	[32–35]
	Mandelic acid and derivatives (O-acetylmandelic acid and O-methylmandelic acid)	[36–38]
	1,1´-Binapthylphosphoric acid	[39–40]
	Camphorsulfonic acid	[41]
	Deoxycholic acid	[42]
	Cyclic phosphoric acid	[29]
	Others (malic acid, lactic acid and derivatives, Mosher's acid, N-derivatized amino acids, etc.)	[43–45]
N-protected amino acids	The same reagents as for acids (brucine, quinine, ephedrine, pseudoephedrine and synthetic chiral bases)	
Unprotected amino acids	Mandelic acid, 10-camphorsulfonic acid, 1,1´-binaphthyl-phosporic acid	[46–48]

Table 2. Resolving agents suitable for resolution of enantiomers of chiral compound without polar acidic or basic functionality [49–64]

Resolving agent	Chiral compounds	Reference
Platinum square planar complexes	Alkenes, allenes, arsines, phosphines, sulfoxides, cyclic ethers, etc.	[49, 50]
π-Acids	Weak chiral bases, aromatic hydrocarbons, heteroaromatic compounds, helicenes	[51, 52]
Cyclodextrins (α, β and γ and their derivatives)	Universal chiral host compounds. However, size and shape compatibility is required and an aromatic moiety is preferred	[53–55]
Crown ethers	Chiral amines and amino acids possessing the primary amino group	[56, 57]
Cyclophanes (resorcarens, cavitands, cryptands, etc.)	Compounds containing aromatic groups	[58]
Clathrates of [tri-O-thymotide (TOT), brucine, sparteine, 2,2´-dihydroxy-1,1´-binaphthyl-diol]	Alkanes, unbranched chiral hydrocarbons, acids and esters bearing small substituents, halogenated hydrocarbons, sulfoxides, phosphine oxides, phosphinates, piperidines	[59–64]

Table 3. Examples of pharmaceuticals resolved using diastereomeric crystallization in the process [5]

Pharmaceutical	Resolving agent
Ampicillin	D-Camphorsulphonic acid
Ethambutol	L-(+)-Tartaric acid
Chloramphenicol	D-Camphorsulphonic acid
Dextropropoxyphene	D-Camphorsulphonic acid
Dexbrompheniramine	D-Phenylsuccinic acid
Fosfomycin	*R*-(+)-Phenethylamine
Thiamphenicol	D-(−)-Tartaric acid (unnatural)
Naproxen	Cinchonidine
Diltiazem	*R*-(+)-Phenethylamine

6.3 Kinetic Resolution

This method relies on the chemical transformation of a racemate in which one of the enantiomers forms a product more rapidly than the other. The very first kinetic resolution of enantiomers was described by Pasteur [65]. This was the resolution of tartaric acid by fermenting yeast. Later, it was found that not only enzymes but also other chiral inductors such as chiral reagents, chiral catalysts, solvents or polarized light beam may effect kinetic resolution. It is important to note that the interaction time must be carefully controlled in kinetic resolution. The reaction has to be stopped at some point short of 100% conversion. Otherwise, both enantiomers of the starting material will be converted into the product, and no resolution is obtained. The slower interacting enantiomer can be obtained in an enantiomerically enriched or even pure form. The product of a kinetic resolution may be either chiral or achiral.

A resolution gives access to both enantiomers, which is desirable if both are required. However, if only one enantiomer is needed, 50% conversion constitutes the maximum yield. Therefore, similar to the above-mentioned crystallization techniques, a combination of resolution–racemization steps may be a very useful approach.

Enzymes are characterized by shape-selective binding which results in enantio-, diastereo-, or chemoselectivity. Nature's catalysts may provide unique advantages of efficiency, stereoselectivity and environmental friendliness. These are perhaps the main arguments for the widespread application of enzymes for kinetic resolutions.

Of the six main classes of enzymes, hydrolases, oxidoreductases and transferases have been the three most useful in kinetic resolution. Among the hydrolases, lipases are extensively used. The molecular machinery of lipases consists of a catalytic triad of the amino acids serine, histidine, and aspartic (or glutamic) acid. The enzyme first transfers the acyl group of an ester to the hydroxyl group of the serine residue to form the acylated enzyme. The acyl group is subsequently transferred to an external nucleophile with the return of the enzyme to its pre-acylated state to start the process again. A variety of nucleophiles can participate in this process: water results in hydrolysis, an amine results in amidation, an alcohol results in esterification or transesterification, and hydrogen peroxide results in the formation of peracid. Another reason which favored the relatively wide applicability of lipases in enzymatic kinetic resolutions is the fact that they act at a water/organic interface and are active in organic solvents [66].

One important disadvantage of natural chiral resolving reagents is their occurrence in mainly one configuration, inherent also for enzymes. However, this drawback can relatively easily be overcome based on the enantiocomplementarity principle. This means that if the enzyme hydrolyzes an (*R*,*S*)-acetate into the (*S*)-alcohol, then an exposure of the (*R*,*S*)-alcohol to the enzyme under acetylating conditions may result in the (*S*)-acetate and (*R*)-alcohol [67].

Other limitations of enzyme-catalyzed kinetic resolutions of enantiomers are that the enzymes are easily denatured, they often require stoichiometric amounts of co-factors, and they are expensive and substrate-specific.

Immobilization of enzymes has been especially important in connection with enzymatic resolutions performed on a commercial scale.

Some of the enzymes are used in order to substitute already well-established classical synthetic schemes. The enzymatic technologies are not always successful in this competition. However, the enantioseparations, together with the production of maize syrup with a high fructose content, human insulin from pig insulin (using trypsin), penicillin analogs (using penicillin cyclase), aspartame using thermolysin, etc. belong to processes where enzymatic technologies prove to be advantageous or simply do not have an alternative.

Several excellent books and reviews are available on the use of enzymes in stereoselective synthesis and resolutions. Therefore, Table 4 only summarizes the enzymes used in chiral separations [68, 69], while Table 5 provides some selected examples of enzyme applications for the kinetic resolution of drug enantiomers [70–83].

Among the chemical methods, the catalytic epoxidation method proposed by Sharpless [84–87] and enantioselective hydrogenation using Noyori catalysts attract the most attention [88] for the kinetic resolution of drug enantiomers.

Table 4. Enzymes used in chiral separation [68, 69]

Enzyme	Substrate
Subtilisin (ex *Bacillus* sp.)	amides, esters
α-Chymotrypsin	amides, esters
Papain	amides, esters
Acetylcholine esterase (AChE)	esters
Pig liver esterase (PLE)	esters
Pig liver acetone powder (crude PLE)	esters
Cholesterol esterase (various sources)	esters
Porcine pancreatic lipase (PPL)	esters
Steapsin (crude PPL)	esters
Microbial lipase ex *Candida cylindracea*	esters
Microbial lipase ex *Pseudomonas cepacia*	esters
Microbial lipase ex *Mucor* sp.	esters
Microbial lipase ex *Candida antarctica*	esters
Transaminases	amines, ketones
Alcohol dehydrogenases	alcohols, including prochiral diols
Aldolases	selected ketones or aldehydes, aldol condensations

Table 5. Selected examples of enzymatic kinetic resolution of drug enantiomer [70-83]

Drug	Enzyme/microbs	Remark	Reference
Diltiazem	Lipase	Resolution will be achieved by preferential hydrolysis of starting racemic p-methoxyphenylglycidate	[70]
Naproxen	Lipase ex *Candida cylindracea*	Racemic naproxen will be transformed to ester (CH_3, CH_2-CH_2Cl) and then hydrolyzed enzymatically	[71]
Ibuprofen	Lipase ex *Candida antarctica* SP 435	Racemic ibuprofen will be transformed to ester (CH_3, CH_2-CH_2Cl) and then hydrolyzed enzymatically	[72]
Ibuprofen	Lipases	Racemic ibuprofen will be transformed to ester (CH_3, CH_2-CH_2Cl) and then hydrolyzed enzymatically	[73]
Ibuprofen	Nitrilase from resting cells of an *Acinctobacter* sp.	Enantioselective hydrolysis of the racemic nitrile precursor of ibuprofen	[74]
β-Blockers	Lipases	Several alternative schemes have been described for the syntheses of optically pure intermediates of β-blockers	[75–77]
β-Blockers	*Rhodococeus* sp.	Epoxidation process which leads to chiral precursors of β-blockers	[78]
β-Blockers	Mycobacteria	Synthesis of (*S*)-epichlorohydrin which is precursor for β-blockers	[79]
β-Blockers	Various microorganisms	Precursor of (*S*)-metoprolol	[80]
ACE-Inhibitors (enalapril and analogs)	Lipase aminopeptidase	Synthesis of optically active precursors	[81]
R-carnitine	Lipase	Enzymatic hydrolysis of corresponding epoxy acid ester to optically active precursor	[82]
β-Lactam antibiotics (penicillins and cephalosporins)	Hydantoinase	The enzyme is used for preparation of D-phenylglycine and D-p-hydroxyphenylglycine, which are chiral intermediates for racemic hydantoins	[82]
β-Lactam antibiotics (penicillins and cephalosporins)	microbial oxidation (microbes?)	n-Butyric acid is oxidized stereoselectively to (R)-β-hydroxybutyric acid	[82]
Propranolol	Lipase	Enantioselective hydrolysis of acetyl precursor	[83]

6.4 Chromatographic Separation of Drug Enantiomers

This section covers three instrumental techniques: gas-chromatography (GC), high-performance liquid chromatography (HPLC) and super/subcritical fluid chromatography (SFC). One additional chromatographic technique which is not intensively used for enantioseparations at present is thin-layer chromatography (TLC).

6.4.1 Gas Chromatography (GC)

GC was the first instrumental technique that allowed the separation of optical isomers [89]. This technique offers relatively high peak efficiency compared with HPLC, but the volatility and thermostability requirements of the analyte limit the application of GC for the separation of drug enantiomers. However, GC is used for the enantioseparation of several drugs and many drug intermediates, natural compounds, food and beverage components, essential oils and perfumes.

Although packed columns have been used in earlier studies, capillary columns containing wall-coated or immobilized chiral selectors are more widely used at present. The chiral selector may be of different nature, such as amino acid derivatives, chiral metal complexes, etc. [90]. At present, cyclodextrins (CD) and their derivatives dominate as chiral stationary phases (CSP) in GC enantioseparations [90, 91]. CD-based CSPs for GC are thermally stable, widely applicable, and easy to use.

Due to the low volatility and thermostability, enantiomers of chiral drugs are basically resolved as their more volatile and stable derivatives obtained by the interaction of parent drugs with isocyanates, trifluoracetic acid, phosgene, etc. Sometimes, derivatization can be performed not only in order to meet the volatility and stability requirements but also in order to improve the chemo- and enantioselectivity of separation. Certainly, a derivatization with a chiral derivatizing reagent and the separation of the covalent diastereomeric compounds on an achiral CSP is also possible, but is essentially not used.

GC enantioseparations of cationic chiral drugs as various derivatives have been reported. To this group of chiral drugs belong sympathomimetics, β-receptor blockers, etc. (Fig. 1) [92]. The need of derivatization is one of the major disadvantages of chiral GC. This makes the method more time-consuming, laborious, and expensive. Chiral drugs and metabolites also may be resolved simultaneously using chiral GC (Fig. 2) [93] but, as already mentioned, a derivatization is commonly required. GC enantioseparation of several drugs of different structural groups in their underivatized form has been also reported [93].

GC is basically a technique for analytical enantioseparations. However, micropreparative separations are also feasible using this technique. The most impressive example of the application of chiral GC for micropreparative enantioseparation of drug enantiomers is the chiral inhalation anesthetic drug, enflurane. According to recent data, the enantiomers of isoflurane may have different pharmacological properties [94]. For the isomeric compound enflurane (Fig. 3) a more intensive metabolism was established for the (*R*)-(–)-enantiomer compared with the (*S*)-(+)-enantiomer [95]. Enflurane is a gas and therefore, the most favorable method for the enantioseparation will certainly be GC. Analytical-scale enantioseparations of this compound have been reported using various CD derivatives as CSPs [97]. The micro-

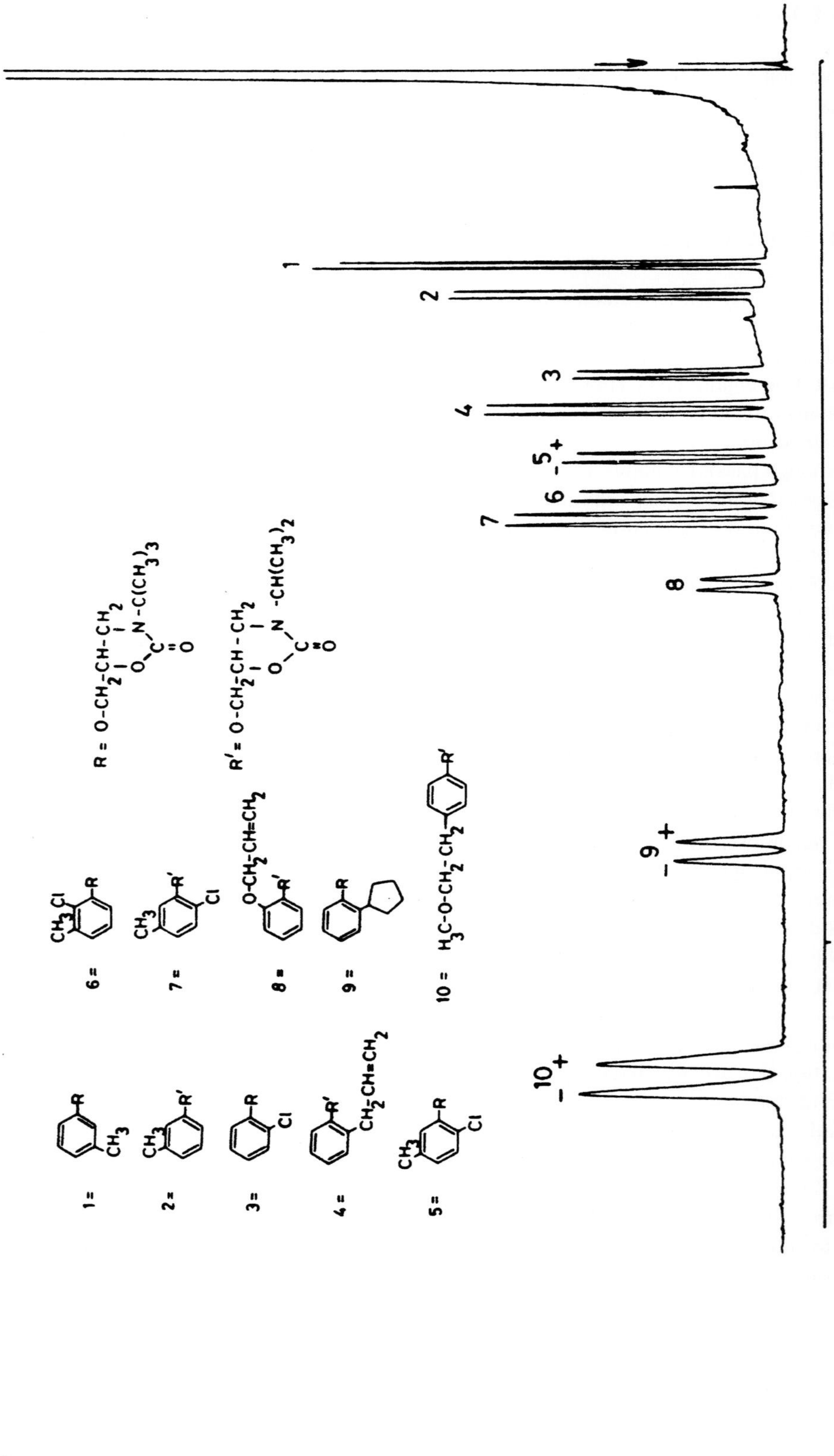
MIN
30
60

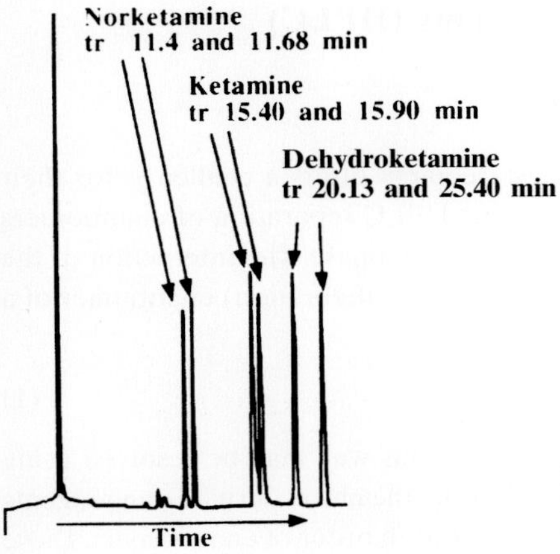

Figure 2. Separation of the enantiomers of ketamine and its metabolites, norketamine and dehydroketamine [93].

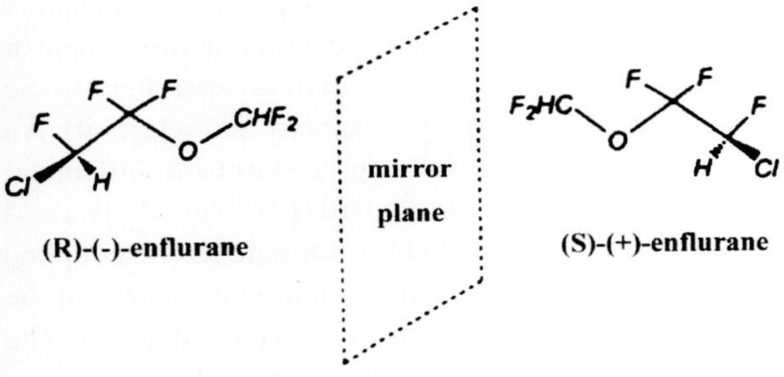

Figure 3. Structure of chiral inhalation anesthetic enflurane [100].

preparative GC separations of the enflurane enantiomers were also reported [97–99]. Preparative separation of enflurane using the simulated moving bed (SMB) principle in GC (GC-SMB) was recently described [100]. The SMB methodology is more commonly used in liquid chromatography (see below); however, as demonstrated in a number of studies, this principle may be successfully applied also to GC [101–103]. Using the experimental set-up described in reference [100], it is possible to prepare 0.01 g of enflurane enantiomers per gram of stationary phase per day. This capacity may be satisfactory for some applications.

Thus, chiral GC possesses historical priority and several conceptual advantages compared with HPLC. However, despite several elegant applications of GC for enantioseparations of chiral drugs, HPLC is the dominating instrumental technique in this field.

◀ **Figure 1.** Separation of the enantiomers of β-receptor blockers as oxazolidin-2-one derivatives. Column: 18-m glass capillary column with XE-60-L-valine-R- phenylethylamide. Peaks: 1, dechlorobupranolol; 2, toliprolol; 3, demethylbupranolol; 4, alprenolol; 5, bupranolol; 6, 2-chlorbupranolol; 7, isopropylbupranolol; 8, oxprenolol; 9, penbutolol; 10, metoprolol [92].

6.4.2 High-Performance Liquid Chromatography (HPLC)

6.4.2.1 Indirect HPLC Enantioseparations

As mentioned above, the different solubility of diastereomers offers a challenge for their separation by diastereomeric crystallization. The indirect HPLC separation of enantiomers relies on their different interaction with a (achiral) stationary phase. The interaction of the enantiomeric pair (*R*, *S*) with one (let us assume it to have *S* configuration) enantiomer of a chiral derivatizing reagent may be expressed as follows:

$$(R, S) + S\,(\text{deriv}) = RS + SS \qquad (1)$$

The diastereomeric compounds *RS* and *SS* obtained by this way may be resolved using achiral HPLC columns, which are less expensive. In addition, the chiral derivatizing reagents available in both configurations allow us to revert the migration order of enantiomers. These are advantages of indirect enantioseparation techniques.

The application of the indirect enantioseparation technique is especially rational in that case when the derivatization, besides affording stereoselectivity, also improves other characteristics of a separation. For instance, chiral derivatizing reagents containing ultraviolet or fluorescent groups may be used not only to introduce stereodifferentiating capabilities to the separation system but also to increase the detection sensitivity, specificity and selectivity. An example of this kind was described in reference [103]. The widely used chiral anti-inflammatory drug ibuprofen and racemic fenoprofen used as internal standard (I. S.) were derivatized with *S*-(–)-1-naphthylethylamine after the addition of ethyl chloroformate as the coupling reagent. The resulting diastereomers were chromatographed at ambient temperature in the reversed-phase mode using C_{18}-material and detected using a fluorescence detector. The suitability of the assay for clinical pharmacokinetics studies was demonstrated.

The useful chiral derivatizing reagents (CDR) for indirect enantioseparations are summarized in a review article by Lindner [104].

The number of reported indirect enantioseparations is currently decreasing compared with direct enantioseparations on CSP. The reason for this is on the one hand the development of more universal, stable and effective CSPs for direct enantioseparations, and on the other hand the inherent limitations of indirect enantioseparation techniques.

One of the major limitations of indirect HPLC enantioseparation is the very strict requirement for the chemical and optical purity of the derivatizing reagent. The level of an enantiomeric impurity contained in a derivatizing reagent introduces *a priori* into the separation system the incorrectness at least at the same level. This is caused by the impossibility, when using achiral stationary phases, to resolve the enantiomeric compounds which are formed from an *S*-analyte and an *R*-derivatizing reagent (*SR*) and the reverse, from *R*-analyte and *S*-derivatizing reagent (*RS*). Other limitations of this technique include the requirement of reactive functional groups of the analyte, additional reaction steps, costs of reagents, time, possible racemization or other chemical transformations of the analyte during the derivatization, the possibility of different reaction kinetics of two enantiomers with a chiral reagent (kinetic resolution), incomplete derivatization, unequal detector response of the diastereomers, etc. [104]. All of these points need to be addressed when an indirect technique is selected for enantiomer analysis [104].

What are the challenges of indirect enantioseparation techniques for preparative purposes? At first, the requirements are more strict for chiral derivatizing reagents for preparative than for analytical applications. In addition, the chiral derivatizing reagent must be inexpensive, easily available, and also recoverable in high chemical and optical purity. The diastereoselectivity of the separation must be high and both the derivatization and the cleavage of the derivatives after a separation should be performed under mild conditions, with high yields. Thus, in principle, this technique can be applied to preparative resolution of drug enantiomers, but it is less probable that it will become a technique with significant impact in this field.

6.4.2.2 HPLC Enantioseparations using Chiral Additives to the Mobile Phase

An enantioseparation in this mode is based on the formation of non-covalent diastereomeric complexes between the enantiomers of an analyte and the chiral additive in the mobile phase (CAMP). Compared with indirect enantioseparations, the CAMP technique has advantages such as the absence of a derivatization step or a higher flexibility (easier change of a chiral additive than a chiral or an achiral packing material). As documented by Davankov [105], the enantiomer migration order with CAMP most likely will be opposite to that observed with the same chiral selector as the stationary phase. The complementary enantioselectivity of enantioseparation with CAMP compared with CSPs is a significant advantage.

Enantioseparations using CAMP in HPLC were first reported in 1979 [106, 107]. Cyclodextrins and their derivatives [108], chiral metal complexes [109] and ion-pair reagents [110] are the most widely used chiral selectors in this technique. Disadvantages such as a need for large amounts of chiral derivatizing reagents, and interference of CAMP signals with the signals for the analytes, limit the application of this approach.

Although some techniques are proposed for the isolation of CAMP and enantiopure analytes after a chromatographic separation [111], this is not a practicable technique for preparative purposes.

6.4.2.3 Direct HPLC Enantioseparations

The fundamental advantage of direct HPLC enantioseparations compared with crystallization and classical enzymatic resolution techniques was clearly expressed by Davankov [112]. As he noted, both the chemical transformation or phase transition are "one-step" processes, whereas chromatography is a "multistep" separation method. This means that the overall resolution of the solutes originates from a large number of stereoselective "one-step" adsorption–desorption cycles. The cumulative nature of the chromatographic separation is the reason that a free energy difference in interactions of enantiomers as small as 0.025 kJ mol^{-1} may in principle be sufficient for baseline chromatographic enantioseparations, whereas the free energy difference required in "one-step" techniques is several orders higher. Further advantages of direct HPLC enantioseparations have been emphasized in an elegant way in the same paper [112]. First, the less strict requirements to the enantioselectivity of chiral selector makes it possible to employ effectively a large number of chiral selectors with relatively low chiral recognition ability. This seems to be the most likely reason for the development of several hundreds of CSPs. More than 150 different CSPs are commercially available. It is possible to obtain both enantiomers in the optically pure form, even when the chiral selector has an optical purity less than 100% [112]. In addition, the absence of irreversible chemical reactions or phase transitions during direct enantioseparations further favors the application of this technique for analytical and preparative purposes.

Thus, direct HPLC enantioseparation has been established as an analytical method of choice for determining the enantiomeric composition. This technique is characterized with high precision, sensitivity, speed, easy automation, tolerance to impurities, and byproducts, etc.

The most important advantages of direct chromatographic enantioseparations for preparative purposes include the presence of resolved enantiomers in original and high optically pure form in different volume fractions of the mobile phase, no loss of chiral selectors, and amost no risk of contamination of desired enantiomers with the chiral selector. Parallel to enantioseparations other impurities may be removed from chiral compounds during this process.

In principle, any chiral compound possessing an ability to interact non-covalently and enantioselectively with chiral molecules may be used more or less successfully as a chiral selector in liquid chromatography. There is a set of characteristics which a chiral selector has to meet, depending on the goal of the separation, the mode, and technique used.

Several books are devoted to the chiral separations in HPLC [69, 93, 104, 108, 110]. Among several hundreds of CSPs described, ligand-exchange [105], Pirkle-type [113], protein and peptide [114], polysaccharide [115], macrocyclic [93, 116], and synthetic polymeric [117–119] CSPs are most widely used.

Direct enantioseparations may be performed in normal-phase as well as reversed-phase HPLC modes. The former offers certain advantages for preparative-scale separations and when the analyte is less soluble or less stable in aqueous solutions. The advantages of the latter process are the possibility of a direct injection of body fluids and a more relevant imitation of stereoselective drug–receptor interactions. Among the above-listed classes of CSPs, Pirkle-type, synthetic polymers and polysaccharide CSPs were originally developed for enantioseparations using organic mobile phases (normal phase) whereas chiral metal complexes, peptides and cyclodextrin CSPs were used in organic–aqueous or pure aqueous buffers. Later, it was established that many of these CSPs (cyclodextrins and their derivatives as well as other most recently developed macrocyclic CSPs, polysaccharides and Pirkle-type CSPs) may be used with both pure organic or organic/aqueous mobile phases [114–116].

The diameter of HPLC columns may vary from several tens of micrometers to several tens of centimeters. Depending on the size of the separation column – nano-, capillary-, narrow-bore – analytical and preparative separations may be distinguished. A current trend in HPLC instrument development provides the hope that in time it may become feasible to use a column of any size with a single instrument.

The obvious advantages of a reduction of the column diameter in analytical enantioseparations include less consumption of packing materials and of high-purity organic solvents, less environmental problems, smaller sample size, and improved analytical characteristics (less dilution of sample, higher plate number and shorter analysis time) of the separation [120].

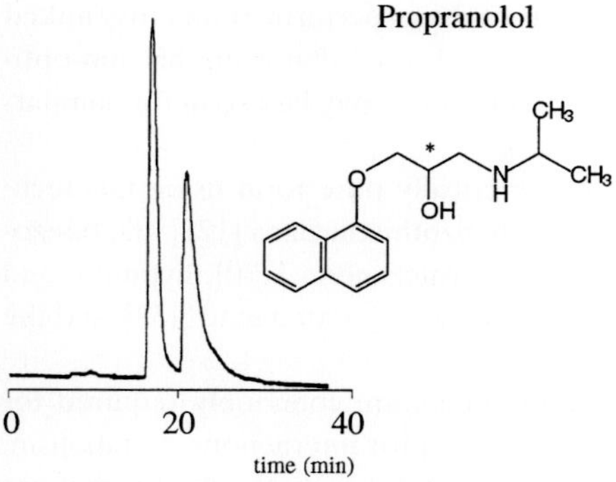

Figure 4. Enantioseparations of propranolol in HPLC using capillary column (100 μm × 25 cm) packed with Chiralcel-OD material [120].

The separation of the enantiomers of the chiral β-blocker propranolol in capillary HPLC using a polysaccharide type CSP is shown in Fig. 4 [120]. The capillary format requires aproximately 10^6 times less stationary and mobile phases, while offering separation characteristics at least adequate or better than that achieved with common-size columns.

Direct HPLC enantioseparation techniques, which are free of many disadvantages of GC, indirect and chiral mobile phase HPLC methods, have gained unequivocal prevalence in bioanalytical studies. Several methods have been advanced so much that they allow enantioselective determination not only of the parent chiral drugs but also of their pharmacologically relevant metabolites [121]. As already mentioned above, a direct injection of biofluids offers several advantages in terms of analysis time and sample recovery. Precolumns packed with achiral or chiral packings, or with the recently developed so-called restricted-access packing materials, may be useful in this case.

In addition to the miniaturization of HPLC enantioseparations, another current trend occurs in the opposite direction, namely the scaling-up of separations. The techniques of preparative-scale enantioseparations using liquid chromatography (LC) are described below.

6.4.2.4 Preparative Resolution of Enantiomers in LC

Chiral pool, diastereomeric crystallization and kinetic resolution currently dominate in the production of pure enantiomers on a commercial scale. Enantioselective synthesis is developing rapidly, but is hardly competitive with the above-mentioned technologies. What and where are the challenges of chromatography to become a competitive tool for a production of enantiomerically pure compounds? The experience with this technique is promising and challenges do exist, at least in certain areas. The following five arguments support this optimistic remark:

1. The chromatographic separation of racemates was originally developed for obtaining pure enantiomers of chiral compounds not accessible by the classical routes. Examples are the preparation of the enantiomers using lactose [122–124], cellulose [125] and starch [126] as chiral packings.

2. The early systematic preparative enantioseparations of drugs performed on cross-linked chiral polyacrylamides in low-pressure LC mode clearly showed that even this non-optimal technique, from the viewpoints of performance and costs, may be useful for comparative biomedical studies of enantiomers of chiral drugs.
3. Several chiral drugs were first obtained in enantiomerically pure form using this technique, such as the diuretics chlorthalidone [127] and benzothiadiazines [128], the benzodiazepinone oxazepam [129], the muscle relaxant chlormezanone [130], hypnotic and anticonvulsant methaqualone [131], the anti-inflammatory drug oxindanac [132], and the antiaromatase fadrozole [133].
4. At an early stage of drug development, both enantiomers are commonly required for comparative biomedical studies (stereoselective drug–receptor interactions, metabolism, pharmacokinetics, drug interactions, toxicity, stability, etc.). When both enantiomers are required, the chromatographic resolution, together with crystallization techniques providing both enantiomers, possess certain advantages. This is relatively difficult to achieve using enzymatic kinetic resolution, chiral pool-based enantioselective synthesis, or asymmetric catalysis. These techniques, especially the latter two, provide higher yields but as a rule only one enantiomer is available.
5. New developments in chiral chromatography (more universal, easily available, stable, tailor-designed CSPs) and technology (recycling, displacement and especially, SMB) makes chromatography a valuable alternative to classical techniques for the preparation of pure enantiomers.

It seems important to note that, similar to any other technique, chromatography suffers from certain drawbacks such as high costs of the stationary and mobile phases, the high dilution of the products, recycling problems of the mobile phase, etc. Below, in some practical examples, the matter of how different technologies of preparative enantioseparation are dedicated to solving the above-mentioned problems will be discussed. The four different techniques such as low-pressure or usual batch separations, displacement-mode, recycling and peak shaving, and SMB can be used with different success for preparative resolution of enantiomers in LC.

6.4.2.4.1 Elution Batch Chromatography

After the first unsuccessful [134] and several successful examples of enantiomer enrichment and separations [122–126] using natural chiral polymers or modifications thereof, chiral polyacrylamides were introduced as chiral sorbents in the mid-1970s [135]. These chiral sorbents were successfully used for the resolution of drug enantiomers which were difficult or even impossible to obtain by synthesis at that time [136]. To these belong the hypnotic drugs methaqualone, glutethimide and hexobarbital, the anticonvulsant mephenytoin, the diuretic drug chlorthalidone and enantiomerically very labile chiral diazepinones. The best resolution was achieved for the enantiomers of the former sedative drug thalidomide (Fig. 5). *R*- and *S*-thalidomide were obtained in amounts which allowed the first study of the pharmacological and toxic properties of the individual enantiomers [137]. The potential of cellulose derivatives for a preparative-scale separations of drug enantiomers has been also demonstrated by the same

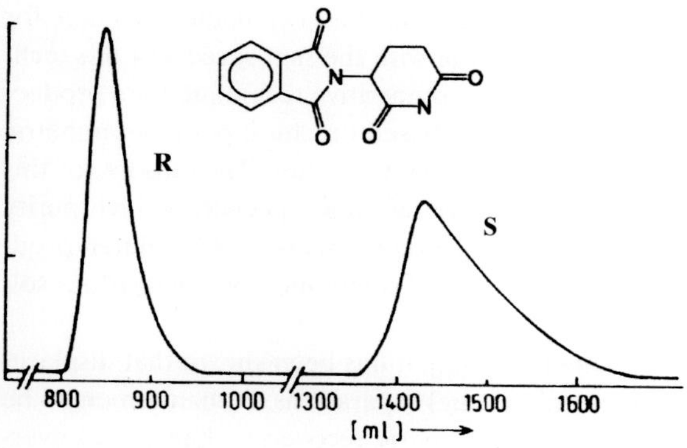

Figure 5. Liquid-chromatographic enantioseparation of 52 mg of thalidomide on the column packed with cross-linked chiral polyacrylamide (65 g) [136].

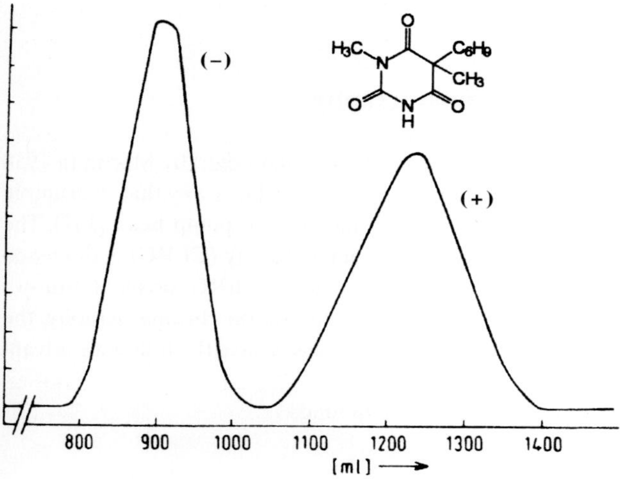

Figure 6. Liquid-chromatographic enantioseparation of hexobarbital on the column (2.5 × 85 cm) packed with cellulose triacetate [136].

group [136]. The material was especially useful for enantioseparations of chiral N-methyl barbiturates (Fig. 6) [136].

The first impressive experience of preparative separations of chiral drug enantiomers by liquid chromatography induced wide interest in this technique, especially in pharmaceutical and fine-chemical industry. Francotte and co-workers [138–140] studied in detail the potential of commercially available relatively inexpensive microcrystalline cellulose triacetate for preparative separations of drug enantiomers in amounts starting from a few milligrams up to several kilograms. These studies are summarized in many original papers and several excellent reviews by this group. Several new cellulose-based packing materials were developed specially for preparative enantioseparations [141, 142]. The chiral drugs resolved into enantiomers are listed in several extensive tables in references [138–140], and cover analgesics, β-blockers, tranquilizers, calcium antagonists, antiaromatases, diuretics, anticonvulsants, nootropics, and several drug candidates.

Thus, LC in the simple elution batch mode recommends itself as a valuable technique for micropreparative separations of drug enantiomers. Along with the first success of this technology, it became clear that, in general, this cannot be a competitive technique for a production-scale (kilograms to tons) enantioselective synthesis based on chiral pools, asymmetric catalysis, kinetic enzymatic resolutions or diastereomeric crystallization. The reasons for this are the above-mentioned high cost of chiral packing materials and especially, of high-purity organic solvents, low loading capacity, i.e., relatively small amounts of highly diluted products, problems related to the recycling of the expensive and sometimes even hazardous solvents, etc.

In a series of papers by Vigh and co-workers [143–145], it has been shown that displacement chromatography may allow micropreparative (1 mg) separations of enantiomers to be performed with somewhat better chemical and enantiomeric recoveries than in the overloaded elution mode. However, this technique does not offer principal solutions to the above-mentioned inherent problems of the elution batch mode, and is not discussed in detail here.

6.4.2.4.2 Closed-Loop Recycling Chromatography

The potential of elution chromatography in a closed cycle was first noted by Martin in 1958 [146]. In this technique, two nearly identical columns are connected in a way that the sample is pumped alternatively through these columns without passing the pump head [147]. The theoretical backgrounds of the closed-loop recycling chromatography (CLRC) is discussed in references [148–151]. Several papers describe the advantages of CLRC for separations of drug enantiomers [138, 139, 148, 149]. Compared with batch elution chromatography, the closed-loop recycling chromatography and peak shaving processes have the following advantages:

- Suitability for the separation of partially resolved compounds.
- Less solvent consumption.
- Higher production rate per time unit.
- Smaller column dimensions can be used to reach a desired goal.
- Less labor-intensive.
- Peak shaving can be used parallel to enantioseparation also to resolve a sample from other impurities.
- Working in the overloaded mode seems more feasible in the closed-loop recycling mode compared with the elution batch mode.

Examples of enantioseparation of selected chiral drugs and drug candidates using CLRC are summarized in Table 6. As an example, enantioseparation of the nootropic drug candidate CGS 16920 is shown in Fig. 7. The volume fraction corresponding to the dashed part on the chromatogram containing a mixture of both enantiomers was recycled and completely resolved, as shown in the figure.

Tailing of the peaks also creates significant problems (high consumption of mobile phases, strong dilutions, etc.) in recycling chromatography. Therefore, if only one enantiomer is desirable, it is preferable to elute it as the first peak.

Table 6. Selected examples of preparative enantioseparation of chiral drugs, drug candidates and drug intermediates using closed-loop recycling chromatography (CLRC)

Chiral compound	Chiral stationary phase	Resolved amount, mg	Reference
Nootropic drug candidate CGS 16920	Tribenzoylcellulose beads	2000	[139]
Chlorthalidone	β-Cyclodextrin bonded to silica	25	[148]
Chlorthalidone	Poly-N-acryloyl-(S)-phenylalanine ethyl ester diol silica	240	[148]
Metoprolol	β-Cyclodextrin bonded to silica	120	[148]
Propranolol	β-Cyclodextrin bonded to silica	5	[148]
Thalidomide	Poly-N-methacryloyl-(R)-cyclohexyl-2-ethylamide	140	[148]
Hetrazepine	Microcrystalline cellulose triacetate	2500000	[18]
Benztriazole derivative	Chiralcel OJ	30000000	[148, 152]
γ-Aryl keto ester	Chiralcel OD	500	[148]

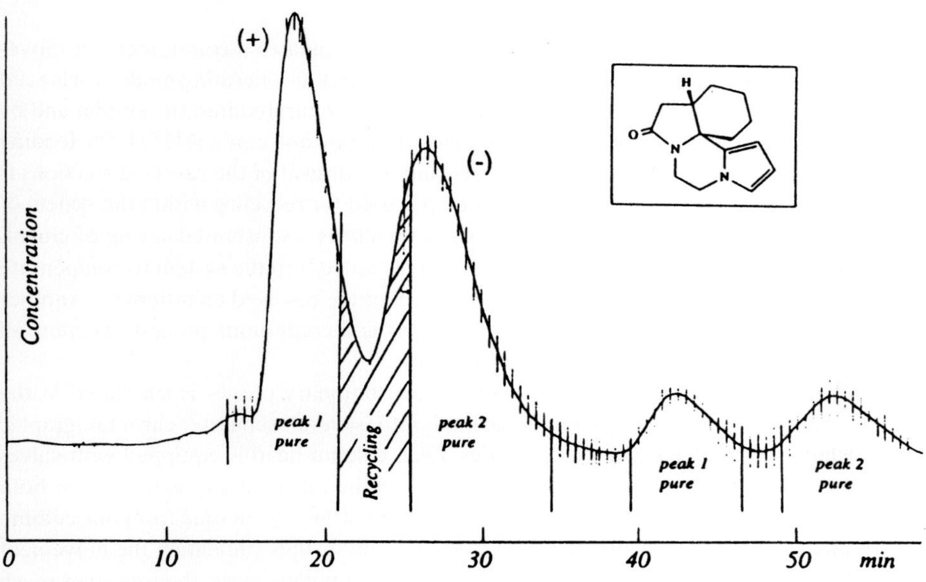

Figure 7. Chromatogram of the resolution of the racemic nootropic drug CGS 16920 (2 g) on the column (5 x 45 cm) packed with tribenzoylcellulose beads with recycling and peak shaving [139].

Thus, application of CLRC to enantioseparations of chiral drugs was a further successful step. The most important advantage of this technique is that by scaling-up analytical separations (which are summarized in a useful form in computerized databases), milligram and gram quantities of chiral drug enantiomers are available without too much effort. These amounts are in most cases sufficient for the drug development stage. The easy access to enantiomerically pure chiral drugs and their synthetic intermediates at an early stage of drug development facilitates the elucidation of alternative production strategies, stability studies, stereoselective analytical monitoring, etc. As shown in references [138, 139, 148, 152], the closed-loop recycling technique may be scaled-up to a production method.

Despite representing a significant step forward, closed-loop recycling chromatography does not offer a conceptually new and effective solutions to problems related to high costs of CSPs, high consumption of organic solvents, dilution of products, relatively low capacity on a production scale, etc. The most effective solutions of these problems are proposed in SMB chromatography, which is considered at present perhaps the most valuable chromatographic alternative for obtaining enantiomerically pure drugs on a production scale (kilograms to tons).

6.4.2.4.3 Simulated Moving Bed Chromatography (SMB)

In paying credit to the fact that the SMB is the most promising chromatographic technique for the production of enantiomerically pure chiral drugs, this technique is discussed below in a rather comprehensive way. Together with the principles, some strategies of method development are also addressed.

SMB is a continuous chromatographic process which simulates a countercurrent movement of the stationary phase and the mobile phase. A continuous operating mode is achieved by continuous feeding of the eluent and the mixture to be separated into the system and by continuous withdrawal of the separated compounds (raffinate and extract) [153]. The feeding of the eluent and of the mixture of enantiomers, and withdrawal of the resolved fractions is achieved by four independent pumps. A fifth pump is used for recycling within the system of the main flow (Fig. 8) [153]. This experimental set-up allows a substantial saving of eluent, because only small amounts of fresh eluent have to be added into the system to compensate for the loss of mobile phase caused by the withdrawal of the resolved enantiomers. Another significant advantage of SMB technology is that this is a continuous process, whereas all other chromatographic techniques are discontinuous.

The countercurrent movement of the mobile and stationary phases is simulated in the following way: the stationary phase is divided into several separate chromatographic columns which are connected in a cyclic series. Each column head is equipped with valves that allow the addition of eluent, feed and removal of the raffinate and extract from both component lines. The inlet and outlet lines will be shifted after a given time from one column to a subsequent column in the direction of the mobile phase, thus simulating the movement of the stationary phase in the opposite direction. After a complete cycle, the four lines reach their initial position. The technological parameters for running the SMB system under optimal conditions can be calculated by the available software, based on several analytical runs.

The strategy for the selection of the CSP/eluent combination for enantioseparations using SMB technology has been described [153–155]. The basic requirements for the mobile phase include selectivity, safety, recycling ability, and costs. One property which is less important in analytical-scale operations but is crucially important in SMB, is the solubility of the sample in the mobile phase. The last requirement means that it is desirable for the CSP to be stable and

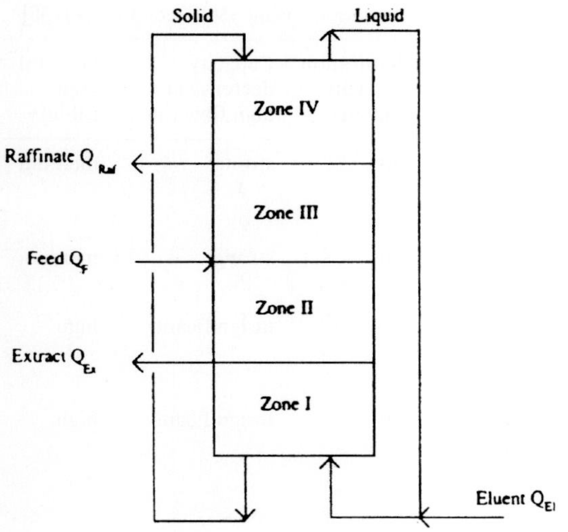

Figure 8. Scheme of a four-zone adsorption process, showing the definition of the four zones with respect to the inlet (eluent and feed) and outlet (extract and raffinate) lines and the countercurrent of mobile and stationary phase [153].

exhibit sufficient (enantio)selectivity in the solvent where the maximum amount of a sample can be dissolved. Together with the productivity of the process, this is also required for running the system under equilibrium conditions. It is almost impossible to automate a SMB separation before it reaches steady-state conditions.

The principal requirements for the CSP in SMB technology are the following: (enantio)selectivity, capacity, and stability. Each of these properties can be characterized from several different points of view. For instance, selectivity includes also the usefulness for a broad range of compounds, i.e., universality, as well as the possibility to predict and adjust a desired selectivity easily. Stability terms cover mechanical stability, stability against solvents, and stability during use. It seems that among the commercially available CSPs, polysaccharide derivatives in various forms (beads or coated on silica) meet these requirements best. Thus, almost 70% of all preparative LC enantioseparations are performed using these CSPs, but this certainly does not mean that there is no niche for the use of other CSPs. At present, detailed characteristics of commercialized CSPs from the viewpoint of their suitability for SMB technology are available [153–155]. Some of these characteristics are summarized in Table 7 [155]. An SMB process-development strategy has been proposed in several papers [153–155]. Before making any decision about the process itself, it is important to set the process requirements such as production-scale, price, purity, yield, product dilution, if one or two enantiomers are required, etc. After setting these requirements, the method development can start according to the following steps [155]:

1. Column screening in order to find a sufficiently selective column.
2. Determination of the stability of the racemate.
3. Elution order of enantiomers.
4. Optimization of mobile/stationary phase combinations.
5. Determination of adsorption isotherms.
6. Simulation and calculation of the separation in different modes.
7. Selection of the production method, calculation of costs.

Table 7. Suitability of commecially available CSP for enantioseparations using SMB-technology [155]

CSP	Selectivity/ Universality	Solvent stability	Saturation capacity, mg/ml	Capacity decrease at a high flow rates	General appli-cability
Microcrystalline cellulose triacetate	medium	medium (alkanes, alcohols, acetonitrile, water)	50	strong	medium
Cellulose beads	medium	medium (alkanes, alcohols, acetonitrile)	50	strong	medium
Cellulose esters/ carbamates	high	medium (alkanes, alcohols acetonitrile, water)	20	insignificant	high
Polyacrylamides	medium	high (alkanes, THF, dichloromethane, ethylacetate, toluene, alcohols, acetonitrile, water)	10	insignificant	high
Brush-type	medium	high (alkanes, THF, dichloromethane, ethylacetate, toluene, alcohols, acetonitrile, water)	8	insignificant	high
Tartardiamide	medium	high (alkanes, THF, dichloromethane, ethylacetate, toluene, alcohols, acetonitrile	8	insignificant	high
Ligand-exchange	low	low (alcohols, acetonitrile, water)	8	–	low
Cyclodextrins	medium	medium (alkanes, alcohols, acetonitrile, water)	5	insignificant	low
Macrocyclic antibiotics	medium	medium (alkanes, THF, alcohols, acetonitrile, water)	8	–	medium

THF, tetrahydrofuran

A method developed according to these schemes is described below [154, 156], using the precursor of the novel Ca-sensitizing drug EMD 53 998 (numbered as EMD 53 986) as an example. This drug precursor was separated into the enantiomers on four different CSPs (Fig. 9). With a comparable α value for all four CSPs ($\alpha > 1.8$), the enantioselective saturation capacity was the highest for the cellulose tris(p-methylbenzoate) beads which contain the highest amount of chiral selector per mass unit. However, the lowest number of plates for this CSP requires a longer column. This leads to unfavorable pressure drop and slower internal flow-rates.

The solubility of EMD 53986 was found to be highest in the mobile phase for the poly (N-acryloyl amino acid ester) CSP.

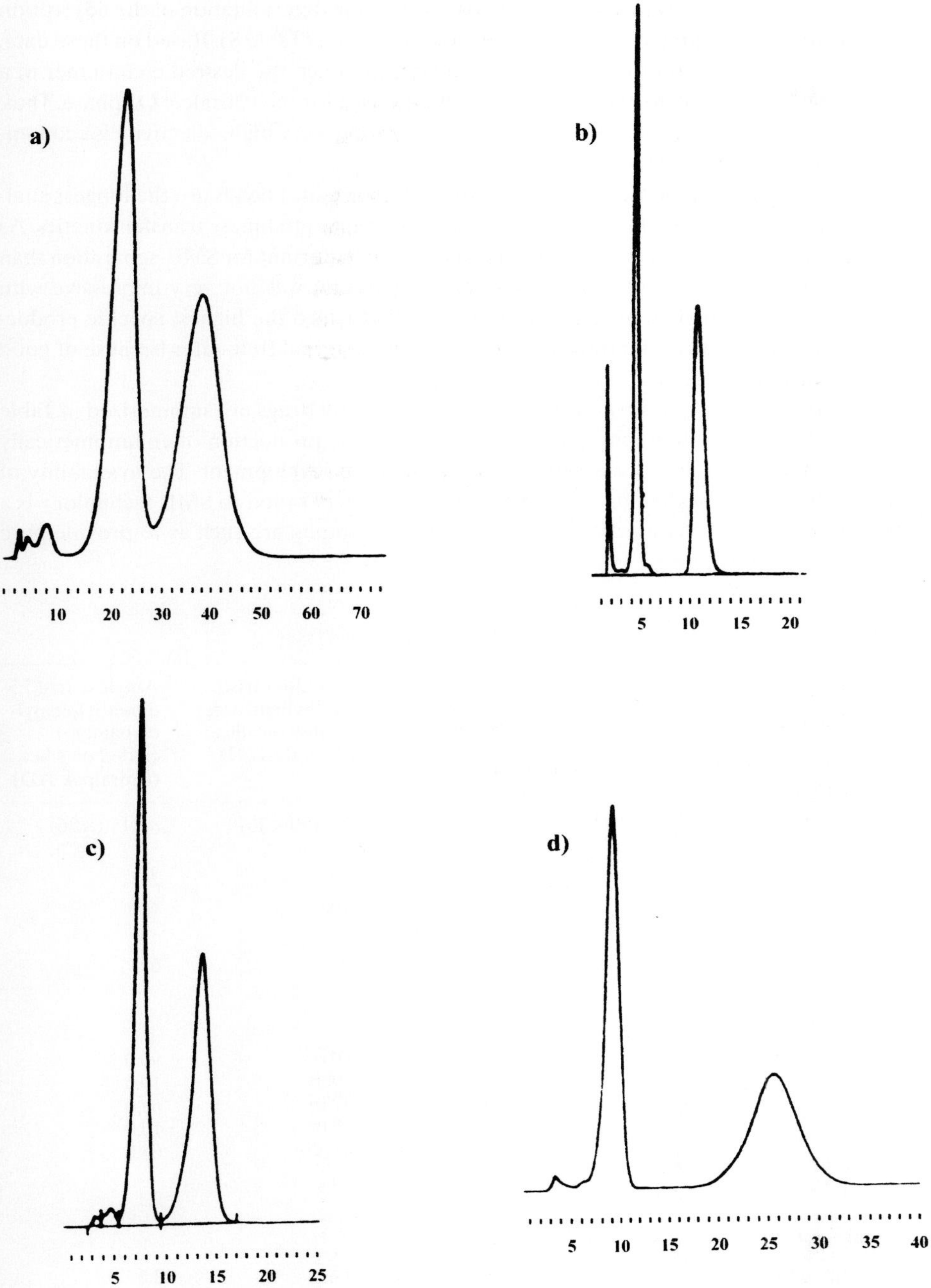

Figure 9. Enantioseparation of the precursor of Ca-sensitizing drug EMD 53998 on: cellulose tris(p-methylbenzoate) beads (a); poly(N-acryloyl-amino acid ester) silica (b); allulose tris(p-methylbenzoate) silica (Chiralal OJ) (c); and anylose tris(3,5-dimethylphenylcarbamate) silica (Chiralpak AD) (d) [153].

After the studies in analytical separation systems and the determination of the adsorption isotherms, the SMB parameters were modeled and simulated (Table 8). Based on these data, it can be concluded that all four CSPs are suitable to produce the desired enantiomer in a purity >99%. The lowest eluent consumption was obtained for the Chiralcel OJ phase. Thus, short retention times with just sufficient, but not necessarily very high, selectivity is economically favorable in the SMB technique.

Specific disadvantages of the cellulose tris(p-methylbenzoate) beads are the longest analysis time, the highest amount of stationary phase, and the limited mass transfer kinetics. As this study showed, a higher peak efficiency is much more important for SMB-separation than a high selectivity. Although the analytical enantioseparation was not very impressive with amylose tris(3,5-dimethylphenylcarbamate), this CSP afforded the highest specific productivity. This was the result of a moderate capacity at high internal flow-rates because of good mass transfer kinetics.

Some examples of preparative enantioseparation of chiral drugs are summarized in Table 9. Overall, SMB chromatography is a powerful tool for the production of enantiomerically pure compounds on a large scale within a short phase of development. The availability of software which allows scaling-up of an analytical enantioseparation to SMB technology is a crucial advantage. The characteristics of SMB chromatography are such as to provide pure

Table 8. SMB parameters for the enantioseparation of EMD 53986 [154]

SMB parameters	Cellulose tris(p-methyl-benzoate) beads	Poly(N-acryloyl-amino acid ester) silica CSP	Cellulose tris(p-methylbenzoate) coated on silica (Chiralcel OJ)	Amylose tris(3,5-dimethylphenyl-carbamate) coated on silica (Chiralpak AD)
Column number and dimensions (mm)	8 × (100 × 26)	8 × (54 × 26)	8 × (50 × 26)	8 × (50 × 26)
Particle diameter (μm)	20–30	20	20	20
Amount of stationary phase (g)	210	90	100	100
Feed concentration (g racemate/L)	5.0	12.0	6.0	6.0
Flow rates (ml/min):				
Recycling	46.50	63.90	70.0	90.0
Feed	5.65	3.32	9.5	10.0
Extract	24.47	37.26	42.0	68.0
Raffinate	7.75	13.40	9.0	10.0
Eluent	36.57	47.34	41.5	68.0
Period time (min)	15.50	7.03	1.5	2.1
Purity desired enantiomer (%)	>99	>99	>99	>99
Eluent consumption (L/g enantiomer)	2.28	2.43	1.8	2.6
Specific productivity (g enantiomer /*d* kg CSP)	98.6	333.2	410	430

Table 9. Selected examples of application of SMB-chromatography for production of enantiomericaly pure compounds

Chiral compound	CSP	Capacity	Remark	Reference
(±)-Phenylethylalcohol	Chiralcel OD	0.98 g/L CSP	5.3 L mobile phase per g enantiomer	[157]
(±)-Threonine	Chirosolve L-proline	30 g was resolved		[158]
(±)-1a, 2, 7, 7a-Tetra-hydro-3-methoxy-napht(2, 3-6) oxirane	Microcrystalline cellulose tri-acetate (CTA)	1.45 g/h kg CSP	The very first real industrial example; 400 ml mobile phase per g enantiomer	[159]
(±) Hetrazepine	Microcrystalline cellulose triace-tate (CTA)	119 g/h kg CSP	94 ml mobile phase per g enantiomer	[160]
(±)-Praziquantel	Cellulose triacetate	4.7 g/h kg CSP	56 ml mobile phase per g enantiomer	[161]
Chiral drug candidate (potent partial agonist at muscarinic receptors)	Chiralpak AD	ca. 30–60 g/h kg CSP	ca. 190–380 ml mobile phase per g enantiomer	[162]
(±)-Tramadol	Chiralpak AD	50 g/h kg CSP	500 ml mobile phase per g enantiomer	[163]
(±)-Binaphthol	Pirkle-type 3,5-DNBPG	2000 g was resolved per day		[164]
(±)-Propranolol	Chiralcel OD	3.83 g/h L CSP	200 ml mobile phase per g enantiomer	[165]
(±)-Guaifenesin	Chiralcel OD	10.0 g/h kg CSP	380 ml mobile phase per g enantiomer	[166]
(±)-Formoterol	Chiralcel OJ	1.2 g/h kg CSP	515 ml mobile phase per g enantiomer	[166]
(±)-Aminoglutethimide	Chiralcel OJ	7.5 g/h kg	740 ml	[166]
(±)-Oxo-oxazolidine	L-Chiraspher	6.8 g/h kg CSP	900 ml	[167]

products in a predictable way, reduce the risk of success, and shorten the time necessary for development. The applicability of SMB technology to a wide variety of compounds may favor the transformation of chromatography from a method of last resort to the method of choice for fast and predictable process development.

SMB allows a drastic reduction of the costs of chiral separations, mainly due to a reduction of chiral stationary phase (50–60% lower than in HPLC) and of eluent consumption (up to 10 times compared with the batch chromatographic process). It allows production scales of 10–100 tons per year, with separation costs as low as US$ 30 per kg of pure enantiomer [155]. The coupling of SMB with racemization and/or enantioselective crystallization techniques is even more promising.

6.4.3 Supercritical (Subcritical) Fluid Chromatography

Super- and subcritical fluid chromatography (SFC) underwent extremely rapid development in the mid-1980s, and was believed to be a technique with a great potential that combined the advantages of GC and HPLC. However, inherent fundamental and technical limitations of this technique became apparent and led to some stagnation in its further development.

In spite of this critical note, the potential of SFC in analytical and preparative-scale enantioseparations has been already illustrated. Technical development in this field may open even more challenges for this technique. The advantage of SFC for preparative separations is that the high-pressure liquid carbon dioxide used as mobile phase can easily be removed from the product. In addition, carbon dioxide is non-hazardous and relatively inexpensive. On the other hand, this mobile phase creates the following problems: the solubility of polar compounds is limited, and alcohols or other polar modifiers have to be used. Although this makes the technical advantage of SFC questionable, the method may offer some advantages for chiral compounds that may dissolve in SFC mobile phases. Selected examples of preparative SFC enantioseparations are summarized in Table 10 [168–171].

Table 10. Selected examples of preparative resolution of enantiomers performed under SFC conditions

Chiral compound	CSP	Mobile phase	Sample size, mg	References
(±)-1-(9-Anthryl)-2,2,2-trifluorethanol	DNBPG (Pirkle-type)	CO_2/2-propanol 82:18	10	[168]
(±)-1-(9-Anthryl)-2,2,2-trifluorethanol	ChyRoSine-A	CO_2/ethanol 96.5:3.5	40	[169]
(±)-Guaifenesin	Chiralcel OD	CO_2/methanol/2-propanol 86:7:7	60	[140]
(±)-Warfarin	Whelk-O-1	CO_2/2-propanol/AcOH 75:25:0.5	200	[170]
(±)-Phosphano-bornadiens	DNBTyr (Pirkle-type)	CO_2/ethanol 92:8 92:12	167 200	[171]
(±)-1,2,5-Triphenyl-phospholane-1-oxide	DNBTyr (Pirkle-type)	CO_2/ethanol 92:8	250	[171]

6.5 Electromigration Techniques

This group includes several related techniques such as classical gel electrophoresis, field-flow fractionation (FFF) and the capillary techniques capillary zone electrophoresis (CZE), capillary electrokinetic chromatography (CEKC) capillary isotachophoresis (CIIP), capillary isoelectric focusing (CIEF) and capillary electrochromatography (CEC).

Classical gel electrophoresis is well-established for the purification and preparation of charged compounds of biological importance such as peptides, nucleic acids, etc. However, it

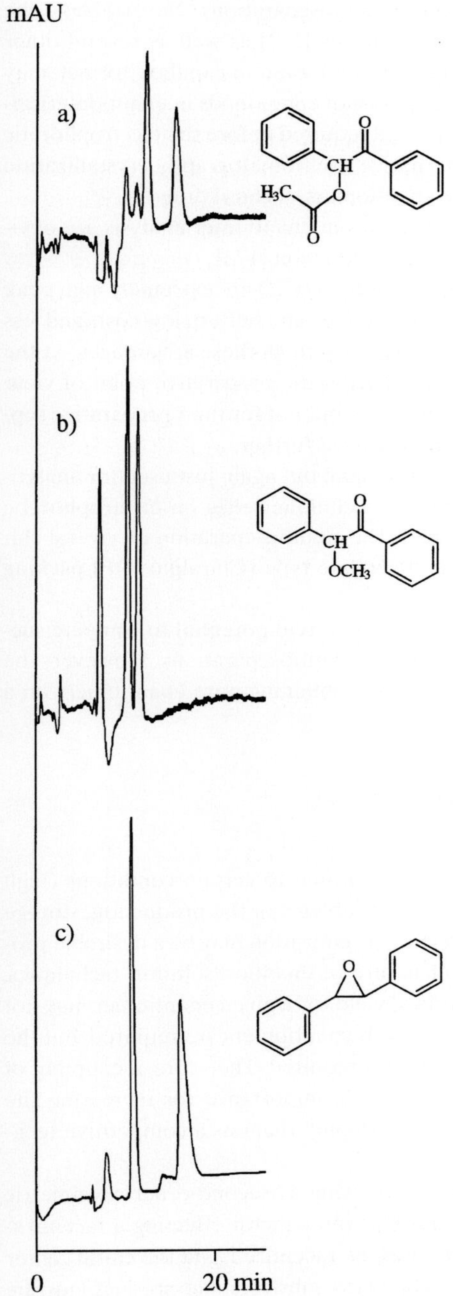

Figure 10. Enantioseparation of (a) benzoin, (b) methylbenzoin, and (c) *trans*-stilbene oxide in capillary electrochromatography using a capillary column (100 μm × 25 cm) packed with chiral poly-pyridylmethylmethacrylate on silica (Chiralpak OP) [174].

is not a technique of choice for preparative-scale enantioseparations. Nevertheless, the newest developments show that classical gel electrophoresis [172] as well as several other electrophoretic techniques (for instance, isoelectric focusing), even in capillary format, may be used for the production of a micropreparative amounts of compounds in enantiomerically pure form. However, significant developments are still required before the electrophoretic techniques will be considered as a competitive alternative to chromatography, crystallization and kinetic resolutions for the production of pure enantiomers of chiral drugs.

The challenges of capillary electromigration techniques in enantiomer analysis are obvious, and in opposition to their potential in enantiomer production [173].

The most important advantages of capillary electrophoresis (CE) are extremely high peak efficiency, small sample size, minute amounts of chiral selectors and buffers, low costs and less environmental problems. A capillary format leads exclusively to all these advantages. At the same time, the capillary format is a clear disadvantage from the preparative point of view. Thus, CE is an excellent technique for enantiomer analysis, but not for their preparative separation. Therefore, at this point the technique is not discussed further.

Another capillary technique, also with significant potential but again just used for analytical enantioseparations, is CEC [173, 174]. This hybrid technique relies on electrophoretic migration and chromatographic separation principles. The enantioseparation of several chiral compounds in non-aqueous CEC using polymethacrylate-type (Chiralpak OP) packing material is shown in Fig. 10 [174].

In summary, capillary electromigration techniques possess real potential to compete successfully with chromatographic methods for analytical enantioseparations. However, the future will show if these techniques may be also used for obtaining pure enantiomers on a micropreparative scale.

6.6 Racemization

Configurational instability of chiral compounds when exposed to certain conditions (high temperature, catalyst, solvent, etc.) is one of the major problems in the production, storage and use of their enantiomerically pure forms. However, racemization may be a desirable process for certain applications. Due to the inherent nature of enantioresolution techniques, both enantiomers may be obtained, but the theoretical yield for a given enantiomer may not exceed 50%. The former is a significant advantage if both enantiomers are required, but the latter is a severe disadvantage if only one enantiomer is required. Therefore, a coupling of enantioresolution techniques to racemization is of crucial importance for increasing the enantiomeric yield in the resolution processes and developing them as a competitive technologies.

The most useful technique is *in situ* racemization, allowing a "second-order asymmetric transformation". However, chemoselectivity is required from a factor effecting a racemization in this case. This means that only the substrate must be racemized, and the chiral center of the product must remain intact. If this is impossible, then substrate and product must be separated before performing racemization of the remaining substrate. Economically, this is not a very attractive scheme. The "second-order" technology becomes especially elegant when racemization and crystallization steps are coupled together, because a substrate–prod-

Table 11. Industrial processes of enantioresolution based on simultaneous crystallization–racemization schemes

Substrate	Racemizing reagent	Product	Reference
(R)-Amino-diazepinone	3 mol % aromatic aldehyde	(S)-Amino-diazepinone crystallized as (S)-camphor-10-sulfonic acid salt	[175]
(R)-Ketone precursor of paclobutrazol	Base 1% (w/w) NaOH	(S)-Ketone precursor of paclobutrazol	[176]
D-α-Amino-ε-caprolactam (precursor for lisine)	Complex of L-substrate with $NiCl_2$	L-α-Amino-ε-caprolactam	[177]
Schiff base of D,L-phenyl-glycine amide	Base, L-mandelic acid	D-Phenylglycine	[178]

uct separation process involves just simple filtration in this case. Certain kinetic compatibility of both steps is required if a "second-order asymmetric transformation" is to work. Commercial-scale processes based on simultaneous crystallization–racemization steps are summarized in Table 11 [175–178].

Several other techniques of so-called crystallization-induced asymmetric transformation of a racemate are described in the literature [179–181], but not all of these can compete with the industrial alternatives.

The acylase-catalyzed resolution of *N*-acetyl-DL-amino acids is a key commercial process. The racemization of the remaining *N*-acetyl-D-amino acid after separation of the L-amino acid must be performed, but adds complexity and cost. Therefore, "*in situ* racemization" with a racemase possessing specificity for N-acylamino acids without affecting the stereochemistry of the product L-amino acids is very desirable. The pentameric enzyme from *Streptomyces* sp. Y-53 specifically catalyzes the racemization of N-acylamino acids without acting on amino acids [182].

Production of optically pure amino acids using their amides as racemic precursors and L-aminopeptidases as catalysts is a well-established commercial process. To overcome the limit of 50% conversion, D-aminopeptidases and amino acid amide racemases were also developed [182].

Another group of precursors for the production of optically pure amino acids include their hydantoin derivatives. An intensive search during the past two decades has led to some effective hydantoin racemases which are already used on pilot-scale processes [183–185].

As these few examples show, the development of effective *in situ* racemization schemes may allow the 50% yield dilemma of enantioresolutions to be successfully overcome.

6.7 Method Selection Strategy

Along with the excellent overview of asymmetric synthetic strategies in Chapter 5, the present chapter shows that many techniques are available for the separation, preparation and production of enantiomerically pure chiral compounds of pharmaceutical interest. The logical question is: how is the most suitable technique for a particular problem to be selected?

Clearly, there is no general prescription as to how an enantioseparation method should be selected, and indeed which is the best technique. The answer is strongly case-dependent, and a decision must be taken based on a multiple strategy, considering the amount and purity of the enantiomers required, whether only one or both enantiomers are desired, the availability of the racemate, other chemicals, solvents, equipments, time, costs, etc.

One principal question which must be answered is which method, synthesis (stereoselective or asymmetric either catalytic or non-catalytic) or enantioresolution, should be used. As experience shows, a synthesis based on chiral pools may be the method of choice for large-scale productions of single enantiomers. Often, only single enantiomers are considered, because the chiral pools are commonly available in one natural configuration, although when this natural configuration is the wrong one, the problem becomes more complicated.

Usually, in the initial stage of the drug development process, there are no clear pharmacological, toxicological, social or economical indications as to whether or not a given drug must be used in a pure enantiomeric form, or in racemic or non-racemic mixtures of the enantiomers. Initially, both enantiomers are required in order to study their properties separately. The enantioresolution techniques which, in principle, provide both enantiomers are advantageous in this case. Another reason why enantioresolution may be the method of choice is that it is in general easier, faster and cheaper to develop than an enantioselective or asymmetric synthetic process. In addition, some chiral compounds may not be easily obtained by synthesis (a lack of suitable functional groups, conformational and configurational instability, etc.).

If the decision is made to use a resolution rather than a synthetic route, the question then arises of whether this should be a classical crystallization, a kinetic resolution, or a chromatographic separation. If the compound either contains ionic groups (and a suitable reagent for diastereomeric salt formation is inexpensive and available) or even better, if it crystallizes as a conglomerate, then crystallization may seem the most suitable technique.

Very often, enzymes are known (for instance esterases) which are able to transform stereoselectively only the required enantiomer of a precursor into the product. Several schemes based on this technique – especially those using immobilized enzymes and microorganisms that also work in organic media – have been established in industrial-scale processes.

If a chiral compound does not contain a suitable functionality and cheap complexing agents (such as native cyclodextrins and other natural compounds) are not available, then chromatographic techniques may be used. The choice regarding which chromatographic technique is used may be relatively simple. Preparative GC offers certain advantages for a few chiral gases of pharmaceutical importance. The applicability of SFC on the preparative scale is limited by instrumental problems or the poor solubility of polar chiral compounds in carbon dioxide. Liquid chromatography in the batch elution mode may offer certain advantages for obtaining chiral pharmaceuticals on the micropreparative scale (milligram to gram), whereas SMB-chromatography is clearly the technique of choice for the large-scale production of enantiomerically pure drugs.

References

[1] Mason, S. *Trends Pharm Sci.* **1**, 20–23 (1986).
[2] Pasteur, L. In: Richardson, G.M. (ed.) *On The Asymmetry of Naturally Occurring Organic Compounds, The Foundations of Stereo Chemistry: Memories by Pasteur, Van't Hoff, Le Bel, and Wislicenus*. American Book Co.: New York, 1901, pp. 1–33.
[3] Ariens, E.J., Soudijn, W., and Timmermans, P.B.M.W.W. *Stereochemistry and Biological Activity of Drugs*. Blackwell Scientific Publications: Oxford, UK 1983.
[4] Wainer, I.W., and Drayer, D.E. *Drug Stereochemistry (Analytical Methods and Pharmacology)*. Marcel Dekker, Inc.: New York, Basel, 1988.
[5] Collins, A.N., Sheldrake, G.N., and Crosby, J. (eds) *Chirality in Industry: The Commercial Manufacture and Applications of Optically Active Compounds*. John Wiley & Sons: Chichester, 1992.
[6] Crossley, R. *Chirality and The Biological Activity of Drugs*. CRC Press, Inc.: Boca Raton, 1995.
[7] Pifferi, G., and Perucca, E. *Eur. J. Drug Metab. Pharmacokinet.* **20**, 15–25 (1995).
[8] Marzo, A. *Arzneim.-Forsch./Drug Res.* **44**, 791–793 (1994).
[9] Stinson, S.C. *C & EN September* 21, 83–104 (1998).
[10] Jacques, J., Collet, A., and Wilen, S.H. *Enantiomers, Racemates and Resolutions.* Wiley-Interscience: New York 1981.
[11] *Chem. Eng. November* 8, 247 (1965).
[12] Davis, C.J., *Chem. Eng. May* 22, 62 (1978).
[13] Crosby, J. In: Collins, A.N., Sheldrake, G.N., and Crosby, J. (eds) *Chirality in Industry: The Commercial Manufacture and Applications of Optically Active Compounds*. JohnWiley & Sons: Chichester, 1992, pp. 1–66.
[14] Wakamatsu, H. *Food Eng. November*, 92 (1968).
[15] Kanoh, S., Hongoh, Y., Motoi, M., and Suda, H. *Bull. Chem. Soc. Jpn.* **61**, 1032–1034 (1988).
[16] Doyle, T.R., and Vogl, O. *J. Am. Chem. Soc.* **111**, 8510–8511 (1989).
[17] Bellucci., G., Berti, G., Ferretti, M., Mastrorilli, E., and Silvestri, L. *J. Org. Chem.* **50**, 1471–1474 (1985).
[18] Jacques, J., and Fouquey, C. *Org. Syn.* **67**, 1–12 (1989).
[19] Webster, F.X., Zeng, X.-N., and Silverstein, R.M. *J. Org. Chem.* **47**, 5225–5226 (1982).
[20] Paquette, L.A., and Gardlik, J.M. *J. Am. Chem. Soc.* **102**, 5016–5025 (1980).
[21] Fizet, C. *Helv. Chim. Acta* **69**, 404–409 (1986).
[22] Himmele, W., and Siegel, H. *Tetrahedron Lett.* **907**, 911–914 (1976).
[23] Markowicz, S.W. *Pol. J. Chem.* **53**, 157–161 (1979).
[24] Faustini, F., DeMunari, S., Panzeri, A., Villa, V, and Gandolfi, C.A. *Tetrahedron Lett.* **22**, 4533–4536 (1981).
[25] Perry, C.W., Brossi, A., Deitcher, K.H., Tautz, W., and Teitel, S. *Synthesis*, 492–494 (1977).
[26] Saigo, K., Kai, M., Yonezawa, N., and Hasegawa, M. *Synthesis*, 214–216 (1985).
[27] Corey, E.J., Trybulski, E.J., Melvin Jr, L.S., Nicolaou, K.C., Secrist, J.A., Lett, R., Sheldrake, P.W., Falk, J.R., Brunelle, D.J., Haslanger, M.F., Kim, S., and Yoo, S.-E. *J. Am. Chem. Soc.* **100**, 4618–4620 (1978).
[28] Marx, M., Marti, F., Reisdorff, J., Sandmeier, R., and Clark, S. *J. Am. Chem. Soc.* **99**, 6754–6756 (1977).
[29] ten Hoeve, W., and Wynberg, H. *J. Org. Chem.* **50**, 4508–4514 (1985).
[30] Shiraiwa, T., Sado, Y., Fujii, S., Nakamura, M., and Kurokawa, H. *Bull. Chem. Soc. Jpn.* **60**, 824–826 (1987).
[31] Larson, E.R., and Raphael, R.A. *J. Chem. Soc. Perkin* **1**, 521–525 (1982).
[32] Suda, H., Motoi, M., Fujii, M., Kanoh, S., and Yoshida, H. *Tetrahedron Lett.* 4565–4568 (1979).
[33] Radler, S., Wermeille, M., and Blaschke, G. *Arzneim. Forsch./Drug Res.* **45**, 1086–1092 (1995).
[34] Blaschke, G., and Walther, B. *Liebigs Ann. Chem.* 561–563 (1987).
[35] Abu Zarga, M., and Shamma, M. *Tetrahedron Lett.*, **21**, 3739–3742 (1980).
[36] Ohgi, T., Kondo, T., and Goto, T. *J. Am. Chem. Soc.* **101**, 3629–3633 (1979).
[37] Corey, E.J., Bakshi, R.K., Shibata, S., Chen, C.-P., and Singh, V.K. *J. Am. Chem. Soc.* **109**, 7925–7926 (1987).
[38] Hecker, S.J., and Heathcock, C.H., *J. Am. Chem. Soc.* **108**, 4586–4594 (1986).
[39] Bey, P., Vevert, J.-P., Van Dorsselaer, V., and Kolb, M. *J. Org. Chem.* **44**, 2732– 2742 (1979).
[40] Blaschke, G., and Doebler, P. *Process for Resolution of Verapamil into its Enantiomers,* Ger. Offen DE 3 723 684 (1989).

[41] Smith, H.E., Neergaard, J.R., de Paulis, T., and Chen F.-M. *J. Am. Chem. Soc.* **105**, 1578–1584 (1983).
[42] Kelly, R.C., Schletter, I., Stein, S.J., and Wierenga, W. *J. Am. Chem. Soc.* **101**, 1054–1056 (1979).
[43] Brown, E., Viot, F., Le Floc'h, Y. *Tetrahedron Lett.* **26**, 4451–4452 (1985).
[44] Jacob, III P. *J. Org. Chem.* **47**, 4165–4167 (1982).
[45] Smith, H., Burrows, E.P., Marks, M.J., Lynch, R.D., and Chen, F.-M. *J. Am. Chem. Soc.* **99**, 707–713 (1977).
[46] Nohira, H., and Ueda, K. *Eur Patent Appl EP* 65867 (1983), *Chem Abstr.* **98**, 161164v.
[47] Viret, J., Patzelt, H., and Collet, A. *Tetrahedron Lett.* **27**, 5865–5868 (1986).
[48] Vriesema, B.K., ten Hoeve, W., Wynberg, H., Kellogg, R.M., Boesten, W.H.J., Meijer, E. M., and Schoemaker, H.E. *Tetrahedron Lett.* **27**, 2045–2048 (1986).
[49] Cope, A.C., Ganellin, C.R., and Johnson Jr, H.W. *J. Am. Chem. Soc.* **84**, 3191–3192 (1962).
[50] Hulshof, L.A., McKervey, M.A., and Wynberg, H. *J. Am. Chem. Soc.* **96**, 3906–3912 (1974).
[51] Block Jr, P., and Newman, M.S. *Org. Synth.* **48**, 120–126 (1968).
[52] Newman, M.S., and Lutz, W.B. *J. Am. Chem. Soc.* **78**, 2469–2473 (1956).
[53] Cramer, F., and Dietsche, W. *Chem. Ber.* **92**, 378–384 (1959).
[54] Benschop, H.P., and Van den Berg, G.R. *J. Chem. Soc. D*, 1431–1432 (1970).
[55] Jin, H.L., Stalcup, A., and Armstrong, D.W. *Chirality* **1**, 137–141 (1989).
[56] Peacock, S.S., Walba, D.M., Gaeta, F.C.A., Helgeson, R.C., and Cram, D.J. *J. Am. Chem. Soc.* **102**, 2043–2052 (1980).
[57] Stoddart, J.F. *Top. Stereochem.* **17**, 207–288 (1987).
[58] Yanagihara, R., Tominaga, M., and Aoyama, A. *J. Org. Chem.* **59**, 6865–6867 (1994).
[59] Gerdil, R., and Allemand, J. *Helv. Chim. Acta* **63**, 1750–1753 (1980).
[60] Gerdil, R. *Top. Curr. Chem.*, **140**, 71–105 (1987).
[61] Toda, F., Tanaka, K., and Ueda, H. *Tetrahedron Lett.* **22**, 4669–4672 (1981).
[62] Toda, F. *Top. Curr. Chem.* **140**, 43–69 (1987).
[63] Toda, F., Tanaka, K., and Mak, T.C.W. *Chem. Lett.* 2085–2088 (1984).
[64] Wilen, S.H., Bunding, K.A., Kascheres, C.M., and Wieder, M.J. *J. Am. Chem. Soc.* **107**, 6997–6998 (1985).
[65] Pasteur, L. *C. R. Acad. Sci.* **46**, 615 (1858).
[66] Johnson, C.R. *Acc. Chem. Res.* **31**, 333–341 (1998).
[67] Wang, Y.-F., and Wong, C.-H. *J. Org. Chem.* **53**, 3127–3129 (1988).
[68] Coffen, D.L. In: Ahuja, S. (ed.) *Chiral Separations, Applications and Technology*. ACS: Washington, DC, USA, 1997, Chapter 4, pp. 69–91.
[69] Ahuja, S. *Chiral Separations, Applications and Technology*. ACS: Washington, DC, USA, 1997.
[70] van Eikeren, P. In: Ahuja, S. (ed.) *Chiral Separations, Applications and Technology*. ACS: Washington, DC, 1997, pp. 9–35.
[71] Wu, S.-H., Guo, Z.-W., and Sih, G.-J. *J. Am. Chem. Soc.* **112**, 1990–1995 (1990).
[72] Blanco, L., Guibe-Jampel, E., and Rousseau, G. *Tetrahedron. Lett.* **29**, 1915–1918 (1988).
[73] Mutsaers, J.H.G.M., and Kooreman, H.J. *Recl. Trav. Chim. Pays-Bas* **110**, 185–188 (1991).
[74] Yamamoto, K., Ueno, Y., Otsubo, K., Kawakami, K., and Komatsu, K. *Appl. Environ. Microbiol.* **56**, 3125–3129 (1990).
[75] Hamaguchi, S., Ohashi, T., and Watanabe, K. *Agric. Biol. Chem.* **50**, 375–380 (1986).
[76] Chen, C.S., Liu, Y., and Marsella, M.J. *Chem. Soc. Perkin Trans.* **I**, 2559–2562 (1990).
[77] Ader, U., and Schneider, M.P. *Tetrahedron Asymm.* **3**, 201–204 (1992).
[78] Bertola, M.A., Koger, H.S., Phillips, G.T., Marx, A.F., and Clussen, V.P. *European Patent Appl* 244, 912 (1987).
[79] Habets-Crützen, A.Q.H., Carlier, S.J.N., De Bont, Jan A.M., Wistuba, D., Schurig, V., Hartmans, S., and Tramper, J. *Enz. Microb. Technol.* **7**, 17–21 (1985).
[80] Robinson, B.W., Bertola, M.A., Koger, H.S., Mary, A.F., and Watt, P.D. *European Patent Appl.* 193, 228 (1986).
[81] Sheldon, R.A. *Chirotechnology (Industrial Synthesis of Optically Active Compounds)*. Marcel Dekker, Inc.: New York 1993, Chapter 9, pp. 343–385.
[82] Bianchi, D., Gabri, W., Cesti, P., Francalanci, F., and Ricc, M. *J. Org. Chem.* **53**, 104–107 (1988).
[83] Matsuo, N., and Ohno, N. *Tetrahedron Lett.* **26**, 5533–5534 (1985).
[84] Finn, M.G., and Sharpless, K.B. In: Morrison, J.D. (ed.) *Asymmetric Synthesis*. Academic Press, Inc.: Orlando 1985, Vol. 5, pp. 207–308.
[85] Hansen, R.M., and Sharpless, K.B. *J. Org. Chem.* **51**, 1922–1925 (1996).
[86] Sharpless, K.B. *Janssen Chim. Acta* **6**, 3–9 (1988).

[87] Gao, Y., Hanson, R.M., Klunder, J.M., Ko, S.Y., Massamane, H., and Sharpless, K.B. *J. Am. Chem. Soc.* **109**, 5765–5780 (1987).
[88] Kitamura, M., Kasahara, I., Manabe, K., and Noyori, R. *J. Org. Chem.* **53**, 708–710 (1988).
[89] Gil-Av, E, Feibush, B., and Charles-Sigler, R. *Tetrahedron Lett.* **10**, 1009–1015 (1966).
[90] Schurig, V. *J. Chromatogr. A* **666**, 111–129 (1994).
[91] König, W.A. *Gas Chromatographic Enantiomer Separation with Modified Cyclodextrins.* John Wiley & Sons: Chichester, England, 1989.
[92] König, W.A. In: Wainer, I.W., and Drayer, D.E. (eds) *Drug Stereochemistry (Analytical Methods and Pharmacology).* Marcel Dekker, Inc.: New York, 1988, Chapter 5, pp. 113–145.
[93] Beesley, T.E., and Scott, R.P.W. *Chiral Chromatography*. John Wiley & Sons: Chichester, 1998, Chapter 6, pp. 141–172.
[94] Hall, A.C., Lieb, W.R., and Franks, N.P. *Br. J. Pharmacol.* **112**, 906–910 (1994).
[95] Garton, K.J., Yuen, P., Meinwald, J., Thummel, E.D., Kharasch, D., and Evan, D. *Drug Metab. Disp.* **23**, 1426–1430 (1995).
[96] Meinwald, J., Thompson, W.R., Pearson, D.L., König, W.A., Runge, T., and Francke, W. *Science*, **251**, 560–561 (1991).
[97] Schurig, V., Grosenick, H., and Green, B.S. *Angew. Chem. Int. Ed. Engl.* **32**, 1690–1691 (1993).
[98] Shitangkoon, A., Staerk, D.U., and Vigh, G. *J. Chromatogr. A* **657**, 387–394 (1993).
[99] Staerk, D.U., Shitangkoon, A., and Vigh, G. *J. Chromatogr. A* **677**, 133–140 (1994).
[100] Juza, M., Di Giovanni, O., Biressi, G., Schurig, V., Mazzoti, M., and Morbidelli, M. *J. Chromatogr. A*, **813**, 333–347 (1998).
[101] Storti, G., Mazzoti, M., Furlan, L.T., Morbidelli, S., and Carra, S. *Sep. Sci. Technol.* **27**, 1889–1916 (1992).
[102] Mazzoti, M., Baciocchi, R., Storti, G., and Morbidelli, M. *Ind. Eng. Chem. Res.* **35**, 2313–2321 (1996).
[103] Lemko, C.H., Gilles, C., and Foster, R.T. *J. Chromatogr.* **619**, 330–335 (1993).
[104] Lindner, W. In: Zief, M., and Crane, L.J. (eds) *Chromatographic Chiral Separations*. (Chromatographic Science Series, 40), Dekker Inc.: New York, 1988, Chapter 4, pp. 91–130.
[105] Davankov, V.A. *Chromatographia* **27**, 475–482 (1989).
[106] LePage, J., Lindner, W., Davies, G., and Karger, B. *Anal. Chem.* **51**, 433–435 (1979).
[107] Hare, P.E., and Gil-Av, E. *Science*, **204**, 1226–1228 (1979).
[108] Sybilska, D., and Zukowski, J. In: Krstulovic, A.M. (ed.) *Chiral Separations by HPLC*. Ellis Horwood Limited: New York, 1989, Chapter 7, pp. 124–172.
[109] Nimura, N. In: Krstulovic, A.M. (ed.) *Chiral Separations by HPLC*. Ellis Horwood Limited: New York, 1989, Chapter 5, pp. 107–123.
[110] Petersson, C., and Heldin, E. In: Subramanian, G. (ed.) *A Practical Approach to Chiral Separations by Liquid Chromatography*. VCH: Weinheim, Germany, 1994, Chapter 9, pp. 279–310.
[111] Cooper, A.D., and Jefferies, T.M. *J. Chromatogr.* **637**, 137–143 (1993).
[112] Davankov, V.A. In: Krstulovic, A.M. (ed.) *Chiral Separations by HPLC*. Ellis Horwood Limited: New York, 1989, Chapter 8, pp. 175–193.
[113] Welch, C.J. *J. Chromatogr. A* **666**, 3–26 (1994).
[114] Allenmark, S.G., and Andersson, S. *J. Chromatogr. A* **666**, 167–179 (1994).
[115] Okamoto, Y., and Yashima, E. *Angew. Chem. Int. Ed.* **37**, 1073–1095 (1998).
[116] Stalcup, A. In: Subramanian, G. (ed.) *A Practical Approach to Chiral Separations by Liquid Chromatography*. VCH: Weinheim, Germany, 1994, Chapter 5, pp. 95–113.
[117] Okamoto, Y., and Hatada, K. In: Zief, M., and Crane, L.J. (eds) *Chromatographic Chiral Separations.* (Chromatographic Science Series, 40), Marcel Dekker Inc: New York, 1988, Chapter 8, pp.199–218.
[118] Sellergren, B. In: Krstulovic, A.M. (ed.) *Chiral Separations by HPLC*. Ellis Horwood Limited: New York, 1989, Chapter 4, pp. 69–92.
[119] Kinkel, J.N. In: Krstulovic, A.M. (ed.) *Chiral Separations by HPLC*. Ellis Horwood Limited: New York, 1989, Chapter 8, pp. 217–269.
[120] Krause, K., Girod, M., Chankvetadze, B., and Blaschke, G. *J. Chromatog. A*, in press (1999).
[121] Noctor, T.A.G. In: Subramanian, G. (ed.) *A Practical Approach to Chiral Separations by Liquid Chromatography*. VCH: Weinheim, 1994, Chapter 12, pp. 357–396.
[122] Henderson, G.M., and Rule, H.G. *J. Chem. Soc.* 1568–1573 (1939).
[123] Lecoq, H. *Bull. Soc. Ray Sci. Liege* **11**, 318–325 (1943).
[124] Prelog, V., and Wieland, P. *Helv. Chim. Acta* **27**, 1127–1134 (1944).
[125] Mayer, W., and Merger, F. *Liebigs Ann. Chem.* **644**, 65–69 (1961).

[126] Lautsch, W., and Heinicke, D. *Kolloid Z.* **154**, 1–4 (1957).
[127] Blaschke, G., and Markgraf, H. *Chem. Ber.* **113**, 2031–2035 (1980).
[128] Blaschke, G., and Maibaum, J. *J. Pharm. Sci.* **74**, 438–442 (1985).
[129] Blaschke, G., Fraenkel, B., Fröhlingsdorf, B., and Marx, A. *Liebigs Ann. Chem.* **8**, 753–756 (1988).
[130] Allenmark, S., and Thompson, R.A. *Tetrahedron Lett.* **28**, 3751–3752 (1987).
[131] Rimböck, K.H., Kastner, F., and Mannschreck, A. *J. Chromatogr.* **329**, 307–310 (1985).
[132] Francotte, E., Stierlin, H., and Faigle, J.W. *J. Chromatogr.* **346**, 321–331 (1985).
[133] Furet, P., Batze, C., Bhatnagor, A., Francotte, E., Rihs, G., and Lang, M. *J. Med. Chem.* **36**, 1393–1400 (1993).
[134] Willstätter, R. *Ber. Dtsch. Chem. Ges.* **37**, 3758–3760 (1904).
[135] Blaschke, G., and Donow, F. *Chem. Ber.* **108**, 1188–1197 (1975).
[136] Blaschke, G. *Angew. Chem.* **92**, 14–25 (1980).
[137] Blaschke, G., Kraft, H.P., Fikentscher, K., and Köhler, F. *Arzneim. Forsch.* **29**, 1640–1642 (1979).
[138] Francotte, E., and Junker-Buchheit, A. *J. Chromatogr.* **576**, 1–45 (1992).
[139] Francotte, E. *J. Chromatogr. A* **666**, 565–601 (1994).
[140] Francotte, E. In: Ahuja, S. (ed.) *Chiral Separations (Applications and Technology).* ACS: Washington, DC, USA, 1997, Chapter 10, pp. 271–308.
[141] Rimböck, K.H., Kastner, F., and Mannschreck, A. *J. Chromatogr.* **351**, 346–350 (1986).
[142] Francotte, E., and Wolf, R.M. *Chirality* **3**, 43–55 (1991).
[143] Vigh, G., Quintero, G., and Farkas, G. *J. Chromatogr.* **484**, 237–250 (1989).
[144] Farkas, G., Irgens, L.H., Quintero, G., Beeson, M.D., Al-Saed, A., and Vigh, G. *J. Chromatogr.* **645**, 67–74 (1993).
[145] Irgens, L.H., Farkas, G., and Vigh, G. *J. Chromatogr. A* **666**, 603–609 (1994).
[146] Martin, A.J.P. In: Coates, Y.J., Noebles, H.J., and Fagerson, I.S. (eds) *Gas Chromatography.* Academic Press: New York, 1958, p. 237.
[147] Biesenberger, J.A., Tan, M., Duvdevan, I., and Maurer, T. *J. Polym. Sci.* **9**, 353–357 (1971).
[148] Dingenen, J., and Kinkel, J.N. *J. Chromatogr. A* **666**, 627–650 (1994).
[149] Chizhkov, V.P. (Russ. Chem. Rev.), *Usp. Khim.* **40**, 161–175 (1971).
[150] Martin, M., Vorillon, F., Eon, C., and Guiochon, G. *J. Chromatogr.* **125**, 17–41 (1976).
[151] Saidel-Morgenstern, A., and Guiochon, G. *Chem. Eng. Sci.* **48**, 2787–2797 (1993).
[152] Dingenen, J. In: Subramanian, G. (ed.) *A Practical Approach to Chiral Separations by Liquid Chromatography.* VCH: Weinheim, New York, 1994, Chapter 6, pp. 115–181.
[153] Devant, M.R., Jonas, R., Schulte, M., Keil, A., and Charton, F. *J. Prakt. Chem.* **339**, 315–321 (1997).
[154] Schulte, M., Ditz, R., Devant, R.M., Kinkel, J.N., and Charton, F. *J. Chromatogr. A* **769**, 93–100 (1997).
[155] Schulte, M., *10th International Symposium on Chiral Discrimination*, Vienna, Austria (1998).
[156] Jonas, R., Klockow, M., Lues, I., and Wurziger, H. *Bioorg. Med. Chem. Lett.* **4**, 2585–2588 (1994).
[157] Negawa, M., and Shoji, F. *J. Chromatogr.* **590**, 113–117 (1992).
[158] Fuchs, G., Nicoud, R.M., and Bailly, M. *Proceedings of the 9th Symposium on Preparative and Industrial Chromatography*, *"Prep 92" INPL*, Nancy, France, 1992, pp. 205–220.
[159] Nicoud, R.M., Fuchs, G., Adam, P., Bailly, M., Küsters, E., Antia, F.D., Reuill, R., and Schmid, E. *Chirality* **5**, 267–271 (1993).
[160] Nicoud, R.M., Bailly, M., Kinkel, R., Devant, R., Hampe, T., and Küsters, E. In: Nicoud, R.M. (ed.) *Simulated Moving Bed: Basics and Applications.* INPL, Nancy, France, 1993, pp. 65–88.
[161] Ching, C.B., Lim, B.G., Lee, E.J.D., and Ng, S.C. *J. Chromatogr.* **634**, 215–219 (1993).
[162] Guest, D.W. *J. Chromatogr. A* **760**, 159–162 (1997).
[163] Cavoy, E., Deltent, M.F., Lehoucq, S., and Miggiano, D. *J. Chromatogr. A* **769**, 49–57 (1997),
[164] Blehaut, J. *10th International Symposium on Chiral Discrimination*, Vienna, Austria (1998).
[165] Gattuso, M.J., and Makino, S. *Poster Presentation on Chiral 94*, Reston, VA, (1994).
[166] Francotte, E.R., and Richert, P. *J. Chromatogr.* **769**, 101–107 (1997).
[167] Seebach, D., Hoffmann, M., Sting, A.R., Kinkel, J.N., Schulte, M., and Küsters, E. *J. Chromatogr. A* **796**, 299–307 (1998).
[168] Macaudiere, P., Caude, M., Rosset, R., and Tambute, A. *J. Chromatogr. Sci.* **27**, 383–394 (1989).
[169] Fuchs, G., Doguet, L., Perrut, M., Tambute, A., and Le Goff, P. *Presentation at the Symposium on Supercritical Fluids*, Boston, MA, USA (1991).
[170] Blum, A.M., and Kumar, M.L. *Poster Presentation at the 4th International Symposium on Chiral Discrimination*, Montreal, Canada, September (1993).
[171] Fuchs, G., Doguet, L., Barth, D., and Perrut, M. *J. Chromatogr.* **623**, 329–336 (1992).

[172] Stalcup, A.M., Gahm, K.H., Gratz, S.R., and Sutton, R.M.C. *Anal. Chem.* **70**, 144–148 (1998).

[173] Chankvetadze, B. *Capillary Electrophoresis in Chiral Analysis*. John Wiley & Sons: Chichester, England, 1997.

[174] Krause, K., Chankvetadze, B., Okamoto, Y., and Blaschke, G. *Electrophoresis*, **20**, 2772–2778 (1999).

[175] Reider, P.J., Davis, P., Hughes, L.D., and Grabowski, E.J.S. *J. Org. Chem.* **52**, 955–957 (1987).

[176] Black, S.N., Williams, L.J., Davey, R.J., Moffatt, F., Jones, R.V.H., McEwan, D.M., and Sadler, D.E. *Tetrahedron* **45**, 2677–2682 (1989).

[177] Sifiniades, S., Boyle, W.J.J., and Van Peppen, J.F. *J. Am. Chem. Soc.* **98**, 3738–3739 (1976).

[178] Kamphuis, J., Boesten, W.H., Kaptein, B., Hermes, H.F.M., Sanke, T., Broxterman, Q.B., Van den Tweel, W.J.J., and Schoemaker, H.E. In: Collins, A.N., Sheldrake, G.N., and Crosby, J. (eds) *Chirality in Industry: The Commercial Manufacture and Applications of Optically Active Compounds*. JohnWiley & Sons: Chichester, 1992, Chapter 8, pp. 187–208.

[179] Belokon, Y.N., Tararov, V.I., Saveleva, T.F., and Belikov, V.M., *Makrom. Chem.* **187**, 1077–1086 (1986).

[180] Chibata, I., Yamada, S., Hongo, C., and Yochioka, R. *European Patent Appl.*, 0. 070. 114 (1982).

[181] Kral, V., Dvorak, D., Javada, J., Stibor, I., Mostecky, J., and Votava, V. *Czechoslovakian Patent*, 257094 (1989), CA: **112**, 35450 (1990).

[182] Kamphuis, J., Boesten, W.H.J., Broxterman, Q.B., Hermes, H.F.M., van Balken, J.A.M., Meijer, E.M., and Schoemaker, H.E. *Adv. Biochem. Eng./Biotechnol.* **42**, 133–186 (1990).

[183] Guivarch, M., Gillonier, C., and Brunie, J.C.. *Bull. Soc. Chem. Fr.* **44**, 91–95 (1980)

[184] Nishida, Y., Nakamichi, K., Nabe, K., and Tosa, T. *Enzyme Microb. Technol.* **9**, 479–483 (1987).

[185] Syldatk, C., and Wagner, F. *Food Biotechnol.* **4**, 87–95 (1990).

7 Biocatalyzed Reactions

Oreste Ghisalba

7.1 Introduction: Biotechnology and Biocatalysis

Biocatalysis or biotransformation encompasses the use of biological systems, whether whole cells, cellular extracts or isolated enzymes, to catalyze the conversion of one compound to another.

For thousands of years, mankind has used biological processes in an empirical way, e.g., to produce alcoholic beverages, bread, fermented foods and dairy products. Thanks to the investigations by Louis Pasteur and Robert Koch in the second half of the 19th century, we learned that these processes are in fact catalyzed by micro-organisms or by microbial enzymes, respectively. This knowledge opened the way for the directed and more rational development of new microbial and enzymatic processes in the food and dairy sector and in the chemical and pharmaceutical industries. Early denominations of this new approach were 'Zymotechnology' or 'Technical Biology'. The term 'Biotechnology' was first used in the year 1917. Already then, in the early times of industrial biotechnology, the leading representatives of the new science promoted the idea of using biological systems to create more efficient, more selective, and environmentally friendlier processes for the conversion of raw materials into industrial products, thereby substituting problematic chemical transformations. The concept of a more sustainable use of the limited resources was thus one of driving forces for the rapid development of industrial biotechnology long before the now world-wide accepted political vision of sustainable development was promoted by national and international conferences and organizations. In the time since World War II, the most significant new achievements of biotechnology were the development and the establishment of microbial large-scale production processes for antibiotics, vitamins, amino acids, nucleotides, and industrial enzymes. The introduction of genetic engineering (since 1973) and cell culture technology for higher organisms drastically enlarged the possibilities of biotechnology and provided an additional innovation potential. Modern biotechnology derived from molecular biology has renewed many traditional methodologies and processes. It is now undoubtedly in the genetics field, where the greatest impact is expected, at least in the short term. Thanks to genetic engineering techniques enzymes from all kinds of biological sources can now be re-cloned and overexpressed in easily mass cultivatable micro-organism or cell cultures. With this, even so far rare enzymes can be produced in large quantities and at affordable costs. In addition, whole-cell biocatalysts or production organisms can now be genetically improved by directed engineering of metabolic pathways. This can be done either by amplification of already present genes (to eliminate metabolic bottlenecks), by genetic modification of such genes, or by the introduction of "new" genes or even whole clusters of genes from other organisms (pathway engineering). With such new engineering approaches on the one hand, the productivity of "classical" biological processes can be further increased, on the other hand, entirely new processes and products can be generated by recombination of metabolic pathway elements from various organisms.

To a significant extent these new options of biotechnology have already been exploited in the health sector (new types of drugs and diagnostic tools), whereas practical applications are emerging at lower speed in other sectors, such as nutrition, environment, raw material supply, or specialty chemicals. New biotechnological production systems based on mammalian cell cultures are in use, e.g., for the manufacture of recombinant therapeutic proteins and antibodies.

A recently published OECD-Report [1] analyzes the state of the art and the future development needs for industrial biotechnology. Some important conclusions in this report summarizing the state of the art are:

- Biotechnological operations are currently used in a wide range of major industrial processes.
- Economic competitiveness has been established for a variety of biotechnological applications to achieve cleanliness.
- Biotechnology-based processes have been successfully integrated into large-scale operations.
- Industrial penetration of biotechnology is increasing as a consequence of advances in recombinant DNA technologies.
- Biotechnological operations have led to cleaner processes with lowered production of wastes and in some cases lower energy consumption.
- The fine chemicals industry is one of the industrial segments where the impact of biotechnology is felt most strongly.

7.2 Advantages of Biocatalysis

To perform (large-scale) synthesis of drugs, materials, or chemicals, one can in principle follow three different approaches with various degrees of complexity:

1. Use a purely chemical strategy with classical organic reactions and/or new special reactions.
2. Use a chemo-enzymatic route combining chemical and biocatalytic steps. In this case the biocatalyst is preferentially used to perform the key reaction(s) requiring high selectivity or specificity.
3. Use a biological total synthesis by fermentation or multi step biotransformation (with growing cells, resting cells, multi-enzyme systems, genetically or metabolically engineered biocatalysts, etc).

Enzymes and whole-cell biocatalysts have several properties that make them attractive for organic synthesis:

- Enzymes are chiral catalysts with high regio- and stereoselectivities towards their natural substrates, but quite often also towards related "unnatural/synthetic" substrates.
- Biocatalysis occurs under mild conditions; there are no requirements for strong bases or acids, heavy metals or other ingredients normally associated with chemical catalysis.

Therefore, biocatalysis offers great chances and advantages for successful applications also in cases where either the substrates or the products of the reaction are chemically labile.

- Biocatalysts optimally operate between 10 and 90 °C (most commonly between 20 and 40 °C), and therefore require only modest energy input.
- Biocatalysis is normally performed in an aqueous environment, but can in many cases also be conducted in solvent mixtures, liquid–liquid two-phase systems, and even in pure organic solvents. This extends the traditional (natural) application range of biocatalysis to more lipophilic starting materials and opens up the way to new reactions. A relevant practical example is the use of esterases and lipases to catalyze esterifications in organic solvents such as vinyl acetate.
- The high regio- and stereoselectivities of enzymes allow the design of synthesis strategies that avoid or limit the use of protecting groups. This is of special interest, e.g., for the chemo-enzymatic synthesis of complex carbohydrates and glycoconjugates.

7.3 Application Range of Biocatalysis

According to the EC-System of the Nomenclature Committee of the International Union of Biochemistry and Molecular Biology IUBMB [2], the enzymes are classified in six main classes (Table 1). To allow a refined classification, the main classes are subdivided into three orders of enzyme subclasses (EC X.a.b.c., e.g., EC 1.1.1.1 for alcohol dehydrogenase). The majority of the described enzymes belong to the EC classes 1, 2 and 3.

Very detailed information on all the 3400 EC-classified enzymes (organized in fact sheets) can be found in the excellent *Enzyme Handbook* [3] edited by GBF (Gesellschaft für Biotechnische Forschung in Braunschweig) in 17 Volumes, so far. The same information is also available in electronic form via BRENDA (Braunschweig Enzyme Databank).

Many thousands of original papers and hundreds of review articles have been written on biotransformations, describing a wide range of synthetic applications in research, development and production. The newcomer in the field is confronted with the difficulties of identifying the suitable biocatalysts, methods and technologies to apply to his or her particular synthesis problem. This report tries to give – from an industrial perspective – an overview on the power of applied biocatalysis and to indicate sources providing more in-depth information. The aim is not to give a comprehensive overview on enzyme applications in general, but to discuss selected "classical" examples (including technological aspects) as well as new trends and future options.

The state of the art is described in many excellent and comprehensive books [4–29] and general review articles [30–70]. The reference list gives only the most prominent and best known reviews. Additional and highly recommendable information sources for the synthetic chemist are the *Warwick Biotransformation Abstracts* and the associated electronic database. Contact address: H.G. Crout, Warwick Biotransformation Club Database, Organic Chemical Institute, University of Warwick, UK (service for members only). The two CD-ROMs *Biotransformation* (K. Kieslich and the Warwick Biotransformation Club) and *BioCatalysis* (H.L. Holland and B. Jones) are available from Chapman & Hall (London, 1996). Practical examples for preparative biotransformations (with checked procedures)

Table 1. The six enzyme classes

IUBMB enzyme class	Reactions catalyzed / special requirements
EC 1 Oxidoreductases	* Reductions/Oxidations at -CH-OH, -C=O, -C=C-, etc. * two-substrate reactions * *co-substrate required*
EC 2 Transferases	* Transfer of functional groups such as C_1, aldehyde, keto, acyl, glycosyl, etc. * two-substrate reactions, *one substrate with activated functional group* → A- (B) + C = A-C + (B)
EC 3 Hydrolases	* Hydrolyses/Condensations of esters, glycosides, ethers, peptides, amides, etc. → A-B + H_2O = AH + BOH * Hydrations → A-B + H_2O = HA-BOH * Transesterifications * two-substrate reactions, one substrate being H_2O * *no co-factors required*
EC 4 Lyases	* Additions/Eliminations * Cleavage of C-C- , C-O-, C-N-bonds * one-substrate reactions →, two-substrate reactions ← * *no co-factors required*
EC 5 Isomerases	* Racemization, cis-trans isomerization, epimerization * one-substrate reactions * *no co-factors required*
EC 6 Ligases	* Formation of C-O-, C-S-, C-N-, C-C-bonds * two-substrate reactions * *co-substrate ATP required*

Enzymes can generally catalyze their reactions in both directions. This is of special interest for the preparative use of hydrolytic enzymes such as lipases/esterases, where the reverse reaction can be performed in organic solvents.

can be found, e.g., in the highly acclaimed handbook edited by Roberts et al. [29], and also in [14].

According to the Warwick Biotransformation Club Database in 1987/88 [60], approximately 65% of all applications reported fell into the classes of esterolytic reactions (ester hydrolysis, synthesis or transesterification) (40%) and dehydrogenase reactions (25%). Next in importance were oxygenase-mediated reactions, peptide and oligosaccharide synthesis, which together comprise 24% of the total. Reports of enzymatic procedures for carbon–carbon bond formation were very few in number (2%). All other reaction types comprised less than 10% of the total.

The reported applications of esterolytic applications used esterases (21%), lipases (63%), and proteases (16%). The most often and widely used biocatalysts were pig liver esterase, pig pancreatic lipase and lipase from *Candida cylindracea*. Among the proteases, α-chymotrypsin, papain, penicillin aminoacylase and subtilisin were used most frequently. In the case of stereospecific reductions of carbonyl compounds the use of whole cells of baker's yeast (*Saccharomyces cerevisiae*) [75–77] accounted for 54% of the total dehydrogenase applications. The dominance of just very few biocatalytic systems, seems to indicate the direction in which most (academic) synthetic chemists looking to use biocatalysts see the best chances to achieve success. The fact that so much has already been achieved in biotransformations with only a very limited fraction of available biocatalysts gives an indication of the huge potential of this field still waiting for investigation and exploitation. The actual scope of bioreactions and the sources for biocatalysts are summarized in Tables 2 and 3.

Table 2. Types of substrate conversions (selection) that have been successfully performed by biocatalysis (compiled from Refs. [4–75]). For more detailed information on the biocatalysts refer to Table 3.

Type of reaction	Type(s) of biocatalyst(s) (selection)
Reductions of	
aromatic carbonyl compounds	BY
α-substituted carbonyl compounds	BY
α-hydroxy ketones	BY, glycerolDH
β- and γ-substituted carbonyl compounds	BY
ketones and α- and β-diketones	BY, An, Cr, Cu, Gc, Ha, Ks, Mi, Mr, Ns, Pd, Pm, Rn, Yl
β-keto acids and esters, γ-keto esters	BY, Ao, Cb, Cg, Cu, Ct, Gc, Hp, Lk, Mj, Pf, Pm, Pv, Rr, Td, glycerolDH
masked carbonyl compounds	BY
activated double bonds (hydrogenation)	BY, An, Gc, Pc
acyclic and cyclic ketones	HLADH, TBADH, other ADHs, HSDHs
reductive amination of keto acids	AADH, GluDH, PheDH
Oxidations/hydroxylations of	
unactivated saturated carbons	Bc, Mi, Mp, Po, Ra, CytP450 enzymes, monooxygenases
(polyunsaturated) fatty acids	SLO
epoxidation and dihydroxylation of alkenes	An, Nc, Po
aromatic compounds (→ unsaturated diols)	Pp (mutant strain), Ce
hydroxylated compounds and aldehydes	Go, Ps
diols (and lactonization)	Bp, Go, Ko, Ps, HLADH
enzyme catalyzed Baeyer-Villiger oxidations	Ac, Ps, CHO
organic sulfides (sulfoxidation)	BY, An, Ceq, Mi, Po, CPO, BSA
Hydrolysis	
esters of acyclic alcohols	An, Bs, CCL, MML, lipases P, P30, and LP-80, PFL, PLE, PPL
hydroxy and enol esters	BY, Bco, Pmi, CCL, lipases A6, P and OF-360, PA, PLE, lipase from Gc
esters of cyclic alcohols	Bs, lipases P and SAM II, CCL, PFL, PPL
esters of cyclic diols and triols	acetylcholinesterase, lipases P and PS, PCL, PLE, PPL, lipase from Mj
esters of acyclic diols	lipases P and MY, LPL, PFL, PLE, PPL
cyanohydrin acetates	Bco, Ct, Pmi, PFL
acyclic monoesters	Sg, CCL, α-chymotrypsin, lipases OF, P, PS, and P30, PFL, PLE, PPL, protease from Ao
acyclic diesters	α-chymotrypsin, lipases OF and MY, PFL, PLE
cyclic mono and diesters	Ag, lipase from Rd, lipase B, PLE, PPL
amides and lactams	Mn, Pp, Pso, Re, PLE
Hydrations of	
nitriles	*Rhodococci*, microbial nitrilases, nitrile hydratases
amides	microbial amido hydrolases
C=C	BY, Ec, Prm, fumarase

Table 2. Continued

Type of reaction	Type(s) of biocatalyst(s) (selection)
Transesterifications of	
(Reversion of hydrolytic reactions)	*esterases, lipases, proteases*
(racemic) alcohols with nonactivated esters	CCL, PPL, lipase from Pc
(racemic) alcohols with activated esters	PPL, pancreatin
irreversible transesterification with vinyl-carboxylates	CCL, PFL, PPL, lipase PS and SAM II, lipase from Pf
Enzymatic ring closure and opening	
lactone formation from hydroxy esters	LPL, PFL
ring opening of anhydrides	lipase P, PFL
Asymmetric glycosylation	microbial β-glucosidases/β-galactosidases
Acylation of amines	CCL, PA
Biocatalytic C–C-formation	BY and other yeasts, AcCoA:AcCoA C-acetyltransferase, tryptophan synthetase, prenyltransferase, oxynitrilases, aldolases, transketolases, sterol cyclase
Additions and eliminations	aminotransferases, decarboxylases
Isomerization of	
sugars	xylose (glucose) isomerase
sugar phosphates	triosephosphate isomerase
Biotransformations of organometallic compounds	
Reductions and oxidations	BY, Ceq, Cb, Cd, Ne, Pf, Rr, HLADH
Hydrolysis and esterification	CCL, CVL, PFL, PLE, lipase P, PA
Multienzymatic approaches	
of special interest for co-factor-dependent systems with recycling enzymes and multistep one-pot enzymatic synthesis	FDH, kinases, glycosyltransferases

Table 3. Frequently used biocatalysts and commercial sources

Whole micro-organisms used as biocatalysts (selection)

Ac: *Acinetobacter calcoaceticus*, Ag: *Arthrobacter globiformis*, An: *Aspergillus niger*, Ao: *Aspergillus oryzae*, As: *Aspergillus* sp., Bc: *Bacillus cereus*, Bco: *Bacillus coagulans*, Bp: *Bacillus polymyxa*, Bs: *Bacillus subtilis*, BY: baker's yeast (*Saccharomyces cerevisiae*), Cb: *Candida boidinii*, Cd: *Corynebacterium dioxydans*, Ce: *Cunninghamella elegans*, Ceq: *Corynebacterium equi*, Cg: *Candida guilliermondii*, Cr: *Candida rugosa* (formerly *Candida cylindracea*), Ct: *Candida tropicalis,* Cu: *Candida utilis,* Ea: *Enterobacter aerogenes,* Ec: *Escherichia coli,* Gc: *Geotrichium candidum,* Go: *Gluconobacter oxidans,* Ha: *Hansenula anomala,* Hp: *Hansenula polymorpha,* Ko: *Klebsiella oxytoca,* Ks: *Kleockera saturnus,* Lc: *Lactobacillus casei,* Lk: *Lactobacillus kefir,* Mi: *Mortierella isabellina,* Mj: *Mucor javanicus,* Mn: *Mycobacterium neoaurum,* Mp: *Mucor plumbeus,* Mr: *Mucor rouxianus,* Nc: *Nocardia corallina,* Ne: *Nocardia erythropolis,* Ns: *Nocardia salmonicor,* Pc: *Phanerochaete chrysosporium,* Pd: *Penicillium digitatum,* Pf: *Pichia farinosa,* Pfr: *Pseudomonas fragi,* Pm: *Pichia membranaefaciens,* Pmi: *Pichia miso,* Po: *Pseudomonas oleovorans,* Pp: *Pseudomonas putida,* Prm: *Proteus mirabilis,* Pc: *Pseudomonas cepacia* (renamed: *Burkholderia cepacia*), Pf: *Penicillium frequentans,* Ps: *Pseudomonas* sp., Pso: *Pseudomonas solanaceum,* Pv: *Proteus vulgaris,* Ra: *Rhizopus arrhizus,* Rd: *Rhizopus delemar,* Re: *Rhodococcus equi,* Rn: *Rhizopus nigricans,* Rr: *Rhodotorula rubra,* Sg: *Streptomyces griseus,* Td: *Torulaspora delbrueckii,* Yl: *Yarrowia lipolytica*

Table 3. Continued

Free enzymes used as biocatalysts (selection)
AADH: α-amino acid dehydrogenase, ADH: alcohol dehydrogenase, BSA: bovine serum albumine, CCL: *Candida cylindracea* lipase (renamed: *Candida rugosa*), CHO: cyclohexanone oxygenase, CPO: chloroperoxidase, CVL: *Chromobacterium viscosum* lipase, CytP450 enzymes (containing cytochrome P450), FDH: formate dehydrogenase, GlycerolDH: glycerol dehydrogenase (from Ea, Gc, *Cellulomonas* sp.), GluDH: L-glutamate dehydrogenase, D-HicDH: D-hydroxyisocaproate dehydrogenase (from Lc), HLADH: horse liver alcohol dehydrogenase, HSDH: Hydroxysteroid dehydrogenase(s), Lipase A6 (from As, Amano), Lipase B (from Pfr), Lipase OF-360 (Meito Sangyo), Lipase P (from Ps, Amano), Lipase P30 (Amano), Lipase LP-80 (Amano), Lipase SAM II (Fluka), LPL: lipoprotein lipase, L-LDH: L-lactate dehydrogenase (from *Lactobacilli*), D-LDH: D-lactate dehydrogenase (from *Lactobacilli*), MML: *Mucor mihei* lipase, Monooxygenases: e.g. camphor MO (from Pp), PA: penicillin G acylase, PheDH: L-phenylalanine dehydrogenase, PCL: *Pseudomonas cepacia* lipase, PFL: *Pseudomonas fluorescens* lipase, PLE: pig liver esterase, PPL: pig pancreatic lipase, SLO: soybean lipoxygenase, TBADH: *Thermoanaerobium brockii* ADH.

Enzyme manufacturers and suppliers (selection)
ABM Chemicals (UK), Aldrich (D), Altus Biologics (USA), Amano Pharmaceutical Co., Ltd. (J), ASA Spezialchemie GmbH (D), Asahi Chemical Industry Co. Ltd. (J), Biocatalysts Ltd. (UK), Biozyme (UK), Boehringer Ingelheim (D), Boehringer Mannheim GmbH/Roche (D), Calbiochem Corporation (USA), Calzyme (USA), Daiwa Kasei K.K. (J), Diversa (USA), Enzymatix (UK), Finnish Sugar (SF), Finnzymes Oy (SF), Fluka Chemie AG (CH), GDS Bulk Enzymes (USA), Genencor International Inc. (SF), Genis Ltd. (Iceland), Genzyme Corporation (UK,USA), Gist-Brocades (now DSM)(NL), Grindsted Products (DK), Hankyu Kyoei Bussan (J), ICN Biomedicals Inc. (USA), Juelich Enzyme Products (D), Kaken Pharmaceutical (J), Kyowa Hakko (J), Meiji Seika Kaisha (J), Meito Sangyo Co. Ltd. (J), Merck (D), Miles Laboratories (D), Nagase & Co. (J), Novo Nordisk (DK), Oriental Yeast Co. Ltd. (J), Osaka Saiken (J), Oxford Glycosystems (UK), Pacific Enzymes (New Zealand), Prozyme (UK), Rapidase (F), Recordati SpA (I), Röhm GmbH (D), Seikagaku Kogyo (J), Serva Feinbiochemica GmbH (D), Sigma Chemical Corporation (USA), Takeda Chemical Industries (J), Thermogen Inc. (USA), Tokyo Kasei (J), Toyobo Co. Ltd. (J), Toyo Jozo (J), United States Pharmaceuticals USA), Unitika Ltd. (J), Wako Chemicals (J), Worthington (USA), Yakult Honsha (J), Yamasa Shoju (J).
A tabular survey of commercially available enzymes and adresses of suppliers can e.g. be found in Ref. [22], pages 963–975.

Sources for microbial and other whole cell biocatalysts – official strain collections (selection)
ATCC: American Type Culture Collection (Manassas VA, USA); CBS: Centraalbureau voor Schimmelcultures (Baarn, NL); DSM: German Collection of Micro-organisms and Cell Cultures (Braunschweig, Germany); IAM: Institute of Applied Microbiology, University of Tokyo (Tokyo, Japan); IFO: Institute for Fermentation, Osaka (Osaka, Japan); JCM: Japan Collection of Micro-organisms, RIKEN (Wako, Saitama, Japan); NCIMB: National Collections of Industrial and Marine Bacteria (Aberdeen, UK); NRRL: Agricultural Research Service Culture Collection, Northern Regional Research Center, USDA (Peoria, Illinois, USA); and others. Comprehensive directories of culture collections, databases on microbes and cell lines, biodiversity related information, etc. can be obtained from the WFCC (World Federation for Culture Collections) World Data Centre for Micro-organisms (WDCM, RIKEN Japan) (http://wdcm.nig.ac.jp) or from the European Culture Collections' Organization (ECCO, Braunschweig, D).

Major driving forces for present and future R&D in biocatalysis are:

- The continued growth in the demand for chiral building blocks and enantiomerically pure compounds, in general. The great majority of applications found in literature are related to enantioselective biocatalysis.
- The rapidly increasing demand for chiral drug building blocks in the pharmaceutical industry, due to major changes in the regulatory procedure.

- Increasing need for regioselective enzyme-catalyzed reactions, e.g., for the selective esterification of di- and multifunctional compounds, the selective hydrolysis of compounds containing multiple ester functions (dicarboxylic acid esters, sugar esters, etc.) and the selective reduction of compounds with multiple carbonyl groups.
- Growing interest in areas of synthesis where biocatalysts offer clear advantages over purely chemical strategies. The most prominent examples are at present: peptide synthesis, protein modification, and synthesis of complex carbohydrates, oligosaccharides and glycoconjugates. The latter area presents many chemical, ecological and economical problems to the classical organic chemistry, due to the multifunctionality calling for sophisticated and laborious protection group strategies. Biotransformation approaches are being used in these fields with increasing frequencies.
- International efforts to promote sustainable development and environmentally friendly technologies [1]

7.4 Screening for Biocatalysts

The first step in the set-up of a viable biocatalytic method or process is the identification of a suitable biocatalyst showing the desired selectivity/specificity, optimal performance and sufficient stability under process conditions. Such biocatalysts may be microbial, plant or animal cells or enzymes derived thereof.

By reviewing literature, it becomes obvious that three main screening strategies have been followed, so far:

1. Testing and evaluating (cheap) commercially available (bulk) biocatalysts or suitable micro-organisms. In most classes of enzymes the selections and the quantities of commercially available preparations are rather limited. Starting from existing knowledge about the enzyme class, the reaction type, and the potential substrate spectrum, such biocatalysts are evaluated for their abilities to chemo-, regio-, stereo, or enantioselectively transform a broad variety of (related) substrates into more or less useful and versatile products, such as chiral building blocks. This approach is predominantly taken in the academic environment and focused on groups of enzymes that are easy to handle in an organic chemistry laboratory. Purified enzymes which do not require co-enzymes, such as hydrolases (lipases, esterases, proteases) are still the preferred biocatalysts for the preparation of optically active compounds. "The discovery of new reactions for well-known commercially available enzymes, and usual transformations of unusual substrates according to well-established reaction processes, eventually non-conventional conditions, still attracts many research groups and is certainly a fast growing area" [68].
2. Searching and screening for suitable novel microbial biocatalysts from natural sources, e.g., by selective enrichment techniques starting from soil or sewage samples, or alternatively and in rare cases, by testing extracts of plant or animal tissues. This more difficult and more demanding and laborious approach is essential when the task is to specifically and selectively synthesize a defined product (drug or drug intermediate) by converting a defined substrate using a particular reaction type and no suitable enzyme or microbe with

the required properties is commercially accessible. In general, microbial enzymes are the better and cheaper choice. The problem is not only to find a biocatalyst that does the requested job, but also rather to find the "best-suited" biocatalyst with respect to operational lifetime, physico-chemical properties, specific catalytic activity, kinetic properties, tolerable substrate/product concentration, etc. In addition, the biocatalyst must be easy and cheap to produce. This screening approach is predominantly taken in the industrial environment and often comprises biocatalysts that require special technical skills to handle. Both approaches 1 and 2 are complementary. Anabolic enzymes are usually substrate-specific, but catabolic enzymes often show broader ranges of specificity.

3. A third approach, related to approach 2, is the search for novel enzymes catalyzing difficult reaction types, where no indications or knowledge on their existence is yet available. Typical examples for this approach from the last decade are the discovery and investigation of enzymes catalyzing Baeyer–Villiger oxidations [53] or Diels–Alder-type reactions [74].

For approaches 2 and 3, often involving biocatalysts from exotic sources (e.g., from microbes that are difficult to grow, or from mammalian tissues), it can be of advantage and recommendable to reclone and overexpress the enzymes in better suitable and easier to handle microbes, such as *Escherichia coli* or *S. cerevisiae*, and/or to improve the biocatalysts by genetic modification (site-directed or random mutagenesis), or by metabolic engineering. A practical introduction to simple cloning methods was written for organic chemists by Schreiber and Verdine [78].

Enzyme screening and bioprocess development were important skills and key factors in the success of the Japanese industrial biotechnology. Based on a working group analysis of the European Federation of Biotechnology completed in 1987, P.S.J. Cheetham came to a very harsh statement on this issue: "in Europe screening is underdeveloped and under-resourced both in absolute terms and by comparison with current activities in Japan, and especially when compared with the resource devoted to genetic manipulation. With hindsight future biological scientists may consider screening to have been a rate-limiting step in the development of biotechnology in Europe in the late 20th century" [79]. This statement is particularly true with respect to screening efforts directed to the discovery of completely new, so far unknown subclasses of enzymes.

Most of the commercialized industrial enzymes are produced from only about 25 species of micro-organisms (fungi, yeasts, bacteria). New enzymes required are first sought from these organisms, as the manufacturers already have the specific knowledge and technology for biomass production and enzyme extraction from these microbes [80]. The disadvantage of this approach is that the isolated enzyme may not show ideal characteristics for the intended application(s). According to data compiled by Foster et al. [71], the following numbers of enzymes are commercially available, most of them however not in bulk quantities: oxidoreductases (90 types of enzymes of different sources); transferases (90), hydrolases (125), lyases (35), isomerases (6), and ligases (5); total number of commercial enzyme types: 351. In order to enlarge this arsenal of available biocatalysts a more systematic screening and exploitation of the existing biodiversity would therefore be highly desirable (see below).

Speculations on the number of species living on earth indicate that the major part (i.e. about 95%) of the microbial world still remains to be discovered and studied (Table 4.)

Table 4. Biodiversity, known species versus estimated total number of species (in thousands)

Class of organisms	Described species	Total number of estimated species		Working figure	% of known species*	Accuracy
		High	Low			
Viruses	4	1000	50	400	1	Very poor
Bacteria	4	3000	50	1000	0.4	Very poor
Fungi	72	2700	200	1500	4.8	Moderate
Algae	40	1000	150	400	10	Very poor
Protozoa	40	200	60	200	20	Very poor
Plants	270	500	300	320	84	Good
Insects	950	100000	2000	8000	12	Moderate
Chordates	45	55	50	50	90	Good

* Described species in relation to working figure.
Source: Global Biodiversity Assessment, UNEP, Cambridge University Press, 1995.

It is generally stated by microbial ecologists that a very large fraction of the microbes, and especially of the bacteria present in soil samples, are "viable but not culturable" (VBNCs), which means that they can neither be isolated nor cultivated, at least not with the existing isolation and cultivation techniques. The VBNC-state can have several reasons: exogenic or circumstantial non-culturability (obligate oligotrophy, obligate symbiosis/syntrophy, presence or absence of oxygen, lack of growth factor or presence of inhibitor); actions of phages, prophages, colicins, etc. present in the cells; irreversible "lethal" damage (the cells are possibly still displaying some viability functions); transient "sublethal" damage or injury with subsequent repair/recovery leading to a culturable cell; survival state as a phase in the life cycle (triggered by environmental stress or starvation). One of the fundamental questions currently discussed among microbiologists is whether and to what extent the VBNC-state is a survival strategy.

Besides the routinely used selective enrichment and isolation techniques, there are various approaches for searching and screening microbial biodiversity: investigation of macro- and microhabitats, isolation of extremophiles, intelligent sampling in megadiversity regions with high density of endemic plant and animal species (to find for example new species of plant-associated microbes), investigation of syntrophic associations of microbes, investigation of "not-culturable" micro-organisms, isolation strategies based on seasonal sampling, isolation and growth media development (exotic C-, N-, P-, S-sources), *in situ* detection and identification of microbes by molecular probing.

A number of smaller enzyme-producing companies focus on thermophilic micro-organisms (and other extremophiles) to identify and produce new types of thermostable enzymes: Unitika, Pacific Enzymes, Genis, Diversa (formerly Recombinant BioCatalysis), and others. One "extremozyme" that has already found commercial application is the heat-stable DNA polymerase from *Thermus aquaticus* (Taq-polymerase) that gave rise to the polymerase chain reaction (PCR). Using PCR, nucleic acids or segments of DNA can by amplified in vitro without having to replace the enzyme after each amplification cycle when the DNA template is denatured by heat. A number of new hyperthermophilic enzymes with temperature optima between 75 and 118 °C have been described in the past few years [81], such as proteases, amylases, glucosidases, galactosidases, xylanases, glucose isomerases, alcohol dehydrogenases, hydrogenases, etc. As thermophilic micro-organisms are in most cases very difficult to grow to higher cell densities, the new enzymes from such sources have for bulk production purposes been recloned and overexpressed either in *E. coli* or *Bacillus* strains.

With modern PCR techniques it has become possible to enrich and investigate the genetic information (DNA) even from the VBNCs present in soil samples. A general approach (followed, e.g., by Diversa) begins with the isolation of nucleic acids directly from soil and environmental samples, from primary enrichment cultures, and from purified micro-organisms across the whole spectrum of biodiversity. In the next stages of the process, gene libraries (genes cloned in suitable expression hosts) are constructed from the purified DNA and subjected to high-throughput automated activity screening systems. The robotic screening of genes by expression from environmental libraries has the potential to increase drastically the speed at which novel and useful biocatalysts can be discovered, even from niches and genera (including VBNCs) that were so far not accessible via the classical microbiological and biochemical techniques. Diversa presently offers more than 340 "novel" robust "CloneZymes" belonging to the most frequently used enzyme EC-subclasses. These enzymes are also marketed for research purposes in the form of CloneZyme Evaluation Libraries (aminotransferases, cellulases/hemicellulases, esterases/lipases, glycosidases, phosphatases). An additional option in this context is the optimization of the novel biocatalysts by "directed evolution" [82].

Screening kits/sets containing samples of the normal commercially available enzymes are also provided by other enzyme suppliers, such as Boehringer Mannheim/Roche (Chirazyme sets for lipases/esterases, aldol reaction kits), Altus Biologics (ChiroScreen Kits TE and EH (based on CLECs, see section 5) for the chiral resolution of alcohols, amines, and esters), Biocatalysts (kits with alcohol dehydrogenases), Enzymatix (lipase biotransformation research kit), and others.

World-wide, comparatively limited screening effort is devoted to discovery of completely new, so far unknown subclasses of enzymes (see also section 6). To make use of the opportunities still hidden in the vast unexplored part of biodiversity, the new approaches using molecular biology techniques can yield exciting and unexpected new types of biocatalysts with interesting synthetic abilities. In order to widen the application spectrum of chemistry and biotechnology, the screening efforts should be enforced in both academic and industrial research.

In contrast to what many people still seem to believe, screening must not necessarily be a frustrating and laborious job for dummies. As is demonstrated by many successful examples from academia and industry, screening can be done with very clever strategies and be very exciting. Some very useful indications in this direction can be found in the excellent proceedings of the International Symposium 'New Frontiers in Screening for Microbial Biocatalysts' held in 1996 [27].

7.5 Technical Aspects of Biocatalysis

To set up a practical bioconversion process screening and optimization work is required on three different levels:

1. Screening for suitable micro-organisms (plant cells, etc.) and/or enzymes with the required catalytic properties and selectivities.
2. Screening for the optimum conditions for culture growth and production of the desired enzyme(s). This can be a very demanding job in cases where the enzyme expression is not constitutive. In this case, the best (and cheapest) conditions for enzyme induction and the right time for harvest must be identified.
3. Screening for the optimum reaction conditions with whole cells, crude or purified enzymes (including technological aspects such as co-factor regeneration, immobilization, type of reactor, etc.).

In order to perform bioreactions, the biocatalyst can be operated under various process conditions:

- Batch application of whole cells in free or immobilized form in aqueous environment.
- Continuous application of whole cells in immobilized form in aqueous environment.
- Application of whole cells in two-liquid phase or multiphase systems or in micelles; biocatalysis at the interface between the aqueous and the organic phase (for poorly soluble substrates).
- Use of acetone-dried or permeabilized cells.
- Batch application of free crude or purified enzymes in aqueous or organic environment.
- Use of polyethyleneglycol (PEG) -modified enzymes in organic solvents.
- Continuous application of immobilized enzymes.
- Use of Enzyme Membrane Reactors (EMR); the method of choice for systems with co-factor recycling or for reactions with expensive enzymes.

For cost reasons, if ever possible, whole-cell biocatalysis is used to perform biotransformations. This is possible when the following criteria are met: (i) no diffusion limitations for substrate(s) and/or product(s); and (ii) no side or follow-up reactions due to the presence of other cellular enzymes. If these conditions are not fulfilled, the use of isolated enzymes – or in special cases of permeabilized cells – is indicated.

Immobilization can be achieved by adsorption or covalent fixation of the biocatalyst to a solid support (e.g. surface-modified polymer or glass beads), by entrapment or by encapsulation in gel beads (e.g., agarose, polyacrylamide, alginate, etc.). Hundreds of immobilization methods have been described and reviewed in the literature [83–89], but only a limited set of methods has found real technical applications. The first large-scale applications of immobilized enzymes were established for the enantioseparation of D- and L-amino acids by Chibata, Tosa and co-workers at Tanabe Seiyaku Company. The Japanese achievements in the large-scale application of immobilized systems are very well documented in an excellent multi-author publication edited by Tanaka, Tosa and Kobayashi [90] (see also section 7). Some enzyme suppliers sell important industrial enzymes not only in the free form (solution or powder) but also immobilized on solid supports.

A more recently launched approach to enzyme stabilization is the Cross-Linked Enzyme Crystals (CLECs) technology developed and marketed by Altus Biologics Inc. (a daughter company of Vertex Pharmaceuticals Inc.). This technology takes well-known soluble enzymes which denature in harsh environments and stacks them in a solid crystal form – adding cross-linking chemical bonds for further strength [91–95]. These "high-performance biocatalysts" retain full activity in harsh conditions of elevated temperature, pH, proteolysis or in the presence of mixed aqueous organic solvents, and eliminate the need for an inert immobilization support. As can be learned from company documents, e.g., the following types of CLECs with potential applications in manufacturing, diagnostics (biosensors) and enzyme drug therapy have been created: asparaginase (from *E. coli*) as therapeutic for childhood leukemia, lipase (*Candida*) as therapeutic for pancreatic insufficiency, lysozyme (hen egg white) as digestive aide, urease (jack bean meal) as diagnostic for measuring urea, luciferase (fire-fly and bacterial) as diagnostic for signal amplification, thermolysin (porcine liver, *Thermoproteolyticus rokko*) for manufacturing aspartame, penicillin acylase (*E. coli*) for manufacturing 6-APA, several lipases (*Candida rugosa*, *Geotrichum candidum*, *Pseudomonas cepacia*), esterases (porcine liver and *Candida rugosa*), elastase (porcine pancreatic) and alcohol dehydrogenases (horse liver and *Thermoanaerobium brockii*) for manufacturing chiral intermediates. CLECs have extremely long storage (at room temperature) and operational lifetimes (at reaction temperature) in comparison with the corresponding free enzymes. This has been demonstrated, e.g., for thermolysin [91, 93] and the synthesis of aspartame and peptides, and for *Candida rugosa* lipase and the resolution of chiral esters [95]. CLECs can be applied in bioreactors at much higher concentration per unit volume than immobilized enzymes. In addition, CLECs-based screening kits are available (see section 4).

An important technical issue is the large-scale applicability of co-factor-dependent enzymatic systems. It is generally accepted that, e.g., NADH-requiring oxidoreductases can easily be used in whole-cell biocatalysis such as baker's yeast-mediated reductions, where the cofactor recycling step is simultaneously performed within the intact cell, driven by the reduction equivalents introduced via the external carbon and energy source (glucose).

Although several useful biochemical methods for the in vitro recycling of co-factors have been established, many chemists still consider the co-factor requirement as a technical and economic obstacle to a more general use of this type of enzymes in their cell-free form. For small-scale synthesis, including some multienzyme systems with co-factor recycling, the simple enclosure of enzymes in dialysis tubes (membrane-enclosed enzyme catalysis, MEEC-technique) has been described [96]. For preparative-scale synthesis, repetitive batch has proved to be an easy to handle technique. The multiple reuse of the enzyme(s) is possible after recovering them from the reacted solution by means of concentration with ultrafiltration equipment. Compared with batch processes, continuous processes often show higher space-time yields. For application of enzymes in continuous processes, and especially for co-factor-dependent systems, the Enzyme Membrane Reactor (EMR) has been developed. The EMR concept (Fig. 1) has been promoted and successfully applied by Kula and Wandrey [97], and others. It was also successfully put into industrial application, e.g., by Degussa, Tanabe Seiyaku, Sepracor, and others (see also section 7).

A classical example for the preparative use of a NADH-dependent enzyme is the enantioselective reduction of 3,4-dihydroxyphenylpyruvic acid to 3,4-dihydroxyphenyllactic acid catalyzed by D-hydroxyisocaproate dehydrogenase [98]. The essential co-factor PEG-NADH is in situ regenerated from PEG-NAD^+ by a second enzyme, formate dehydrogenase (FDH), using formate as the hydrogen donor. By coupling to water-soluble polyethyleneglycol (PEG) with a molar mass of 20000, the co-factor can be retained together with the enzymes by an ultrafiltration membrane, and the whole process can be performed continuously in an EMR.

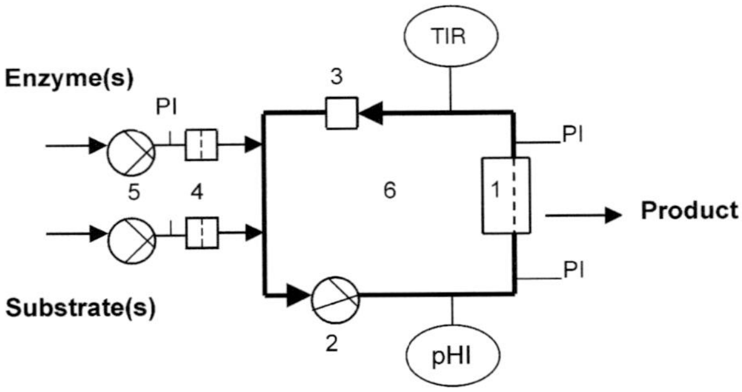

Figure 1. Enzyme Membrane Reactor (EMR) for bulk production - schematic representation. 1, ultrafiltration module; 2, peristaltic pump; 3, septum; 4, sterile filter; 5, metering pumps; 6, reactor loop (thermostated); pHI, pH indication; TIR, temperature indication and regulation; PI, pressure indication.

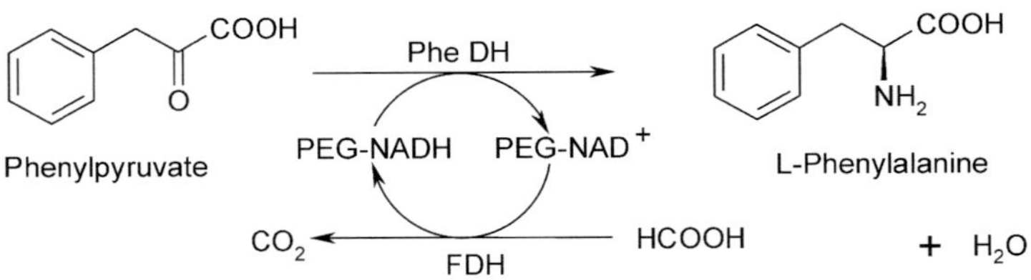

Figure. 2. Enzymatic synthesis of L-phenylalanine by reductive amination starting from phenylpyruvate (using EMR technology).

As much as 600000 cycles have been reported for the FDH/formate-driven PEG-NADH-recycling system in the case of L-phenylalanine production (Fig. 2) by reductive amination with phenylalanine dehydrogenase [99].

At the time when the EMR concept was introduced, most of the co-factors were still very expensive compounds and therefore contributed to a large fraction to the overall process costs. Therefore, for application in continuous processes, the molecular weight enlargement by covalent coupling to PEG (in order to allow very high recycling numbers) was a *conditio sine qua non* for the economic viability of such processes. In the meantime, however, thanks to the rapid progress in fermentation technologies, at least NAD(H) and ATP have become affordable compounds, at least when purchased in bulk (i.e., multi-kilogram) quantities. Under these conditions, the co-factor cost drops to 1 US$ per gram (or even less).

A more recent EMR-process for the enantioselective reduction of 2-oxo-4-phenylbutyric acid (OPBA) to (*R*)-2-hydroxy-4-phenylbutyric acid (HPBA, e.e. > 99.9%) with NADH-dependent D-lactate dehydrogenase (D-LDH) from *Staphylococcus epidermidis* and FDH/formate for co-factor recycling uses free NADH instead of PEG-NADH [100, 101] (Fig. 3). Within the selected residence time of 4.6 h, a cycle number for the co-factor of 1000 can easily be achieved. Instead of NADH, the less expensive NAD^+ is used. A second reason for the application of the native co-enzyme is the fact that the activity of D-LDH is reduced to one tenth when PEG-enlarged co-enzyme is used instead of the native co-enzyme. HPBA produced by this method with a space-time yield of 165 $g\,l^{-1}\,d^{-1}$ is of considerably higher enantiomeric purity and the process shows competitive economy in comparison with chemical

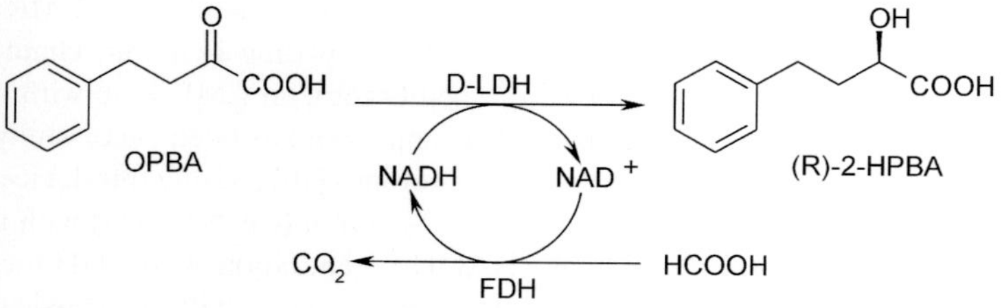

Figure 3. Enantioselective enzymatic reduction of 2-oxo-4-phenylbutyric acid to *(R)*-2-hydroxy-4-phenylbutyric acid (using EMR technology).

enantioselective hydrogenation approaches [101]. On the other hand, performance, product quality and economics of this EMR process are comparable with an alternative biocatalytic approach using whole cells of reductase-containing *Proteus vulgaris* or *Proteus mirabilis* [101]. HPBA is a versatile building block, e.g., for the angiotensin-converting enzyme (ACE) inhibitor benazepril/cibacen and related pharmaceuticals.

An attractive variation of the EMR concept is the "Charged Ultrafiltration Membrane Enzyme Reactor" promoted by the group of Kulbe and Chmiel [102, 103]. This concept applies negatively charged ultrafiltration membranes to retain efficiently the native co-factor in the reactor when NAD(P)(H)-dependent enzymes coupled to co-factor recycling systems are used. The successful application of this approach was described, e.g., for the simultaneous conversion of glucose/fructose mixtures into gluconic acid and mannitol via the enzymes glucose dehydrogenase and mannitol dehydrogenase (both NAD(H)-dependent enzymes; co-factor turnover number 150000), or for the enantioselective conversion of sulcatone to the bark beetle pheromone (*S*)-sulcatol (2-methyl-2-heptene-6-ol) using the alcohol dehydrogenase from *Thermoanaerobium brockii*. In the latter case, the co-factor NADPH was regenerated by the same enzyme by simultaneous conversion of isopropanol (hydrogen donor) to acetone (co-factor turnover number 4400).

In a 1996 report on reaction engineering, the Wandrey group reached similar conclusions with respect to the use of native co-enzymes instead of PEG-enlarged ones in EMR-systems [104]: "With native co-factors quite high cycle numbers of up to 4500 can be reached with charged ultrafiltration membranes." And: "at least in the case of NAD^+ and ATP the co-factor costs are no longer economically limiting. Also for more expensive co-factors (e.g. $NADP^+$), enzyme reaction engineering will help make appropriate processes economically feasible." An additional "upgrade" of this technology to further improve the co-factor retention is the use of nanofiltration instead of ultrafiltration membranes [105].

Until quite recently, NADP(H)-dependent enzymes were economically not very attractive, as no suitable co-factor regeneration system was available. To overcome this limitation, the NAD(H)-dependent FDH from the methylotrophic bacterium *Pseudomonas* sp. 101 was successfully engineered by multipoint site-directed mutagenesis. The finally generated NADP(H)-specific mutant enzyme shows about 60% of the activity of the wild-type FHD with NAD(H). A first practical application of this mutant enzyme for co-factor regeneration was described for the enantioselective reduction of acetophenone with the NADPH-dependent alcohol dehydrogenase from *Lactobacillus* sp. [106]. This new option might open up the way to future technical applications of interesting NADP(H)-dependent enzymes.

As a new option, for the bioconversion of poorly soluble substrates the classical EMR-concept can be extended to an "Emulsion Membrane Reactor", comprising a separate chamber for emulsification (with a hydrophilic ultrafiltration membrane), an EMR-loop with a normal ultrafiltration module, and a circulation pump. This approach has been successfully demonstrated for the enzymatic reduction of poorly soluble ketones [107]. Using this device, e.g., for the enantioselective reduction of 2-octanone to (*S*)-2-octanol (e.e. >99.5%) with a carbonyl reductase from *Candida parapsilosis* under NADH-regeneration with FDH/formate, the total turnover number was increased by a factor 9 as compared with the classical EMR.

Typical performance data of selected EMR-processes (under optimal conditions) are listed in Table 5.

Table 5. Applications of Enzyme Membrane Reactors (EMR) of commercial interest [108, 109]

Enzyme	Reaction	Product	Space-time yield ($g\,l^{-1}\,d^{-1}$)	Enzyme consumption (U per kg)	Co-factor-cycles
Aminoacylase	Hydrolysis of DL-methionine	L-Methionine	1200	1000	
Fumarase	Hydration of fumarate	L-Malic acid	1000		
(R)-Oxynitrilase	C–C-bond formation with CN and benzaldehyde	Mandelic nitrile	2400	7000	
Aspartase	Amination of fumarate	L-Aspartic acid	400	5000	
α-Chymotrypsin	C–N-bond formation MalTyr-OEt and Arg-NH_2	Kyotorphin	25600	175 mg/kg	
L-HicDH/FDH*	Reduction of α-ketoisocaproate	L-Hydroxy isocaproate	411	730 L-HicDH 2280 FDH	69400
L-LeuDH/FDH*	Reductive amination of α-ketoisocaproate	L-Leucine	250	260 L-LeuDH 300 FDH	75000
L-LeuDH/FDH*	Reductive amination of trimethylpyruvate	L-tert. Leucine	640	930 L-LeuDH 2280 FDH	125000
L-PheDH/FDH*	Reductive amination of phenylpyruvate	L-Phenylalanine	456	1500 L-PheDH 150 FDH	600000
Acylase/ L-PheDH/FDH*	Reductive amination and deacetylation of acetamido cinnamate	L-Phenylalanine	277	1170 Acylase 1770 L-PheDH 400 FDH	200000

* With PEG-enlarged co-factor NAD(H).
FDH, Formate dehydrogenase; L-HicDH, L-Hydroxyisocaproate dehydrogenase; LeuDH, Leucine dehydrogenase; PheDH, Phenylalanine dehydrogenase.

7.6 Biocatalysis in Research and Development: On the Road to New Bioprocesses

In the past, a small number of academic groups have greatly contributed to the discovery of synthetically very useful novel microbial biocatalysts, and also to the set-up of quite a number of industrial processes. The most prominent Japanese example is the Laboratory of

Table 6. Bioprocesses established by the Laboratory of Applied Microbiology of Kyoto University in collaboration with industry. (Compiled from Refs. [32, 38, 52, 73, 110–116] and from institute documentations)

Product	Enzyme (source)	Yield $g\,l^{-1}$	(mol%)
D-*p*-Hydroxyphenylglycine	Dihydropyrimidinase/Hydantoinase (*Bacillus* sp.)	5	(74)
D-Phenylglycine	Dihydropyrimidinase/Hydantoinase (*Bacillus* sp.)	6	(91)
L-Tyrosine	β-Tyrosinase (*Erwinia herbicola*)	61	
L-Dopa	β-Tyrosinase (*Erwinia herbicola*)	>100	
L-Tryptophan	Tryptophanase (*Proteus rettgeri*)	100	(95)
L-Cysteine	Cysteine desulfhydrase (*Enterobacter cloacae*)	50	(86)
L-Cysteine	Cysteine synthase (*Bacillus sphaericus*)	70	(82)
D-Cysteine	β-Chloro-D-alanine chloride-lyase (*Ps. putida*)	22	(88)
L-Cystathionine	Cystathionine-γ-synthase (*Bacillus sphaericus*)	42	(92)
L-Serine	Serine hydroxymethyltransferase (*Hyphomicrobium sp.*)	52	
Ethyl *(R)*-4-chloro-3-hydroxybutanoate	Aldehyde reductase (*from Sporobolomyces salmonicolor,* overexpressed in *E. Coli)*	300	(100)
Acrylamide	Nitrile hydratase (*Pseudomonas chloroaphis*)	400	(100)
Acrylamide	Nitrile hydratase (*Rhodococcus rhodochrous*)	650	(100)
Methacrylamide	Nitrile hydratase (*Pseudomonas chloroaphis*)	200	
Crotonamide	Nitrile hydratase (*Pseudomonas chloroaphis*)	200	
Nicotinamide	Nitrile hydratase (*Rhodococcus rhodochrous*)	1465	(100)
Acrylic acid	Nitrilase (*Rhodococcus rhodochrous*)	380	(100)
p-Aminobenzoic acid	Nitrilase (*Rhodococcus rhodochrous*)	110	(100)
Pyrazinoic acid	Nitrilase (*Rhodococcus rhodochrous*)	434	(100)
Nicotinic acid	Nitrilase (*Rhodococcus rhodochrous*)	172	(100)
6-Hydroxynicotinic acid	Hydroxylase (*Comamonas acidovorans*)	120	(96)
6-Hydroxypicolinic acid	Hydroxylase (*Alcaligenes faecalis*)	116	(97)
D-Malic acid	Maleate hydratase (*Arthrobacter* sp.)	87	(72)
Pyrogallol	Gallic acid decarboxylase (*Citrobacter* sp.)	23	(100)
Theobromine	Oxygenase (*Pseudomonas putida*)	20	(92)
D-Pantoyl lactone	Carbonyl reductase (*Candida parapsilosis*)	350	(100)
D-Pantoic acid	Aldonolactonase (*Fusarium oxysporum*)	700	(95)
Co-enzyme A	Multistep enzyme system (*Brevibacterium ammoniagenes*)	115	(95)
Adenosylmethionine	AdoMet synthetase (*Saccharomyces sake*)	12	(45)
Adenosylhomocysteine	AdoHcy hydrolase (*Alcaligenes faecalis*)	74	(97)
FAD	FAD pyrophosphorylase (*Arthrobacter globiformis*)	18	(28)
Pyridoxyl-5′-phosphate	PMP oxidase (*Pseudomonas fluorescens*)	0.15	(98)
NADH	Formate dehydrogenase (*Arthrobacter* sp.)	30	(90)
NADPH	Glucose dehydrogenase (*Gluconobacter suboxydans*)	73	(100)
Dihomo-γ-linolenic acid	Multistep conversion (*Mortierella alpina*)	4.1	
Arachidonic acid	Multistep conversion (*Mortierella alpina*)	13	
Eicosapentaenoic acid	Multistep conversion (*Mortierella alpina*)	1.8	

Applied Microbiology at Kyoto University with Hideaki Yamada (now retired) and his coworkers and colleagues Kumagai, Shimizu, and others [32, 38, 52, 73, 110–116]. In the past 40 years, this laboratory has discovered, studied and promoted in collaboration with industrial partners an impressive number of synthetically useful microbial enzymes. More than 30 industrialized processes have emerged or been enabled from the work of this laboratory (see Table 6).

The reported product end concentrations in these processes clearly disprove the very common and often-heard prejudice that biological reactions can be performed at low concentrations only. The reported value of 1465 $g\,l^{-1}$ reaction volume in the case of the nicotinamide fed batch process seems to represent the highest product end concentration of a bioreaction ever achieved. The nicotinamide process (see Fig. 13) is commercialized under license by Lonza (new production plant in Gangzhou, China). Some of the processes listed in Table 6 are discussed in more detail in section 7.

Table 7. Novel microbial biocatalyts found in target-oriented screening programs by Japanese research groups, selection. (From Refs. [73, 112] and references therein)

Product	Enzyme	Enzyme source
D-Amino acids [114–116]	D-Hydantoinase and D-Decarbamoylase	*Pseudomonas putida*, *Bacillus* sp. *Blastobacter* sp., *Agrobacterium* sp.
L-Amino acids [115, 116]	L-Hydantoinase and L-Decarbamoylase	*Pseudomonas putida* *Alcaligenes xylosoxidans*
trans-4-Hydroxy-L-proline	Dioxygenase	*Dactylosporangium sp.*
cis-3-Hydroxy-L-proline	Dioxygenase	*Streptomyces* sp.
Ethyl *(S)*-4-chloro-3-hydroxybutanoate	Carbonyl reductase	*Candida magnoliae*
2*S*,3*R*-3-(4-Methoxyphenyl)-glycidic acid methyl ester	Lipase	*Serratia marcescens*
(S)-1-Methyl arylalkylamines	*(S)*-Arylalkyl acylamidases	*Nocardia erythropolis, Cellulomonas fimi*
(R)-1-Methyl arylalkylamines	*(R)*-Arylalkyl acylamidases	*Pseudomonas putida*
chiral Epoxides	Alkene monooxygenase	*Nocardia corallina*
chiral 2,3-Dichloro-1-propanol	Halohydrin hydrogen-halide lyase	*Alcaligenes* sp., *Pseudomonas* sp.
chiral 3-Chloro-1,2-propanediol	Halohydrin hydrogen-halide lyase	*Alcaligenes* sp., *Pseudomonas* sp.
(S)-1,2-Pentanediol	Alcohol dehydrogenase and Reductase	*Candida parapsilosis*
Trehalose	Maltooligosyl trehalose synthase	*Arthrobacter* sp.
Adenine arabinoside	Nucleoside phosphorylase	*Enterobacter aerogenes*
Ribavirine	Nucleoside phosphorylase	*Erwinia carotovora*
5-Methyluridine	Nucleoside phosphorylase	*Erwinia carotovora*
Dideoxyadenine	Nucleoside phosphorylase	*Escherichia coli*
5,8,11-*cis*-Eicosatrienoic acid	Multistep conversion	*Mortierella alpina*

A considerable number of other novel microbial enzymes of high industrial interest were identified and described by the Kyoto University Laboratory and other Japanese groups in the years since 1985 [112] (see Table 7)

Tables 6 and 7 clearly demonstrate that very interesting novel microbial biocatalysts with high industrial potential can be found by intelligent screening using selective enrichment and

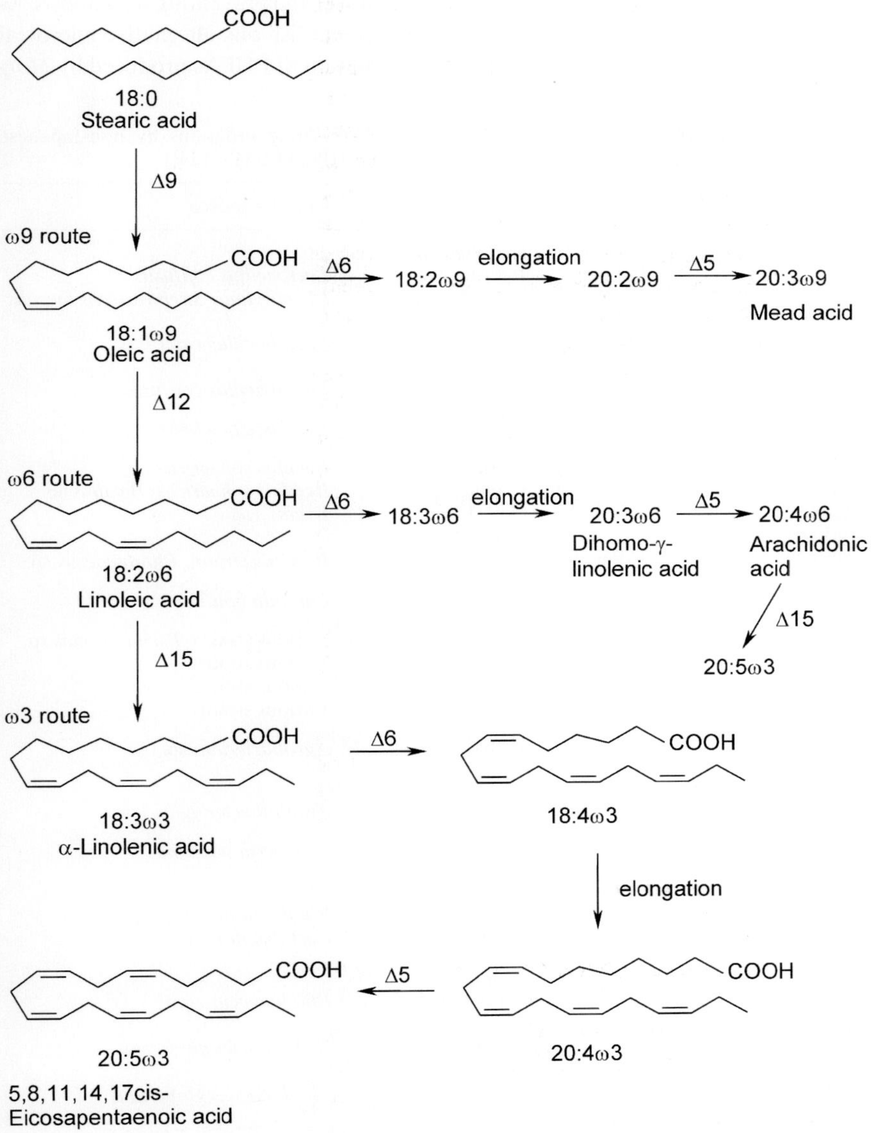

Figure 4. Microbial production of PUFAs with mutant strains of *Mortierella alpina*, starting from 18:1ω9, 18:2ω6 and 18:3ω3 (ω9, ω6 and ω3 routes, respectively).

isolation techniques. This is a very important strength of the Japanese Biotechnology. Some of the above processes in Table 7 are already commercialized on large scale (e.g., 2*S*,3*R*-3-(4-methoxyphenyl)-glycidic acid methyl ester, or D- and L-amino acids via the hydantoinase/carbamoylase route; see section 7). Hydroxyprolines are interesting pharmaceutical building blocks (for antiphlogistics, carbapenem antibiotics and ACE inhibitors).

C_{20} polyunsaturated fatty acids (PUFAs; see Tables 6 and 7 and Fig. 4) exhibit unique biological activities, such as lowering the plasma cholesterol level and preventing thrombosis. As food sources rich in PUFAs are limited to a few seed oils and fish oils, alternative microbial sources of PUFAs are of high interest. The species distribution of PUFAs produced by *Mor-*

Table 8. Novel microbial enzymes found in target-oriented screening programs by non-Japanese research groups in academia and industry, selection. (From Refs. [108, 112, 117–124])

Product	Enzyme	Enzyme source
Chiral 2-Hydroxy acids [108]	2-Hydroxy acid dehydrogenases	
	D-Lactate dehydrogenase	*Lactoacillus confusus*
	D-Mandelate dehydrogenase	*Lb. curvatus, Streptococcus faecalis*
	D-Hydroxyisocaproate dehydrogenase*	*Lactobacillus casei*
	L-2-Hydroxyisocaproate dehydrogenase*	*Lactobacillus confusus*
Chiral Hydroxyketones [108]	Acetoin dehydrogenase	*Lactobacillus kefir*
L-Alanine [108]	L-Ala dehydrogenase	*Bacillus sphaericus*
L-Amino acids [108] (L-Leu, L-Met, L-Ala, L-Phe L-tertiary Leu)	L-Leu dehydrogenase	*Bacillus sphaericus*, *Bacillus* sp. *E. coli, etc.*
L-Phenylalanine [108]	L-Phe dehydrogenase	*Brevibacterium, Rhodococcus* sp.
NADH from NAD [108]	Formate dehydrogenase	*Candida boidinii*
D-Amino acids [118–120]	D-Hydantoinases and	Various strains, *Pseudomonas* sp.
	D-Decarbamoylases	Various strains
L-Amino acids [118–120]	L-Hydantoinases and	Various strains
	L-Decarbamoylases	Various strains
5-Hydroxypyrazine-2-carboxylic acid [121]	Nitrilase and dehydrogenase	*Agrobacterium* sp.
Carbacephem [122]	*o*-Phthalyl amidase	*Xanthobacter agilis*
(R)-2-(4-Hydroxyphenoxy)-propionic acid [123]	Hydroxylase	*Beauveria bassiana*
(S)-*p*-Chlorophenylethanol	Alcohol dehydrogenase	*Rhodococcus erythropolis*
(R)-*p*-Chlorophenylethanol [124]	Alcohol dehydrogenase	*Lactobacillus kefir*
(S)-1-Phenylethylamine and analogs/homologs [125]	*(S)*-N-Acetyl-1-phenylethyl-amine amidohydrolase	*Rhodococcus equi*
(R)-1-Phenylethylamine and analogs/homologs [126]	*(R)*-N-Acetyl-1-phenylethyl-amine amidohydrolase	*Arthrobacter aurescens*
(S)-2-amino-1-phenyl-4-pentene and analogs/homo logs [127]	*(S)*-N-Acetyl-2-amino-1-phenyl-4-pentene amidohydrolase	*Rhodococcus globerulus*

* Alternative name: L- and D-2-hydroxy-4-methyl-pentanoate dehydrogenase)

tierella alpina can be modulated by mutagenesis strategies, by knocking out specific desaturases [112].

Ethyl (*S*) and (*R*)-4-chloro-3-hydroxybutanoates (see Tables 6 and 7) are promising chiral building blocks for L-carnitine (*R*-form), 3-hydroxy-3-methylglutaryl (HMG)-CoA-reductase inhibitors and 1,4-dihydropyridine-type β-blockers (*S*-form).

Interesting new biocatalysts were also found and successfully developed by research groups in the Western hemisphere (see Table 8). In the academic sector, e.g., German groups achieved good screening results: Kula and Hummel (Jülich, Düsseldorf) described several new amino acid and hydroxy acid dehydrogenases with high industrial potential [108, 117]. Most of these co-factor-dependent dehydrogenases have been evaluated and used in EMR-processes. Wagner and Syldatk (Braunschweig, Stuttgart) made very valuable contributions in the field of hydantoinases and carbamoylases [118, 119]. In the industrial sector, most of the pharmaceutical and specialty chemicals companies have at least one specialized and interdisciplinary biotransformation group at their disposal. All of these industrial laboratories use commercial biocatalysts to do their jobs, some of them, however, are also heavily engaged in the screening for novel types of biocatalyts.

For the enantioselective preparations of chiral synthons, the most interesting oxidations are the hydroxylations of unactivated saturated carbons or carbon–carbon double bonds in alkene and arene systems, together with the oxidative transformations of various chemical functions. Of special interest is the enzymatic generation of enantiopure epoxides. This can be achieved by epoxidation of double bonds with cytochrome P450 mono-oxygenases, ω-hydroxylases, or biotransformation with whole micro-organisms. Alternative approaches include the microbial reduction of α-haloketones, or the use of haloperoxidases and halohydrine epoxidases [128]. The enantioselective hydrolysis of several types of epoxides can be achieved with epoxide hydrolases (a relatively new class of enzymes). These enzymes give access to enantiopure epoxides and chiral diols by enantioselective hydrolysis of racemic epoxides or by stereoselective hydrolysis of *meso*-epoxides [128, 129].

Interesting new enzymes can of course also be isolated from non-microbial sources. A recent example of a useful novel enzyme or plant origin is the peptide amidase from orange flavedo (Fig. 5) discovered by Steinke and Kula [130, 131]. This enzyme, which has an extremely wide substrate range, is useful for C-terminal enzymatic deprotection in peptide synthesis under very mild conditions. Substrates of this peptide amidase are: protected and unprotected peptide amides, N-protected amino acid amides. The enzyme is stereoselective with regard to the C-terminal position; only L-amino acid amides are accepted as substrates, with the exception of proline. Other interesting enzymes of plant origin are the oxynitrilases (see below).

It is easy to predict that enzymatic protection and deprotection strategies will become increasingly important in future organic synthesis strategies. Novel approaches for drug research use structural information on the complex biological drug targets, and therefore call for compounds with increased complexity and higher functionality; they thus create additional synthetic challenges. The existing enzymatic options for protection and deprotection – mainly using hydrolytic enzymes (lipases, esterases, acylases, proteases) – have recently been reviewed by Waldmann and co-workers [132, 133] (see also Chapter B8 in [22]) for hydroxyl, amino, mercapto, and carboxyl groups in the chemistry of peptides, carbohydrates, alkaloids,

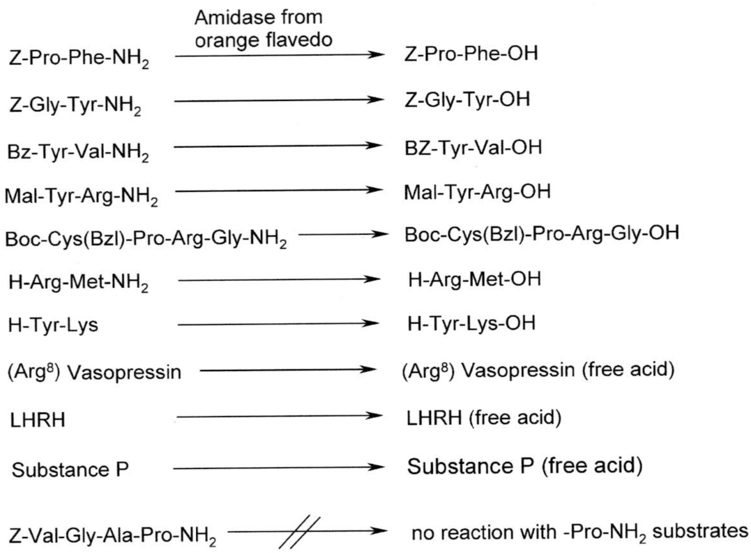

Figure 5. Peptide amidase for the C-terminal peptide deprotection under mild conditions (selected examples).

and steroids. A lipase-labile linker, the 4-acyloxy-carboxybenzyl group, has been successfully developed and used in solid-phase synthesis for combinatorial chemistry [134].

The stereocontrolled enzymatic formation of C–C bonds is a very attractive option, due to the general interest for this fundamental reaction in organic chemistry. Compared with other types of enzymatic reactions, the number of reports on asymmetric C–C bond formation is still small, and the number of practical examples in preparative scale is very limited. So far, baker's yeast-mediated acyloin condensation that was discovered some 80 years ago has found many practical synthetic applications. However, this seems mainly due to the easy availability of this cheap biocatalyst and also to the operational simplicity of its application. Many enzymes can catalyze condensation reactions, which finally lead to the formation of a C–C bond. A highly interesting option is the oxynitrilase-catalyzed enantioselective cyanohydrin formation by addition of cyanide to aldehydes. The long-known reaction catalyzed by the almond oxynitrilase (EC 4.1.2.10) has been reinvestigated [135–140] and applied to various preparative purposes directed to the synthesis of chiral α-hydroxy acids and β-amino alcohols. (*R*)-Cyanohydrins are formed by addition of HCN to aldehydes or ketones under the catalysis of almond (*R*)-oxynitrilase in organic solvents. The use of (*S*)-oxynitrilase from *Sorghum bicolor* allows the formation of the corresponding (*S*)-cyanohydrins [138, 141].

Simple carbohydrates (cellulose, starch, glucose, fructose, inulin, chitin, etc.) extracted from biomass, oligo- and polysaccharides of microbial or algal origin and derivatives thereof play a very important role in many sectors ranging from textiles and specialty chemicals to cosmetics, and food and feed industries. Large-scale carbohydrate-based processes in these sectors are well established and based on relatively simple chemical or biotechnological transformations (see also section 7). However, when it comes to carbohydrate-based pharmaceuticals, increased structural complexity and more sophisticated synthesis strategies are required. In principle, such strategies are available from high-quality academic research; however, their practical realization on an industrial scale is still hampered by many economical and technological limitations (see below).

One of the most classical reactions for the formation of C–C bonds is the aldolic condensation. In nature, such stereocontrolled reactions are catalyzed by enzymes of the class of lyases (EC 4). The majority of these enzymes can be found in the biosynthesis of carbohydrates, and are used for the synthesis of natural and unnatural carbohydrates. Aldolases (and transketolases) have been intensively investigated and their scope of applications has been evaluated and reviewed by the groups of Whitesides and Wong [16, 21, 142–145] (see also Chapter B4 in [22]). Aldolases can be divided into three main types:

- Type A enzymes use dihydroxyacetone phosphate (DAHP) as the nucleophilic substrate and, by reacting with an aldehyde, form a ketose 1-phosphate sugar.
- Type B aldolases use pyruvic acid as the nucleophile to form a 3-deoxy-2-keto acid product.
- Type C aldolases (e.g. deoxyribose 5-phosphate aldolase (DERA)) use acetaldehyde as the nucleophilic substrate to form a 2-deoxyaldose.

Aldolases accept a wide range of aldehydes in place of their natural substrates and permit the synthesis of carbohydrates such as azasugars, deoxy sugars, deoxythio sugars, fluorosugars, and C_8 or C_9 sugars. In the case of D-fructose-1,6-diphosphate aldolase (FDP aldolase, Type A), more than 75 aldehydes have been identified as substrates [143].

N-Acetylneuraminic acid aldolase (a type B aldolase) catalyzes the cleavage of N-acetylneuraminic acid (Neu5Ac) to N-acetylmannosamine (ManNAc) and pyruvate (Pyr). The reverse reaction can be applied to synthesize Neu5Ac (Fig. 6), which plays an important physiological role as the terminal sugar residue in mammalian glycoproteins and glycolipids [146]. This reverse reaction was known for a long time and some research groups used it for the preparation of Neu5Ac on a small scale. The careful kinetic analysis of the system and the use of EMR allowed the scaling-up to multikilogram scale (with a space-time yield of 650 g l^{-1} d^{-1}) in a collaboration between academic research and industry [147, 148]. The reaction can also be performed with immobilized Neu5Ac aldolase [149]. A number of companies have patented or are developing Neu5Ac-based drugs, e.g., sialidase inhibitors or antiallergic agents. Alternative accesses to Neu5Ac are the isolation from natural sources (Chinese

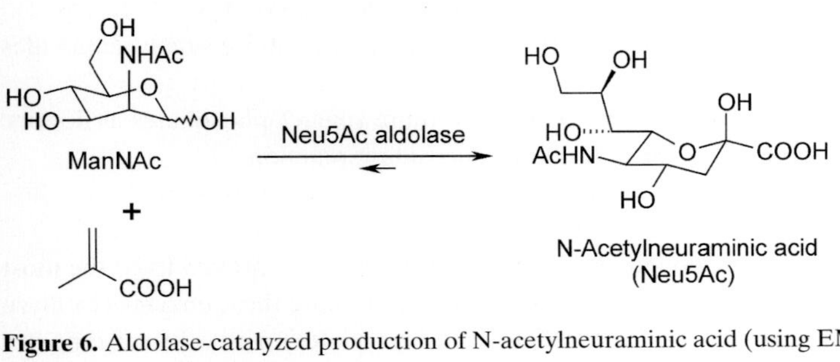

Figure 6. Aldolase-catalyzed production of N-acetylneuraminic acid (using EMR technology).

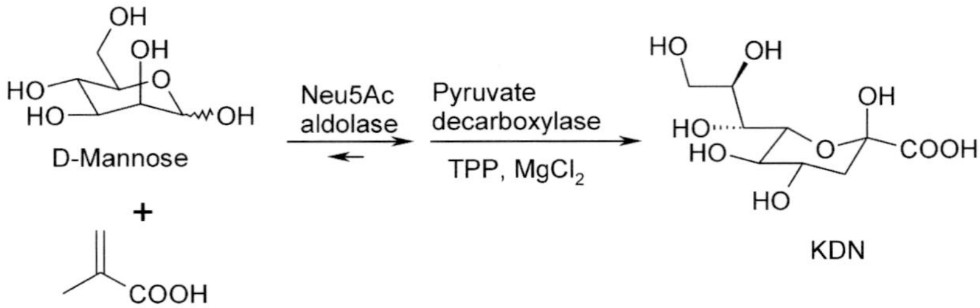

Figure 7. Aldolase- and pyruvate decarboxylase-catalyzed synthesis of KDN.

birds nests, cow's milk, egg yolk) or the enzymatic degradation of the microbial Neu5Ac polymer colominic acid.

As Neu5Ac aldolase accepts a range of substrates (more than 60 are known) in place of ManNAc, this enzyme can also used to synthesize Neu5Ac derivatives [149]. For example, if ManNAc is replaced by D-mannose, Neu5Ac aldolase can be used for the preparative-scale synthesis of KDN (3-deoxy-β-D-*glycero*-D-*galacto*-2-nonulosonic acid; Fig. 7) [150].

Recent research has revealed that complex carbohydrates and glycoconjugates play important roles in biological processes such as fertilization, embryogenesis, neuronal development, and cellular proliferation. They are also important as blood group determinants and for the secretion and circulation half-life of glycoproteins. Because of these important roles of carbohydrates in health and disease (including cancer, rheumatoid arthritis and other autoimmune diseases, peptic ulcers, inflammatory conditions, and infectious diseases), new fields – glycobiology/glycotechnology and the development of carbohydrate-based drugs – are becoming increasingly important and a major focus of current biological research [144, 151]. A number of companies are involved or specialized in carbohydrate research (Alpha-Beta, Bayer AG, Carrington Labs, Cytel, Glaxo Wellcome, GlycoMed, Greenwich Pharmaceuticals, Johnson & Johnson, MECT, Neose Pharmaceuticals, Novartis, Oxford GlycoSystems, Sanofi/Organon, Searle, Snowbrand, and others). The synthesis of complex carbohydrates is a very challenging task. A very elegant approach to design straightforward synthesis routes with a drastically reduced need for laborious, costly and time-consuming protection and deprotection of functional groups, is the introduction of biotechnology into carbohydrate chemistry, taking advantage of the high selectivity, specificity, and catalytic power of enzymes. A number of excellent reviews by Wong, Whitesides and others describe the present state of enzyme application in carbohydrate chemistry [152–157].

Several strategies are available for the enzymatic oligosaccharide synthesis [144, 155]:

- The use of the glycosyltransferases of the Leloir pathway, which require sugar nucleotides as donors.
- The use of the non-Leloir pathway enzymes, which require sugar-1-phosphates as donors.
- The use of the reverse glycosidase or of the transglucosidase reactions.
- The use of phosphorylase (e.g. sucrose phosphorylase).

Of these, the sugar nucleotide-dependent glycosyltransferases have proved to be the most suitable for the synthesis of complex oligosaccharides. This is because these enzymes catalyze stereo- and regiospecific reactions (high specificity and high yields) with a broad range of complex acceptor structures. Therefore, due to space constraints only this approach is shortly outlined in this review.

In nature, at least 100 different glycosyltransferases are responsible for the biosynthesis of the various complex carbohydrates and glycoconjugates. Of interest for the in vitro synthesis of glycoconjugates and glycans are the following transferase families: fucosyl-, galactosyl-, mannosyl-, N-acetylgalactosaminyl-, N-acetylglucosaminyl-, and sialyltransferases. In each of these families there are several enzymes with different linkage and acceptor specificities (for relevant examples, see Table 9). Natural sources for glycosyltransferases are, e.g., human milk and mammalian tissues.

Each glycosyltransferase of the Leloir pathway uses an individual sugar nucleotide as donor. Eight types of sugar nucleotides are essential for glycotechnology directed to mammalian systems: UDP-glucose (UDP-Glc), UDP-galactose (UDP-Gal), UDP-N-acetylglucosamine (UDP-GlcNac), UDP-N-acetylgalactosamine (UDP-GalNac), UDP-glucuronic acid (UDP-GlcU), GDP-mannose (GDP-Man), GDP-fucose (GDP-Fuc), and CMP-N-acetylneuraminic acid (CMP-Neu5Ac). The glycosyltranferases, as well as the sugar nucleotides, are still very expensive and (if at all) commercially available only in small quantities. GDP-Fuc is at present by far the most expensive sugar nucleotide, followed by CMP-Neu5Ac and UDP-GalNAc, whereas UDP-Glc, UDP-GlcNAc, and UDP-GlcU are among the "cheaper" ones. The high costs represent a severe limitation to the industrial use of the methodology. With the so far known chemical and enzymatic methods, however, skilled scientists can synthesize all of these sugar nucleotides in 1 g to 100 g scale. In view of the high costs, it is indicated to use the precious co-factors for glycosyltransferase reactions in the presence of a suitable in situ recycling system. Many such approaches can be found in literature and in the cited reviews (for an example, see below).

Table 9. Important examples of glycosyltransferases [157]

Glycosyltransferase (source)	Donor sugar nucleotide	Product formed
N-acetylgalactosaminyltransferase		
(Fucα1-2)Gal α1-3-N-Acetyl-galactosaminyltransferase (from human milk)	UDP-GalNac	GalNAcα1-3(Fucα1-2)Gal-R
Fucosyltransferase		
β-Galactoside α1-2-fucosyltranferase (from porcine submaxillary gland)	GDP-Fuc	Fucα1-2Galβ1-3(4)GlcNAc-R
Galactosyltransferase		
β-N-Acetylglucosaminide β1-4-galactosyl transferase (from human milk)	UDP-Gal	Galβ1-4GlcNAc-R
(Fucα1-2)Gal α1-3-galactosyltransferase	UDP-Gal	Galα1-3(Fucα1-2)Gal-R
Sialyltransferase		
β-Galactoside α2-3-sialyltransferase (from porcine submaxillary gland)	CMP-Neu5Ac	Neu5Acα2-3Galβ1-3GalNac-R
α-N-Acetylgalactosaminide α2-6-sialyl-transferase (from porcine submaxillary gland)	CMP-Neu5Ac	Neu5Acα2-6GalNAc1α-Thr/Ser Galβ1-3

Efficient enzymatic synthesis approaches have been published for some of these rare sugar nucleotides. CMP-Neu5Ac, which is formed from CTP and Neu5Ac by CMP-Neu5Ac synthetase (CMP-sialic acid synthetase), is the activated form of Neu5Ac used in the natural biosynthesis of sialyl oligosaccharides. This enzyme can be isolated from several mammalian tissues such as calf brain and used for CMP-Neu5Ac synthesis [158]. However, this is a very laborious task as the enzyme concentration in such tissues is very low and very variable depending on the age and the quality of the tissue. The enzyme can also be found in colominic acid-forming bacteria, e.g., in polysaccharide capsule-forming variants of *E. coli*. A recombinant enzyme from *E. coli* has been successfully used for the gram-scale synthesis of CMP-Neu5Ac [149, 159]. An alternative approach uses the CMP-Neu5Ac synthetase from *E. coli* K-235 [160] or from an overproducing spontaneous mutant *E. coli* K-235/K1 [161]. Under optimal growth conditions with a designed medium, the mutant strain produces up to 1000 U of CMP-Neu5Ac synthetase per kg wet biomass (reported values for calf brain are 30–70 U per kg). Careful reaction engineering for the enzymatic preparative-scale synthesis of CMP-Neu5Ac (based on the kinetic study of the purified enzyme) opens a way to the more economic use of the biocatalyst [162] (Fig. 8).

So far, the most efficient protocols for the chemoenzymatic "large-scale" (multigram) syntheses of GDP-Fuc and analogs, starting from commercially available fucose, have recently been published by Baisch and Öhrlein [163]. In this seven-step synthesis, the crucial last step is the enzymatic deprotection of the triacetylated fucose moiety with acetylesterase from orange peel (Fig. 9).

A new chemoenzymatic synthesis of UDP-GalNAc consists of a multi-enzyme catalyzed generation of UDP-galactosamine from uridine-5'monophosphate (UMP) and sucrose in a repetitive batch mode, followed by a chemical acetylation step [164].

A large number of pharma-relevant oligosaccharide sequences and glycoconjugates for biological evaluation have been synthesized by many academic and industrial research groups following the enzymatic and chemo-enzymatic strategies [16, 21, 22, 151–157].

Multistep enzymatic transformations can be done with combinations of compatible aldolases, glycosyltransferases and recycling systems. A practical and successful example for this concept is the multienzymatic synthesis of the sialyl Lewis^x tetrasaccharide (sialyl Le^x;

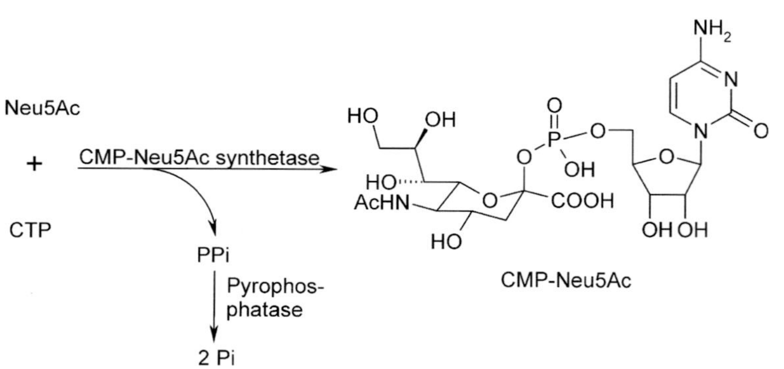

Figure 8. Enzymatic activation of N-acetylneuraminic acid to CMP-Neu5Ac.

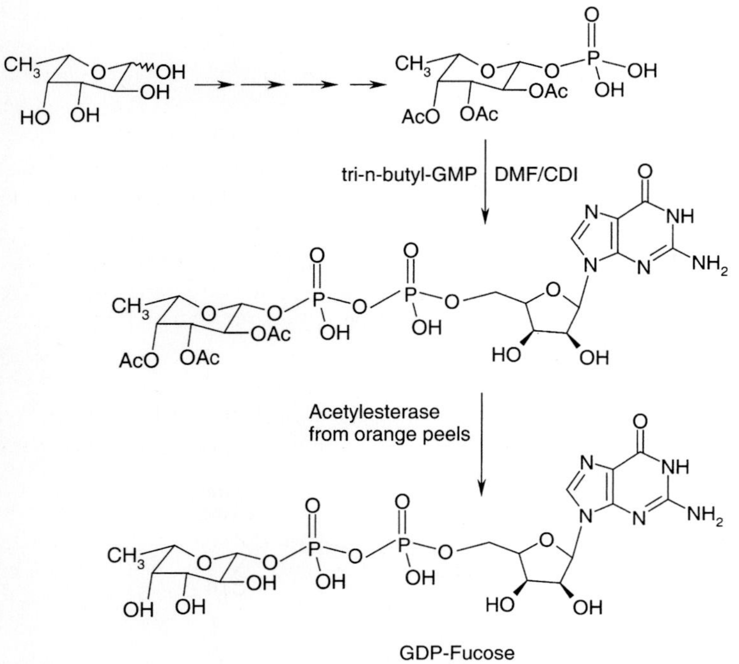

Figure 9. Preparation of GDP-fucose.

Fig. 10) [165]. The Cytel corporation in San Diego has been able to scale-up this process for the routine production of kilogram quantities in each run of reactions [144]. The sialyl Le^x tetrasaccharide and its analogs are of pharmaceutical interest in the context of new treatments for chronic inflammatory diseases (psoriasis, rheumatoid arthritis), tumors, and tissue injuries following a heart attack, stroke, or organ transplant. Sialyl Le^x and its analogs are therefore enthusiastically pursued by several pharmaceutical companies. Strategies for the solid-phase chemoenzymatic synthesis of glycopeptides (and glycolipids) containing sialyl Le^x have also been designed and reported [144, 156, and references therein].

The all-important prerequisite for the future industrial development of carbohydrate chemistry and glycotechnology is the establishment of economically affordable accesses to all necessary types of (recombinant) glycosyltranferases and sugar nucleotides. With the increasing availability of sugar nucleotides and recombinant glycosyltransferases, the glycosyltransferase-mediated glycosylation coupled with in situ regeneration of the sugar nucleotides will become the most effective method for large-scale stereocontrolled synthesis of oligosaccharides and glycoconjugates. This highly stereo- and regioselective enzymatic route is very attractive in comparison with classical chemical approaches, as no protection/deprotection steps are necessary, and no side products are formed.

A new industrial avenue to oligosaccharide synthesis may be opened by an approach using whole-cell biotransformations. Researchers from Kyowa Hakko Co. (Japan) have developed systems for the large-scale production of UDP-Gal and globotriose (Fig. 11) from inexpen-

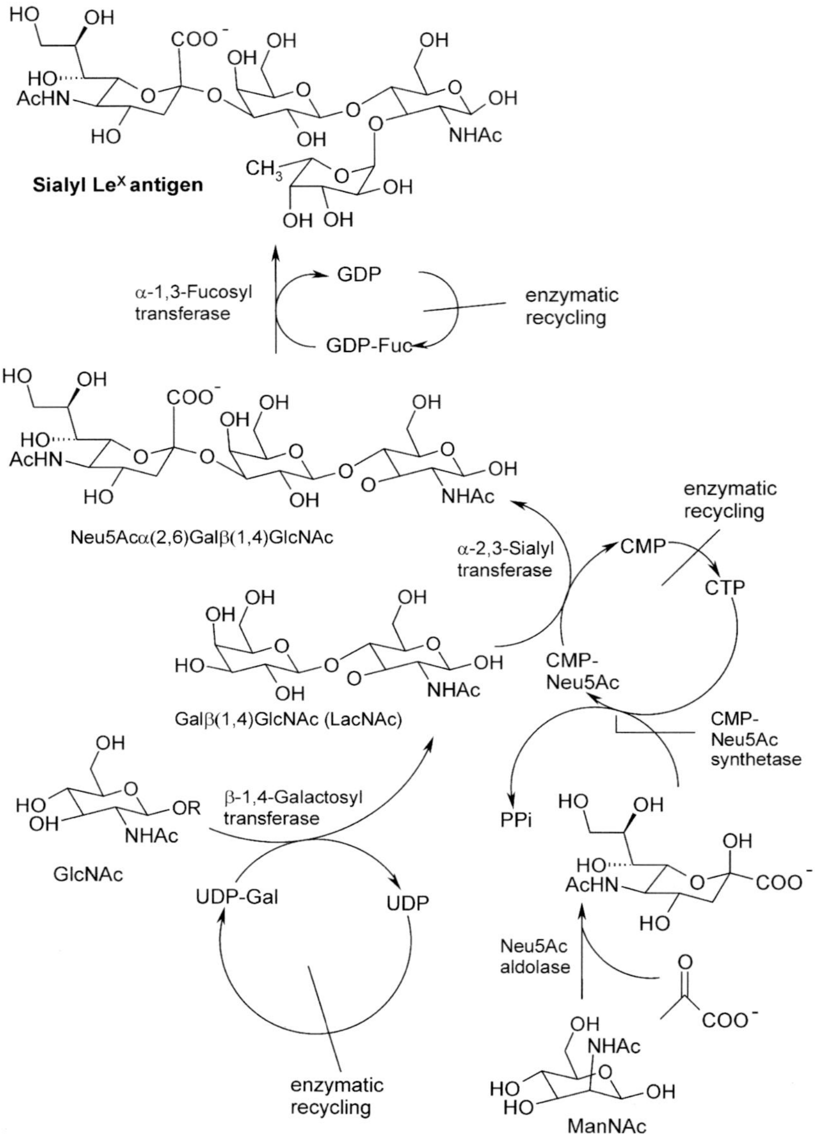

Figure 10. Multienzymatic synthesis of sialyl Lewis^x tetrasaccharide (simplified scheme).

sive starting materials by coupling metabolically engineered bacteria [166]. The production system for UDP-Gal combines a metabolically engineered *E. coli* NM522/pNT25/pNT32 (overexpressing the UDP-Gal biosynthetic genes galactose-1-phosphate uridyltransferase (*galT*), galactokinase (*galK*), glucose-1-phosphate uridyltransferase (*galU*), and pyrophosphatase) with *Corynebacterium ammoniagenes* DN510 producing uridine 5′-triphosphate

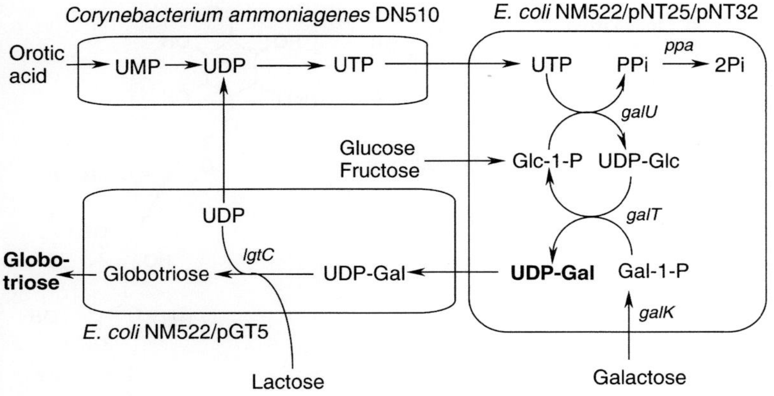

Figure 11. Production of UDP-Gal and globotriose using metabolically engineered microbial mixed cultures.

(UTP), the substrate for UDP-Gal biosynthesis, from the inexpensive precursor orotic acid. This system is reported to accumulate 44 g l^{-1} (72 mM) UDP-Gal after 21 h of reaction starting from orotic acid (110 mM), galactose (250 mM), and glucose or fructose (energy source). The cells are permeabilized with polyoxyethylene octadecylamine and xylene added to reaction mixture. No formation of UDP-Glc is observed. When another recombinant *E. coli* strain NM522/pGT5 expressing the α1-4-galactosyltranferase gene (*lgtC*) from *Neisseria gonorrheae* is coupled with this UDP-Gal production system, 188 g l^{-1} (372 mM) globotriose (Galα1-4Galβ1-4Glc) are produced after 36 h of reaction starting from orotic acid (47 mM), galactose (722 mM), lactose (468 mM), and fructose. In this process, orotic acid is recycled about eight times via UDP and UTP. The biocatalyst concentrations in these systems are 150 g l^{-1} (wet weight) for the DN510 cells and 50 g l^{-1} for the NM522/pNT25/pNT32 and NM522/pGT5 cells. Globotriose is a trisaccharide portion of globotriosylceramide, a receptor of verotoxin produced by some bacterial strains such as *E. coli* O157.

The strategy of producing sugar nucleotides by combining metabolically engineered *E. coli* with a nucleoside 5'-triphosphate-producing micro-organism, and the concept of producing oligosaccharides by coupling such systems with glycosyltransferases, have a great potential for application to the manufacture of other sugar nucleotides and oligosaccharides. Several other bacterial glycosyltransferase genes have already been cloned and overexpressed in *E. coli*. Such recombinant strains would be suitable for the set-up of analogous production systems for various sugar nucleotides and oligosaccharides.

Chemoenzymatic strategies were recently also applied for the preparative-scale synthesis of the very rare and expensive research biochemicals D-*myo*-inositol-1-phosphate, and D- and L-1,3,4,5-*myo*-inositol tetrakisphosphates [127, 167]. The approach uses commercial lipase from *Pseudomonas* sp. for the regio- and enantioselective acetylation (in vinyl acetate) of 4,6-di-O-benzyl-*myo*-inositol to 1D-1-acetoxy-4,6-di-O-benzyl-*myo*-inositol (intermediate for D-IP_1), and lipase from *Candida antarctica* for the enantioseparation of racemic 2,6-dibenzyl-*myo*-inositol by enantioselective C5-acetylation (in the synthesis of D- and L-1,3,4,5-IP_4) (Fig. 12).

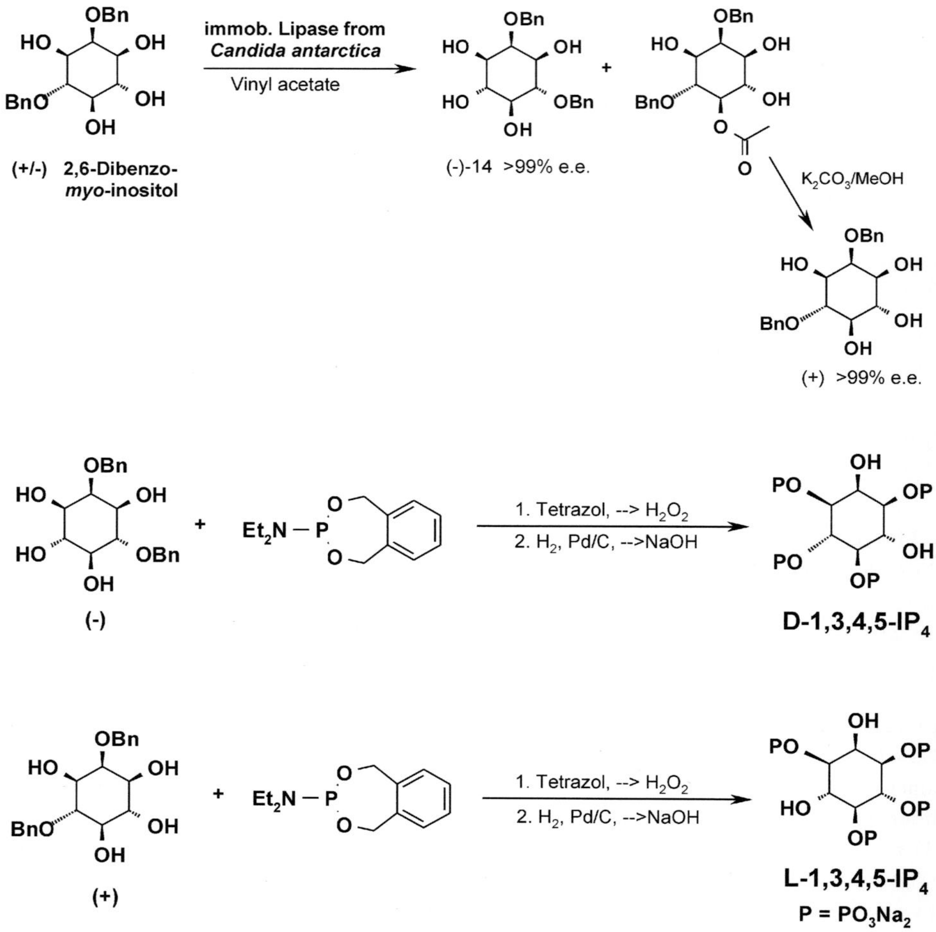

Figure 12. Chemo-enzymatic approach to D- and L-1,3,4,5-inositol phosphates.

7.7 Large-Scale Biocatalysis

Wandrey [168] and Tramper [169] have defined criteria for the optimization and the viability of biochemical conversions in comparison with chemical processes. Process design includes the choice of an appropriate substrate, of a catalyst, and of methods of downstream processing in order to obtain the desired product in a defined purity. The aims of process selection and optimization (especially in the pharmaceutical sector) are to find process conditions defined by high conversion of the substrate, high selectivity of the reaction, high optical purity of the product, high space-time yield (productivity) of the process, and low enzyme and co-enzyme consumption per unit mass of product. Of high importance are also:

- The number of reaction steps involved, environmental aspects → "greenness".
- *Biocatalysis versus chemical conversion?* When the number of reaction steps can be drastically reduced, chances are in favor of biocatalytic strategies. Enantio- and stereoselectivity can be strong points in favor of biocatalysis.
- *Reactor configuration: standard versus tailored?* Decision is based on the kinetics of the biocatalytic process; availability and experience; desired mixing; mass transfer properties:
  a) Standard case: stirred-tank reactor for batch operation → in industry great reluctance to change to other configurations.
  b) Potential alternatives: plug-flow reactor, packed-bed reactor, fluidized-bed reactor, two-liquid-phase reactor, membrane reactor (especially for co-factor-requiring free enzymes), etc.
- *Mode of operation: (fed)-batch versus continuous?* Parameters for decision: kinetics, stability and form of the biocatalyst; desired substrate conversion, product concentration (solubility); need of process control, inhibitory/toxic effect of substrate or product. Examples for fed-batch bioprocesses: acrylamide process (Nitto Chemicals), L-carnitine process (Lonza). Examples for continuous operation: production of L-tert. leucine and other amino acids (Degussa).
- *Level of process integration?* Parameters for decision: number of process steps (subsequent reactions); nature of the reactions.

One of the major difficulties in promoting biological reactions in the academic environment – and sometimes also in the industrial environment – is to obtain access not only to the vast research literature and to comprehensive reviews in the field but also to a sufficient number of suitable industrial case studies and success stories. This is of great importance to demonstrate the power of biocatalysis for large-scale conversions and to convince hardcore chemists and chemical engineers of the practical feasibility and advantages of industrial bioprocesses. At present, the number of available detailed bioprocess descriptions and comparative case studies is still relatively small, and in addition this information is not easily accessible for non-specialists in the field. Therefore, this section is devoted to the compilation of valuable information on some of the most important bioprocesses, including indications on system productivity, production scales, specific advantages, etc.

Large-scale applications of biocatalysis are found in the fields of drug and drug intermediate manufacturing, production of selected bulk chemicals, processing of food and beverages, and waste treatment. A classical example is the Reichstein process for the synthesis of ascorbate (vitamin C) involving the transformation of sorbitol to L-sorbose with *Gluconobacter suboxidans*. This process – established in 1933 – is still in use, but attempts are being made to substitute it by a new process based on the direct one- or two-step transformation of glucose to ascorbate using metabolically engineered micro-organisms.

Many of today's large-scale applications use immobilized enzymes. In particular, the Japanese industry has for a long time pioneered this sector of biotechnology. In the multi-author publication edited by Tanaka, Tosa and Kobayashi [90] and in a more recent review [89], a number of well-established continuous production processes with immobilized biocatalysts are described in detail (for a selection, see below). Complementary information on some of these Japanese bioprocesses as well as additional case studies for other bioprocesses has been compiled by Cheetham [170].

7.7.1 Immobilized Enzymes [89, 90]

Examples of these include:

- Optical resolution of racemic amino acids (methionine, phenylalanine, tryptophan, valine) by the action of L-specific amino acylase from *Aspergillus oryzae* (Tanabe Seiyaku Co., Ltd.). The theoretical productivity of 1000 liter immobilized amino acylase columns ranges from 214 kg per day for L-Ala to 715 kg per day for L-Met.
- Production of an optically active diltiazem intermediate (2*R*, 3*S*)-methoxyophenylglycidate methyl ester ((–)-MPGM) from racemic MPGM by the action of lipase from *Serratia marcescens* in a toluene aqueous biphasic system (Tanabe Seiyaku Co., Ltd.). For the continuous production of (–)-MPGM, a hollow fiber bioreactor was set up in collaboration with Sepracor Inc. The introduction of this enzymatic step allowed the shortening of the diltiazem synthesis from nine down to five steps.
- Production of 6-amino penicillanic acid (6-APA) from penicillin G or V by the action of penicillin amidase from *E. coli*, *Bacillus megaterium*, or *Bovista plumba* (Toyo Jozo Inc., Asahi Chemical Industry Co., Ltd., Fujisawa Pharmaceutical Co., Gist-Brocades/DSM, Novo-Nordisk, Pfizer, and others). Annual world production of 6-APA: 6000 tons, used for the manufacture of semisynthetic penicillins.
  In comparing the chemical and enzymatic routes, the former route comprises three reaction steps with "dirty" and expensive chemicals/organic solvents under relatively extreme conditions. The enzymatic route requires only one reaction step under mild conditions; the enzyme is immobilized (operating lifetime: $>$1000 h; productivity: up to 2000 kg 6-APA per kg immobilized enzyme). In this case, the biocatalytic route has largely replaced the chemical route! [169, 170].
- Production of 7-aminocephalosporanic acid (7-ACA) (and of 7-aminodesacetoxy cephalosporanic acid (7-ADCA)) from cephalosporins by the action of cephalosporin C amidase from *Pseudomonas* sp. (Toyo Jozo Inc., Asahi Chemical Industry Co., Ltd., and others). Annual world production of 7-ACA: 1000 tons and 7-ADCA: 500 tons, used for the manufacture of semisynthetic cephalosporins.
- Production of low-lactose milk by hydrolysis of the lactose with β-galactosidase (lactase) from *Aspergillus oryzae* or *Aspergillus niger* (Snow Brand Milk Products, Co, Ltd. in Japan, Centrale del Latte in Milan, Italy, Valio Oy in Finland, Drouin Cooperative Butter Factory in Australia, and others). Typical plant sizes range from 10 000 to 250 000 liters per day [170]. Lactose is the major sugar present in milk. Its hydrolysis to glucose and galactose is of considerable importance for the food and dairy industry, as a large fraction of the world population is lactose-intolerant.
- Production of cocoa butter-like fats by enzymatic transesterification with the extracellular lipase from *Rhizopus niveus* (Fuji Oil Co., Ltd).
- Production of high-fructose syrup (HFS) from glucose (enzymatically degraded starch) by the action of glucose isomerases from various micro-organisms (immobilized enzyme from Novo Nordisk A/S, Gist-Brocades/DSM, and other companies). The annual world production of HFS, for use in soft drinks like Pepsi Cola and Coca Cola, is in the order of 10 million tons (dry basis). Fructose is significantly sweeter than glucose. No effective chemical isomerization methods are possible, and other sources of fructose, e.g. the hydrolysis of inulin, are not yet effective. Therefore, the enzymatic isomerization technology is applied [170].

7.7.2 Immobilized (Resting) Microbial Cells [89, 90]

Examples of these include:

- Production of L-aspartic acid from fumaric acid by stereoselective addition of ammonia under the action of the intracelluar aspartase in *E. coli* (Tanabe Seiyaku Co., Ltd.). When a 1000-liter column is used, theoretical yield of L-aspartic acid is 3.4 tons per day (and even considerably higher for mutant strains and plasmid pNK101-harboring strains). A similar industrial process using the immobilized *E. coli* aspartase (instead of the whole cells) was established earlier by Kyowa Hakko Kogyo, Co., Ltd.. L-Aspartate is mainly used as a building block for the manufacture of the sweetener aspartame [170].
- Production of L-malic acid from fumaric acid by stereoselective hydration under the action of the intracellular fumarase in *Brevibacterium ammoniagenes* (Tanabe Seiyaku Co., Ltd). A 1000-liter column bioreactor can typically yield 30 tons of L-malic acid per month.
- Production of L-alanine by decarboxylation of L-aspartic acid under the action of the intracellular L-aspartate-β-decarboxylase in *Pseudomonas dacunhae* (Tanabe Seiyaku Co., Ltd). A 1000-liter pressurized column bioreactor can typically yield 5 tons of L-alanine per month.
- Production of D-aspartic acid (and L-alanine) from DL-aspartic acid by decarboxylation of the L-aspartic acid under the action of the intracellular L-aspartate-β-decarboxylase in *Pseudomonas dacunhae* (Tanabe Seiyaku Co., Ltd). A 1000-liter pressurized column bioreactor can typically yield 9.5 tons of D-aspartic acid and 5.1 tons of L-alanine per month. D-Aspartic acid is used as an important component of the semisynthetic penicillin, aspoxicillin.
- Production of L-isoleucine from ethanol and α-keto butyric acid or α-amino butyric acid using a multistep bioconversion with *Brevibacterium flavum* under "native immobilization" and biotin-free conditions (Mitsubishi Petrochemical Co., Inc.). Productivity of this system is 200 $mmol\ l^{-1}\ d^{-1}$.
- Production of HFS by isomerization of glucose under the action of the intracellular glucose isomerase in *Streptomyces phaechromogenes* (Nagase Biochemicals Co., Ltd).
- Production of acrylamide (Fig. 13) by hydration of acrylonitrile under the action of the intracelluar nitrile hydratase in *Rhodococcus rhodochrous* (Nitto Chemical Industry Co., Ltd., fed-batch process). The annual production amounts to >30000 tons (see also Table 6). Acrylamide is one of the most important commodity chemicals and is required in large quantities as the pre-polymer of polyacrylamide that is widely used in polymer and flocculent applications. The advantages of this hydratase approach in comparison with the classical chemical nitrile hydration are: higher product end concentration, quantitative yields, no formation of acrylic acid, no need for copper catalyst, and only five chemical/technical operations instead of seven [73, 112, 113, 171]. An analogous process for nicotinamide is being commercialized by Lonza (see also section 6).
- Production of palatinose from sucrose under the action of the intracellular α-glucosyltransferase in *Protaminobacter rubrum* (Mitsui Seito Co., Ltd.). Palatinose is a useful substitute for sucrose with low insulin stimulation; annual production 4000 tons.
- Production of fructo-oligosaccharides (low-calorie sweetener) from HFS under the action of the intracellular β-fructofuranosidase in *Aspergillus niger* (Meiji Seika Co. Ltd). The annual production amounts to 4000 tons.

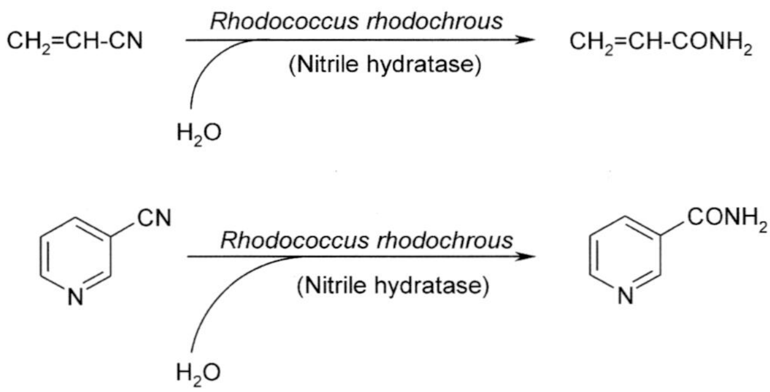

Figure 13. Biocatalytic production of acrylamide and nicotinamide.

- α-, β-, and γ-Cyclodextrin production from enzymatically hydrolyzed starch (glucose) under the action of the intracellular cyclodextrin glucosyltransferase in *Bacillus circulans* (Mercian Co., Ltd). The annual production amounts to 150 tons.
- Production of cis,cis-muconic acid by partial biodegradation of benzoic acid (three enzymatic steps) with a mutant of *Pseudomonas putida* (Mitsubishi Kasei Corporation). Productivity is $>70\ g\ l^{-1}\ d^{-1}$. This process still requires further development.
- Production of p-hydroquinone by hydroxylation oxidation of phenol with *Mycobacterium* sp. (Mitsubishi Gas Chemical Co., Inc.). Productivity is $75\ g\ l^{-1}\ d^{-1}$. This process needs further development to become competitive with the established chemical process.

In Japan, continuous production processes using immobilized living cells are currently introduced in such classical areas as beer (Kirin Brewery Co.) and sake brewing (Ohzeki, Co. Ltd), vinegar production (Kewpie Jyozo, Co. Ltd), and in the production of soy sauce (Kikkoman Co.).

7.7.3 Additional Established Bioprocesses Documented by Case Studies [170, 172–173]

Examples of these include:

- D-p-Hydroxyphenylglycine, D-phenylglycine and their derivatives are important side-chain precursors for semisynthetic penicillins and cephalosporins. The industrial production process for these amino acids involves the chemical synthesis of D/L-hydantoins, the enantiospecific hydrolysis of the corresponding D/L-5-monosubstituted hydantoins to N-carbamoyl-D-amino acids catalyzed by a microbial D-hydantoinase (e.g. from *Bacillus brevis*, *Agrobacterium radiobacter*, *Blastobacter* sp., or *Pseudomonas putida*), and decarbamoylation either with nitrite under acidic conditions or with D-carbamoylase (Fig. 14) [73, 112, 169]. The enantiospecific cleavage of the racemic hydantoin derivatives is performed under simultaneous racemization of the unwanted enantiomer, thus allowing a theoretical yield of 100%. This is an excellent example for a high degree of process integration. The process is performed by Kanegafuchi (in Singapore), Ajinomoto, Ricordati, and

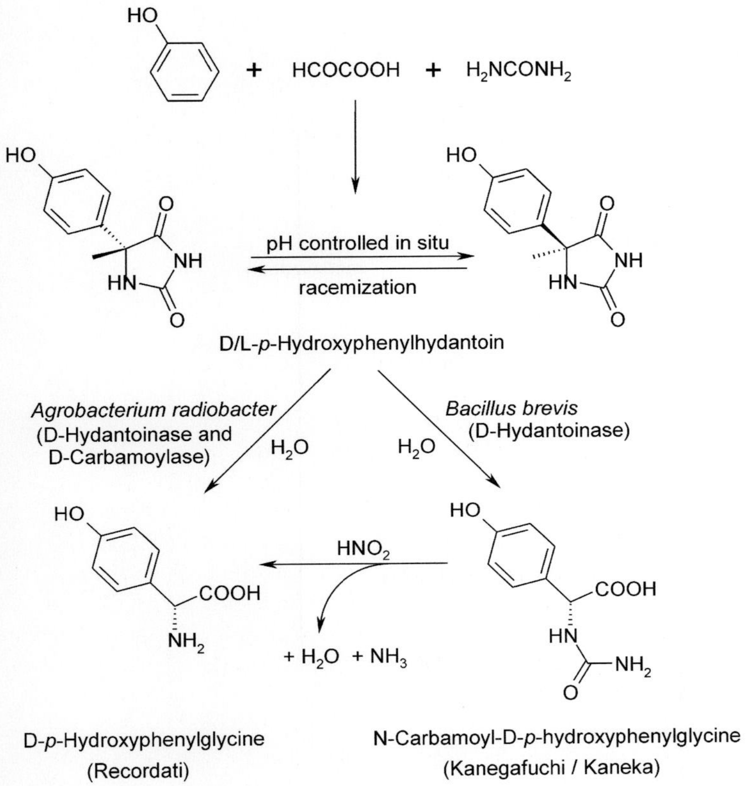

Figure 14. The hydantoinase route to D-*p*-hydroxyphenylglycine.

others, with either immobilized enzymes or whole cells; annual world production is several thousand tons.

- L-specific hydantoinases can be applied in an analogous way for the production of L-glutamate, L-lysine, L-valine, L-leucine, L-tryptophan, and L-phenylalanine, so wider commercial applications for hydantoinase technology can be expected in the future [73, 118–120].
- Production of the artificial low-calorie sweetener aspartame from Z-L-aspartate and D/L-phenylalanine methylester by peptide bond formation with immobilized thermolysin from *Bacillus thermoproteolyticus* (Tosoh Corp., Ajinomoto, Toyo-Soda, DSM, annual world production approx. 10 000 tons). Aspartame is about 200 times as sweet as sucrose, and is used in drinks such as Coca Cola and Pepsi Cola Light. In contrast to the older chemical process, the enzymatic process can – due to the L-selectivity of the enzyme – use the cheaper D/L-phenylalanine methylester instead of the pure L-form. The enzymatic process (Fig. 15) yields α-aspartame exclusively, whereas the chemical route yields a mixture of α-aspartame and bitter-tasting β-aspartame, thus requiring an additional separation step.

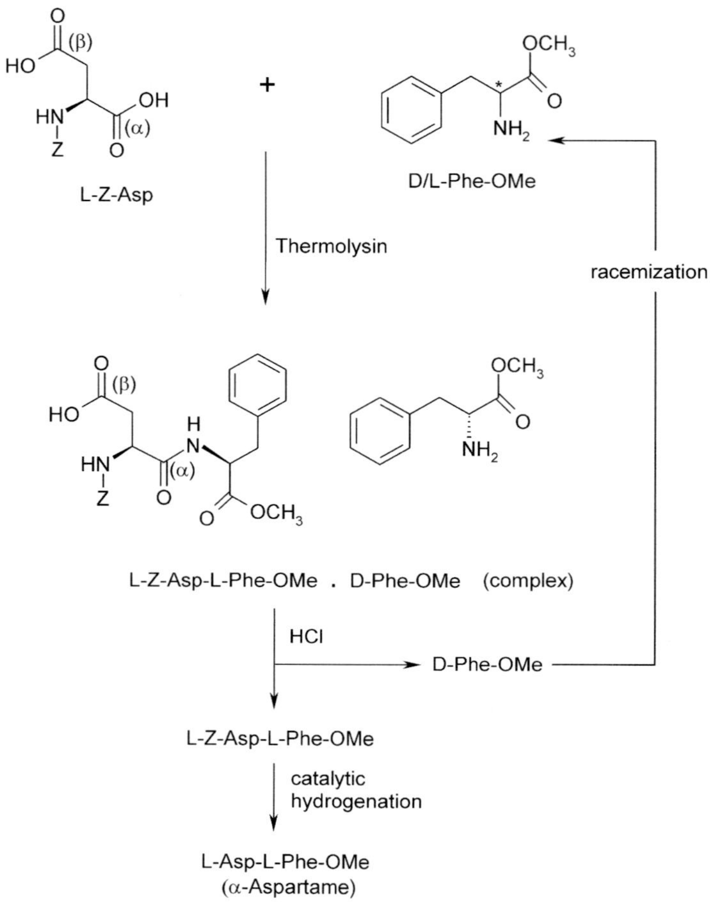

Figure 15. Enzymatic synthesis of aspartame.

- L-Lysine, an essential amino acid, is used in very large quantities to supplement human foods and animal feeds. Traditionally, L-Lysine is produced by fermentation processes. An alternative route developed by Toray Ind. involves the chemical synthesis of D/L-α-amino-ε-caprolactam followed by the selective hydrolysis of the L-α-amino-ε-caprolactam catalyzed by intracellular lactamase in *Cryptococcus laurentii*, to give L-Lysine. The process can be improved by adding a second micro-organism, *Achromobacter obae*, containing α-amino-ε-caprolactam racemase. Thus, quantitative yields of L-lysine are obtained.
- Substituted nicotinic acid derivatives, such as 6-hydroxy nicotinic acid are important building blocks for the synthesis of pesticides and pharmaceuticals, acting as specific inhibitors of NAD/NADP-dependent enzymes. In a process developed by Lonza (Switzerland), 6-hydroxy nicotinic acid is produced by whole-cell microbial hydroxylation of nicotinic acid with *Achromobacter xylosoxidans* (yield: >100 g l^{-1}) (Fig. 16). A similar process for the production of 6-hydroxypicolinic acid (98 g l^{-1}) and derivatives from picolinic acid (derivatives), based on *Alcaligenes faecalis*, has also been established [174, 175].

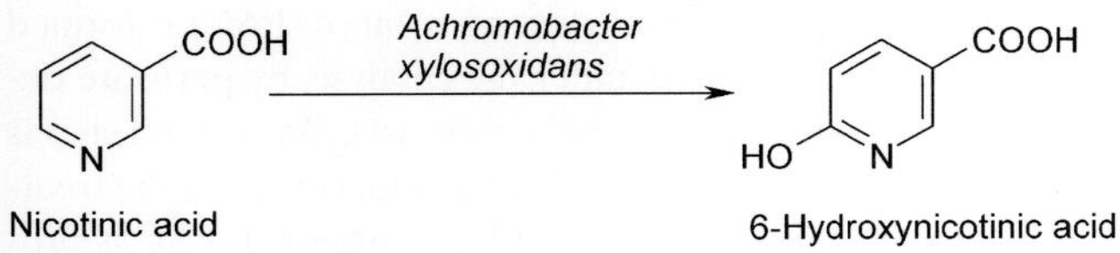

Figure 16. Microbial hydroxylation of nicotinic acid.

- Production of L-carnitine by microbial hydroxylation of γ-butyro-betaine (continuous process, yield >130 g l^{-1} d^{-1}, or repeated batch process) with a mutant strain of *Agrobacterium/Rhizobium* lacking L-carnitine dehydrogenase. This process (Fig. 17), developed by Lonza, operates on a 140 tons per year scale. L-Carnitine is used in pharmaceuticals as a thyroid inhibitor, as a slimming aid, and in sports foods and drinks [174, 176, 177].
- Production of L-tert. leucine by enantioselective reductive amination of trimethylpyruvate (2-oxo-3,3-dimethylbutanoate) with leucine dehydrogenase under PEG-NADH-recycling with formate/formate dehydrogenase (Degussa, EMR-technology, 100-kg scale) (see also Table 5) [104, 178]. L-tert. leucine is an interesting new chiral synthon which is now appearing in pharmaceutical development compounds and drug candidates.

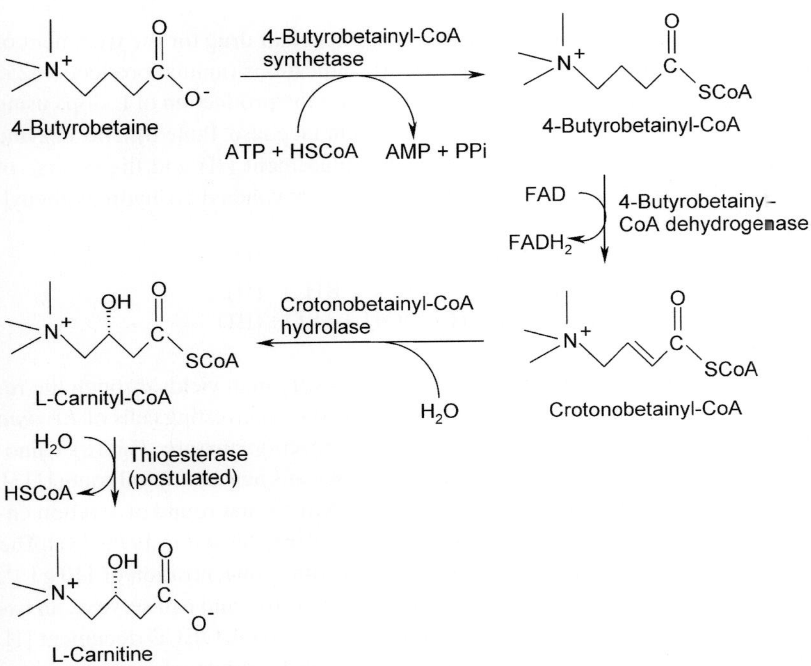

Figure 17. Microbial production of L-carnitine by hydroxylation 4-butyrobetaine using whole cells of strain HK4 (*Agrobacterium/Rhizobium*).

- Production of L-ephedrine. In a first reaction step phenylacetylcarbinol (PAC) is formed by "condensation" of benzaldehyde and pyruvic acid under the catalysis by pyruvate decarboxylase in whole cells of *Saccharomyces cerevisiae* (baker's yeast). The second step is the chemical conversion of PAC into L-ephedrine. L-Ephedrine is widely used in the treatment of asthma and hay fever as a bronchodilating agent and decongestant. PAC is produced by this method on large scale (10 m^3 reactor with 200 kg of yeast) by Knoll (BASF) and Malladi Drugs & Pharmaceuticals (India). The reaction can also be performed with substituted aromatic aldehydes, e.g., with *m*-hydroxybenzaldehyde, giving access to ephedrine analogs.
- Production of (*S*)-2-chloropropanoic acid from racemic 2-chloropropanoic acid by enantioselective degradation of the *R*-enantiomer with whole cells of *R*-dehalogenase-containing *Pseudomonas putida*. (*S*)-2-chloropropanoic acid is a building block for a wide range of herbicides and is produced by ICI (2000 tons per year). The side product (*S*)-lactic acid is also of commercial interest.
- Production of (*R*)-glycidyl butanoate and (*R*)-glycidol by PPL (porcine pancreatic lipase)-catalyzed enantioselective hydrolysis of racemic glycidyl butyrate. Both chiral products are required for various commercial pharmaceuticals. The method has been commercialized by Andeno-DSM and others.
- Production of (*S*)-phenylethylamine by *R*-selective methoxyacetylation of racemic phenylethylamine with lipase from *Burkholderia plantari* or *Pseudomonas* sp. in MTBE (BASF, 100-ton scale) [179].

L-Dopa, a metabolic precursor of dopamine, is a very important drug for the treatment of parkinsonism, but is also of interest in other therapeutic indications (annual production 250 tons). A very interesting recently industrialized bioprocess is the production of L-dopa using β-tyrosinase (tyrosine phenol lyase) in a resting cell system (see also Table 6). This enzyme catalyzes a variety of reactions: α,β-elimination (I), β-replacement (II) and the reverse of α,β-elimination (III). R,R′ = phenyl-, hydroxyphenyl-, dihydroxyphenyl-, trihydroxyphenyl-

$$RCH_2CHNH_2COOH + H_2O \rightarrow RH + CH_3COCOOH + NH_3 \quad (I)$$
$$RCH_2CHNH_2COOH + R'H \rightarrow R'CH_2CHNH_2COOH + RH \quad (II)$$
$$R'H + CH_3COCOOH + NH_3 \rightarrow R'CH_2CHNH_2COOH + H_2O \quad (III)$$

L-Tyrosine and related amino acids can be synthesized in very high yields through the reverse of α,β-elimination (III). In the case of L-dopa synthesis by the resting cells of *Erwinia herbicola*, high concentrations of L-dopa are obtained in the reaction mixture (Fig. 18). Ajinomoto Co. Ltd started the commercial production of L-dopa via this new biological route [113]. The new access to L-dopa was first described in 1969 [180]. After a first round of reaction engineering and strain development, titers of around 50 g l^{-1} were achieved in 1975 [181]. The now commercialized fed batch process [182], with a final product concentration of 110 g l^{-1}, has many advantages over the classical chemical process (Table 10), and can serve as an excellent example for a sustainable production process in the sense of the OECD document [1].

The enzymatic process (60 m^3 scale) shows a fivefold improved productivity in comparison with the chemical process, and with a significant reduction in time required to complete the process.

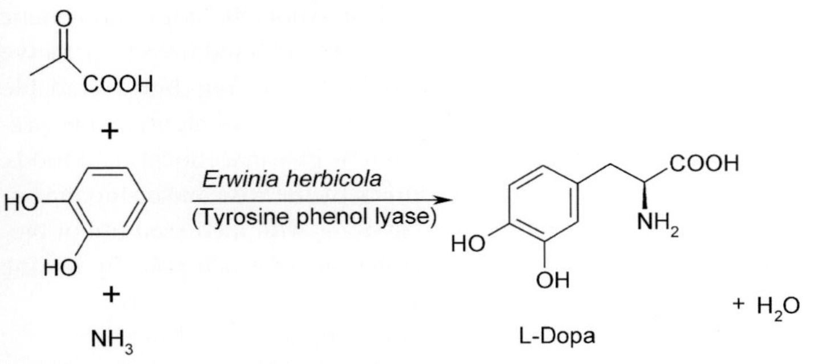

Figure 18. The new synthetic process for L-dopa.

Table 10. Comparison between the enzymatic and the chemical processes for the production of L-dopa. (Data from Ajinomoto Co. Ltd., [182])

	Enzymatic synthesis	Chemical synthesis
Main staring materials	Pyrocatechol, pyruvic acid, and ammonia	Vanillin, hydantoin, hydrogen, acetic acid anhydride
Number of individual reactions	1	8
Reaction byproducts	Water	Ammonia, carbon dioxide, acetic acid
Optical separation	Not necessary	Separation of reaction intermediates (acetyl-D/L-veratroylglycine) with the enzyme acylase and racemization of the D-compound
Production facilities (reaction and isolation)	Versatile equipment can be used for both the enzymatic reaction and the isolation process	Special plant is required for both the synthesis reactions and the separation process
Time required for production	Approx. 3 days	Approx. 15 days
Downstream processing	Simple → crystallization, ultrafiltration, recrystallization; no organic solvent extraction	Complex, stepwise
Types of amino acids present in the product as impurities in minor amounts	L-Tyrosine	L-Tyrosine, 3-methoxy L-tyrosine, 3,4,6-trihydroxyphenylalanine

7.8 Outlook

In 1987, the Brundtland Commission defined sustainable development as "development that meets the needs of the present without compromising the ability of future generations to meet their needs".

On the way to sustainable production strategies, the chemical industry must play a key role and must therefore undergo important structural changes by introducing more and more biological processes and biobased thinking into its R&D effort. Biotechnology and genetic engineering greatly enlarge the arsenal of organic synthesis methods, and help to preserve natural resources and ecosystems, to protect the environment by producing (bio)degradable products, to avoid or reduce pollution and repair damages to the environment, and to preserve and increase the quality of life by the introduction of new pharmaceuticals and foods. This can be achieved by the utilization of renewable resources, by the wise and economic use of non-renewable resources, and by the protection of ecosystems with increased use of biological processes. The necessary changes in technologies can only be achieved by linking economical, ecological, and social goals.

Two conclusions in the earlier-mentioned OECD-Report [1] read:

- "The key area of opportunity is improved and novel biocatalysts. The search is on to examine the remaining unexplored biological diversity which presumably holds a great wealth of biocatalytic potential and of new bioactive compounds and biomaterials. The introduction of biotechnology into many industrial processes will be increasingly dependent on the development of recombinant biocatalysts.
- Bioprocess engineering and integrated bioprocessing also remain as critical factors for the commercialization of biotechnology."

References

[1] Biotechnology for Clean Industrial Products and Processes – Towards Industrial Sustainability. OECD: Paris, 1998.

[2] Webb, E.C. *Enzyme Nomenclature – Recommendations (1992) of the Nomenclature Committee of the International Union of Biochemistry and Molecular Biology IUBMB*. Academic Press: San Diego, 1992.

[3] Schomburg, D., et al. (eds) *Enzyme Handbook*. Springer Verlag: Berlin, Vol. 1 (1990) to Vol. 17 (1998).

[4] Kieslich, K. (ed.) *Microbial Transformations of Non-Steroid Cyclic Compounds*. Thieme: Stuttgart 1976

[5] Jones, J.B., Sih, C.J., and Perlman, D. (eds) *Applications of Biochemical Systems in Organic Chemistry*. Wiley Interscience: New York, 1976.

[6] Rosazza, J.P.N. (ed.) *Microbial Transformation of Bioactive Compounds*, Vols. 1 and 2. CRC Press: Boca Raton, 1982.

[7] Fogarty, W.M. (ed.) *Microbial Enzymes and Biotechnology*. Applied Science Publishers: London, 1983.

[8] Kieslich, K. (ed.) *Biotransformations*. Biotechnology Vol. 6a, Verlag Chemie: Weinheim, 1984.

[9] Porter, R., and Clark, S. (eds) *Enzymes in Organic Synthesis*. Ciba Foundation Symposium 111. Pitman: London, 1985.

[10] Tramper, J., van der Plas, H.C., and Linko, P. (eds) *Biocatalysts in Organic Synthesis*. Studies in Organic Chemistry 22. Elsevier: Amsterdam, 1985.

[11] Schneider, M.P. (ed.) *Enzymes as Catalysts in Organic Synthesis*. Proceedings of the NATO Advanced Research Workshop. Reidel: Dordrecht, 1986.

[12] Laane, C., Tramper, J., and Lilly, M.D. (eds) *Biocatalysts in Organic Media*. Elsevier: Amsterdam, 1987.

[13] Moody, G.W., and Baker, P.B. (eds) *Bioreactors and Biotransformations*. Elsevier Applied Science Publishers: London, 1987.

[14] Davies, H.G., Green, R.H., Kelly, D.R., and Roberts, S.M. (eds) *Biotransformations in Preparative Organic Chemistry: the Use of Isolated Enzymes and Whole Cell Systems in Synthesis*. Academic Press: London, 1989.

[15] Abramowicz, D.A. (ed.) *Biocatalysis*. Van Nostrand Reinhold: New York, 1990.

[16] Whitesides, G.M., and Wong, C.-H. *Enzymes in Organic Synthesis*. Pergamon Press: Oxford, 1990.

[17] Gopping, L.G., Martin, R., Pickett, J.A., Bucke, C., and Bunch, A.W. (eds) *Opportunities in Biotranformations*. Elsevier: London, 1990.
[18] Gerhartz, W. *Enzymes in Industry, Production and Application*. VCH Publishers: Weinheim, 1990.
[19] Holland, H.L. *Organic Synthesis with Oxidative Enzymes*. VCH Publishers: Weinheim, 1992.
[20] Halgas, J. *Biocatalysts in Organic Synthesis*. Studies in Organic Chemistry 46. Elsevier: Amsterdam, 1992.
[21] Wong, C.-H., and Whitesides, G.M. *Enzymes in Synthetic Organic Chemistry*. Tetrahedron Organic Chemistry Series Vol. 12. Pergamon: New York, 1994.
[22] Drautz, K., and Waldmann, H. (eds) *Enzyme Catalysis in Organic Synthesis – A Comprehensive Handbook*, Vols. 1 and 2. VCH: Weinheim, 1995.
[23] Fogarty, W.M. *Microbial Enzymes and Biotechnology*. Applied Science Publishers: London, 1993.
[24] Faber, K. *Biotransformations in Organic Chemistry*. Springer Verlag: Berlin, 2nd Edition, 1995.
[25] Godfrey, T., and West, S. (eds) *Industrial Enzymology – Application of Enzymes in Industry*, 2nd Edition. Macmillan: London, 1996.
[26] Koskinen, A.M.P., and Klibanov, A.M. *Enzymatic Reactions in Organic Media*. Chapman & Hall: London, 1996.
[27] Kieslich, K., van der Beek, C.P., de Bont, J.A.M., and van den Tweel, W.J.J: (eds) *New Frontiers in Screening for Microbial Biocatalysts*. Studies in Organic Chemistry 53, Elsevier: Amsterdam, 1998.
[28] Kelly, D.R. (ed.) *Biotransformations I*, Biotechnology, Vol. 8a (second, completely revised edition). VCH: Weinheim, 1998; Vol.8b will be published in 1999.
[29] Roberts, S.M., Wiggins, K., Casy, G., Phythian, S., Todd, C., and Nielsen, M-B. *Preparative Biotransformations. Whole Cell and Isolated Enzymes in Organic Synthesis*. Wiley: Chichester, 1992 (with updates 1993, 1994, 1995, 1996, 1997).
[30] Abbott, B.J. *Adv. Appl. Microbiol.* **20**, 203–257 (1976).
[31] Whitesides, G.M. Large-Scale Organic Synthesis Using Cell-free Enzymes. In: *Applications of Biochemical Systems in Organic Chemistry*. Jones, J.B., Sih, C.J., and Perlman, D. (eds) Wiley Interscience: New York, 1976, pp. 901–927.
[32] Yamada, H., and Kumagai, H. *Pure Appl. Chem.* **59**, 1117–1127 (1978).
[33] Kieslich, K. Microbial Transformations – Type Reactions. In: *Antibiotics and Other Secondary Metabolites; Biosynthesis and Production*. FEMS. Symposium No.5, Academic Press, 1978, pp. 57–85.
[34] Fischli, A. Chiral Building Blocks in Enantiomer Synthesis Using Enzymatic Transformations. In: *Modern Synthetic Methods*. Scheffold, R. (ed.) Otto Salle Verlag: Frankfurt/Main, 1980, pp. 268–350.
[35] Kieslich, K. Steroid Conversions. In: *Economic Microbiology, Microbial Enzymes and Bioconversions*. Vol. 5. Rose, A.H. (ed.) Academic Press: London, 1980, pp. 369–465.
[36] Rose, H. History and Scientific Basis of Commercial Exploitation of Microbial Enzymes and Bioconversions. In: In: *Economic Microbiology, Microbial Enzymes and Bioconversions*. Vol. 5. Rose, A.H. (ed.) Academic Press: London, 1980, pp. 1–47.
[37] Jones, J.B., Enzymes in Synthetic Organic Chemistry. In: *Enzymic and Non-Enzymic Catalysis*. Dunnill, P., Wiseman, A., and Blakebrough, N. (eds) Ellis Horwood: Chichester, 1980, pp. 54–83.
[38] Esaki, N., Soda, K., Kumagai, H., and Yamada, H. *Biotechnol. Bioeng.* **22**, Suppl. 1, 127–141 (1980).
[39] Whitesides, G.M., and Wong, C.-H. *Aldrichimica Acta* **16**, 27–34 (1983).
[40] Sih, C.J., and Chen, C-S. *Angew. Chemie, Int. Ed. Engl.* **23**, 570–578 (1984).
[41] Neidleman, S.L. *Biotechnol. Genet. Eng. Rev.* **1**, 1–38 (1984).
[42] Whitesides, G.M., and Wong, C.-H. *Angew. Chemie, Int. Ed. Engl.* **24**, 617–638 (1985).
[43] Simon, H., Bader, J., Gunther, H., Neumann, S., and Thanos, J. *Angew. Chemie, Int. Ed. Engl.* **24**, 539–553 (1985).
[44] Best, D.J. Applications of Biotechnology to Chemical Production. In: *Molecular Biology and Biotechnology*. Walker, J.M., and Gingold, E.B. (eds) The Royal Society of Chemistry: London, 1985, pp. 228–267.
[45] Jones, J.B. Enzymes as Chiral Catalysts. In: *Asymmetric Synthesis*. Vol. 5. Morrison, J.D. (ed.) Academic Press: Orlando, 1985, 309–344.
[46] Jones, J.B. *Tetrahedron* **42**, 3351–3403 (1986).
[47] Kieslich, K. *Arzneim.-Forsch.* **36**, 774–778, 888–892, 1006–1009 (1986).
[48] Butt, S., and Roberts, S.M. *Natural Products Reports* 489–503 (1986).

[49] Breitgoff, D., Essert, T. Laumen, K., and Schneider, M.P. Hydrolytic Enzymes in Organic Synthesis – Chiral Building Blocks from Prochiral Substrates. In. *F.E.C.S. Int. Conf. Chem. Biotechnol. Biol. Act. Nat. Prod., 3rd*, Vol. 2, VCH: Weinheim, 1987, pp. 127–147.
[50] Akiyama, A., Bednarsky, M., Kim, M-J., Simon, E.S., Waldmann, H., and Whitesides, G.M. *Chem. Brit.* 645–654 (1987).
[51] Soda, K., and Yonaha, K. Application of free Enzymes in Pharmaceutical and Chemical Industries. In. *Biotechnology.* Vol. 7a, Rehm H.J., and Reed. G. (eds) VCH: Weinheim, 1987, pp. 605–652.
[52] Yamada, H., and Shimizu, S. *Angew. Chemie, Int. Ed. Engl.* **27**, 622–642 (1988).
[53] Walsh, C.T., and Chen, Y-C.J.. *Angew. Chemie, Int. Ed. Engl.* **27**, 333–343 (1988)
[54] Holland, H.L. *Chem. Rev.* **88**, 473–485 (1988).
[55] Schneider, M.P and Laumen. K. Enzymes in Organic Synthesis – Chiral Building Blocks from Racemic and Prochiral Substrates. In: *Bioflavour '97*, Schreier, P. (ed.) Walter de Gruyter: Berlin, 1988, pp. 483–529.
[56] Akiyama, A., Bednarsky. M., Kim, M-J., Simon, E.S., Waldmann, H., and Whitesides, G.M. *CHEMTECH*, 627–634 (1988).
[57] Toone, E.J., Simon, E.S., Bednarsky, M.D., and Whitesides, G.M. *Tetrahedron* **45**, 5365–5422 (1989).
[58] Wong, C.-H. *Science* **244**, 1145–1152 (1989).
[59] Turner, N.J. *Natural Products Reports* **5**, 625–644 (1989).
[60] Crout, D.H.G., and Christen M. Biotransformations in Organic Synthesis. In: *Modern Synthetic Methods*, Vol. 5. Scheffold. R. (ed.) Springer-Verlag: Berlin, 1989, pp.1–114.
[61] Jones, J.B. *Pure Appl. Chem.* **62**, 1445–1448 (1990).
[62] Davies, H.G., Green, R.H., Kelly, D.R., and Roberts, S.M. *Biotechnology* **10**, 129–152 (1990).
[63] Davies, H.G., Green, R.H., Kelly, D.R., and Roberts, S.M. *Crit. Rev. Biotechnol.* **10**, 129–153 (1990).
[64] Kamphuis, J, Boesten, W.H.J., Broxtermann, Q.B., Hermes, H.F.M., van Balken, J.A.M., Meijer, E.M., and Schoemaker, H.E. *Adv. Biochem. Eng. Biotech.* **42**, 133–186 (1990).
[65] Crosby, J. *Tetrahedron* **47**, 4789–4846 (1991).
[66] Kulla, H.G. *Chimia* **45**, 81–85 (1991).
[67] Boland, W., Frössl, C., and Lorenz, M. *Synthesis*, 1049–1972 (1991).
[68] Santaniello, E., Ferraboschi, P., Grisenti, P., and Manzocchi, A. *Chem. Rev.* **92**, 1071–1140 (1992).
[69] Crout, D.H.G., Davies, S., Heath, R.J., Miles, C.O., Rathborne, D.R., and Swoboda, B.E.P. *Biocatalysis* **9**, 1–30 (1994).
[70] Coffen, D.L. Enzyme-Catalyzed Reactions. In: *Chiral Separations – Applications and Technology.* Satinder, A. (ed.) American Chemical Society: Washington, 1997, pp. 59–91.
[71] Foster, K.A., Frackman, S., and Jolly, J.F. Production of Enzymes as Fine Chemicals. In: *Enzymes, Biomass, Food and Feed.* Reed, G., and Nagodawithana, T.W. (eds) *Biotechnology*, Vol. 8a (second, completely revised edition). VCH: Weinheim, 1998, pp. 73–120.
[72] Patel, R.N. *Adv. Appl. Microbiol.* **43**, 91–140 (1997).
[73] Shimizu, S., Ogawa, J., Kataoka, M., and Kobayashi, M. *Adv. Biochem. Engin. Biotechnol.* **58**, 45–87 (1997).
[74] Laschat, S. *Angew. Chemie. Int. Ed. Engl.* **35**, 289–291 (1996).
[75] Servi, S. *Synthesis*, 1–24 (1990).
[76] Csuk, R., and Glänzer, B.I. *Chem. Rev.* **91**, 49–97 (1991).
[77] Kometani, T., Hidefumi, Y., and Matsuno, R. *J. Mol. Catal. B: Enzymatic* **1**, 45–52 (1996).
[78] Schreiber, S.L., and Verdine, G.L. *Tetrahedron* **47**, 2543–2562 (1991).
[79] Cheetham, P.S.J. *Enzyme Microb, Technol.* **9**, 194–213 (1987).
[80] Trevan, M.D. Enzyme Technology: towards Usable Catalysts. In: *Molecular Biology and Biotechnology.* Walker, J.M. and Gingold, E.B. (eds) The Royal Society of Chemistry: London, 1985, pp. 191–227.
[81] Adams, M.W.W., and Kelly, R.M. *C&EN, Dec. 18, 1995.*
[82] Robertson, D.E., Mathur, E.J., Swanson, R.V., Marrs, B.L., and Short, J.M. *SIM News* **46**, 3–8 (1996).
[83] Zaborski, O.R. (ed.) *Immobilized Enzymes.* CRC Press: Cleveland Ohio, 1973.
[84] Chang, M.S. *Biomedical Applications of Immobilized Enzymes and Proteins*, Vols. 1 and 2, Plenum Press: New York, 1977.
[85] Chibata, I., Tosa, T., and Sato, T. Application of Immobilized Enzymes in Pharmaceutical and Chemical Industries. In. *Biotechnology.* Vol. 7a, Rehm H.J., and Reed. G. (eds) VCH: Weinheim, 1987, pp. 653–684.

[86] Chibata, I. (ed.) *Immobilized Enzymes – Research and Development*. Kodansha: Tokyo, 1978.
[87] Mattiasson, B. (ed.) *Immobilized Cells and Organelles*, Vols. 1 and 2. CRC Press: Boca Raton, 1983.
[88] Mosbach. K. (ed.) Immobilized Enzymes. *Methods Enzymol.* **44**, 1976.
[89] Tosa, T., and Shibatani, T. *Ann. N.Y. Acad. Sci.* **750**, 364–375 (1995).
[90] Tanaka, A., Tosa, T., and Kobayashi, T. (eds) *Industrial Application of Immobilized Biocatalysts*. Dekker: New York, Basel, 1993.
[91] St. Clair, N.L., and Navia, M.A. *J. Am. Chem. Soc.* **114**, 7314–7316 (1992).
[92] Borman, S. *C&EN*, **70**, 40–43 (1992).
[93] Persichetti, R.A., St. Clair, N.L., Griffith, J.P., Navia, M.A., and Margolin, A.L. *J. Am. Chem. Soc.* **117**, 2732–2737 (1995).
[94] Lalonde, J. *Chemtech*, 38–45, February (1997).
[95] Lalonde, J.J., Govardhan, C., Khalaf, N., Martinez, A.G., Visuri, K., and Margolin, A.L. *J. Am. Chem. Soc.***117**, 6845–6852 (1995).
[96] Bednarsky, M.D., Chenault, H.K., Simon, E.S., and Whitesides, G.M. *J. Am. Chem. Soc.* **109**, 1283–1285 (1987).
[97] Kula, M-R., and Wandrey, C. *Methods Enzymol*, **136**, 9–21 (1987).
[98] Pabsch, K., Petersen, M., Rao, N.N., Alfermann, A.W., and Wandrey, C. *Rec. Trav. Chim. Pays-Bas* **110**, 199–205 (1991).
[99] Hummel, W., Schütte, H., Schmidt, E., Wandrey, C., and Kula, M-R. *Appl. Microbiol. Biotechnol.* **26**, 409–416 (1987).
[100] Schmidt, E., Ghisalba, O., Gygax, D., and Sedelmeier, G. *J. Biotechnol.* **24**, 315–327 (1992) and **31**, 233 (1993).
[101] Schmidt, E., Blaser, H.U., Fauquex, P.F., Sedelmeier, G., and Spindler, F. Comparison of Chemical and Biochemical Reduction Methods for the Synthesis of (R)-2-Hydroxy-4-phenylbutyric Acid. In: *Microbial Reagents in Organic Synthesis*, Servi, S. (ed.) Kluwer Academic Publishers: Amsterdam, 1992, pp. 377–388.
[102] Howaldt, M.W., Kulbe, K.D., and Chmiel, H. *Ann. N.Y. Acad. Sci.* **589**, 253–260 (1990).
[103] Kulbe, K.D., Howaldt, M.W., Schmidt, K., Röthig, T.R., and Chmiel, H. *Ann. N.Y. Acad. Sci.* **613**, 820–826 (1990).
[104] Kragl, U., Kruse, W., Hummel, W., and Wandrey, C. *Biotechnol. Bioeng.* **52**, 309–319 (1996).
[105] Seelbach, K., and Kragl, U. *Enzyme Microb. Technol.* **20**, 389–392 (1997).
[106] Seelbach, K., Riebel, B., Hummel, W., Kula, M-R., Tishkov, V.I., Egorov, A.M., Wandrey, C., and Kragl, U. *Tetrahedron Lett.* **37**, 1377–1380 (1996).
[107] Liese, A., Zelinsky, T., Kula, M-R., Kierkels, H., Karutz, M., Kragl, U., and Wandrey, C. *J. Mol. Catal. B: Enzymatic* **4**, 91–99 (1998).
[108] Hummel, W., and Kula, M-R. *Eur. J. Biochem.* **184**,1–13 (1989).
[109] Kragl. U., Vasic-Racki, D., and Wandrey, C. *Chem.-Ing.-Tech.* **64**, 499–509 (1992).
[110] Nagasawa, T., and Yamada, H. *Trends Biotech. (TIBTECH)*, **7**, 153–158 (1989).
[111] Mauger, J., Nagaswa, T., and Yamada. H. *Tetrahedron*, **45**, 1347–1354 (1989).
[112] Ogawa, J., and Shimizu, S. *Trends Biotech. (TIBTECH)*, **17**, 13–21 (1999).
[113] Yamada, H., Screening of Novel Enzymes for the Production of Useful Compounds. In: *New Frontiers in Screening for Microbial Biocatalysts*, Kieslich, K., van der Beek, C.P., de Bont, J.A.M., and van den Tweel, W.J.J: (eds) Studies in Organic Chemistry 53, Elsevier: Amsterdam, 1998, pp. 13–17.
[114] Yamada, H., Takahashi, S., Kii, Y., and Kumagai, H. *J. Ferment. Technol.* **56**, 484–491 (1978).
[115] Ogawa, J., Shimizu, S. *J. Mol. Catal. B: Enzymatic*, **2**, 163–176 (1997).
[116] Yamanaka, H., Kawamoto, T., and Tanaka, A. *J. Ferment. Bioeng.* **84**, 181–184 (1997).
[117] Hummel, W., Schütte, H., and Kula, M-R. *Ann. N.Y. Acad. Sci.* **434**, 194–205 (1984).
[118] Syldatk, C., Möller, R., Pietzsch. M, and Wagner. F., In: *Biocatalytic Production of Amino Acids and Derivatives*, Rozzell, J.D., and Wagner, F. (eds) Hanser: Munich, 1992, pp. 129–176.
[119] Syldatk, C. et al. *Biotechnol. Lett.* **14**, 99–104 and 105–110 (1992).
[120] Morin, A., Hummel, W., and Kula, M-R. *J. Gen. Microbiol.* **133**, 2101–1207 (1987).
[121] Wieser, M., Heinzmann, K, and Kiener, A. *Appl. Microbiol. Biotechnol.* **48**, 174–176 (1997).
[122] Black, T.D., Briggs, B.S., Evans, R., Muth, W.L., Vangala, S., and Zmijewski, M.J. *Biotechnol. Lett.* **18**, 875–880 (1996).
[123] Dingler, C., Ladner, W., Krei, G.A., Cooper, B., and Hauer, B. *Pestic. Sci.* **46**, 33–35 (1996).
[124] Hummel, W. *Adv. Biochem. Eng. Biotechnol.* **58**, 145–184 (1997).
[125] Brunella, A., Graf, M., Kittelmann, M., Laumen, K., and Ghisalba, O. *Appl. Microbiol. Biotechnol.* **47**, 515–520 (1997).

[126] Graf, M., Brunella, A., Kittelmann, M., Laumen, K., and Ghisalba, O. *Appl. Microbiol. Biotechnol.* **47**, 650–657 (1997).
[127] Laumen, K., Brunella, A., Graf, M., Kittelmann, M., Walser, P., and Ghisalba, O., New Biocatalytic Approaches for the Synthesis of Chiral Drugs, Intermediates, and Substrates. In: *Trends in Drug Research II*, van der Goot, H. (ed.) Pharmacochemistry Library, Vol. 29. Elsevier: Amsterdam, 1998, pp. 17–28.
[128] Archelas, A., and Furstoss, R. *Annu. Rev. Microbiol.* **51**, 491–525 (1997).
[129] Archelas, A., and Furstoss, R. *Trends Biotech. (TIBTECH)* **16**, 108–116 (1998).
[130] Steinke, D., and Kula, M-R. *Angew. Chemie, Int. Ed.* **29**, 1139–1140 (1990).
[131] Kammermeier-Steinke, D., Schwarz. A., Wandrey, C., and Kula, M.-R. *Enzyme Microb. Technol.* **15**, 764–769 (1993).
[132] Reidel, A, and Waldmann, H. *J. Prakt. Chem.* **335**, 109–127 (1993).
[133] Schelhaas, M., and Waldmann, H. *Angew. Chemie, Int. Ed. Engl.* **35**, 2056–2083 (1996).
[134] Sauerbrei, B, Jungmann, V., and Waldmann, H. *Angew. Chemie, Int. Ed. Engl.* **37**, 1143–1146 (1998).
[135] Smitskamp-Wilms, E., Brussee, J., van der Gen, A., van Scharrenburg, and G.J.M. Sloothaak, J.B. *Recl. Trav. Chim. Pays. Bas* **110**, 209–215 (1991).
[136] Effenberger, F., Ziegler, T., and Förster, S. *Angew. Chemie, Int. Ed.* **26**, 458–460 (1987).
[137] Effenberger, F., Hörsch, B., Weingart, F., Ziegler, T., and Kühner, S. *Tetrahedron Lett.* **32**, 2605–2608 (1991).
[138] Effenberger, F. *Angew. Chemie, Int. Ed. Engl.* **33**, 1555–1564 (1994).
[139] Kragl, U., Neidermeyer, U., Kula, M-R., and Wandrey, C. *Ann. N.Y. Acad. Sci.*, **613**, 167–175 (1990).
[140] Niedermeyer, U., and Kula, M.R. *Angew. Chemie*, **102**, 424–425 (1990).
[141] Effenberger, F., Hörsch, B., Förster, S., and Ziegler, T. *Tetrahedron Lett.* **31**, 1249–1252 (1990).
[142] Bednarski, M.D., Simon, E.S., Bischofberger, N., Fessner, W-D., Kim, M-J., Lees, W., Saito, T., Waldmann, H., and Whitesides, G.M. *J. Am. Chem. Soc.*, **111**, 627–636 (1989).
[143] Bednarski, M.D., Applications of Enzymatic Aldol Reactions in Organic Synthesis. In: *Applied Biocatalysis*, Vol. 1, Blanch, H.W., and Clark, D.S. (eds) Marcel Dekker: New York, 1991, pp. 87–116.
[144] Wang, P.G., Fitz, W., and Wong, C.-H. *CHEMTECH*, 22–32, April 1995.
[145] Takayama, S., McGarvey, G.J., and Wong, C.-H. *Annu. Rev. Microbiol.* **51**, 285–310 (1997).
[146] Schauer, R. *Trends Biochem. Sci. (TIBS)*, **10**, 357–360 (1985).
[147] Kragl. U., Gygax, D., Ghisalba, O., and Wandrey, C. *Angew. Chemie, Int. Ed. Engl.* **30**, 827–828 (1991).
[148] Kragl, U., Kittelmann, M., Ghisalba, O., and Wandrey, C. *Ann. N.Y. Acad. Sci.*, **750**, 300–305 (1995).
[149] Liu, J.L.-C., Shen, G.J., Ichikawa, Y., Rutan, J.R., Zapata, C., Vann, W.F., and Wong, C.-H. *J. Am. Chem. Soc.* **114**, 3901–3910 (1992).
[150] Sugai, T., Kuboki, A., Hiramatsu, S., Okazaki, H., and Ohta, H. *Bull. Chem. Soc. Jpn.* **68**, 3581–3589 (1995).
[151] Hughes, S. *Scrip Magazine*, 28–31, April 1994.
[152] Toone, E.J., Simon, E.S., Bednarski, M.D., and Whitesides, G.M. *Tetrahedron* **45**, 5365–5422 (1989).
[153] David, S., Augé, C., and Gautheron, C. *Adv. Carbohydr. Chem. Biochem.* **49**, 175–237 (1991).
[154] Drueckhammer, D.G., Hennen, W.J., Pederson, R.L., Barbas, C.F., Gautheron, C.M., Krach, T., and Wong, C.-H. *Synthesis* 499–525 (1991).
[155] Ichikawa, Y., Look, G.C., and Wong, C.-H. *Anal. Biochem.* **202**, 215–238 (1992).
[156] Wong, C.-H., Halcomb, R.L., Ichikawa, Y., and Kajimoti, T. *Angew. Chemie, Int. Ed. Engl.* **34**, 412–432, 521–546 (1995).
[157] Nilsson, K.G.I., Enzymatic Synthesis of Complex Carbohydrates and Their Glycosides. In: *Applied Biocatalysis*, Vol. 1, Blanch, H.W., and Clark, D.S. (eds) Marcel Dekker: New York, 199??, pp. 117–177.
[158] Simon, E.S., Bednarsky, M.D., and Whitesides, G.M. *J. Am. Chem. Soc.* **110**, 7159–7163 (1988).
[159] Shames, S.L., Simon, E.S., Christopher, C.W., Schmid, W., and Whitesides, G.M. *Glycobiology* **1**, 187–191 (1991).
[160] Kittelmann, M., Klein, T., Kragl, U., Wandrey, C, and Ghisalba, O. *Ann. N.Y. Acad. Sci.* **672**, 444–450 (1992).
[161] Kittelmann, M., Klein, T., Kragl, U., Wandrey, C., and Ghisalba, O. *Appl. Microbiol. Biotechnol.* **44**, 59–67 (1995).

[162] Kragl, U., Klein, T., Vasic-Racki, D., Kittelmann, M., Ghisalba, O. and Wandrey, C. *Ann. N.Y. Acad. Sci.* **799**, 577–583 (1996).

[163] Baisch, G., and Öhrlein, R. *Bioorg. Med. Chem.* **5**, 383–391 (1997).

[164] Bülter, T., Wandrey, C, and Elling, L. *Carbohydrate Res.* **305**, 469–473 (1997).

[165] Ichikawa, Y., Liu, J.L.-C., Shen, G.-J., and Wong, C.-H. *J. Am. Chem. Soc.* **113**, 6300–6302 (1991).

[166] Koizumi, S., Endo, T., Tabata, K., and Ozaki, A. *Nature Biotechnology* **16**, 847–850 (1998).

[167] Laumen, K., and Ghisalba, O. *Biosci. Biotech. Biochem.* **58**, 2046–2049 (1994).

[168] Wandrey, C., Reaction Engineering for Enzyme-Catalyzed Biotransformations. In: *Enzyme Catalysis in Organic Synthesis – A Comprehensive Handbook*, Vol. 1, Drautz, K., and Waldmann, H. (eds) VCH: Weinheim, 1995, pp. 89–155.

[169] Tramper, J. *Biotechnol. Bioeng.* **52**, 290–295 (1996).

[170] Cheetham, P.S.J., Case Studies in Applied Biocatalysis – From Ideas to Products. In: *Applied Biocatalysis*, Cabral, J.M.S., Best, D., Boross, L, and Tramper, H. (eds) Harwood Academic Publishers: Chur, 1994, pp. 47–108.

[171] Kobayashi, M., Nagasawa, T., and Yamada, H. *Trends Biotech. (TIBTECH)*, **10**, 402–408 (1992).

[172] Adlercreutz, P., Iborra, J.L., Schmidt, E., and Pedersen, S., Applications. In: *Applied Biocatalysis*, Cabral, J.M.S., Best, D., Boross, L, and Tramper, H. (eds) Harwood Academic Publishers: Chur, 1994, pp. 109–156.

[173] Tramper, J., Applied Biocatalysts: From Product Request to Idea to Product. In: *Applied Biocatalysis*, Cabral, J.M.S., Best, D., Boross, L, and Tramper, H. (eds) Harwood Academic Publishers: Chur, 1994, pp. 1–45.

[174] Birch, O.M., Brass, J.M., Kiener, A., Robins, K., Schmidhalter, D., Shaw, N.M., and Zimmermann, T. *Chimica oggi/Chemistry Today*, 9–13, September, 1995.

[175] Glöckler, R., and Roduit, J-P. *Chimia*, **50**, 413–415 (1996).

[176] Hoeks, F.W.J.M.M., Kulla, H., and Meyer, H-P. *J. Biotechnol.* **22**, 117–128 (1992).

[177] Hoeks, F.W.J.M.M., Mühle, J., Böhlen, L., and Psenicka, I. *Biochem. Engin. J.*, **61**, 53–61 (1996).

[178] Bommarius, A.S., Schwarm, M., Stingl, K., Kottenhahn, M., Huthmacher, K., and Drauz, K. *Tetrahedron Asymmetry* **6**, 2851–2888 (1995).

[179] Balkenhohl, F., Ditrich, K., Hauer, B., and Ladner, W. *J. Prakt. Chem.* **339**, 381–384 (1997).

[180] Kumagai, H., Matsui, H., Ohgishi, H., Ogata, K., Yamada, H., Ueno, T., and Fukami, H. *Biochem. Biophys. Res. Commun.* **34**, 266–270 (1969).

[181] Yamada, H., and Kumagai, H. *Adv. Appl. Microbiol.* **19**, 249–288 (1975).

[182] Enei, H., Nakazawa, E., Tsuchida, T., Namerikawa, T., and Kumagai, H. *Japan Bioindustry Letters*, **13**, No. 1, 2–4, April, 1996.

8 Selective Glycosidation Reactions and Their Use in Medicinal Chemistry

Giuseppe Capozzi, Stefano Menichetti and Cristina Nativi

Abbreviations

All	Allyl
Ac	Acetyl
Bn	Benzyl
Bz	Benzoyl
Cp	Cyclopentadienyl
DAST	Diethylaminosulfur trifluoride
DBU	1,8-Diazobicyclo[5.4.0]undec-7-ene
DIBALH	Di-iso-butylaluminum hydride
DPTBS	Diphenyl-tert-butylsilyl
MCA	Chloroacetyl
MS	Molecular sieves
NBS	N-bromosuccinimide
NIS	N-iodosuccinimide
Piv	Pivaloyl
Pht	Phthaloyl
PTSA	Para-toluenesulfonic acid
TBDMS	Tert-butyldimethylsilyl
TBPA	Tri(4-bromophenyl)ammoniumyl hexachloroantimonate
TfO^-	Trifluoromethanesulfonate
TIPS	Tri-iso-propylsilyl
TMS	Trimethylsilyl
TNCP	Tetrachlorophthalimido

8.1 Introduction

Carbohydrates, known to organic chemists for more than a century, have always occupied a confined area of interest [1–4]. Recently, the discovery of saccharides as abundant moieties on cellular surfaces, where they play a fundamental role in cellular communication processes or constitute the hook-up for proteins, bacteria and antibodies, has prompted a general upsurge of interest in carbohydrate chemistry, which nowadays represents a link between organic chemistry and medicinal chemistry [5–9].

Sugar molecules are components of antibodies, antitumor and antiviral agents [10]; moreover, they have an especially important role as source of biological information in glycolipids and glycoproteins, so that often non-glycosylated proteins or lipids do not express their biological activities any longer [7, 11, 12].

Another fundamental discovery in carbohydrate chemistry regards the role of sialyl Lewis^X in inflammatory diseases [13–17]. Sialyl Lewis^X is a terminal tetrasaccharide (α-Neu-5-Ac-(2′3)-β-D-Gal-(1′4)-[α-L-Fuc-(1′3)]-β-Glc-NAc) of glycoproteins and glycopeptides displayed on the surface of white blood cells. The cell-surface receptors P-, E- and L-selectins, which are present on vascular endothelium and on platelets, are able to recognize and bind the tetrasaccharide sialyl Lewis^X, following the action of inflammatory factors [18, 19].

Cell-surface carbohydrates are also involved in tumor progression: changes in their sequence or configuration seem to be involved in the state of metastasis [11].

Because carbohydrates have such a relevant biochemical role, many efforts have been devoted to the elucidation of the nature of carbohydrates in glycoconjugates as well as to elucidate the biological information contained in sugar chains; to this purpose a new subdiscipline called "glycobiology" was recently established.

The renewed interest in carbohydrate chemistry has also prompted the research for new and more efficient methods to supply complex sugar-containing compounds, whose isolation from natural sources is generally insufficient. In the synthetic process, a crucial point is the regio- and stereocontrolled formation of the glycosidic bonds between two sugar moieties, or between a saccharide and an aglycone residue. For example, in the formation of a glycosyl bond between galactose (**A**) and glucose (**B**) (Fig. 1), **A** can be linked to four different positions of **B**, namely C-2, C-3, C-4, and C-6, resulting in four isomers. Moreover, since the galactose residue can have α or β configuration at the anomeric carbon, and can exist in the furanose or the pyranose form, the number of possible isomers rises to 16 [7].

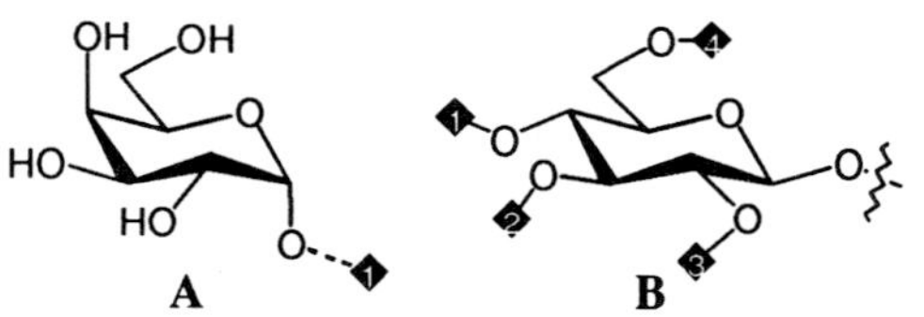

Figure 1. Galactose (**A**) and glucose (**B**) involved in sugar chain formation.

Since each oligosaccharide has a specific biological role which can be deeply altered by any possible change in configuration and in points of attachment of the sugar chain, as well as in substitution of sugar moieties, it is evident that a regio- and stereoselective glycosidation is mandatory to control the key events for healthy cellular development or interruption of established diseases [11].

Important advances have been achieved in this field and valuable monographs [20–28] and reviews [29–37] are available, providing the organic chemist with up-to-date methods concerning selective glycosyl syntheses.

8.2 Selective Protection of Carbohydrates

Except for a very limited number of cases, in order to address the regioselectivity issue and to achieve an efficient glycosyl synthesis, the first problem to face is a suitable derivatization of all those hydroxyl groups not involved in the glycosyl bond generation. Moreover, the introduction of differentiated protecting groups which can be selectively removed is a desirable feature. Glycosyl donors or acceptors protected as ether derivatives are the most widely used synthetic intermediates which allow an easy and systematic manipulation of hydroxyls in a carbohydrate moiety, without affecting the optically purity of the molecule. The subject has been extensively reviewed [38–46], and a complete listing of protecting groups is beyond the scope of the present chapter. Among the rich repertoire of groups and methods, we have selected three widespread methodologies that have received recent developments and improvements namely: (i) Protection as acetals and ketals; (ii) activation of hydroxyls as stannoxanes; and (iii) protection as silylethers, mainly focusing on practical aspects that are of concern for the synthetic chemist. Our aim is to provide guidelines that permit readers to find their way through such an overwhelming field.

8.2.1 Acetal Derivatives

The temporary protection of 1,2 or 1,3-hydroxy groups as cyclic acetals has been frequently employed. Cyclic acetals are easily prepared in high yields, they are stable to strong bases, to metal hydrides and to mild oxidants, and they can be easily cleaved in high yields [39]. *O*-Isopropilidene and *O*-benzylidene derivatives are the most commonly used protected sugars. For instance, 1,2:5,6-di-*O*-isopropylidene-D-glucofuranose (**1**), [47] 1,2:3,4-di-*O*-isopropylidene-D-galactopyranose (**2**) [47] and 1-*O*-methyl-4,6-*O*-benzylidene-α-D-glucopyranose (**3**) [48] are useful starting materials for the synthesis of a large number of carbohydrate-containing molecules [40]. As a general rule, aldehydes tend to form six-membered rings from hydroxyls in 1,3 relative position, while ketones tend to engage hydroxyl groups in 1,2 relative position to give five membered rings (Fig. 2).

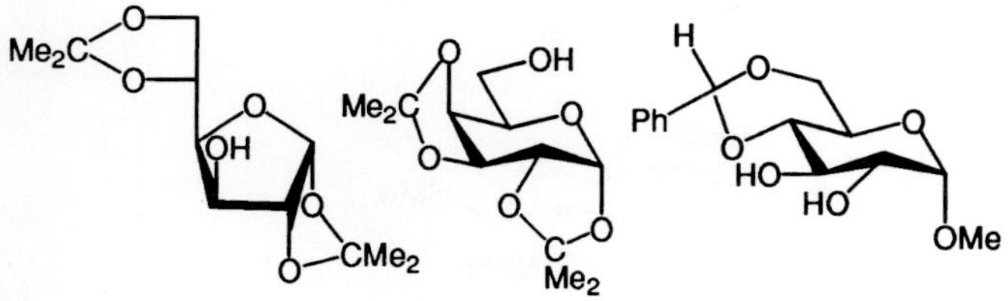

1 **2** **3**

Figure 2. O-Diisopropylidene acetals **1**, **2** and O-benzylidene acetal **3**.

Isopropylidene acetals are satisfactorily prepared by reacting the carbohydrate moiety with stoichiometric amounts of dimethoxypropene in the presence of catalytic *p*-toluenesulfonic acid (PTSA), a convenient variant of the classical procedure consisting in the use of acetone and PTSA in the presence of a drying agent [39].

Benzylidene acetals are generally prepared following two standard procedures that is, by reacting a dihydroxy compound: (i) with benzaldehyde in tetrahydrofuran in the presence of stoichiometric amounts of Lewis acid; and (ii) with benzaldehyde dimethylacetal in dimethylformamide in the presence of a catalytic amount of PTSA [39].

With regard to the preparation of benzylidene acetals, there are two major drawbacks. Whenever possible, both the 1,2 and the 1,3 acetals are formed, leading to a mixture of products. Furthermore, the equilibrium of acetal formation is disfavored by the water produced in the reaction. These disadvantages are largely overcome by reacting benzaldehyde dimethylacetal in tetrahydrofuran in the presence of stoichiometric amounts of zinc chloride–diethylether complex; the use of benzaldehyde dimethylacetal instead of benzaldehyde itself favors the formation of 1,3-dioxanes and facilitates the purification of final products, while the use of zinc chloride–diethylether complex instead of PTSA avoids the presence of added water in the reaction medium, increasing the yield of the desired products.

Free sugars are normally restored from *O*-isopropylidene acetals by acidic treatment. Instead, *O*-benzylidene protecting groups can be cleaved with a number of reagents which do not return the free sugar itself, but selectively deprotect only one of the two hydroxyl groups. An efficient regioselective reductive cleavage of a benzaldehyde acetal is accomplished under mild conditions by diisobutylaluminum hydride (DIBALH) [39] or lithium aluminum hydride and aluminum trichloride at room temperature [49], to afford the benzyl ether of the most substituted hydroxyl and the unprotected hydroxyl group of the least substituted position as shown in Scheme 1 for the reduction of **4** to **5**. The reversed regioselectivity (the benzyl ether of the least substituted alcohol) is obtained either by sodium cyanoborohydride [50] or by borane–trimethylamine complex in the presence of aluminum trichloride [51] (see Scheme 1). When two benzylidene groups are present in the molecule, selective deprotection has been achieved under very mild conditions, by a borane-tetrahydrofuran complex in the presence of dibutylboron triflate [52] (see section 2.4). The selective partial deprotection of non-symmetric benzylidene acetals represents a substantial advance in protecting groups chemistry and it has found extensive use in oligosaccharide synthesis [41].

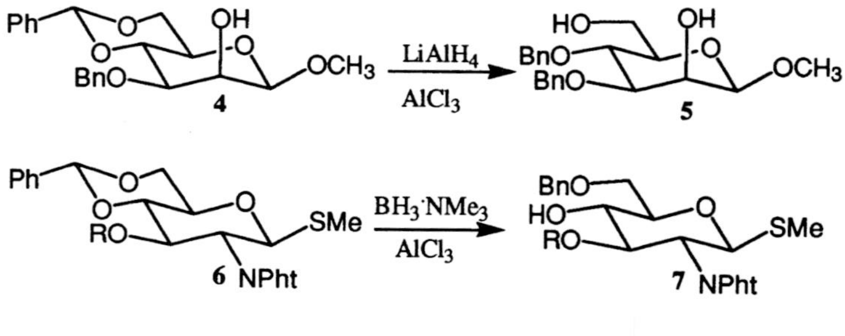

8.2.2 Activation via Stannoxanes

Selective activation of hydroxyl groups by the stannylation procedure is an established practice to efficiently manipulate diols and polyols [53, 54]. Among the various reagents and methods, activation through dialkylstannylene acetals is noteworthy. This two-step protection method involves the conversion of dihydroxy substrates into cyclic dialkyl dioxastannanes which, in turn, are treated with suitable nucleophiles, to regioselectively afford the corresponding mono alkyl, silyl or acyl derivatives under neutral conditions.

Dialkylstannylene acetals which are usually prepared by azeotropic removal of water in the presence of dibutyltin oxide, can be used to selectively protect only one of two secondary hydroxyl groups or a primary, in the presence of a secondary, hydroxyl group. When more than two free hydroxyl groups are present on the same molecule, the tin atom can bridge several different pairs of oxygen atoms. Nevertheless only one derivative is generally obtained [53].

Many aspects of the regioselective manipulation of polyols through dialkylstannylene acetals have been studied and some interesting modifications have improved this procedure [53, 55]. For example, the regioselective formation of monobenzyl, monoallyl and monomethyl ethers, which normally proceeds at very slow speed, is markedly enhanced when the reaction of benzyl and allyl bromides or methyl iodide on dialkylstannylene derivatives of polyhydroxy compounds is carried out in the presence of stoichiometric amounts of quaternary ammonium halides [37,56,57]. Several examples of this modified procedure, such as the regioselective mono-*O*-alkylation of disaccharide glycosides (Scheme 2), have been reported [58].

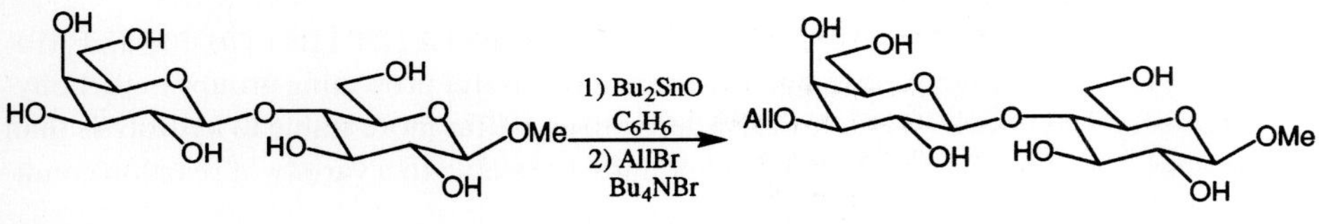

For small-scale manipulations, and in those cases where prolonged heating would affect the substrate, the use of the commercially available dibutyltin dimethoxide in place of the less reactive dibutyltin oxide, is advisable [59] (Scheme 3).

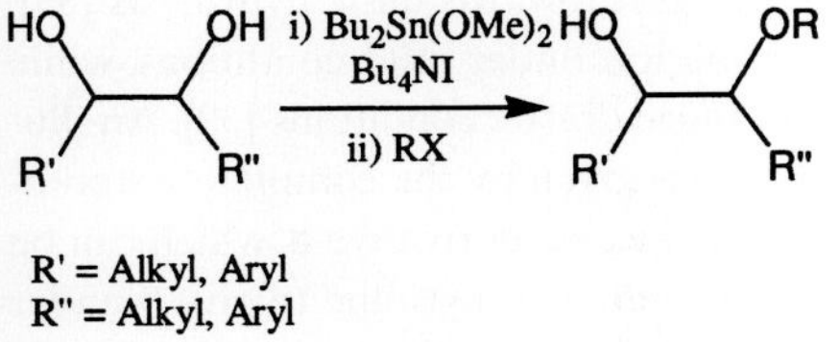

A remarkable application of the stannylation procedure is the organotin-mediated monoacylation of diols with reversed chemoselectivity, by which monoesterification of unsymmetrically substituted diols at the most substituted hydroxyl group can be achieved with acyl chlorides [60]. In the reported mechanism [61], this unusual reversal of chemoselectivity rests on a fast intramolecular transesterification equilibrium in which the dibutylstannylene acetal plays the double role of reagent and catalyst. The knowledge of the reaction mechanism allows for adjustment of experimental conditions to achieve remarkable selective level which can be higher than 99% using appropriate reagents (Scheme 4).

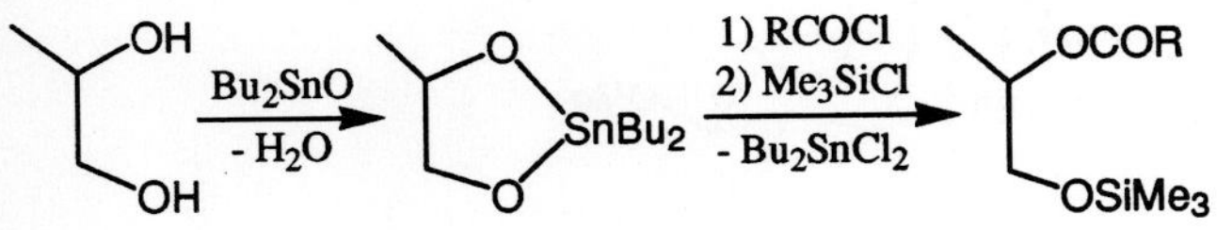

8.2.3 Silyl Derivatives

Another class of widely employed protecting groups is represented by silyl ethers [38, 39, 62–64]. The trimethylsilyl (TMS) moiety was the silyl group originally employed [63, 64], but because of the low stability of TMS ethers in acidic or basic media [66, 67], many other silyl derivatives are nowadays preferred in synthetic organic chemistry.

All trialkylsilyl ethers are deprotected with fluoride ion (tetrabutylammonium fluoride or cesium fluoride in tetrahydrofuran, or hydrofluoric acid in acetonitrile). The advantage of using fluoride-cleaving reagents is due to their compatibility with a wide variety of functional and protecting groups [38, 39, 67]. Trialkylsilyl ethers are generally cleaved by acidic or basic treatment, but their stability and easy deprotection can be tuned by varying substituents on silicon. The presence of bulky groups on silicon increases the stability of the corresponding silyl ethers and facilitates selective protection of polyhydroxylated molecules, depending on their steric requirements. The use of bulky trialkylsilyl ethers is nowadays unavoidable. As a matter of fact, a silyl protecting group is employed at some intermediate step in virtually every total synthesis.

t-Butyldimethylsilyl (TBDMS) [68, 69], diphenyl-*t*-butylsilyl (DPTBS) [70–72] and tri-*i*-propylsilyl (TIPS) [73, 74] have become the three most useful protecting groups in carbohydrate chemistry. TBDMS, DPTBS and TIPS derivatives are far more stable to hydrolysis than TMS analogs. They are often crystalline compounds, are stable to a variety of reaction conditions [75–82], and do not possess additional stereogenic centers.

Best results are obtained with TIPS ethers in terms of regioselectivity, yields and ease of cleavage [83–87]. The large steric bulk of the tri-*i*-propylsilyl group ensures high selectivity in the protection of primary versus secondary hydroxyl groups, and the least hindered versus the most hindered secondary hydroxyls. TIPS ethers stability to acidic hydrolysis is intermediate between that of TBDMS and of DPTBS, but it is the largest towards basic hydrolysis [88]. TIPS ethers are quantitatively deprotected with fluoride ion under mild conditions, while DPTBS ethers are often cleaved in lower yields under more drastic conditions [39]. An illuminating example of the merits of TIPS protecting group is given by the completely stereoselective protection of the hydroxyl group at C-3 of the galactose derivative **8**, which can be satisfactorily achieved by tri-*i*-propylsilyl triflate in the presence of pyridine to give **9** under

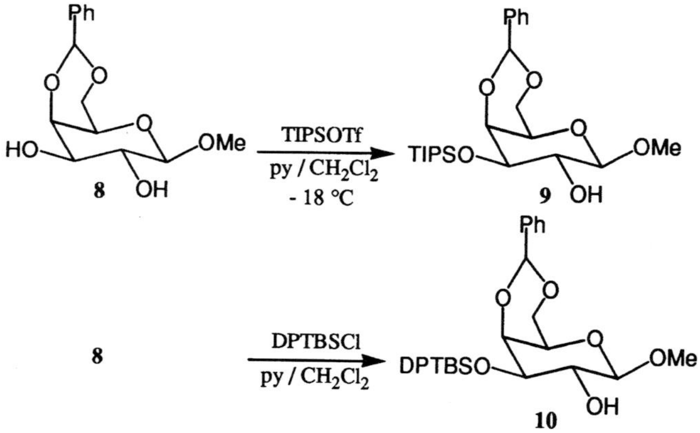

very mild conditions (see section 2.4) (Scheme 5). Correspondingly, the reaction of **8** with *t*-butyldimethylsilyl triflate affords a mixture of regioisomers. On the other hand, DPTBS chloride reacts with **8** affording regioselectively the 2-hydroxy-3-*O*-*t*-butyldiphenylsilyl ether (10), but is subsequently cleaved in low yields.*

8.2.4 Experimental Procedures

Methyl 3,4-di-O-benzyl-β-D-mannopyranoside (**12**): [49]

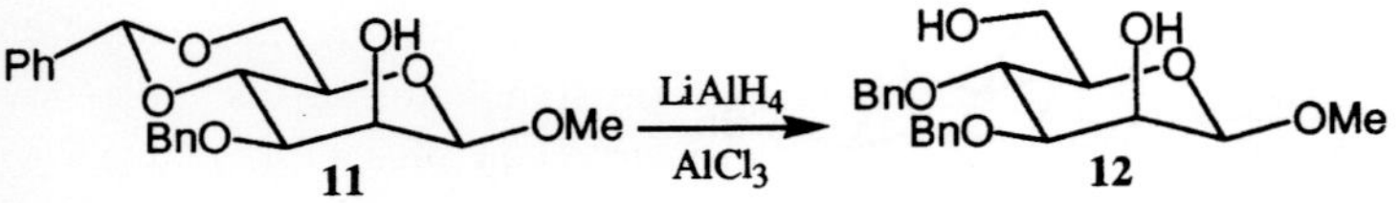

11 **12**

A mixture of **11** (100 mg), $LiAlH_4$ (10 mg) and $AlCl_3$ (34 mg) was refluxed in 1:1 ether-dichloromethane (10 mL) for 2 h. After work-up, the syrupy **12** (84 mg; 83.5%) was obtained as a single product $[\alpha]_D$ −27.9° (c 0.39, chloroform). Flash chromatography on silica gel, R_f 0.45 (dichloromethane/methanol, 19:1).

Methyl 6-O-benzyl-2-deoxy-3-O-(phenylcarbamoyl)-2-phthalimido-1-thio-β-D-glucopyranoside (**14**): [51]

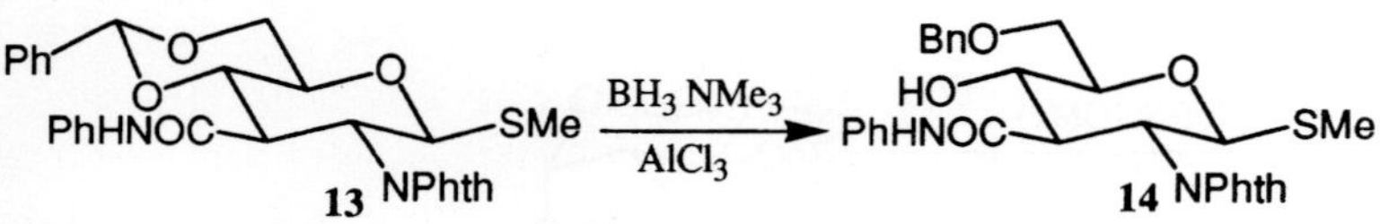

13 **14**

Compound **13** (3.583 g, 6.556 mmol) was treated with $BH_3 \cdot NMe_3$ (3.80 g, 52 mmol) and $AlCl_3$ (7.00 g, 52 mmol), stirred at r.t. for 18 h and then diluted with Et_2O. Ice/water and 2 N aqueous HCl (10 mL) were added and the mixture was filtered through Celite®. The filtrate was extracted twice with Et_2O and the combined extracts were processed as usual. MeOH was added to the residue, and the mixture was exposed to high vacuum. This treatment was repeated three times. Column chromatography of residue on silica gel (toluene/EtOAc, 3:1) afforded 2.580 g (72%) of 14: m.p. 148–150 °C (Et_2O); R_f 0.29; $[\alpha]_D$ +19.4° (c 0.8).

Methyl 2,3:4,6-di-O-benzylidene-α-D-mannopyranoside (**16**): [52]

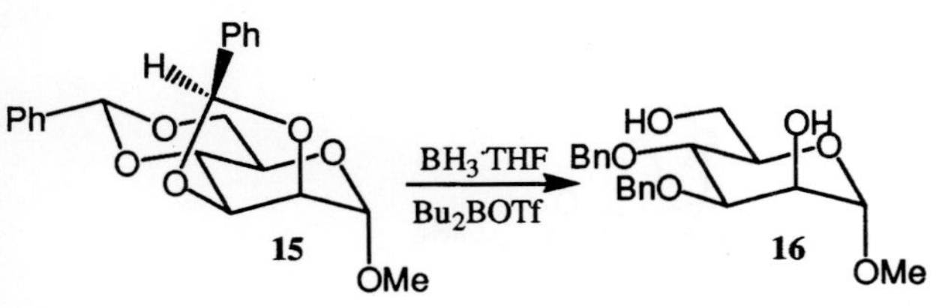

15 **16**

---

* Capozzi, G., Falciani, C., Menichetti, S., and Nativi, C., unpublished results.

A solution of 1 M BH_3 in THF (10 mL) was added to a 50 mL dry flask containing compound 15 at 0 °C and the solution was stirred for 5 min. A solution of 1 M Bu_2BOTf in CH_2Cl_2 (1 mL) was then added slowly to the clear solution. After 1 h at 0 °C, TLC showed that the starting material had disappeared. Triethylamine (0.5 mL) was then added to the reaction mixture followed by careful addition of methanol until the evolution of H_2 had ceased. The reaction mixture was co-distilled with methanol three times before being put on the silica gel column. Elution with 1:1 hexane/ethyl acetate gave the pure 3,4-dibenzyl derivative **16**.

Benzyl 3-O-allyl-β-D-galactopyranoside (**18**): [57]

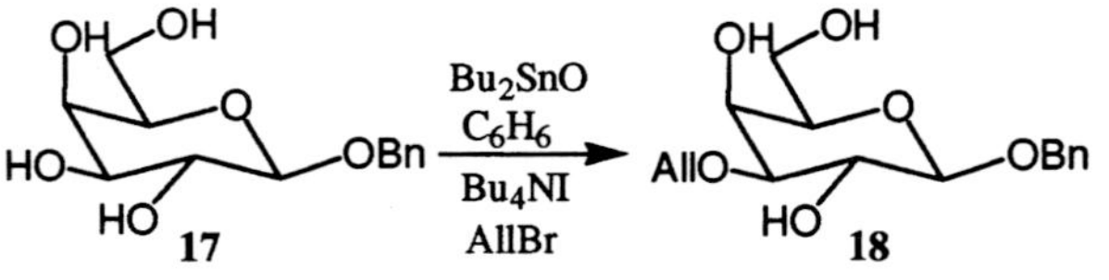

A mixture of the glycoside **17** (1 mmol) and dibutyltin oxide (1 mmol) in benzene was refluxed for 16 h with azeotropic removal of water. The solution was evaporated to ca. 25 mL, tetrabutylammonium iodide (1 mmol) and allyl bromide (0.5 mL) were added, and the mixture was heated at 60 °C for 8 h. Evaporation to dryness gave a residue which was processed by column chromatography on silica gel (chloroform/methanol, 95:5). Crystals (**18**, 85%), m.p. 105 °C (Et_2O), $[\alpha]_D$ −20° (2.0, dichloromethane).

Methyl 3-O-benzyl-4,6-O-benzylidene-α-D-glucopyranoside (**20**): [59]

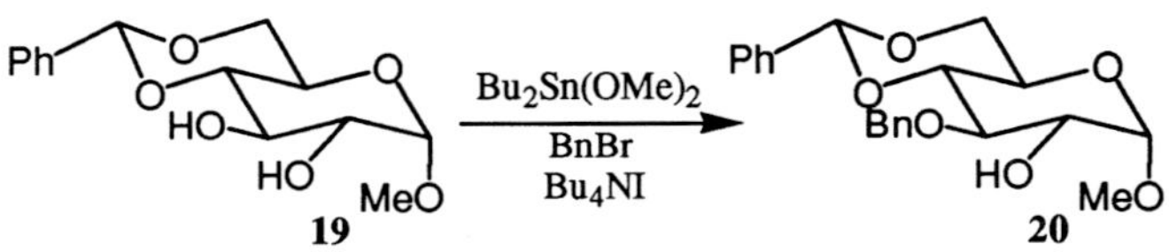

The diol **19** (1 mmol) was taken up in dry benzene (or toluene) (40 mL) and the dibutyltin dimethoxide (1.1 mmol) added in one portion. Benzene (20 mL) was then removed by distillation and the tin acetal solution allowed to cool to r.t. To the solution was added benzyl bromide (1.1 mmol) and tetra-N-butylammonium iodide (1.5 mmol) and the temperature was then raised to 50 °C and maintained until the reaction was complete by TLC. The reaction mixture was quenched by the addition of water (10 mL) and standard worked-up. The crude was purified by flash chromatography on silica gel to give **20** (83%).

Methyl 4,6-O-benzylidene-3-O-tri-iso-propyl-β-D-galactopyranoside (**9**):*

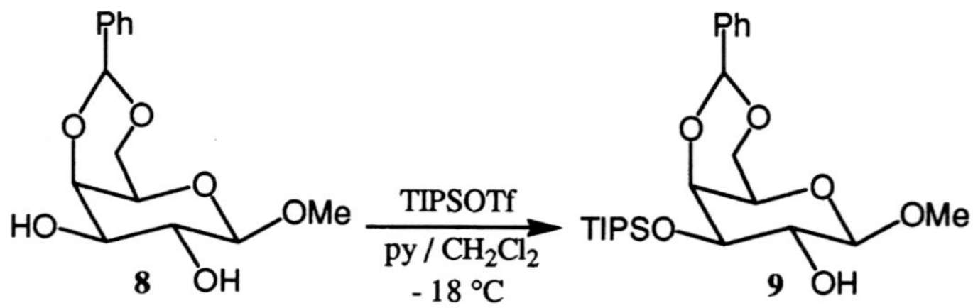

* Capozzi, G., Falciani, C., Menichetti, S., and Nativi, C., unpublished results

To a solution of diol **8** in dichloromethane (0.3 mL) was added pyridine (0.5 mL) and the reaction mixture, cooled to −18 °C, was treated dropwise with 1.2 equiv. of TIPSOTf. During the addition the reaction temperature was carefully controlled. When the starting material was consumed (checked by TLC), the solution was evaporated to dryness to give a residue which was processed by column chromatography on silica gel (petroleum ether/EtOAc, 3:1). Monosilyl ether **9** (146 mg, 95%) was obtained as a glassy solid.

8.3 *O*-Glycosidation Reaction

The second issue to address is the stereoselectivity of the glycosyl linkage.

Unfortunately, a universal method for selective *O*-glycosidation does not yet exist. In fact, even small changes of reaction conditions, nature of donor, acceptor or promoter can dramatically compromise the success of the selected glycosidation protocol. However well-adaptable procedures are nowadays available, affording *O*-glycoside chains with a remarkable α/β selectivity.

In the following sections some of the most frequently used methods for stereoselective *O*-glycosidation will be discussed, with particular attention to their applications in the synthesis of relevant biologically active molecules.

8.3.1 The Koenigs–Knorr and Related Methods

In 1901, Koenigs and Knorr published an efficient method for the preparation of alkyl and aryl *O*-glycosides [89]. The Koenigs–Knorr reaction has for a long time been the best method achievable for the synthesis of *O*-glycosides. This method has been widely studied and improved in the years [31, 90, 91], and although valuable alternatives are now established, it still is the method of choice when more sophisticated procedures fail.

In the Koenigs–Knorr method, glycosyl halides are the glycosyl donors. The procedure consists of the treatment of glycosyl bromides or chlorides, easily obtained as pure α anomers [3], with alcoholic or phenolic acceptors in the presence of an excess of a heavy metal salt [92]. Silver carbonate or silver oxide were originally used, but at present silver triflate or mercury (II) salts are the most frequently promoters employed [93].

Due to the lability of some glycosyl halides, the α-chloro or α-bromoglycosyl donors can be prepared "in situ" (activation step) and reacted with suitable acceptors (glycosidation step) in the presence of halophilic promoters. The reaction proceeds via the oxocarbenium ion **23** (Scheme 6), which is transformed into the corresponding *O*-glycoside **24**.

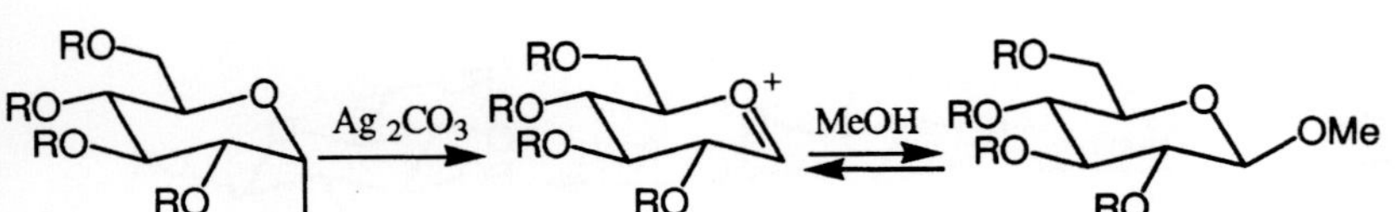

23 **24**

A disadvantage affecting the original method is the low selectivity often observed in the glycosyl linkage formation, due to the intermediacy of the oxocarbenium ion. Fortunately, it has been found that with the appropriate choice of the substituent at C-2 and its relative stereochemistry to C-1 the stereochemistry of the glycoside bond can be conveniently controlled. As a matter of fact, when the hydroxyl group at the C-2 position of the haloglycoside is protected as an ester, the initially formed oxocarbenium ion is stabilized by the anchimeric assistance of the ester carbonyl group. The dioxolanium ion intermediate **25** subsequently undergoes a nucleophilic attack by the acceptor. Because of this neighboring group participation, the cyclic intermediate **25** reacts at the anomeric center from the opposite side with respect to the substituent that is *trans* to substituent at C-2 affording the corresponding β-*O*-glycoside **26** (Scheme 7).

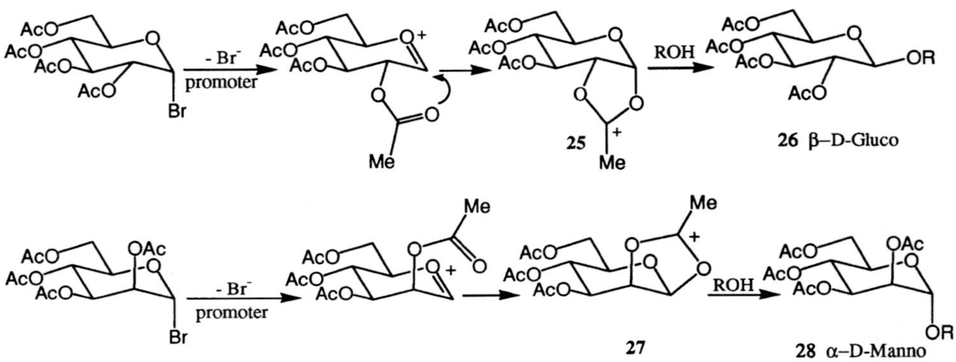

As shown in Scheme 7, in the D-*gluco* series the glycosidation gives the corresponding β-D-*gluco* derivative **26**. In the D-*manno* series, the C-2 carbon has the opposite configuration so that the corresponding dioxolanium intermediate **27** resides on the β-site of the mannoside and the subsequent opening affords the α-D-*manno*pyranoside **28** [94]

In some interesting modifications of the Koenigs–Knorr method, bromide or chloride leaving groups are replaced with fluoride [95–97] or alkyl- and arylthio groups [98–100]. Although known for many years, glycosyl fluorides have only recently become seen as useful donors; that is, when efficient methods for their preparation and fluorophilic promoters for their conversion into reactive glycosylating agents were introduced [101-103]. Although glycosyl fluorides are less reactive than the corresponding chlorides and bromides, their value as glycosyl donors is demonstrated by the wide number of reactions they undergo when suitably activated [3].

Similar to glycosyl fluorides, thioglycosides – first introduced in glycosidation reactions by Ferrier [104, 105] – offer efficient temporary protections of the anomeric center. They can be introduced at an early stage of the synthesis and selectively activated to form active glycosyl donors [106] (Scheme 8).

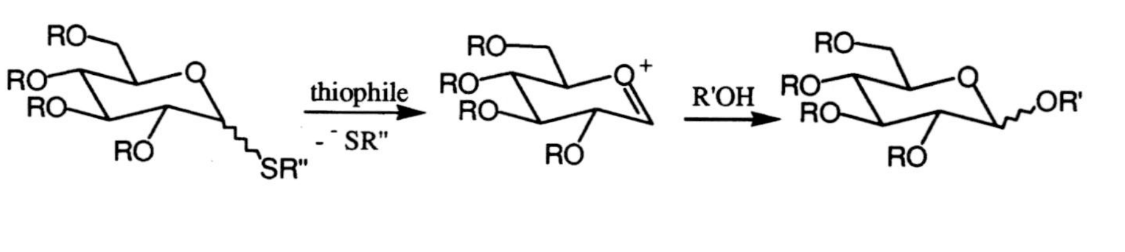

A variety of thiophilic promoters have been employed for the activation of thioglycosides. Heavy-metals salts, such as mercury and copper salts, were initially used, but recently more efficient activators have been introduced [22]. Among several different variants, a conceptually novel method is worth mentioning which involves a single-electron transfer from sulfur to the activating species tri(4-bromophenyl)ammoniumyl hexachloroantimonate ($TBPA^{+\cdot}$) [32] (Scheme 9). The radical cation $GlySR^{+\cdot}$ formed undergoes cleavage to afford the thiyl radical ($RS^{\cdot}$) and the Gly^{+} cation, which is stabilized by the neighboring oxygen atom and trapped "in situ" by the acceptor.

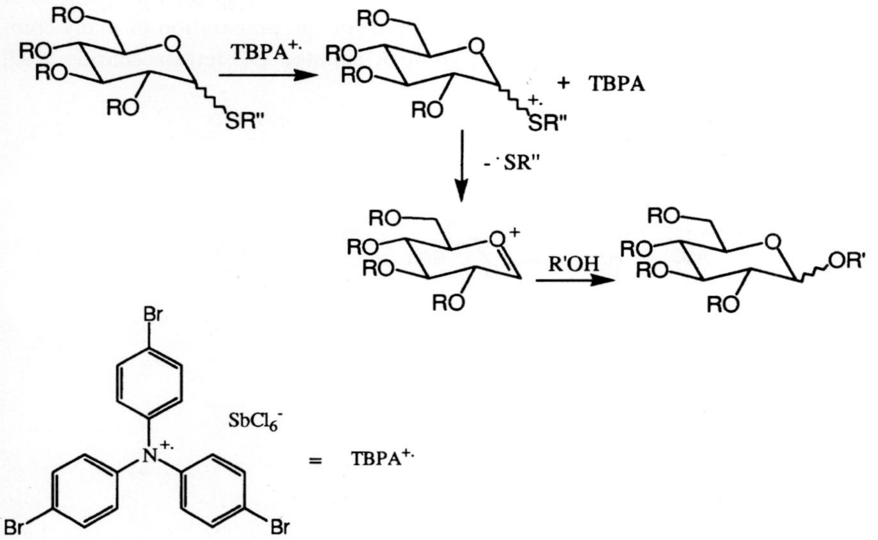

Thioglycosides were developed and widely used by Hanessian [98]. Recently, he introduced the use of glycosyl 2-thiopyridylcarbonates which by *remote activation* (as opposed to activation of a substituent directly attached to the anomeric center) in the presence of an excess of alcohol, afford the desired glycoside in good yield and with good selectivity (Scheme 10). As depicted in Scheme 10, 1,2-*cis* or 1,2-*trans*-glycosides can be obtained depending on the nature of the C-2 substituent in the donor. With the O-benzyl protected donor, the glycosidation is α-selective while employing a participating group like the acetoxy group, at C-2 in the 2-thiopyridylcarbonate donor, a 1,2-*trans*-disaccharide is successfully prepared [107].

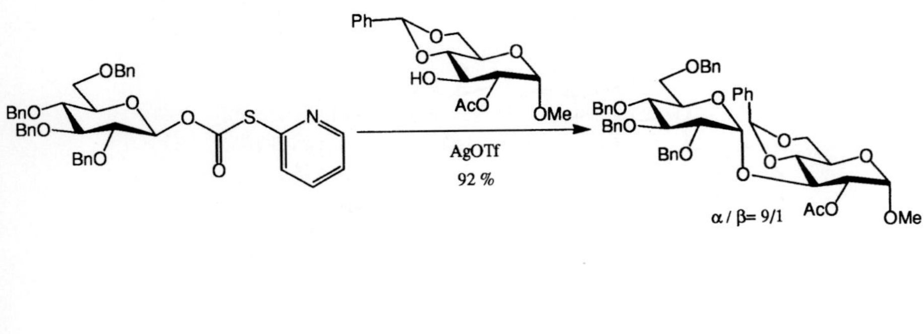

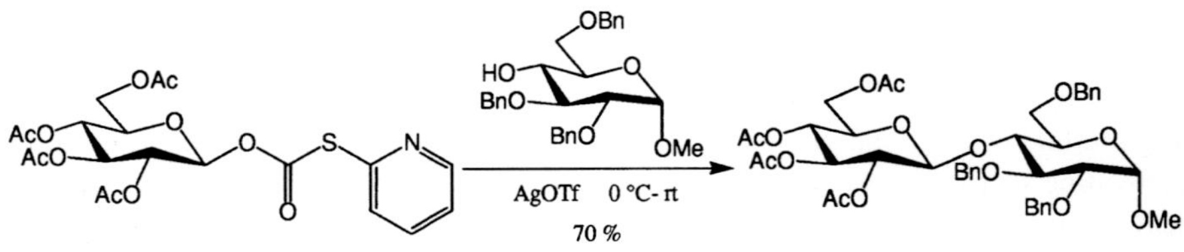

The combination of the chemistry of thioglycosides with that of glycosyl fluorides, known as "two-stage activation procedure", proved to be an efficient strategy for the synthesis of oligosaccharides and it has been successfully employed for the preparation of many complex molecules, [22] such as Avermectin B_{1a} [108] or sulfated Le^{a} tetrasaccharide [109] (Scheme 11).

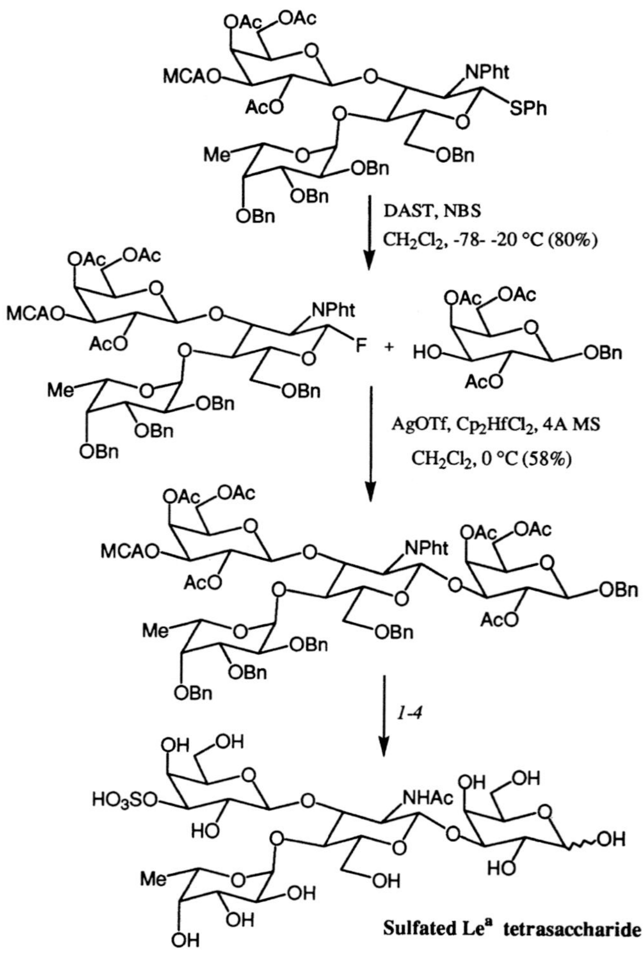

8.3.2 The Glycal Method

1-Glycals are extensively studied carbohydrate derivatives [110, 111] that were first obtained by Fisher and Zach at the beginning of the 20th century [112]. Although for many decades carbohydrates have been successfully used as chiral building blocks for the synthesis of natural products, glycals have often been considered unattractive "chirons" for these purposes [113]. At the present time however, the major efforts devoted to the enantioselective preparation of complex carbohydrate-containing molecules found in the glycal are coming to fruition in the synthesis of the almost ubiquitous *C*- and *O*-glycosides as well as 2-deoxyglycosides.

Glycals can be synthesized under mild conditions, and with good yield [114–117]. They are stable in basic media, but under thermal or acidic conditions undergo Ferrier rearrangement [118]* (Scheme 12).

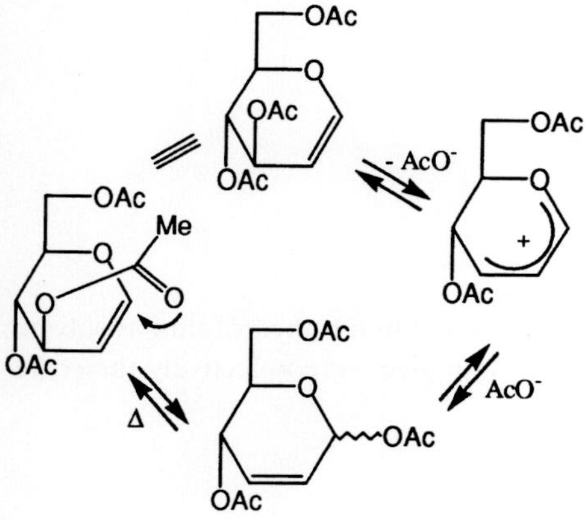

The addition of electrophiles to the double bond of glycals is the most common reaction used in the synthesis of C- and O-glycosides. Among the examples reported and extensively reviewed [31, 113], the transformation of glycals into the corresponding 1,2-epoxisugars and the use of the latter as glycosyl donors is worthy of mention. The opportunity to use 1,2-anhydro sugars as glycosyl donors was first recognized by Lemieux [120, 121] but, perhaps due to the poor availability of these oxiranes, the method has never been suitably developed [122]. More recently, the introduction of dimethyldioxiranes (**29**) (see Scheme 13) as epoxidant reagent of olefins [123, 124] allowed stereoselectively isolateable epoxides to be produced from a wide number of glycals**, promoting the successful exploitation of 1,2-anhydro sugars as glycosyl donors [126, 127].

* The rearrangement commonly involves acylated glycals and to a lesser extent their O-alkylated analogs. Although the Ferrier rearrangement is generally considered a complication, it can have attracting synthetic utility and offer ways to introduce hydrogen, fluorine or other groups on position 1 and 3 of the sugar ring [3, 119].

** More recently, 1,2-anhydro sugars have also been obtained in good yields by the use of anhydrous *m*CPBA [125].

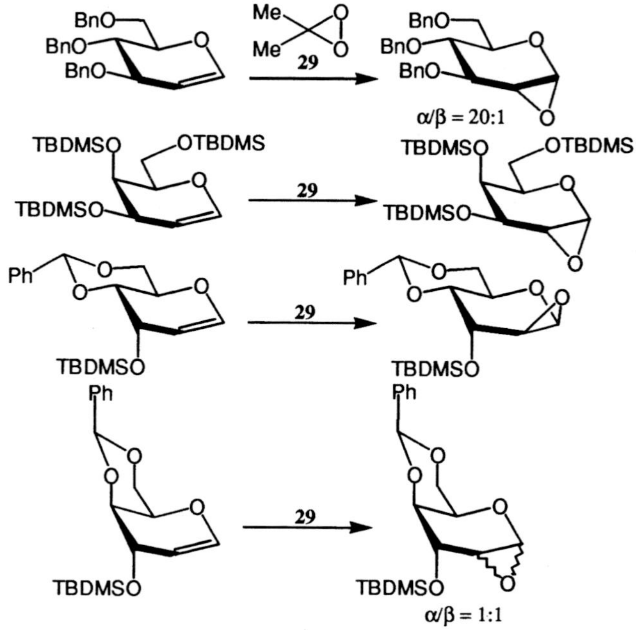

In the presence of anhydrous zinc chloride as catalyst and in tetrahydrofuran as solvent, the α dioxirane **30** reacts with a glycosyl acceptor [A], to give stereoselectively the corresponding β-*O*-glycoside **31** [122, 128] (Scheme 14).

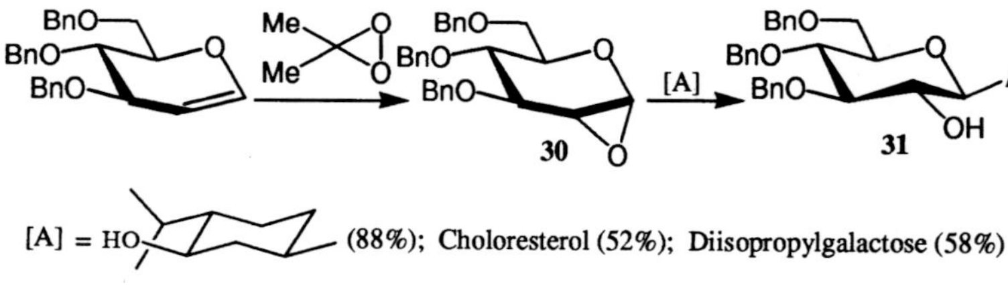

As shown in Scheme 14, with alcohol acceptors, including those which the hydroxyl group is part of a monosaccharide, the epoxide ring opening occurs *anti* with respect to the epoxidic oxygen atom, affording the corresponding 2-hydroxy *gluco*-β-*O*-glycosides **31**. The finding that the glycosyl acceptor for the glycal epoxide can be itself a glycal is however the real advantage of this method (Scheme 15).

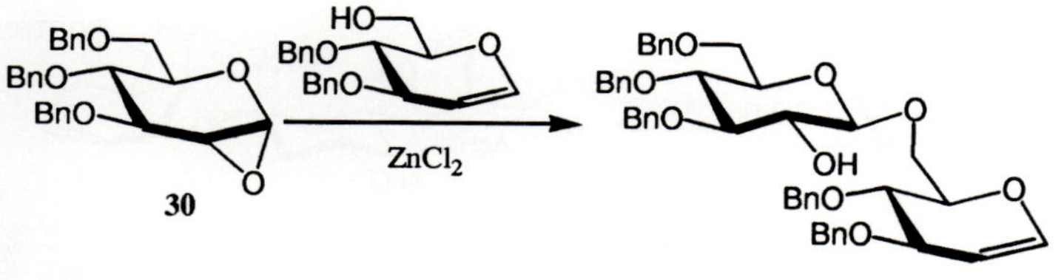

30

The repeatability of the present strategy made the use of glycal epoxide one of the most versatile ways to produce β-oligosaccharide, and it is the central reason for the successful extension of the glycal approach to solid-phase synthesis. In solid-phase glycosylations, the donor is fixed to a solid matrix and reacts repeatedly with glycal acceptors to give an oligosaccharide which, in the final step, is removed from the solid support [131, 132] (Scheme 16).

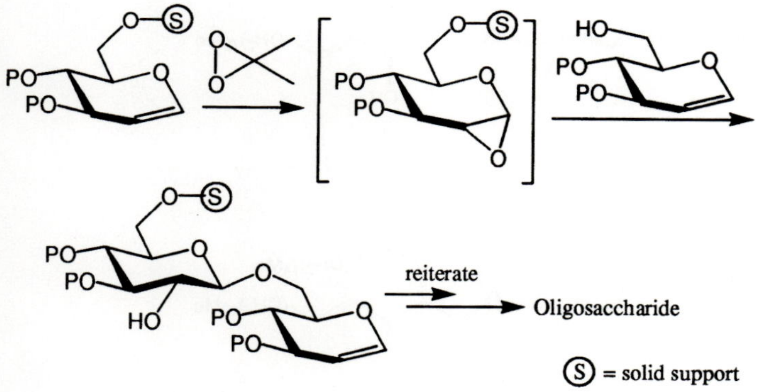

The glycal epoxide method was also applied to the synthesis of α-*O*-glycosides [133] and of 2-deoxy-β-*O*-glycosides [134]. Although interesting results were reported in both cases, other strategies seem to be preferable for these purposes (see sections 3.3 and 3.5).

The possibility of using the glycal method in the construction of biologically relevant molecules has been widely employed. The synthesis of gangliosides was indeed realized relying on the use of galactal-derived **32** (Scheme 17) which was glycosylated with the sialyl donor **33** [135]. The 3-sialylated galactal obtained, was treated with 1,8-diazobicyclo[5.4.0]undec-7-ene (DBU), providing the 3,4-spirolactone glycal **34**. Epoxidation and glycosylation with the ceramide precursor **35** gave the glycoside **36** which was converted to the sodium salt of GM_4 **37** (Scheme 17).

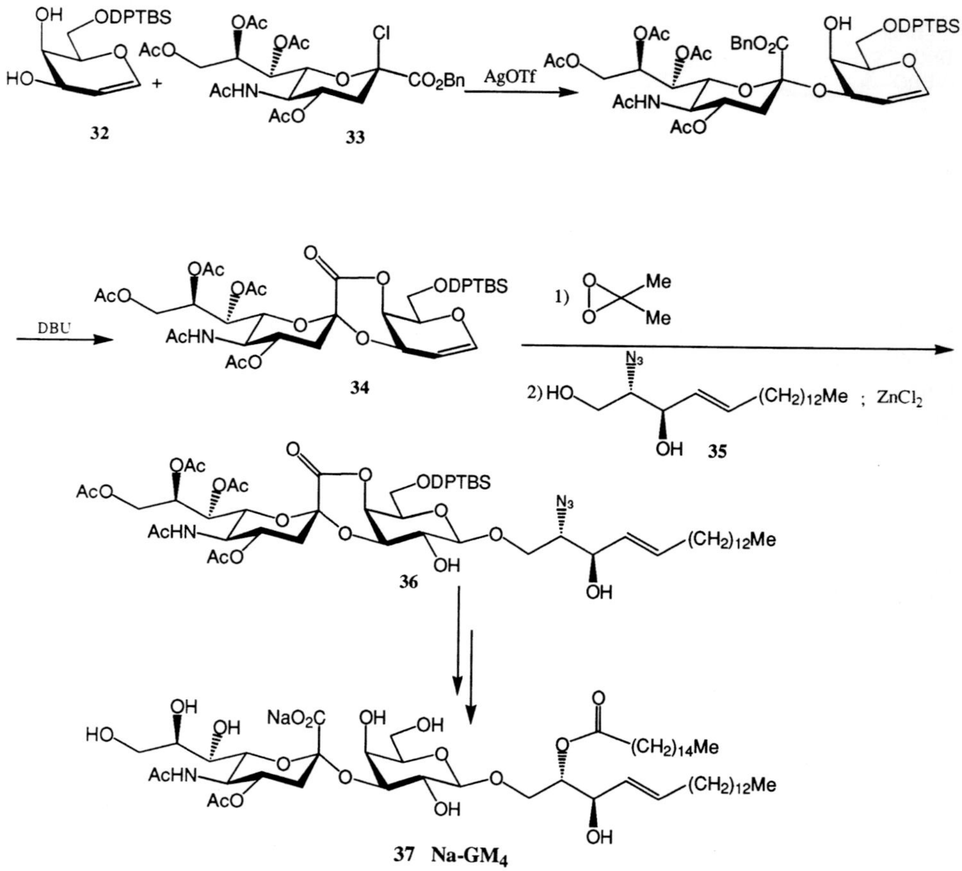

The glycal assembly logic was also applied to the synthesis of Lewis determinants [136], blood group determinants [137] and tumor antigens, as for the synthesis of **38**, a hexasaccharide glycosphingolipid [138] which is a breast tumor associated antigen of biological relevance (Fig. 3).

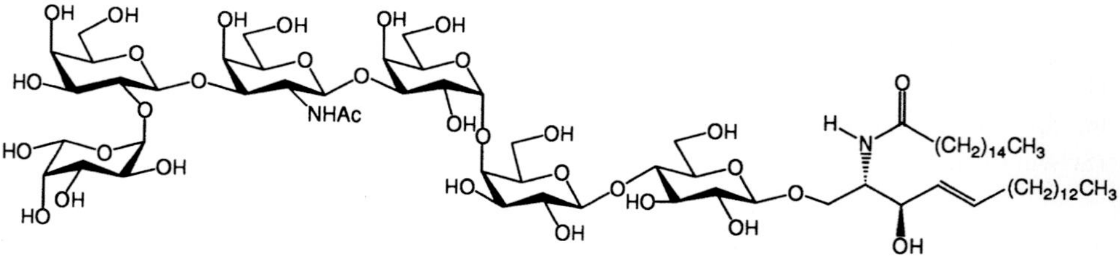

38

Figure 3. Structure of MBr1 antigen.

The synthetic planning for this challenging molecule recognizes the two trisaccharides **39** and **40** as principal substructures. The glycosidation of **39** (donor) with **40** (acceptor) produces the hexasaccharide **41** suitable for the introduction of the ceramide via its glycal linkage. Worthy of mention with regard to this strategy is the conciseness in achieving a complex system such as **39** and **40**, which are obtained from the readily available galactal **42** and glucal **43** respectively. Surprisingly the formation of the α-linkage in the synthesis of **40** and the β-glycosidation of **39** with **40** are realized reacting as donors a fluoroglycoside in the first case, [138] and a thioglycoside in the second, rather than the respective epoxides [138] (Scheme 18).

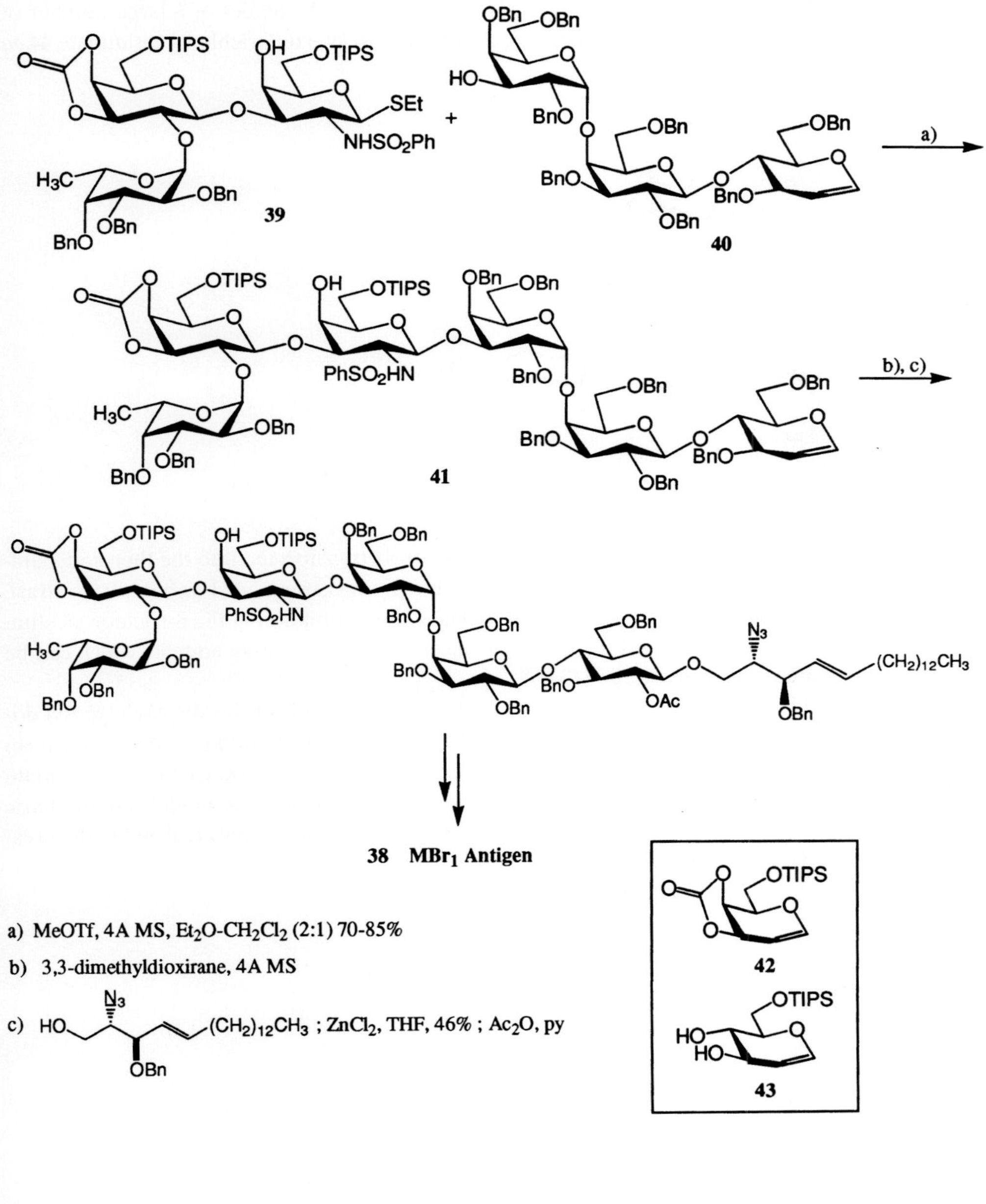

8.3.3 The Trichloroacetimidate Method

A valuable alternative to the Koenigs–Knorr-type methods is represented by the treatment of pyranose or furanose derivatives with bases. This generates anomeric alcohoxides which are known to react with suitable nitriles systems, providing stable O-alkyl imidates [36]. A successful and widely used glycosylation method, developed following this type of reaction, relies on the addition of alkoxides to the electron-poor trichloroacetonitrile, to provide the corresponding trichloroacetimidate derivatives, in which the anomeric oxygen has been transformed into a good leaving group [91]. The anomeric alchoxides of a large number of sugars were reported to react with trichloroacetonitrile to give the trichloroacetimidate **44** *eq* in a reversible way (Scheme 19).

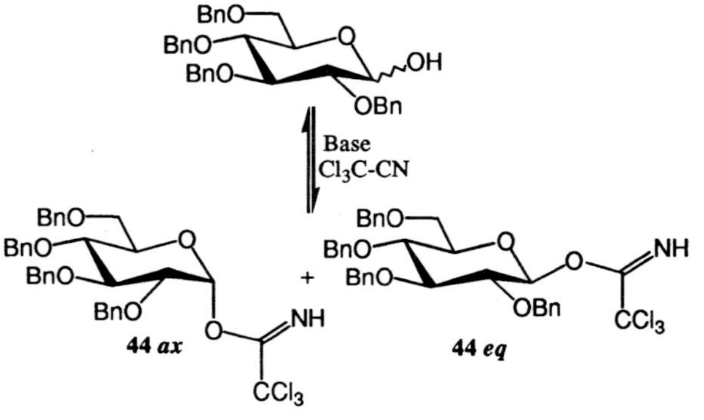

In basic medium, the kinetic derivative **44** *eq* is slowly transformed into the thermodynamically more stable **44** *ax*, which is obtained as single product using NaH as base. In contrast, the use of a weak base such as potassium carbonate allows isolation of the β-trichloroacetimidate in good yield. The β- and the α-trichloroacetimidates are stable adducts which can be directly isolated and easily stored [139].

Under mild acid treatment these anomeric oxygen-activated species display glycosyl donor properties leading, in the presence of acceptors, to the corresponding *O*-glycosides, in an irreversible manner (Scheme 20). Thus, as reported in Scheme 20, the trichloroacetimidate method consists of a first a*ctivation step* (formation of the trichloroacetimidate) under basic conditions and a second *glycosidation step* (synthesis of the O-glycoside) realized in the presence of catalytic amounts of an acidic promoter.

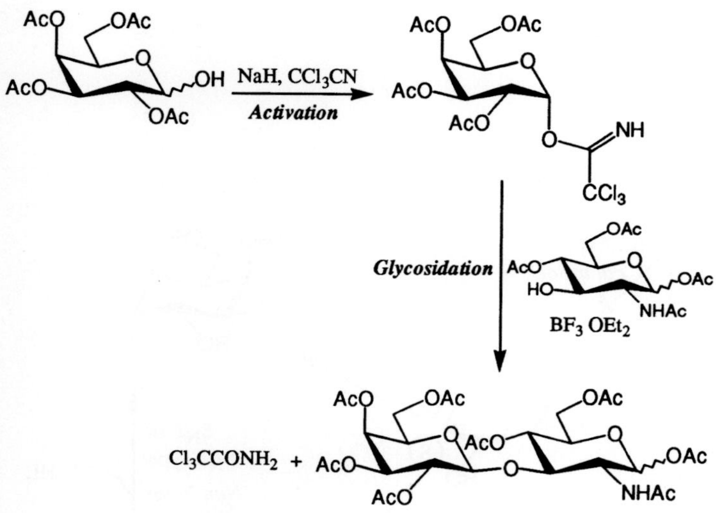

The stereochemical control in the trichloroacedimidate method is generally substrate-dependent and mainly subordinated to the strength of the anomeric effect, as well as to the neighboring group participation of the substituent at C-2. Thus, the α-*manno* and β-*gluco* glycosides are formed more easily than the corresponding α-*gluco* and β-*manno* derivatives (see Scheme 7; section 3.1). When non-participating protective groups are present at C-2 of the donor, the solvent effect becomes dominating in the anomeric stereocontrol. In non-polar solvents, S_N2-type reactions generally occur, hence α-trichloroacetimidates yield β-glycosides while β-trichloroacetimidates give α-glycosides. In polar solvents, S_N1-type reactions occur which for kinetic and thermodynamic reasons preferentially afford α-*galacto*-type glycosides (Scheme 21). The same type of selectivity is also observed for sugars of the *manno* and *gluco* series.

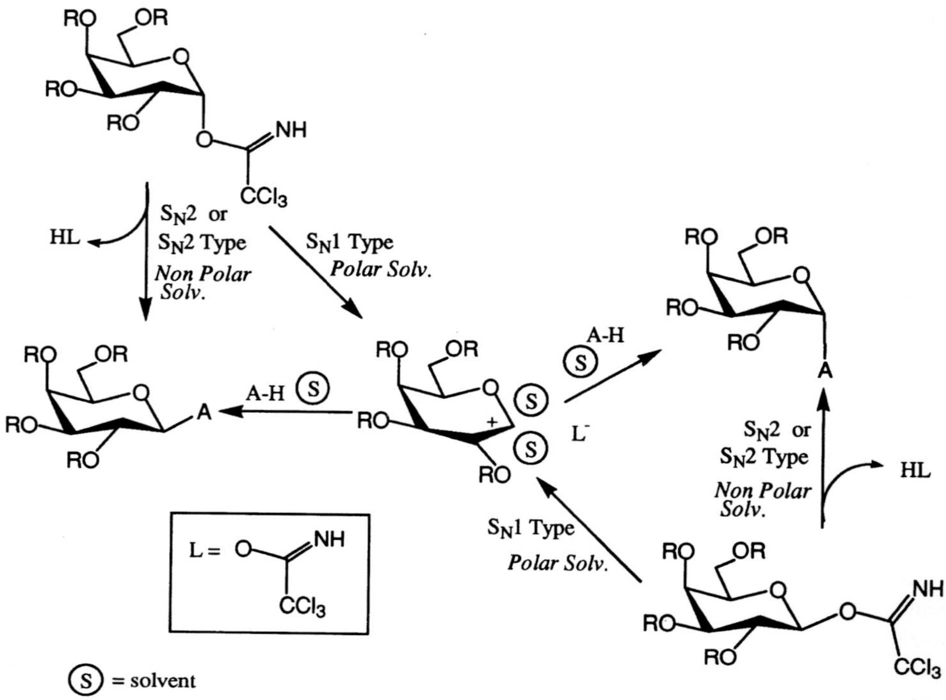

A dramatic effect has recently been observed in the case of nitriles as participating solvents in glycoside reactions [140, 141]. In these solvents, a high preference for β glycosidation was noticed.

The mild conditions required and the applicability of the present method to complex carbohydrate-containing molecules, made the choice of the trichloroacetimidate as donor highly popular; two representative examples, such as the synthesis of the GD_3 [142] and of the GM_3 [143] gangliosides, are reported in Schemes 22 and 23.

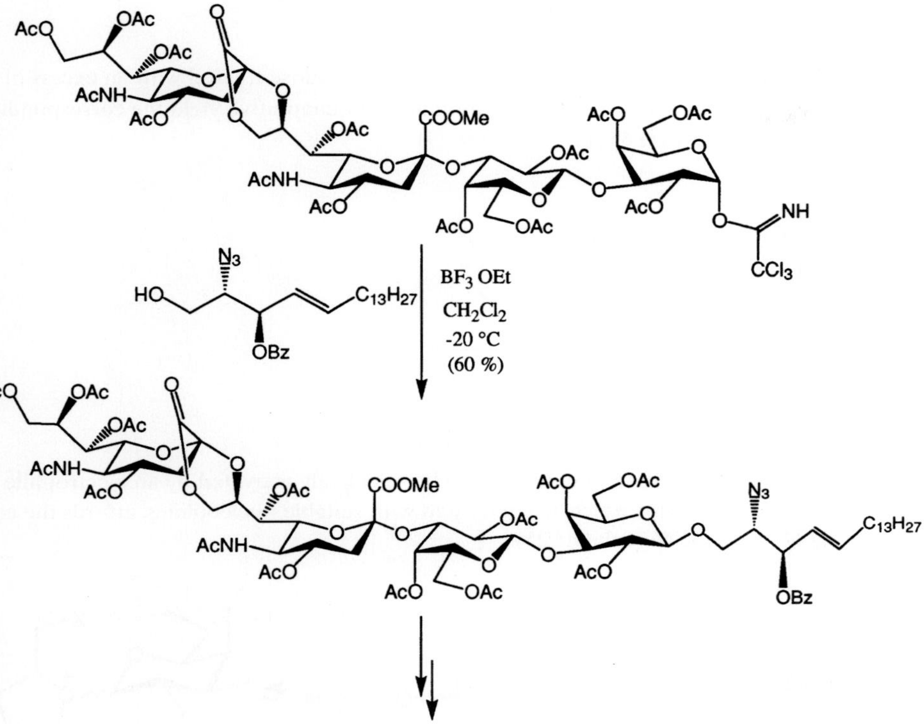

GD$_3$

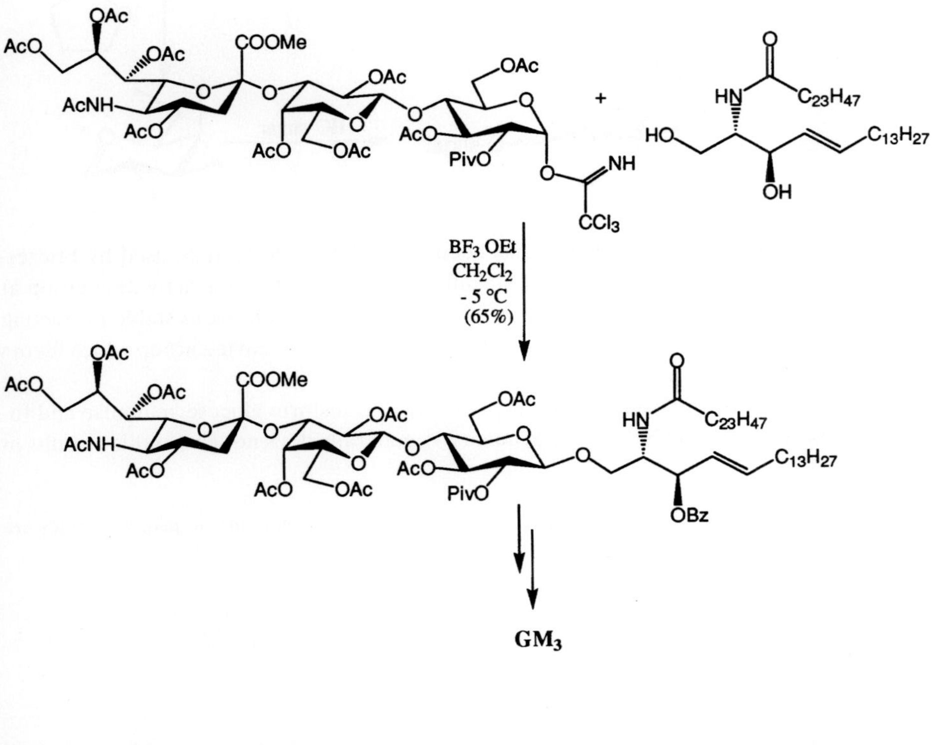

GM$_3$

8.3.4 The *n*-Pentenyl Method

Under standard Fisher conditions for glycosidation*, an aldose treated with an excess of *n*-pentenyl alcohol under acid catalysis, affords in almost quantitative yield the corresponding *n*-pentenylglycoside (NPG) **45** [144] (Scheme 24).

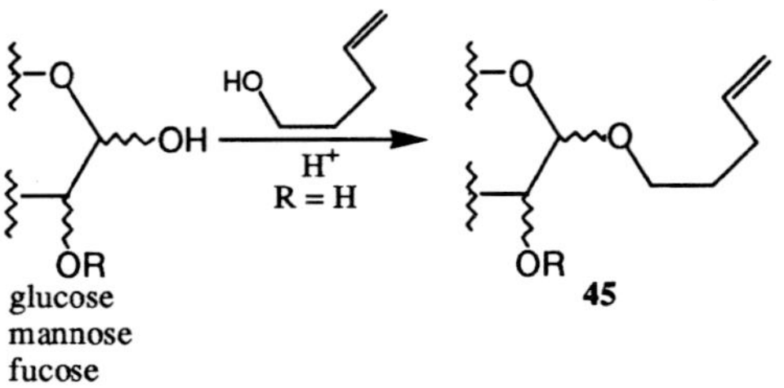

The *n*-pentenylglycoside **45** so obtained can be selectively activated by an electrophile to give rise to a reactive donor which, when treated with suitable nucleophiles, affords the corresponding *O*-glycoside **46** (Scheme 25).

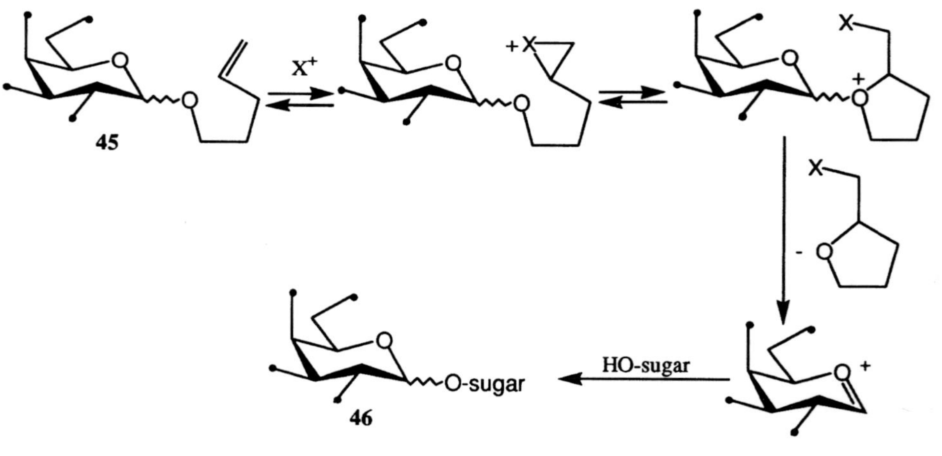

The role of NPGs as glycosyl donors has been studied and successfully used by Frieser-Reid and co-workers [23, 145]. The major advantage of NPGs is that the activating group at the anomeric center can be introduced at an early stage in the synthesis, as stable protecting group of the anomeric hydroxyl, and after activation, it provides a leaving group which favors the coupling of the donor with an acceptor.

The Fisher glycosidation, which is an efficient way to transform glucose, mannose and fucose into the corresponding NPGs (see Scheme 24), generally does not give similar results in

* When sugars are treated with alcohols in the presence of catalytic amounts of acid, glycosides are formed.

the case of galactose or glucosamine. In the latter two instances, the standard Koenigs–Knorr (a) or the glycosyl acetate (b) procedures can be satisfactory followed [23] (Scheme 26).

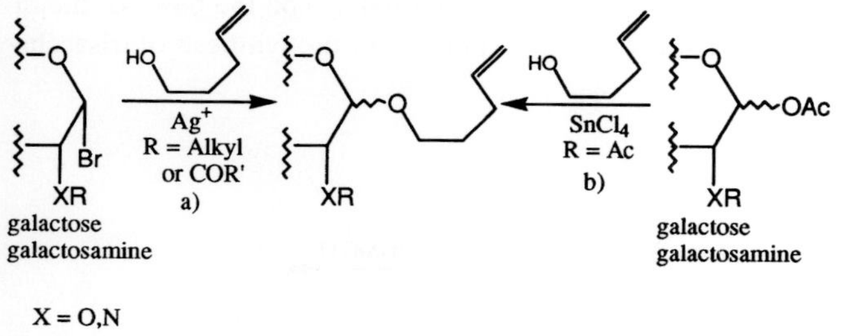

The coupling of NPGs with acceptors can be realized in two different ways [146]: a direct coupling is obtained under the action of an halonium ion (*path a*), while in particular situations the conversion into the glycosyl bromide, followed by a Koenigs–Knorr coupling is preferable (*path b*) [147] (Scheme 27).

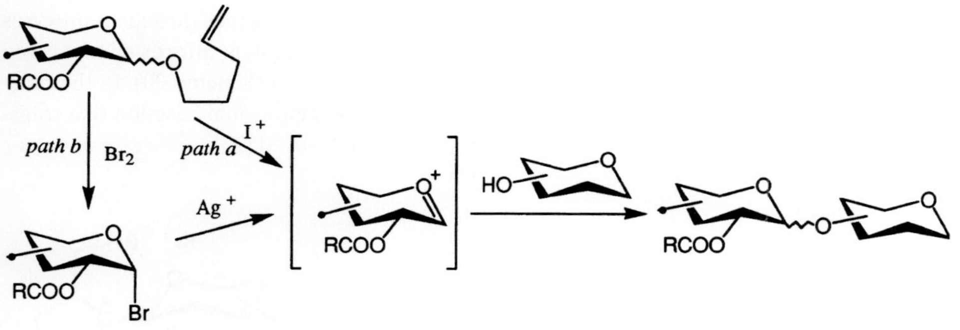

For direct coupling, which is the most widely used procedure, the promoter of choice is *N*-iodosuccinimide (NIS) in the presence of triethylsilyl trifluoromethanesulfonate [148]. Under these conditions the reaction is very fast, while the use of NIS or *N*-bromosuccinimide without acid catalyst slows down the coupling rate. The effect of triflic acid or silyl triflate on NIS is shown in Scheme 28.

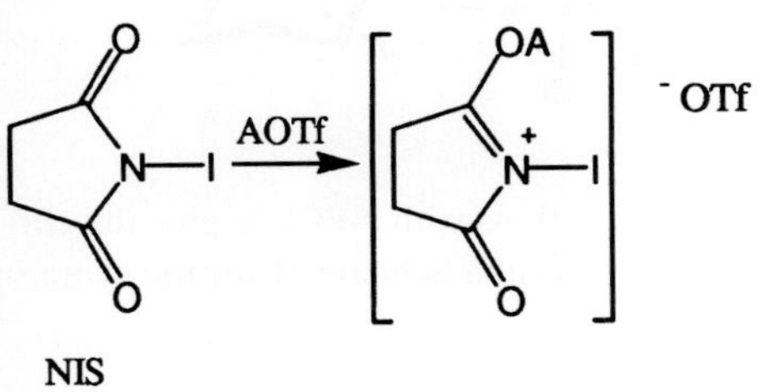

A major improvement of the NPGs method involved the discovery of the armed–disarmed protocol for oligosaccharide assembly [149]. This finding came from the observation that acyl-protected NPGs (disarmed) reacted more slowly than the corresponding alkyl-protected (armed) derivatives. This difference, which can be rationalized on the bases of the inductive effect of the C-2 group, has been successfully exploited for the synthesis of trisaccharides (Scheme 29).

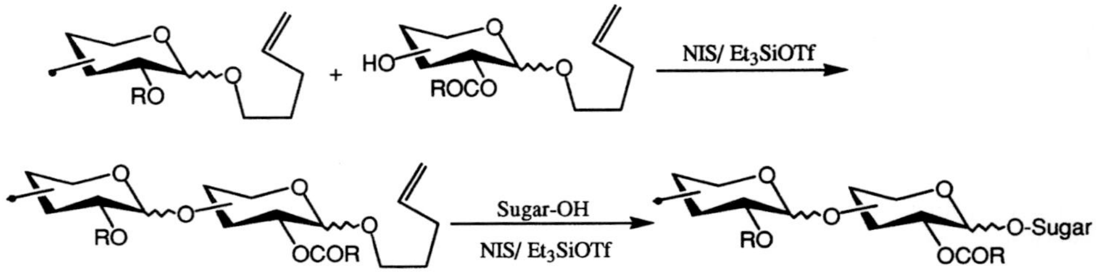

In the case of monosaccharides with protecting groups which does not allow the application of the armed–disarmed process, two NPGs can be coupled together through a previous treatment of one of the two NPG moiety with bromine. This treatment affords an undesired α-bromo derivative (**47**) and the dibromo intermediate (**48**) [150] (Scheme 30). In the presence of an excess of bromide ion (by adding Et_4NBr), the intramolecular reaction that transforms **49** into **50** can be prevented, and the formation of **48** favored [23].

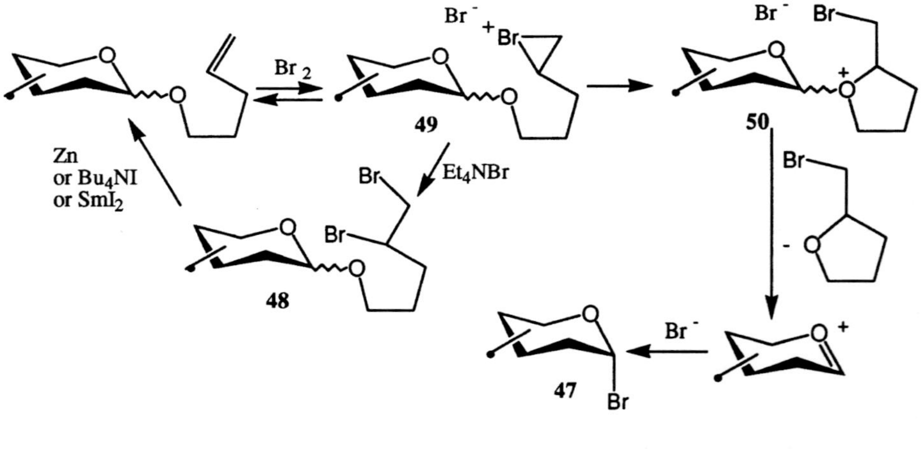

The dibromide **48** so obtained can undergo coupling reactions with NPGs to give the corresponding disaccharides. An example of this strategy is shown in Scheme 31 for the synthesis of an oligosaccharide of the *manno* series.

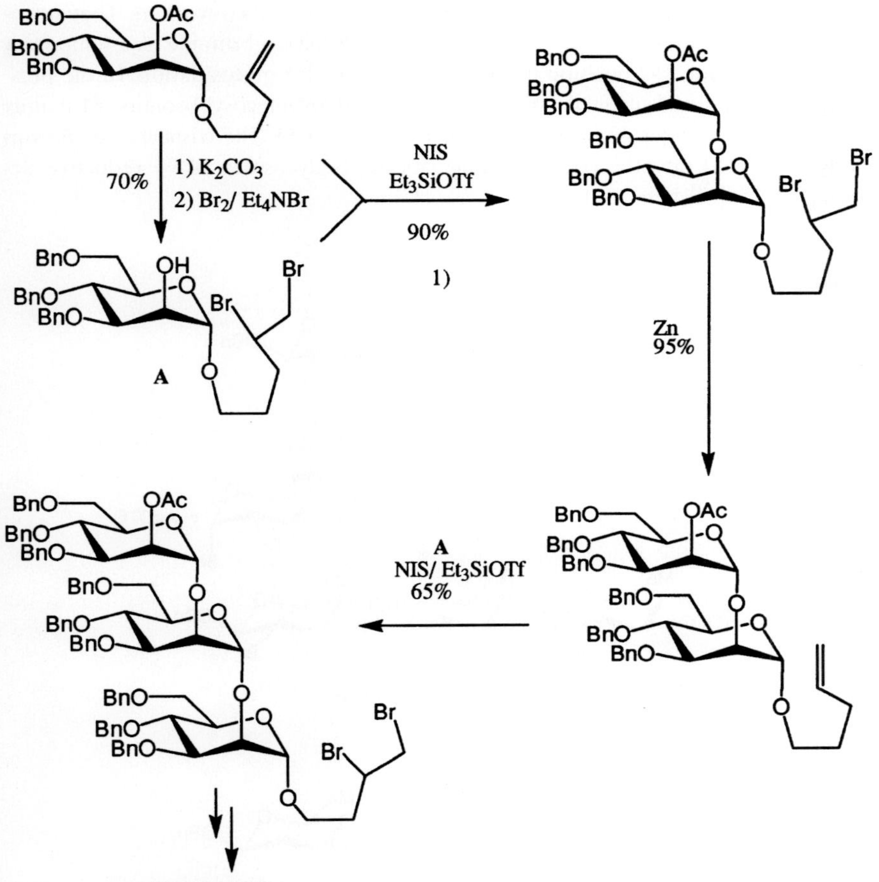

After glycosidation, the dibromodisaccharide formed (see Scheme 31) can be further activated by reductive debromination with zinc, samarium iodide or iodide ion [23] to give the corresponding NPG which is ready to react with another molecule of dibromide **A**.

8.3.5 2-Deoxyglycosides

2-Deoxyglycosides are a class of derivatives of wide biological relevance. They frequently appear in naturally occurring substances such as anthracycline, cardiac glycosides, aureolic acid antibiotics, erythromycins and many other glycosylated macrolides [151, 152]. Although many efforts have been devoted to optimize 2-deoxy glycosyl synthesis, the lack of assistance from proximal substituents and the low stability of precursors make the preparation of pure anomers, particularly the β-anomer, a major challenge [31, 153, 154].

A possible solution to the problem was advanced by Thiem and co-workers. They introduced the use of 2-bromo-2-deoxyglycosyl bromides which have a bromide as a temporary participating group at the C-2 position to promote a β-selective glycosidation. In the presence of catalytic amounts of silver triflate, the 2-bromo-2-deoxyglycosyl bromide **51** mainly gives the desired β-glycosides **52**, which are transformed into **53**. The trisaccharide **53** was subsequently converted into the corresponding 2-deoxy-β-glycosides **54** by reductive debromination [155, 156] (Scheme 32).

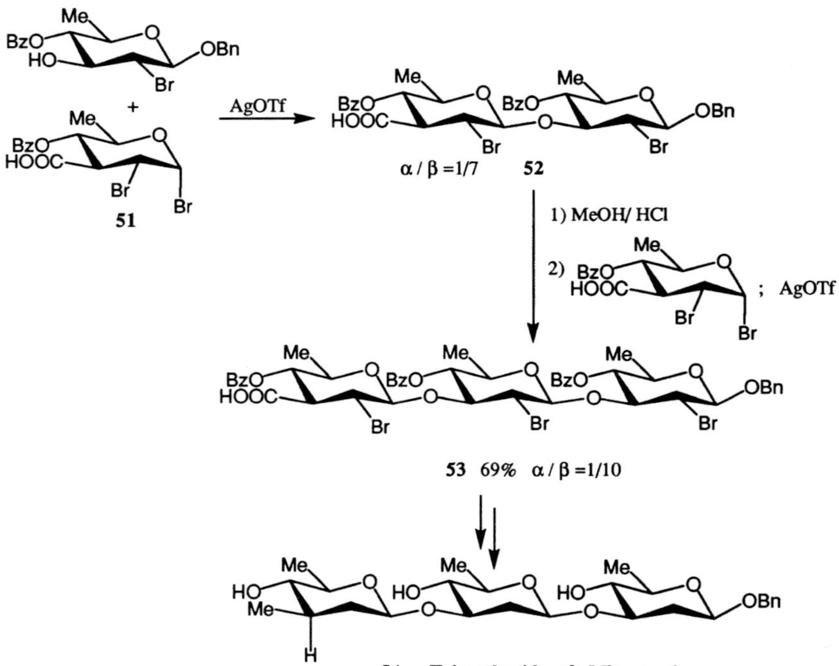

54 Trisaccharide of Mitramycin

2-Iodo-glycosides can be a valuable alternative to the above-mentioned 2-bromo derivatives. Easily prepared via addition reactions to glycals of NIS or $I(coll)_2ClO_4$ and alcohols [30, 157], these donors were successfully used in the synthesis of aureolic acid oligosaccharides [155, 156, 158].

The easily removable arylthio and arylseleno groups were also widely employed as temporary participating groups at the C-2 position*. Although many synthetic ways have been reported for the preparation of 2-thio and 2-seleno donors [159, 160], the widely used 2-deoxy-2-phenylthioglycosides can be prepared in high yield from the corresponding phenylthioglycosides *via* 1,2-migration with diethylaminosulfur trifluoride (DAST). The glycosidation of the 2-thioglycosyl fluoride obtained, is realized using $SnCl_2$ and affords α- or β-derivatives, depending on the solvent [161] (Scheme 33).

* 2-Thio and 2-selenoglycosides are commonly transformed into the corresponding 2-deoxy derivatives, in high yield and without affecting the stereochemistry of the anomeric center, by hydrogenolysis using Raney Ni or etheroatom radical elimination [32, 151].

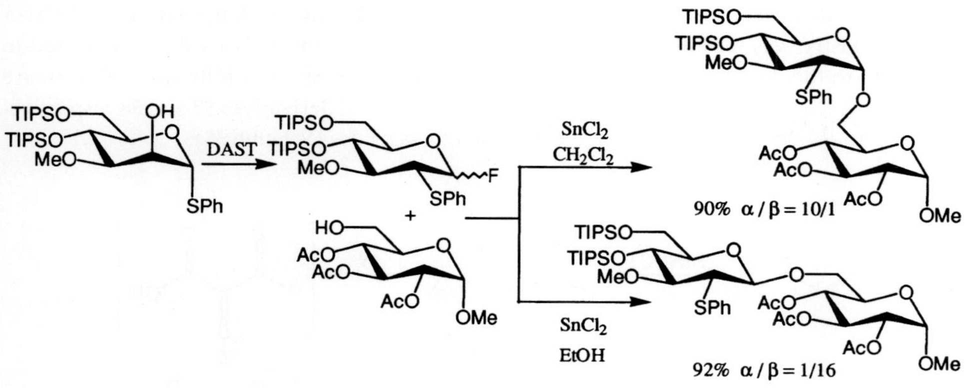

The electrophilic addition of arylbis(arylthio)sulfonium salts to glycals in the presence of alcohols, directly provide 2-thio-β-glycosides with good stereoselectivity [162]. Stereoselective syntheses of β-glycosides were also obtained from the reaction of 1,2-*trans* acetoxyselenides, prepared by treatment of glycals with phenylselenyl chloride and silver acetate, in the presence of trimethylsilyl triflate [163]. Combined application of this method and the trichloroacetimidate method (see section 3.3) was suggested for the synthesis of functionalized precursors of the olivomycin C-D-E trisaccharide [164] (Scheme 34).

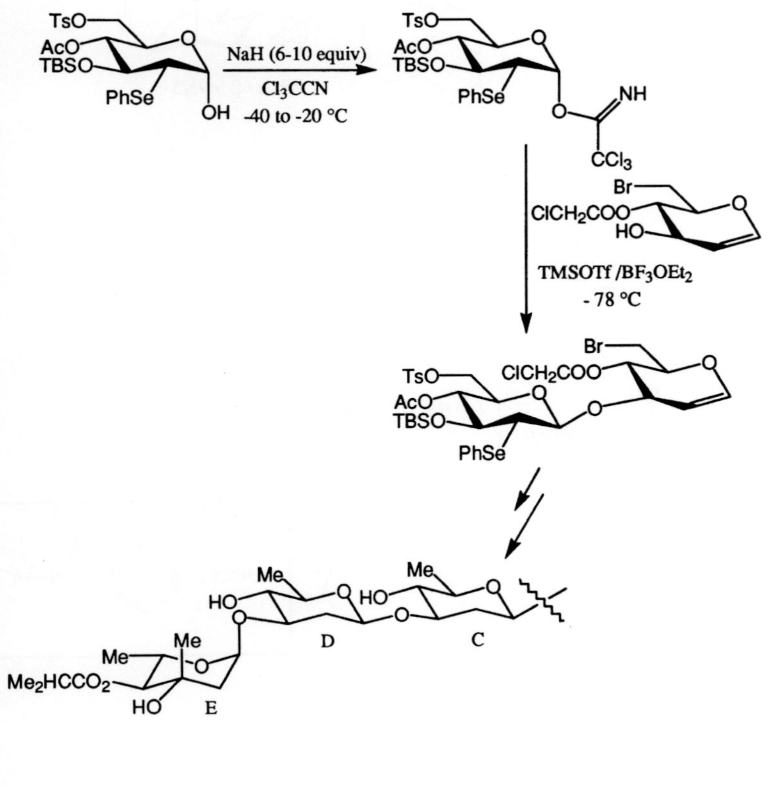

A protocol for highly stereoselective glycosyl syntheses, without using conventional glycoside chemistry, was recently published. Substituted or unsubstituted glycals are reported to cycloadd to the diacylthione **55** or to the *ortho* thioquinone **56** [165–168], which in turn are easily generated in situ from the parent phthalimidesulfenyl derivatives **57** and **58** respectively [169], to give regiospecifically, stable bicyclic *O*-glycosides (Scheme 35).

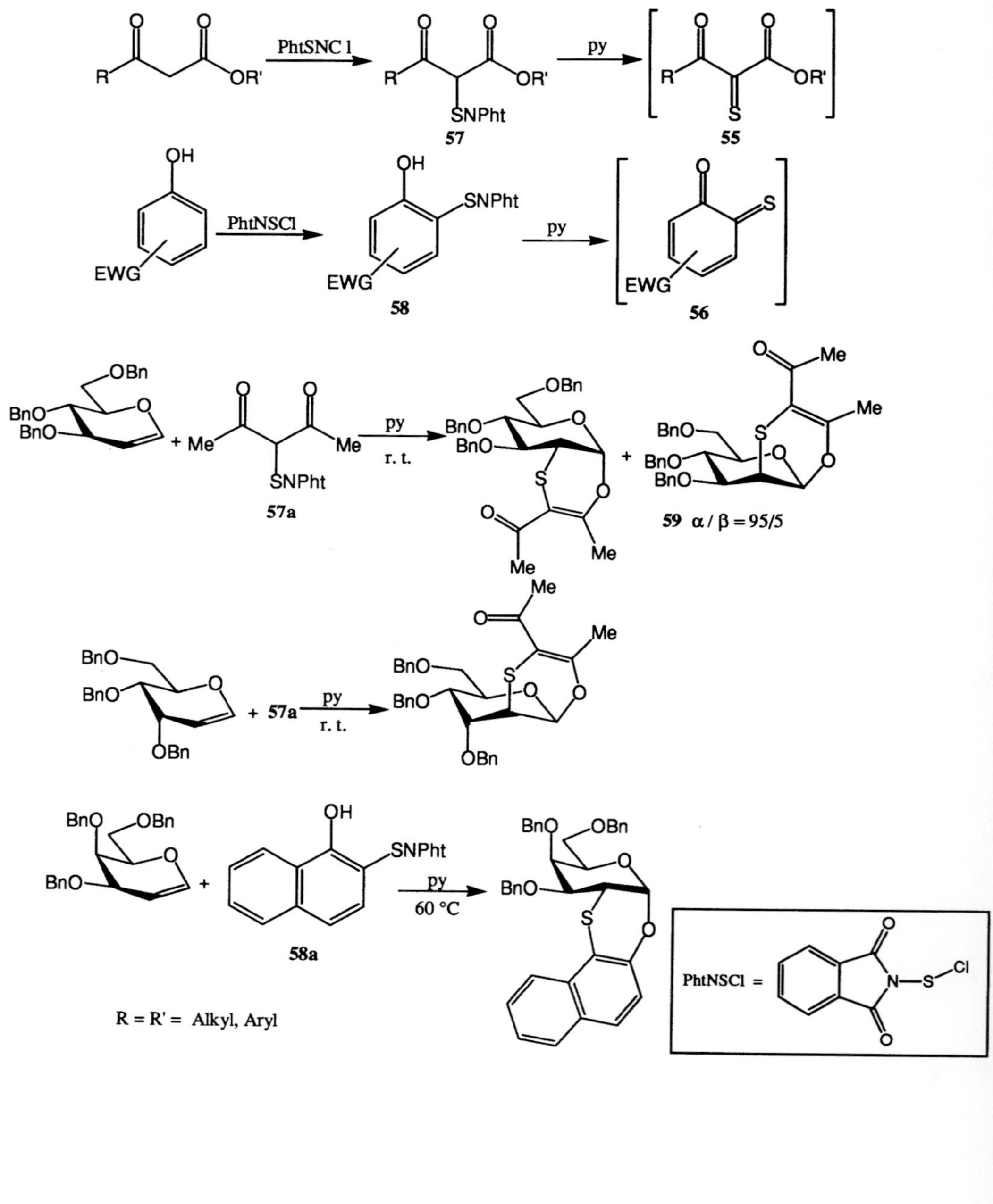

Since the cycloadditions selectively afford α or β-2-thio-*O*-glycosides depending on the nature of the glycal (see Scheme 35), the subsequent desulfurization of these latter allows the isolation of the corresponding α- and β-2-deoxy-*O*-glycosides as pure anomers [167, 170] (Scheme 36).

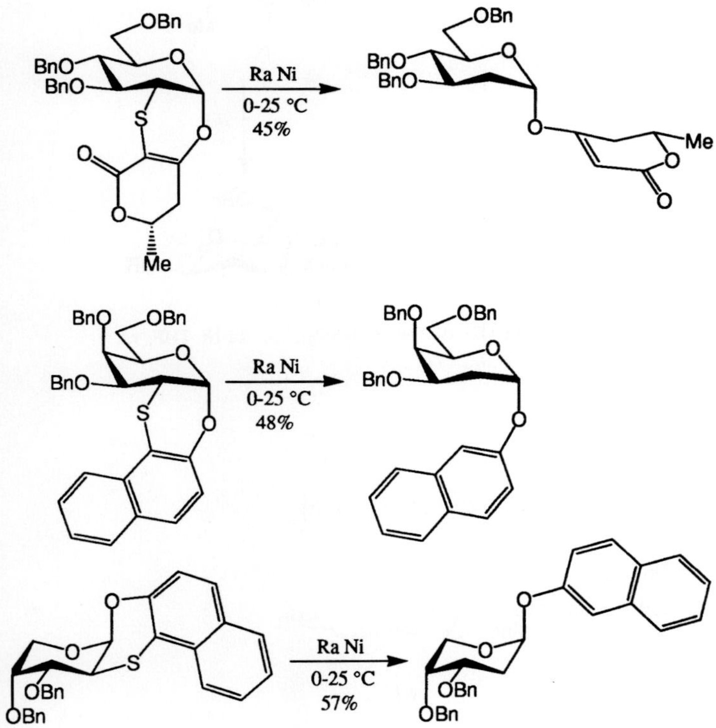

The stable 1,4-oxathiin **59** was also employed for the total stereoselective synthesis of β-*O*-glycosides [172]. The quantitative reduction and acetylation of **59** afforded the "active" species **60***, able to react, under mild conditions and in good yield, with oxygenated nucleophiles. In fact, in the presence of catalytic amounts of methyl triflate or trimethylsilyl triflate, **60** undergoes the attack of the acceptor from the opposite side of the oxathiin-fused ring (i.e., the β-site) providing the corresponding 2-thio-β-*O*-glycoside as single isomer (Scheme 37). Desulfurization of the latter produces the desired 2-deoxy-β-*O*-glycoside.

---

* For other methods to activate the cycloadduct **59**, see Refs [173, 174].

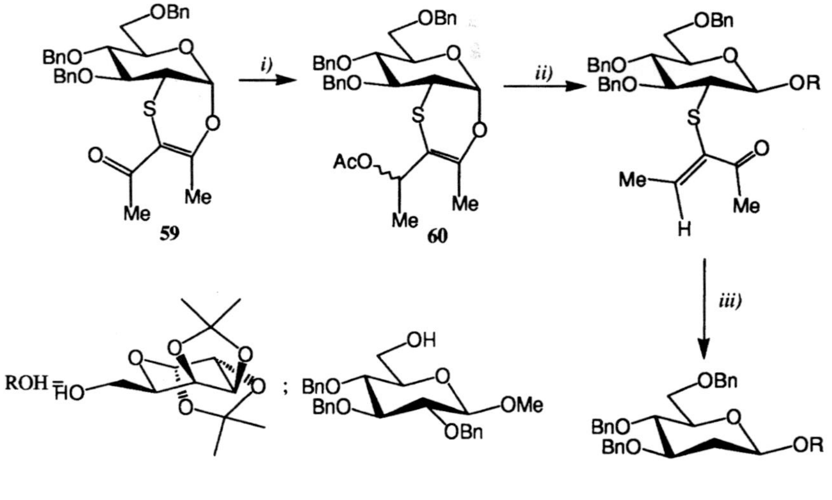

i) $LiAlH_4$- THF, Ac_2O, py, 98% ; *ii)* ROH (2 eq), MeOTf (0.2 eq), CH_3NO_2 ; *iii)* Ra-Ni THF, r.t.

8.3.6 Experimental Procedures

2,3,4-Tri-O-methyl-S-phenyl-D-glucopyranoside (**62**): [175, 176]

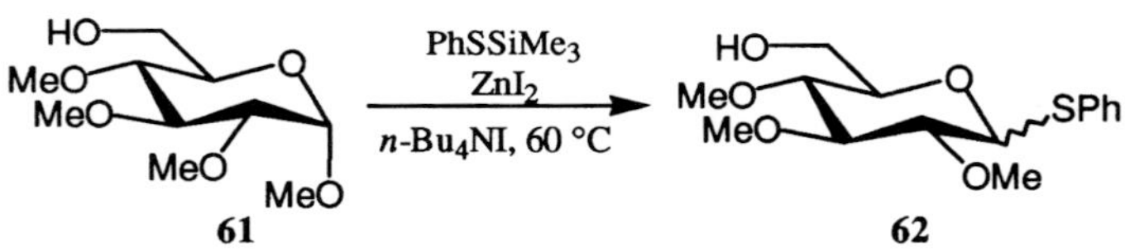

Methyl α-pyranoside **61** (1.0 mmol) was dissolved in $ClCH_2CH_2Cl$ (5 mL) containing Me_3-SiSPh (5.0 mmol). ZnI_2 (3.0 mmol) and n-Bu_4NI (1.20 mmol) were added and the mixture was heated under argon at 60 °C with stirring. After 1–2 h, the mixture was filtered and diluted with CH_2Cl_2, and the filtrate was washed with 10% $Ba(OH)_2$ (three times) and processed as usual to give a yellow syrup. Flash column chromatography gave pure 1-thiopyranoside 62 as α/β mixture in 70% yield.

*6-Deoxy-2,3-di-O-methyl-4-O-*tert *butyldimethylsilyl-S-phenyl-D-allopyranoside* (**64**): [176]

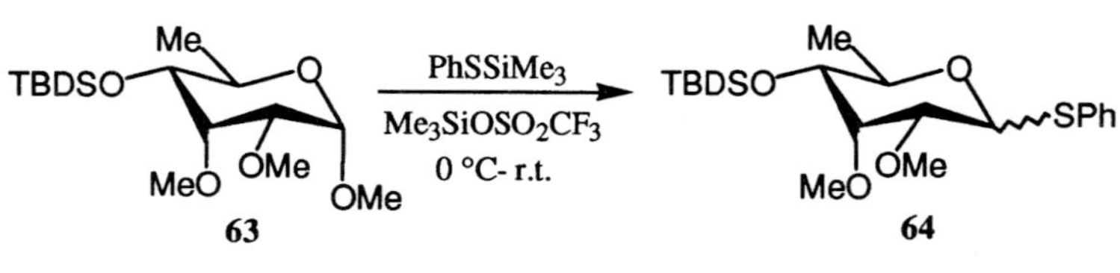

Methyl α-pyranoside (**63**) was dissolved in CH_2Cl_2 (3 mL) containing Me_3SiSPh (5.0 mmol) under argon, and the solution was cooled to 0 °C. $Me_3SiOSO_2CF_3$ (1.2 mmol) was added, and the mixture was allowed to warm to room temperature. After 1–2 h of stirring, the mixture was worked-up, the reaction mixture filtered and diluted with CH_2Cl_2. The filtrate was washed with 10% $Ba(OH)_2$ (three times) and processed as usual to give a yellow syrup. Flash column chromatography gave pure 1-thiopyranoside 64 as α/β mixture in 74% yield.

1,2-Anhydro-3,4,6-tri-O-benzyl-α-D-glucopyranose (**66**): [122]

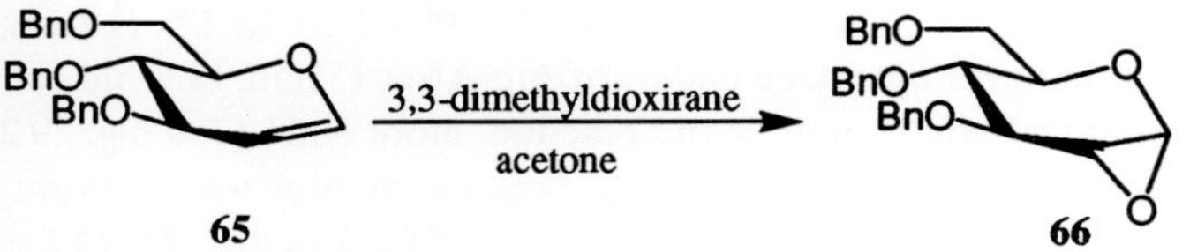

The glycal **65** (0.1 mmol) was dissolved in 1.0 mL of CH_2Cl_2 and the resulting solution was cooled to 0 °C. A solution of 3,3-dimethyldioxirane (1.2 equiv., ca. 0.05 M) was added dropwise. The reaction mixture was stirred at 0 °C for 1 h or until TLC indicated complete consumption of the glycal. The solution was evaporated with a stream of dry N_2 and the residue was dried in vacuo to afford the 1,2-anhydro sugar **66** in quantitative yield $[\alpha]_D$ +29.2° (0.96, chloroform).

O-(3,4,6-Tri-O-benzyl-β-D-glucopyranosyl)-(1→6)-1,5-anhydro-3,4-di-O-benzyl-2-deoxy-D-arabino-hex-1-enopyranose (**68**): [122]

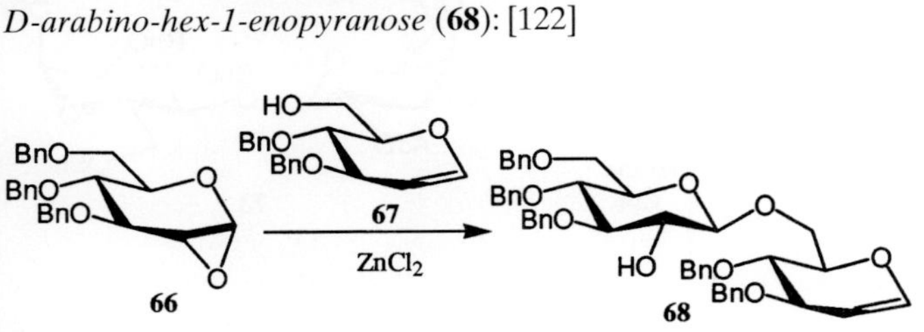

Epoxide **66** (49.4 mg, 0.114 mmol) was dissolved in 0.15 mL of THF, and the resulting solution was cooled to −78 °C. A solution of glycal **67** (55.8 mg, 0.171 mmol) in 0.15 mL of THF was added followed by the dropwise addition of 0.25 mL of a 1.0 M solution of $ZnCl_2$ in ether. The mixture was stirred at −78 °C for 1 h and then allowed to warm to room temperature for 18 h, the mixture was added to 25 mL of saturated $NaHCO_3$ which was then extracted with 3×10 mL of EtOAc. The combined organic extracts were dried over $MgSO_4$, filtered and concentrated. The residue was chromatographed on silica gel (eluent hexane/EtOAc, 7:3) to give 48.7 mg of **68** (56%, 81% based on **67**) along with 30 mg of unreacted **67** $[\alpha]_D$ −5.9° (2.97, chloroform).

2,3,6-Tri-O-benzyl-α-D-glucotrichloroacetimidate (**70**): [177]

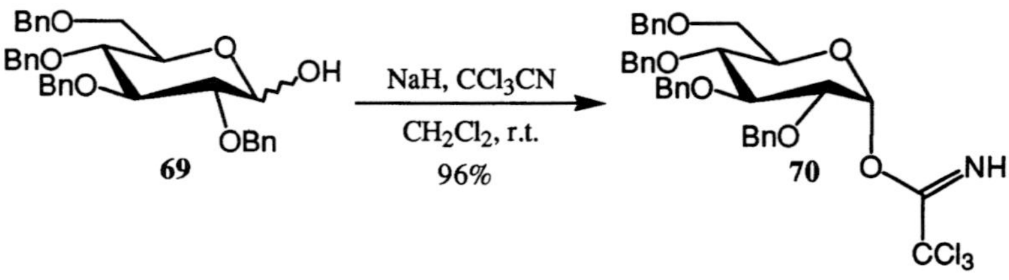

To a solution of compound **69** (12.0 g, 22.2 mmol) in dry CH_2Cl_2 (100 mL), Cl_3CCN (10 mL, 99 mmol) and NaH (50 mg, 2.08 mmol) were added under stirring. After 15 min, TLC indicated $\alpha/\beta \approx 1/3$. For anomerization and completion of the reaction, more NaH (700 mg, 29.2 mmol) was added. After 2 h, the mixture was filtered (Celite), and the solution was evaporated in vacuo. Short-column chromatography (petroleum ether/: Et_2O, 3:2) yielded **70** (14.6 g, 96%) as a colorless oil, which crystallized slowly; m.p. 77 °C (petroleum ether/ Et_2O, 1:1); $[\alpha]_D$ +61.5° (1, chloroform).

1,2-Di-O-acetyl-4,6-O-benzylidene-β-D-galactopyranosyl(1→4)-1,2,3,6-tetra-O-benzyl-α-D-glucopyranoside (**73**): [139]

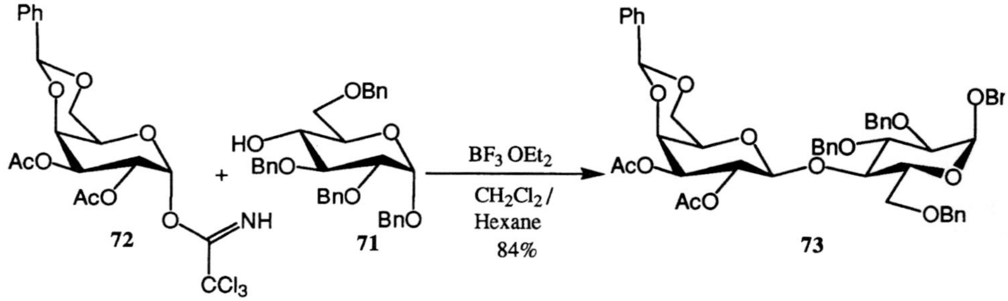

To a solution of the acceptor **71** (10.87 g, 20 mmol) and of the trichloroacetimidate 72 (15.90 g, 32 mmol) in dry *n*-hexane (25 mL) under argon, $BF_3 \cdot OEt_2$ (4.0 mL, 32.5 mmol) was added at 0 °C. After stirring at room temperature for 30 min, the mixture was poured into ice-cold saturated aqueous $NaHCO_3$ with vigorous stirring. The aqueous layer was extracted with Et_2O (2 × 50 mL), and the combined organic layers were washed with water, dried ($MgSO_4$) and evaporated. Chromatography (petroleum ether/EtOAc, 65:35) and crystallization from EtOAc/hexane yielded **73** (14.7g, 84%): m.p. 135–137 °C, $[\alpha]_D$ +52.4° (1, chloroform).

Pent-4-enyl 2,3,4,6-tetra-O-benzoyl-β-D-glucopyranoside (**75**): [178]

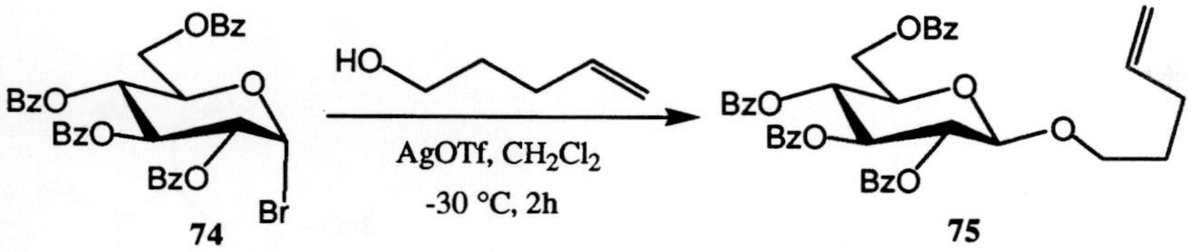

To a mixture of 4-penten-1-ol (0.5 mL, 4.84 mmol), AgOTf (1.2 g, 4.67 mmol) and powdered, activated 4A molecular sieves (1 g) in CH_2Cl_2 (10 mL) was added at −30 °C, through a syringe, a solution of **74** (2.5 g, 3.79 mmol) in CH_2Cl_2 (4 mL) during 10 min. The reaction mixture was stirred at −30 °C for 2 h, quenched with saturated aqueous $NaHCO_3$, and then filtered through Celite. The molecular sieves were washed with CH_2Cl_2 (20 mL). The organic phase was washed with saturated aqueous $NaHCO_3$ (20 mL), dried and concentrated. The residue was purified by flash chromatography (petroleum ether/EtOAc, 4:1) to afford **75** (1.71 g, 68%); m.p. 113–114 °C, $[\alpha]_D$ +18.7° (1, chloroform).

*1-O-[N-(Benzoyloxycarbonyl)-*L*-serine benzyl ester]-3,4,6-tri-O-benzoyl-2-deoxy-2-tetrachloro phthalimido-β-D-glucopyranoside* (**78**): [179]

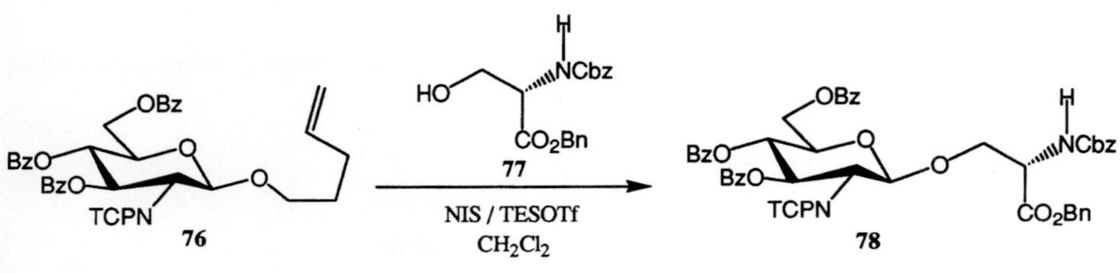

To **76** (0.35 g, 0.43 mmol) and **77** (0.10 g, 0.30 mmol) (both dried by azeotroping together with toluene) in 3 mL of dichloromethane was added *N*-iodosuccinimide (0.124 g, 0.553 mmol) and triethylsilyl triflate (58 μl, 0.26 mmol). After stirring for 30 min at room temperature the glycosyl donor had been consumed and the reaction was quenched with 1.5 mL 10% aq. $Na_2S_2O_3$ and 1.5 mL of saturated aq. $NaHCO_3$ soln. The mixture was stirred for an additional 5 min before separating the layers and extracting the aqueous portion with dichloromethane (3 × 10 mL). The concentrated solution was purified via flash chromatography eluting with a gradient of 25–30% EtOAc/petroleum ether, affording **78** as a white foam (0.25 g, 76%); eluent EtOAc/petroleum ether, 30:70; $[\alpha]_D$ +54.8° (0.98, chloroform).

1-O,2-S-(2-Acetyl-1-methyl-1,2-ethenediyl)-3,4,6-tris-O-(phenylmethyl)-2-thio-α-D-gluco-pyranose (**59**): [165]

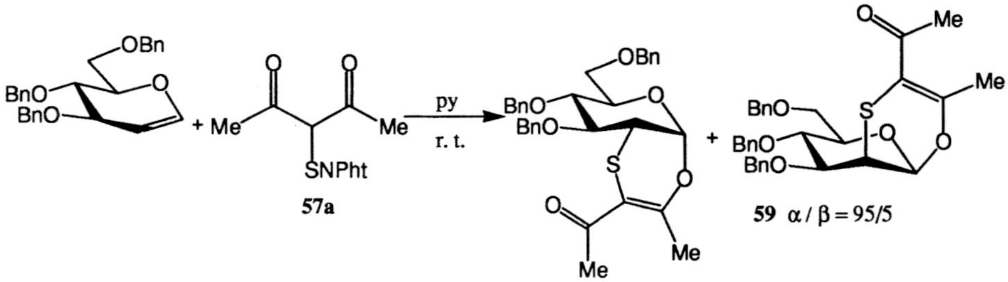

Thiophthalimide **57a** (1.32 g, 4.80 mmol) was dissolved in 10 mL of dry chloroform under nitrogen atmosphere. To the solution was added 3,4,6-tri-*O*-benzylglucal (1.01 g, 2.4 mmol) followed by dry pyridine (0.8 mmol) by syringe. The reaction mixture turned orange immediately and over 4-day period, it acquired a deep red color. After 4 days the reaction had processed almost completely (by NMR analysis). The mixture was washed three times with a saturated solution of NH_4Cl. The organic fraction was dried with anhydrous Na_2SO_4 and concentrated. The resulting crude was put through a flash silica gel column and the final product **59** was isolated to give 1.10 g (83%); m.p. 115–116 °C (*n*-heptane). Starting material (tri-*O*-benzylglucal, 0.156 g, 15%) was also recovered from the chromatography.

1-O-(Methyl-2,3,4-tri-O-benzyl-β-D-glucopyranosyl)-3,4,6-tri-O-benzyl-2-deoxy-2-S-(3-thio-pent-3-en-2-one) (**62**): [169]

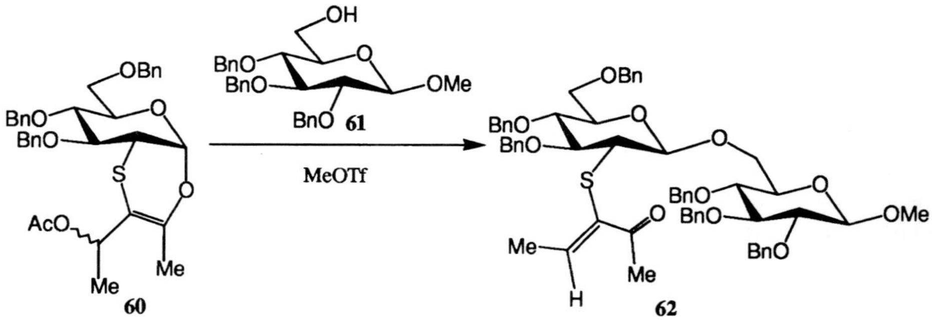

The acetyl derivative **60** (129 mg, 0.219 mmol) was dissolved in 1.5 mL of dry dichloromethane under nitrogen atmosphere; a solution of **61** (2 equiv., 0.438 mmol) in dry dichloromethane (0.5 mL) and methyl triflate (0.2 equiv.) were subsequently added to the reaction mixture. The reaction is quenched with two drops of pyridine at the turning of the color of the mixture from pale yellow to red. After evaporation of the solvent, the crude was purified by flash silica gel chromatography affording **62** (48%) as β-isomer; m.p. 98–100 °C; $[\alpha]_D$ +7.1° (0.3, chloroform).

8.4 Conclusion

In this chapter we have discussed the relevance of glycosides in medicinal chemistry and the problems connected to the regio- and stereoselective synthesis of O-glycosides. With regard to the issue of regioselectivity, three diffuse selective protections of hydroxyls groups have been examined and have focused our attention on recent improvements. Among the many methods available for stereoselective O-glycosyl bond formation, we have reviewed some wide-ranging protocols that have been employed to achieve molecular targets of biological relevance.

We hope that the present chapter can provide readers with guidelines that will enable them to find their way through such an overwhelming discipline.

References

[1] For basic references on carbohydrates see [2–4].
[2] Kennedy, J.F., and White, C.A. *Bioactive Carbohydrates: In Chemistry, Biochemistry and Biology*. Wiley: New York, 1983.
[3] Collins, P.M., and Ferrier, R.J. *Monosaccharides*. Wiley: New York, 1995.
[4] Hanessian, S. *Total Synthesis of Natural Products: The 'Chiron' Approach*. Pergamon Press: Oxford, 1983.
[5] Sharon, N., and Lis, H. *Le Scienze* 46–53 (1993).
[6] Ashwell, G., and Morell, A.G. *Adv. Enzymol.* **41**, 99 (1974).
[7] Kobata, A. *Acc. Chem. Res.* **26**, 319–324 (1993).
[8] Borman, S. *C&EN* Sept. 30, 36–40 (1996).
[9] Watkins, W.M. *Pure & Appl. Chem.* **63**, 561–568 (1991).
[10] Witczak, Z.J. *Curr. Med. Chem.* **1**, 392–405 (1995).
[11] Lemieux, R.U. *Chem. Soc. Rev.* **18**, 347–374 (1989).
[12] Quiocho, F.A. *Pure Appl. Chem.* **61**, 1293–1306 (1989).
[13] Phillips, M.L., Nudelman, E., Gaeta, F.C.A., Perez, M., Singhal, A.K., Hakamori, S.-I., and Paulson, J.C. *Scienze* **250**, 1130–1132 (1990).
[14] Walz, G., Aruffo, A., Kolanus, W., Bevilacqua, M., and Seed, B. *Science* **250**, 1132–1135 (1990).
[15] Lasky, L.A. *Scienze* **258**, 964–969 (1992).
[16] Lucas, R., Mages, S., De Leys, R., Scheerlink, J.-P., Rampelberg, M., and De Baetselier, P. *Science* **263**, 814–817 (1994).
[17] Varki, A. *Glycobiology* **3**, 97 (1993).
[18] Springer, T.A., and Lasky, L.A. *Nature* **349**, 196–197 (1991).
[19] Erbe, D.V., Watson, S.R., Presta, L.G., Walitzky, B.A., Foxall, C., Brandley, B.K, and Lasky, A. *J. Cell. Biol.* **120**, 1227 (1993).
[20] *Carbohydrates-Synthetic Methods and Applications in Medicinal Chemistry*. Ogura, H., Hasegawa, A., and Suami, T. (eds) VCH: Weinheim, 1992.
[21] Dondoni, A., and Marra, A. Thiazole- based one-carbon extension of carbohydrate derivatives. In: *Preparative Carbohydrate Chemistry*. Hanessian, S. (ed.) Marcel Dekker: New York, 1997, pp. 173–205.
[22] Nicolaou, K.C., and Ueno, H. Oligosaccharide synthesis from glycosyl fluorides and sulfides. In: *Preparative Carbohydrate Chemistry*. Hanessian, S. (ed.) Marcel Dekker: New York, 1997, pp. 313–338.
[23] Fraser-Reid, B., and Madsen, R. Oligosaccharide synthesis by *n*-pentenyl glycosides. In: *Preparative Carbohydrate Chemistry*. Hanessian, S. (ed.) Marcel Dekker: New York, 1997, pp. 339–356.
[24] Hanessian, S. Glycoside synthesis based on the remote activation concept: an overview. In: *Preparative Carbohydrate Chemistry*. Hanessian, S. (ed.) Marcel Dekker: New York, 1997, pp. 381–388.
[25] Beau, J.-M., and Gallagher, T. *Top. Curr. Chem.* **187**, 1–54 (1997).
[26] Nicotra, F. *Top. Curr. Chem.* **187**, 55–84 (1997).

[27] Kirschning, A., Bechthold, F.-W., and Rohr, J. *Top. Curr. Chem.* **188**, 1–84 (1997).
[28] Gigg, J., and Gigg, R. *Top. Curr. Chem.* **154**, 77–140 (1990).
[29] Ogawa, T. *Chem. Soc. Rev.* 397–407 (1994).
[30] Danishefsky, S.J., and Bilodeau, M.T. *Angew. Chem. Int. Ed. Engl.* **35**, 1380–1419 (1996).
[31] Toshima, K., and Tatsuta, K. *Chem. Rev.* **93**, 1503–1531 (1993).
[32] Sinaÿ, P. *Pure Appl. Chem.* **63**, 519–528 (1991).
[33] Okamoto, K., and Goto, T. *Tetrahedron* **46**, 5835–5857 (1990).
[34] Lemieux, R.U. *Chem. Soc. Rev.* 423–452 (1978).
[35] Schmidt, R.R., and Kinzy, W. *Adv. Carbohydr. Chem. Biol. Chem.* **50**, 21 (1994).
[36] Schmidt, R.R. *Angew. Chem. Int. Ed. Engl.* **25**, 212 (1986).
[37] Banoub, J., Boullanger, P., and Lafont, D. *Chem. Rev.* **92**, 1167–1195 (1992).
[38] Greene, T.W., and Wuts, P.G.M. *Protective Groups in Organic Synthesis*. John Wiley: New York, 1991.
[39] Kocienski, P.J. *Protecting Groups*. Thieme: Stuttgart, 1994.
[40] Calinaud, P. and Gelas, J. Synthesis of isopropylidene, benzylidene, and related acetals. In: *Preparative Carbohydrate Chemistry*. Hanessian, S. (ed.) Marcel Dekker: New York, 1997, pp. 3–34.
[41] Horton, D., and Norris, P. Dialkyl dithioacetals of sugars. In: *Preparative Carbohydrate Chemistry*. Hanessian, S. (ed.) Marcel Dekker: New York, 1997, pp. 35–52.
[42] Garegg, P.J. Regioselective cleavage of *O*-benzylidene acetals to benzyl ethers. In: *Preparative Carbohydrate Chemistry*. Hanessian, S. (ed.) Marcel Dekker: New York, 1997, pp. 53–68.
[43] Foster, A.B. Cyclic acetals derivatives of sugars and alditols. In: *The Carbohydtrates: Chemistry and Biochemistry,* Vol. 1A. Pigman, W., and Horton, D. (eds) Academic Press: New York, 1972, pp. 391–402.
[44] Stoddart, J.F. *Stereochemistry of Carbohydrates*. John Wiley: New York, 1971.
[45] De Belder, A.N. *Adv. Carbohydr. Chem.* **34**, 179 (1977); (i) Bols, M. *Carbohydrate Building Blocks*. John Wiley: New York, 1996, pp. 19–32.
[47] Schmidt, O.T. *Methods Carbohydr. Chem.* **2**, 318 (1963).
[48] Evans, M.E. *Carbohydr. Res.* **21**, 473 (1972).
[49] Lipták, A., Imre, J., Harangi, J., and Nánási, P. *Tetrahedron* **38**, 3721–3727 (1982).
[50] Garegg, P.J., Hultberg, H., and Oscarson, S. *Carbohydr. Res.* **108**, 97 (1982).
[51] Ito,Y., Nunomura, S., Shibayama, S., and Ogawa, T. *J. Org. Chem.* **57**, 1821–1831 (1992).
[52] Jiang, L., and Chan, T.-H. *Tetrahedron Lett.* **39**, 355–358 (1998).
[53] David, S., and Hanessian, S. *Tetrahedron* **41**, 643–663 (1985).
[54] Pereyre, M., Quintard, J.P., and Raham, A. *Tin in Organic Synthesis*. Butterworth: London 1987; Grindley, T.B. *Adv. Carbohydr. Chem. Biochem.* **53**, 17–142 (1998).
[55] David, S. Selective *O*-substitution and oxidation using stannylene acetals and stannyl ethers. In: *Preparative Carbohydrate Chemistry*. Hanessian, S. (ed.) Marcel Dekker: New York, 1997, pp. 69–84.
[56] Alais, J., and Veyrières, A. *J. Chem. Soc., Perkin Trans.* **1**, 377–381 (1981).
[57] David, S., Thieffry, A., and Veyrières, A. *J. Chem. Soc., Perkin Trans.* 1, 1796–1801 (1981).
[58] Alais, J., Maranduba, A., and Veyrières, A. *Tetrahedron Lett.* **24**, 2383–2386 (1983).
[59] Boons, G.J., Castle, G.H., Clase, J.A., Griece, P., Ley, S.V., and Pinel, C. *Synlett* 913–914 (1993).
[60] Reginato, G., Ricci, A., Roelens, S., and Scapecchi, S. *J. Org. Chem.* **55**, 5132–5139 (1990).
[61] Roelens, S. *J. Org. Chem.* **61**, 5257–5263 (1996).
[62] Pierce, A.E. *Silylation of Organic Compounds*. Pierce Chemical Co.: Rockford, Ill, 1968.
[63] Olah, G.A., and Narang, S.C. *Tetrahedron* **38**, 2225 (1982).
[64] Muzart, J. *Synthesis* 11 (1993).
[65] Colvin, E.W. *Silicon in Organic Synthesis*. Butterworth: London, p. 178, 1981.
[66] Sommer, L.H. *Stereochemistry, Mechanism and Silicon*. McGraw-Hill: New York, 1965, p. 138.
[67] Clark, J.H. *Chem. Rev.* **80**, 429–452 (1980).
[68] Stork, G., and Hudrlik, P.F. *J. Am. Chem. Soc.* **90**, 4462–4465 (1968).
[69] Corey, E.J., and Venkatesvarlu, A. *J. Am. Chem. Soc.* **94**, 6190–6191 (1972).
[70] Hanessian, S., and Lavallee, P. *Can. J. Chem.* **53**, 2975–2977 (1975).
[71] Nicolaou, K.C., Pavia, M.R., and Seitz, S.P. *J. Am. Chem. Soc.* **103**, 1224 (1981).
[72] Pecquet, F., and D'Angelo, J. *Tetrahedron Lett.* **23**, 2777–2780 (1982).
[73] Ogilvie, K.K., Thompson, E.A., Quilliam, M.A., and Westmore, J.B. *Tetrahedron Lett.* **33**, 2865–2868 (1974).
[74] Corey, E.J., and Rücker, C. *Tetrahedron Lett.* **23**, 719–722 (1982).
[75] Prakash, C., Saleh, S., and Blair, I.A. *Tetrahedron Lett.* **30**, 19 (1989).

[76] Metcalf, B.W., Burkhart, J.P., and Jund, K. *Tetrahedron Lett.* **21**, 35–36 (1980).
[77] Palmisano, G., Danieli, B., Lesma, G., and Mauro, M. *J. Chem. Soc., Chem. Commun.* 1564–1565 (1986).
[78] Nicolaou, K.C., and Magolda, R.L. *J. Org. Chem.* **46**, 1506–1508 (1981).
[79] Wood, W.W., and Rashid, A. *Tetrahedron Lett.* **28**, 1933–1936 (1987).
[80] Morton, H.E., and Guidon, Y. *J. Org. Chem.* **50**, 5379–5382 (1985).
[81] Hanessian, S., Delorme, D., and Dufresne, Y. *Tetrahedron Lett.* **25**, 2515–2518 (1984).
[82] Robins, M.J., Samano, V., and Johnson, M.D. *J. Org. Chem.* **55**, 410 (1990).
[83] Corey, E.J., Cho, H., Rücker, C., and Hua, D.H. *Tetrahedron Lett.* **22**, 3455–3458 (1981).
[84] Nakatsuka, M., Ragan, J.A., Sammakia, T., Smith, D.B., Uehling, D.B., and Schreiber, S.L. *J. Am. Chem. Soc.* **112**, 5583 (1990).
[85] Rousseau, G., Slougui, N. *Tetrahedron* **41**, 2653–2664 (1985).
[86] Park, T.K., Kim, I.J., Hu, S., Bilodeau, M.T., Randolph, J.T., Kwon, O., and Danishefsky, J. *J. Am. Chem. Soc.* **118**, 11488–11500 (1996).
[87] Frye, S.V., and Eliel, E.L. *Tetrahedron Lett.* **27**, 3223 (1986).
[88] Cunico, R.F., and Bedell, L. *J. Org. Chem.* **45**, 4797–4798 (1980).
[89] Koenigs, W., and Knorr, E. *Ber.* **34**, 957–962 (1901).
[90] Paulsen, H. *Angew. Chem. Int. Ed. Engl.* **21**, 155–224 (1982).
[91] Schmidt, R.R. *Pure & Appl. Chem.* **61**, 1257–1270 (1989).
[92] Overend, W.G. Glycosides. In *The Carbohydtrates: Chemistry and Biochemistry*, Vol. 1A. Pigman, W., and Horton, D. (eds) Academic Press: New York, 1972, pp. 279–353.
[93] Bonkhov, A.-F., and Zaikov, G.E. *Chemistry of the O-Glycosidic Bond: Formation and Cleavage*, Pergamon Press: Oxford, 1979.
[94] Paulsen, H. *Chem. Soc. Rev.* 15–45 (1983).
[95] Nicolaou, K.C., Chucholowski, A., Dolle, R.E., and Randall, J.L. *J. Chem. Soc., Chem. Commun.* 1155–1156 (1984).
[96] Hashimoto, S., Hayashi, M., and Noyori, R. *Tetrahedron Lett.* **25**, 1379–1382 (1984).
[97] Dolle, R.E., and Nicolaou, K.C. *J. Am. Chem. Soc.* **107**, 1691 (1985).
[98] Hanessian, S., Bacquet, C., and Lehong, N. *Carbohydr. Res.* **80**, C17 (1980).
[99] Fügedi, P., Garegg, P.J., Lönn, H., and Norberg, T. *Glycoconjugate J.* **4**, 97–108 (1987).
[100] Nicolaou, K.C., Dolle, R.E., Papahatjis, D.P., and Randall, J.L. *J. Am. Chem. Soc.* **106**, 4189–4192 (1984).
[101] Roush, W.R., Lin, X., and Straub, J.A. *J. Org. Chem.* **56**, 1649–1655 (1991).
[102] Halcomb, R.L., and Danishefsky, S.J. *J. Am. Chem. Soc.* **111**, 6661–6666 (1989).
[103] Dushin, R.G., and Danishefsky, S.J. *J. Am. Chem. Soc.* **114**, 3471–3475 (1992).
[104] Ferrier, R.J., Hay, R.W., and Vethaviyasar, N. *Carbohydr. Res.* **27**, 55 (1973).
[105] Ferrier, R.J., and Haines, S.R. *Carbohydr. Res.* **127**, 157 (1984).
[106] Williams, R.M., and Stewart, A.O. *Tetrahedron Lett.* **24**, 2715–2718 (1983).
[107] Lou, B., Huynh, H.K., and Hanessian, S. Oligosaccharide synthesis by Remote Activation: O-Protected Glycosyl 2-thiopyridylcarbonate Donors. In: *Preparative Carbohydrate Chemistry*. Hanessian, S. (ed.) Marcel Dekker: New York, 1997, pp. 431–448.
[108] Nicolaou, K.C., Dolle, R.E., Papahatjis, D.P., and Randall, J.L. *J. Am. Chem. Soc.* **106**, 4189 (1984).
[109] Nicolaou, K.C., Bockovich, N.J., and Carcanague, D.R. *J. Am. Chem. Soc.* **115**, 8843–8844 (1993).
[110] Ferrier, R.J., *Adv. Carbohydr. Chem.* **20**, 67 (1965).
[111] Ferrier, R.J., *Adv. Carbohydr. Chem. Biochem.* **24**, 199 (1969).
[112] Fisher, E., and Zach, K. *Sitz. Ber. Kgl. Preuss. Akad. Wiss.* **16**, 311 (1913).
[113] Tolstikov, A.G., and Tolstikov, G.A. *Russian Chem. Rev.* **62**, 579–601 (1993).
[114] Eitelman, S.J., Hall, R.H., and Jordaan, A. *J. Chem. Soc., Perkin Trans.* **1**, 595 (1978).
[115] Sharma, M., and Brown, R.K. *Can. J. Chem.* **44**, 2825 (1966).
[116] Lancelin, J., Morin-Allory, L., and Sinay, P. *J. Chem. Soc., Chem. Commun.* 355 (1984).
[117] Somsák, L., and Németh, I. *J. Carbohydr. Chem.* **12**, 679–684 (1993).
[118] Ferrier, R.J., and Prasad, N., *J. Chem. Soc.* 570–575 (1969).
[119] López, J.C., Gómez, A.M., Valverde, S., and Fraser-Reid, B. *J. Org. Chem.* **60**, 3851–3858 (1995).
[120] Lemieux, R.U. *Can. J. Chem.* **31**, 949 (1953).
[121] Lemieux, R.U., and Hubor, G. *J. Am. Chem. Soc.* **78**, 4117 (1956).
[122] Halcomb, R.L., and Danishefsky, S.J. *J. Am. Chem. Soc.* **111**, 6661–6666 (1989).
[123] Curci, R., Fiorentino, M., Troisi, L., Edwards, J.O., and Pater, R.H. *J. Org. Chem.* **45**, 4758 (1980).
[124] Montgomery, R.E. *J. Am. Chem. Soc.* **96**, 7820 (1974).

[125] Chiappe, C., Lo Moro, G., and Munforte, P. *Tetrahedron* **53**, 10471–10478 (1997).
[126] Danishefsky, S.J., and Bilodeau, M.T. *Angew. Chem. Int. Ed. Engl.* **35**, 1380–1419 (1996).
[127] Seeberger, P.H., Bilodeau, M.T., and Danishefsky, S.J. *Aldrichimica Acta* **30**, 75–92 (1997).
[128] For previous examples of this type of reaction see: [129, 130].
[129] Tietze, L.-F., Niemeger, U., and Marx, P. *Tetrahedron Lett.* **18**, 3441 (1977).
[130] Wyss, P.C., Kiss, J., and Arnold, W. *Helv. Chem. Acta* **58**, 1847 (1975).
[131] McClure, K.F., Randolph, J.T., Ruggeri, R., and Danishefsky, S.J. *Science* **260**, 1307 (1993).
[132] Randolph, J.T., McClure, K.F., and Danishefsky, S.J. *J. Am. Chem. Soc.* **117**, 5712 (1995).
[133] Liu, K.K.-C., and Danishefsky, S.J. *J. Org. Chem.* **59**, 1895–1897 (1994).
[134] Liu, K.K.-C., and Danishefsky, S.J. *J. Org. Chem.* **59**, 1892–1894 (1994).
[135] Gervay, J.M., Peterson, T., and Danishefsky, S.J. *J. Org. Chem.* **58**, 5465 (1993).
[136] Behar, V., and Danishefsky, S.J. *Angew. Chem. Int. Ed. Engl.* **33**, 1468 (1994).
[137] Danishefsky, S.J., Behar, V., Randolph, J.T., and Lloyd, K.O. *J. Am. Chem. Soc.* **117**, 5701–5711 (1995).
[138] Park, T.K., Kim, I.K., Hu, S., Bilodeau, M.T., Randolph, J.T., Kwon, O., and Danishefsky, S.J. *J. Am. Chem. Soc.* **118**, 11488–11500 (1996).
[139] Schmidt, R.R., and Jung, K.-H. Oligosaccharide Synthesis with Trichloroacetimidates. In: *Preparative Carbohydrate Chemistry*. Hanessian, S. (ed.) Marcel Dekker: New York, 1997, pp. 283–312.
[140] Schmidt, R.R. In: *Carbohydrates-Synthetic Methods and Application in Medicinal Chemistry*. Hasegawa, A., Ogura, H, and Suami, T. (eds) Kodanasha Scientific: Tokyo, 1992, pp. 66–70.
[141] Schmidt, R.R., Behrendt, M., and Toepfer, A. *Synlett* 694 (1990).
[142] Ishida, H., Ohta, Y., Tsukada, Y., Kiso, M., and Hasegawa, A. *Carbohydr. Res.* **446**, 75 (1993).
[143] Numata, M., Sugimoto, M., Ito, Y., and Ogawa, T. *Carbohydr. Res.* **203**, 205 (1990).
[144] Brooks, L.A., and Snyder, H.R. *Org. Synth. Coll.* **3**, 698 (1955).
[145] Fraser-Reid, B., Mootoo, D.R., Konradsson, P., Udodong, U.E., Andrews, C.W., Ratcliffe, A.J., Wu, Z., and Yu, K.L. *Pure & Appl. Chem.* **61**, 1243–1256 (1989).
[146] Madsen, R., and Fraser-Reid, B. *Modern Methods in Carbohydrate Chemistry*. Khan, S.H., and O'Neill, R.A. (eds) Harwood Academic Publishers: Amsterdam, Chapter 7, 1996.
[147] Roberts, C., Madsen, R., and Fraser-Reid, B. *J. Am. Chem. Soc.* **117**, 1546 (1995).
[148] Konradsson, P., Mootoo, D.R., McDevitt, R.E., and Fraser-Reid, B. *J. Chem. Soc. Chem. Commun.* 270 (1990).
[149] Mootoo, D.R., Konradsson, P., Udodong, U., and Freser-Reid, B. *J. Am. Chem. Soc.* **110**, 5583–5584 (1988).
[150] Merritt, J.R., and Fraser-Reid, B. *J. Am. Chem. Soc.* **114**, 8334–8336 (1992).
[151] Remers, W.A. In: *The Chemistry of Antitumor Antibiotics*. Wiley-Interscience: New York, 1979, pp. 133–175.
[152] Henderson, F.G. *Digitalis* Fish, C., and Surawicz, R. (eds) Grune & Stratton: New York, 1969, pp. 3–21.
[153] Thiem, J., and Klaffke, W. *Top. Curr. Chem.* **154**, 285–332 (1990).
[154] Roush, W.R., and Lin, X.-F. *J. Am. Chem. Soc.* **117**, 2236 (1995).
[155] Thiem, J., and Gerken, M. *J. Org. Chem.* **50**, 954 (1985).
[156] Thiem, J., and Schöttmer, B. *Angew. Chem. Int. Ed. Engl.* **26**, 555 (1987).
[157] Horton, D., Priebe, W., Sznaidman, M. *Carbohydr. Res.* **205**, 71 (1990).
[158] Thiem, J., Gerken, M., Schöttmer, B., and Weigand, J. *Carbohydr. Res.* **164**, 327 (1987).
[159] Ito, Y., and Ogawa, T. *Tetrahedron Lett.* **28**, 2723 (1987).
[160] Preuss, R., and Schmidt, R.R. *Synthesis* 694 (1988).
[161] Nicolaou, K.C., Hummel, C.V., Bockovich, N.J., and Wong, C.-H. *J. Chem. Soc., Chem. Commun.* 870 (1991).
[162] Grewal, G., Kaila, N., and Franck, R.W. *J. Org. Chem.* **57**, 2084–2092 (1992).
[163] Perez, M., and Beau, J.-M. *Tetrahedron Lett.* **30**, 75 (1989).
[164] Roush, W.R., Sebesta, D.P., and James, R.A. *Tetrahedron* **53**, 8837–8852 (1997).
[165] Capozzi, G., Menichetti, S., Nativi, C., Rosi, A., Valle, G. *Tetrahedron* **48**, 9023–9029 (1992).
[166] Capozzi, G., Franck, R.W., Menichetti, S., Mattioli, M., Nativi, C., Valle, G. *J. Org. Chem.* **60**, 6416–6426 (1995).
[167] Capozzi, G., Falciani, C., Menichetti, S., Nativi, S., Franck, R.W. *Tetrahedron Lett.* **36**, 6755–6758 (1995).
[168] Tietze, L.F., and Kettschau, G. Hetero Diels–Alder Reactions in Organic Chemistry. In: *Stereoselective Heterocyclic Synthesis I*. Metz, P. (ed.) Springer-Verlag: Berlin, 1997, pp. 80–82.
[169] Capozzi, G., Falciani, C., Menichetti, S., and Nativi, C. *Gazz. Chim. It.* **126**, 227–232 (1996).

[170] Capozzi, G., Dios, A., Franck, R.W., Geer, A., Marzabadi, C., Menichetti, S., Nativi, C., and Tamarez, M. *Angew. Chem. Int. Ed. Engl.* **35**, 777–779 (1996).
[171] Capozzi, G., Falciani, C., Menichetti, S., Nativi, C., and Raffaelli, B. *Chemistry. A European Journal*, **5**, 1748–1754 (1999).
[172] Capozzi, G., Mannocci, F., Menichetti, S., Nativi, C., and Paoletti, S. *Chem. Commun.* 2291–2292 (1997).
[173] Diep, V., Dios, A., Geer, A., Franck, R.W., Li, B.Q., Marzabadi, C.H., Tamarez, M., Win, W.W., Capozzi, G., Menichetti, S., and Nativi, C. *209th National ACS Meeting*, Anaheim, CA; April, 1995; CARB 033.
[174] Franck, R.W., and Marzabadi, C.H. *J. Org. Chem.* **63**, 2197–2208 (1998).
[175] Hanessian, S., and Giundon, Y. *Carbohydr. Res.* **86**, C3 (1980).
[176] Nicolaou, K.C., Seitz, S.P., and Papahatjis, D.P. *J. Am. Chem. Soc.* **105**, 2430–2434 (1983).
[177] Schmidt, R.R., and Stumpp, M. *Liebigs Ann. Chem.* 1249 (1983).
[178] Madsen, R., and Fraser-Reid, B. *J. Org. Chem.* **60**, 772 (1995).
[179] Debenham, J.S., Debenham, S.D., and Fraser-Reid, B. *Biorg. & Med. Chem.* **4**, 1909–1918 (1996).

9 Chemistry of Antisense Oligonucleotides

Christian R. Noe and Lucius Kaufhold

Abbreviations

ACN	Acetonitrile
AE	Absorptive endocytosis
B	Nucleobase (adenine, guanine, cytidine, thymine) or a nucleobase with attached protecting groups.
$BnCl_2$	2,4-Dichlorophenylmethyl-
bp	Base pairs
CAP-site	GTP is attached with the triphosphate at the 5´ end of an mRNA chain in a 5´,5´-condensation to form the structure 3´-G-5´ppp5´-N-3´p.. generally known as a cap.
CE	Capillary electrophoresis
CGE	Capillary gel electrophoresis
CHOL	Cholesterol
CIP	Cahn–Ingold-Prelog nomenclature system for absolute configuration
CPG	Controlled pore glass
CTAB	Hexadecyltrimethylammonium bromide
CZE	Capillary zone electrophoresis
dA, dG, dC	Deoxyadenosine, deoxyguanosine, deoxycytosine
DBU	1,8-Diazabicyclo[5.4.0]undec-7-en
DEAE	Diethylaminoethyl-
DMF	Dimethylformamide
DMT	Bis(4-methoxyphenyl)phenylmethyl-, (syn. Dimethoxytrityl-)
DNP	2,4-Dinitrophenyl-
DOPE	Dioleylphosphatidylethanolamine
DOTMA	N-[1-(2,3-dioleyloxy)propyl]-N, N, N-trimethylammonium chloride
EDTA	Ethylenediamine tetra-acetic acid
EM	Electrophoretic mobility
EOF	Electro-osmotic flow
ESI-MS	Electrospray ionization mass spectrometry
Fmoc	9-Fluorenylmethoxycarbonyl-
FPE	Fluid phase endocytosis
ICAM-1	Intracellular Adhesion Molecule-1
IGS	Internal guiding sequence
MALDI-TOF	Matrix-assisted laser desorption ionization
MEKC	Micellar electrokinetic chromatography

MeNPOC	5′-O-9-(α-methyl-6-nitropiperonyl)oxycarbonyl)-
MS	Mass spectroscopy
$N(iPr)_2$	Diisopropylamino-
NCEt	2-Cyanoethyl-
NHS	N-hydroxysuccinimide esters
OA	Oleic acid
OD	Optical density
ODN	Oligodeoxynucleotide
ORN	Oligoribonucleotide
PCR	Polymerase chain reaction
PNA	Peptide nucleic acid
RME	Receptor-mediated endocytosis
RNase	Ribonuclease
RT	Room temperature
SDS	Sodium dodecylsulfate
SODN	Oligodeoxynucleotide phosphorothioate
SRU	Structural repeating unit
*t*Boc	tert-Butyloxycarbonyl-
THF	Tetrahydrofuran
THP	Tetrahydropyranyl-
ΔT_m/mod.	Difference in transition temperature of an oligonucleotide duplex per number of modified monomer

9.1 Introductory Remarks

With the launch of the first antisense drug "Fomivirsen sodium" in 1998, a sometimes seriously questioned innovative drug concept with gene therapeutic background has at once become an established drug therapy. This situation has generated a unique momentary opportunity to look at this topic. Bearing in mind that antisense action may include prevention of synthesis of pharmacologically relevant enzymes and receptors, this new type of drugs opens – apart from novel disease targets – an additional level for enzyme- and receptor-oriented drug research.

9.2 Historical Aspects

9.2.1 Nucleic Acids and Molecular Biology in Drug Research

The first step towards understanding the molecular basis of genetic information transfer was taken by the discovery of the nucleic acids by Miescher [1] in 1869. However, it took rather a long time until the biological significance of this class of compounds was recognized. At the beginning of the 20th century the heterocyclic bases adenine (A), guanine (G), cytosine (C) and thymine (T) were characterized, which are structural components of DNA [2–5]. At the same time, Ascoli described uracil (U), which is present in RNA [6]. After discovery of the sugar components D-ribose and D-2-deoxyribose, classification of ribonucleic acid (RNA) and deoxyribonucleic acid (DNA) could take place [7–11].

After proof of the functional role of DNA had been provided by Avery in 1944 [12], the "race" to solve the "riddle of life" began. Finally, in 1953 the key for understanding the mechanism of genetic information transfer was found by Watson and Crick by their ingenious concept of base pairing [13]. Their work was based last but not least on Franklin's X-ray analysis [14] and the work of Wilson [15]. The eminent role of DNA, which contains all information required for the synthesis of all proteins in a living organism, has led to its nickname, the "molecule of life".

The findings on the significant role of nucleic acids in cell biology had immediate impacts on pharmaceutical research. The first disease targets were obvious: unwanted processes involving DNA and RNA action are central in cancer and viral diseases. Thus, starting in the 1950s, the field of antimetabolites emerged as an important field of research and development resulting in a significant number of (still relevant) nucleoside drugs (e.g., 5-fluorouracil (5-FU), aciclovir, zidovudine, didanosine) and a huge array of structural variations of nucleosides, which never reached the level of application (among them Belikova's chloroethylamine and nitrogen mustard substituted 2-deoxyribonucleotides ([16] in which oligonucleotides are mentioned as potential drugs for the first time). Antimetabolites, as well as alkylating agents, may be seen as a "first generation" of drugs acting at the genetic level. In the meantime, their mode of action has been intensively studied and it is well established that – apart from their function as "fake building blocks" in a nucleic acid chain – their mechanism of action includes the whole enzymatic array controlling nucleic acid synthesis. From a structural point of view, these drugs are mostly either modifications of nucleobases or nucleosides. Nucleotide drugs, which by definition contain a phosphate moiety attached to the nucleoside, do not appear among the presently used compounds.

In 1973, some 20 years after the decoding of the genetic code, Cohen and Boyer succeeded in gene recombination, thus establishing the era of gene technology. Barely another event has changed the landscape of pharmaceutical research in a comparable manner. Only three years after the appearance of the first scientific paper, the company Genentec was founded and has become the model of a new "culture" in the pharmaceutical field, a characteristic feature of which is a very early integration of scientific research projects and results into a commercial environment. In the meantime, about 1500 "biotech" companies exist – estalished to a great extent in the USA with the support of venture capital – dealing with commercialization of molecular biology research. By the end of the 1990s – about 25 years after this paradigmatic event – exaggerated short-term expectations from gene technology have been reduced to a realistic level. Nevertheless, pharmaceutical research without use of molecular biology has become uncompetitive nowadays. There are at present four main fields, which have significantly changed or were newly established in the "biotech" era:

- Biotechnology
- Pharmacology
- Diagnostics
- Gene therapy.

Recombinant drugs, e.g., human insulin, plasminogen activator (tPA), interferons, interleukins, growth factors and others, constitute a steadily growing subset of drugs in the market. The impact of molecular biology on pharmacology has been much less spectacular for the public, but its role cannot be overestimated. Tools have been developed to study physiological and pathophysiological processes, e.g., monoclonal antibodies or antisense oligonucleotides, which in fact have found here their first successful application. They are extensively used as "primers" in the polymerase chain reaction (PCR). "Genomics" aims at the systematic discrimination between physiological and pathophysiological states of cell function at the level of nucleic acids. The field of diagnostics has been, without any doubt, dramatically changed by molecular biology in general, and by application of nucleic acids in particular. The "Human Genome Project" provides the basis for gene diagnosis. A novel, nucleic acid-based chip technology is rapidly emerging providing the technical basis for this extremely complex task. Not surprisingly, this field is developing as a further option for a broad application of antisense oligonucleotides.

9.2.2 Gene Therapy

In gene therapy, genes or specific nucleic acid sequences of genes are the direct target of therapeutic attack. Particular broad attention has been paid to somatic gene therapy, in which therapeutically relevant DNA segments are inserted into genomic DNA and are expressed in the human body. They exert their effects either after transcription at the RNA-level [17, 18] or after translation at the protein level. In the beginning, there had been overexpectations in somatic gene therapy for the treatment of cancer and genetically caused diseases, and this had resulted in a subsequent wave of skepticism. Presently, a realistic view prevails of a very important new therapeutic strategy, which will need some time for its successful implementation. "Gene surgery" may certainly be seen as an utopic target of these efforts, by which missing genes are properly inserted into the genome and impaired genes are either deleted or repaired.

The idea of influencing gene expression by nucleic acid segments has, however, not been restricted to somatic gene therapy. As early as 1978 it was shown by Zamecnik and Stephenson that a synthetic oligodeoxynucleotide (ODN) can block expression of a single gene. They used a synthetic 13-mer ODN to block replication of Rous sarcoma viruses [19]. Although they are in principle gene therapeutics, such antisense oligonucleotides are much less elegant, when compared with somatic gene therapy with the active principle generated by a biological process in the body. Thus, it is not surprising that the fundamental discovery that RNA may also exert a catalytic activity [20] and that mRNA can be selectively cleaved by ribozymes [21, 22] added a strong impetus to antisense research. An intrinsic catalytic activity destroying the therapeutic target detected by sequence specific recognition provided the required "extra" incentive.

9.3 Scope of the Article

The attempt will be made in this article to present a comprehensive survey on the chemistry of oligonucleotides and their bio-isosteres, specifically referring to their potential use as "antisense therapeutics". In most cases, such antisense drugs might be classified as "nucleic acid antagonists", because they antagonize different steps of gene expression. They are able to recognize specific sequences of cellular nucleic acids. Their primary effect is exerted through specific base pairing (of Watson–Crick or Hoogsteen type) supported by other supramolecular effects such as base stacking or interaction with tertiary structures. The compounds interfere with translation, transcription, degradation or other related processes of DNA and RNA pathways. Antisense oligomers are not necessarily nucleotides containing phosphate linkages, nor are their structures confined to naturally occurring base and sugar moieties of nucleic acids.

Since antisense compounds are a specific type of regular drugs, emphasis will be laid to apply the criteria of medicinal chemistry. Pharmacological and clinical aspects of antisense drug research will be treated only in brief. The significant overlap of techniques applied in antisense drug development and in related fields of nucleic acid research requires some further limitations to avoid a too broad coverage of the topic. Therefore the following aspects will *not* be treated:

- Modified oligonucleotides exhibiting base pairing with non-"biological" molecules, generating an isolated (artificial) system of recognition and replication [23].
- Nucleic acids sequences interacting with proteins. By definition, these nucleic acids, called aptamers are not "antisense" molecules.
- Monomeric nucleosides and nucleotides that interfere with cellular pathways of nucleic acids without making use of specific sequential information (antivirals and cancer antimetabolites).
- Nucleotides acting at enzymes or receptor sites (e.g., adenosine triphosphate-receptors).

9.4 Criteria in the Development of Antisense Drugs

9.4.1 Structural Aspects

Basically, any natural oligonucleotide sequence exerting a specific pharmacological effect might be considered to be itself a lead structure for the development of an antisense drug. This generates at first sight a rather simple situation of target and drug being structurally related and well defined. Nevertheless, considerable knowledge in nucleic acid chemistry is required as outlined in the following section:

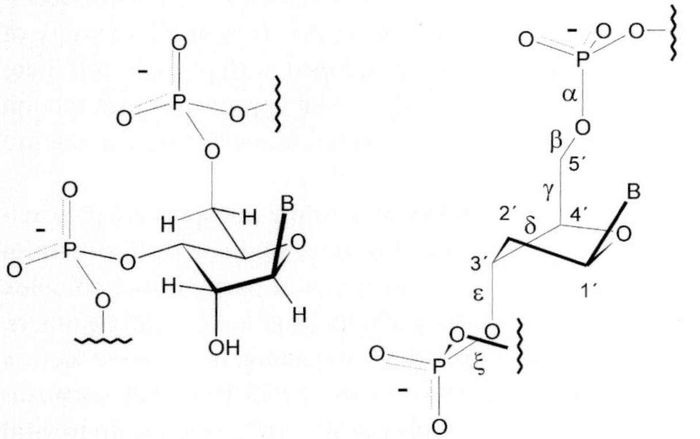

Figure 1. Conformation of nucleotides in double-stranded helices of RNA (type A) and DNA (B type).

The famous double helix, at the first glance, seems to be a comparably "simple" and rather uniform structure of a biopolymer. In fact however, not only may the double helix adopt several different conformations, the best known being the A-, B- and Z-types, but also single-stranded oligonucleotides, depending on their sequence and derivatization and conditions such as humidity, ion strength, counter ion charge, pH, buffer salts, solvent and last but not least temperature, display numerous types of polymorphism, the study of which has become an important field of biophysical research.

There are a few rules about these conformations, which are of rather general importance and ought to be considered, when dealing with antisense chemistry. One of them refers to the geometry of the sugar phosphate backbone, the conformation of which can be defined by the torsion angels (α, β, γ, δ, ε and ξ). Angle δ describes the orientation around C5´-C4´-C3´-O3´ and discriminates between A-type and B-type double helices (Fig. 1). While the other torsion angles remain in the same range for both types [24], angle δ in A-type helices, which are found in double-stranded RNA, usually adopts *sc*+ conformation (+60°), while in B type helices, most frequently adopted by DNA, an *ap* conformation (180°) for angle δ is found. Electronegative atoms like oxygen or fluorine at the 2´-position of the sugar ring favor C3´-endo conformation of the sugar ring [26] probably due to a preferred gauche orientation of the 2´-subsitutent and the ring oxygen, thus leading to A-type structures. A-typ RNA:RNA duplices are usually more stable than DNA:DNA duplices. A comprehensive summary on RNA, DNA and oligonucleotide polymorphism can be found in the excellent book of Saenger [25].

It is not surprising that achievement of chemical synthesis of natural oligonucleotides prompted also search for structural modifications. Study of scope and limitation of novel modifications to undergo base pairing with natural nucleic acids has been a first-line research target, together with attempts to involve oligonucleotide stability. Since oligonucleotide drugs, presently in the stage of clinical research, are very "close" to the natural structures (see section 6), it is evident that there is still a tremendous potential for structural optimization of such drugs. The theoretical part of this type of research has shifted in the meantime towards the study of artificial pairing systems or of alternatives to natural base pairing systems that might have played a role in evolution [23]. Much of this work is beyond the scope of the present article.

With respect to pharmacokinetics it is obvious that biotransformation of natural or only slightly modified natural oligonucleotides – mainly by action of nucleases – is a major obstacle towards application of such compounds. It is not surprising that first of all a variety of modifications of the labile phosphodiester bond has been developed, with phosphorothioate and methylphosphonate groups being hitherto the most successful approaches (see section 6). This report will also include structures such as peptide nucleic acids (PNAs), which are promising bio-isosteres of natural lead structures (see section 10.1).

Due to the polarity and high molecular weight of ODNs, membrane transport is a pharmacokinetic parameter of particular importance in antisense drug development. Both structural modifications of antisense compounds themselves and application of partly rather complex transport systems have been used successfully to improve this important factor. Among others, promising results have been obtained by the use of liposomes and nanoparticles (see section 17). Bearing in mind that antisense specificity is a molecular target specificity and by no means a cellular or tissue specificity, the use of even more complex drug carrier systems up to viral capsids provides an opportunity to include drug targeting into formulation of antisense drugs.

9.4.2 Definition of "Antisense" and Size of Oligonucleotides

Double-stranded DNA is made up of two strands. One of them is named the "coding strand". This contains the triplet code for the protein-sequence to be expressed. The other strand is the "complementary strand", which contains the complementary nucleotides in antiparallel arrangement. It is usually not expressed in the course of a specific gene expression. A mere look at a double helix – without referring to gene expression – does not allow us to decide which of the strands is the coding strand. During gene expression a gene is first transcribed from DNA to mRNA and then translated into a specific protein. In this process, by definition, the term "sense" connotes the mRNA-sequence of bases that is translated into the protein.

The "antisense" sequence contains the complementary bases in an antiparallel arrangement. Thus, the definition of "antisense compounds" refers to their interaction with transcribed RNA. In a broader sense the term "antisense action" is also used to describe inhibition of mRNA processing by molecules binding to sites formed by tertiary structures of single-stranded mRNA and interaction of molecules with duplex DNA by sequence specific insertion into the grooves.

Nucleic acids, due to their triplet code and use of a limited number of different building blocks, provide an extremely specific target for attack by antisense oligonucleotides. From Table 1 it can be seen that – based on the laws of statistics – an oligomer sequence of 17 bases should appear only once in the mRNA of a single cell, thus theoretically providing an almost incredible specificity for a targeted mRNA-sequence. These calculations are however restricted by the following factors:

- Coding gene sequences exhibit by no means a statistical variety of bases. Certain mRNA sequence patterns appear in the genome much more frequently than predicted by laws of statistics.
- Growing length of an oligomer does not only increase its target specificity, but will also increase binding energy of "drug–target"interaction.

As a direct consequence, there will be an increased readiness accept mismatched binding of the drug to undesired sequences. As a general rule it is presently accepted that in order to

Table 1. Selectivity related to the number of bases in an oligonucleotide and the total number of mRNA-Segments. In an average eucaryotic cell 10000 to 20000 different mRNA-sequences can be found. This sums up to approximately 1.5×10^7 to 3.0×10^7 base pairs. A short oligomer of 10 bases should then be found in 15 to 30 different mRNA-strands. Modified from [29]

Length of oligomer	Example	Abundance of random sequence
1 base	A	1 in 4 bases
2 bases	CG	1 in 16 bases
3 bases	GUC	1 in 64 bases
4 bases	GGGA	1 in 256 bases
5 bases	CCUCC	1 in 1.0×10^3 bases
10 bases	CCAACUUCGC	1 in 1.0×10^6 bases
13 bases	GCAUUCCGAUGAC	1 in 1.0×10^7 bases
15 bases	GGAGCCUCUUCGAUG	1 in 1.0×10^9 bases
16 bases	CUACUACCUGCCCGUC	1 in 4.3×10^9 bases
17 bases	CAGCCGUAUUGACGAGA	1 in 1.7×10^{10} bases

provide sufficient selectivity and acceptable handling, antisense oligonucleotides should be made up of about 15–25 base pairs [27, 28].

9.4.3 Molecular Targets of Antisense Action

From Fig. 2 it can be seen that different states of nucleic acids and different phases of their cellular pathways have been chosen as targets for antisense attack.

The most obvious result of antisense attack is translational arrest [27, 28]. Some antisense oligomers form triple helices with double stranded DNA via Hoogsteen or reversed-Hoogsteen base pairing (Fig. 2: A). This "early" phase of inhibition of transcription seems promising in principle, there are, however, many restrictions in this approach, one of them being the fact that formation of triple helices is slower and stability is lower than with duplex formation. Attempts have been made to increase triplex stability using intercalators covalently linked to antisense oligomers [30]. Duplex formation of the antisense molecule with mRNA (Fig. 2: B) is the prevailing strategy in antisense drug development. Intervention with transcription, splicing and mRNA transport has been reported. A particularly well-suited target for inhibition of mRNA are the site of translation initiation (AUG site) [28] (Fig. 2: D) and the CAP-site (Fig. 2: E) [31]. The 3′-region, which is not translated, does not seem to be a suited target sequence.

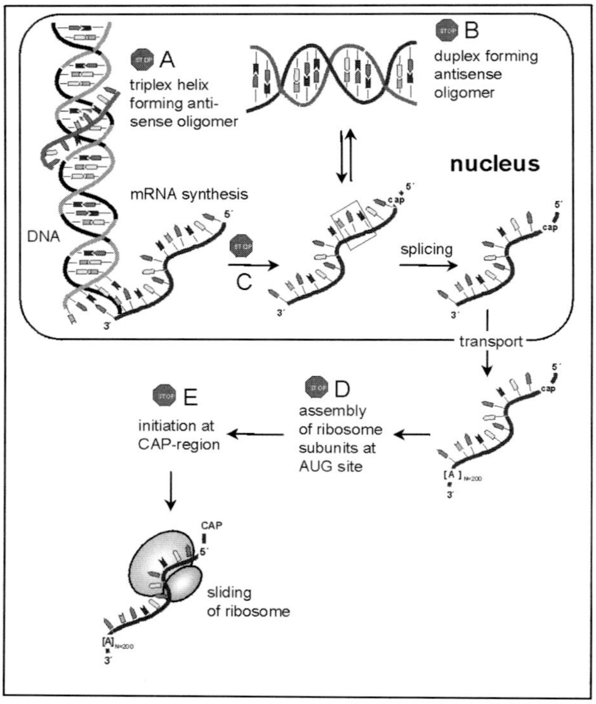

Figure 2. RNA processing and translation; intracellular action of antisense oligomers.

Table 2. Eucaryotic genes with known naturally occurring antisense RNAs of significant length (i.e. greater than 150 nucleotides). From [32]

Gene (sense/antisense)	Overlap (nucleotides)	Species	Sense gene properties
N-Myc/-Ncym	746–769	Human	Proto-oncogene, possible transcription factor
r-erbAα-2/Rev-ErbAα	269	Rat	DNA-binding protein
Thymidylate synthase/rTS	522	Human	DNA precursor synthesis, mRNA binding protein
c-myc	~400	Human	Proto-oncogene
RAD 10/ASR-106	>600	Yeast	DNA repair
ERCC-1/ASE-1	170	Human	DNA repair
Cytochrome P450c27/25	291	Human Rat	Vitamin D3 and cholesterol-27 hydroxylation, mitochondrial
α1(1) Collagen	~235–240, 467	Chicken	Structural protein
Chorion	~223, ~381	Silkmoth	Structural protein
Gonadotropin-RH	~500	Rat	Peptide hormone
Basic fibroblast growth factor	900	*Xenopus*	Growth factor
RPS12	250, 280	Human	Ribosomal protein

Antisense is by no means an "artificial" concept. Apart from basic coding/non-coding strand interactions, a series of cases has been found in nature, in which gene regulation is effected by antisense-type nucleic acid/nucleic acid interactions. Frequently, control of translation and accelerated depletion of target mRNA is effected by antisense ODNs in eucaryotic organisms like yeast, insects or mammals. In Table 2 a series of reported natural sense/antisense interaction is listed. In case given, sequences involved in such interactions are also potential elements of therapeutic antisense action.

9.5 Solid Support Synthesis of Oligonucleotides

It is frequently forgotten that the synthesis of DNA and RNA has been one of the major topics in synthetic chemistry. Khorana's outstanding work in this field, not least the chemical synthesis of tRNA, was duly awarded with a Nobel Prize. With the tremendous development in techniques of molecular biology, today the synthesis of genes and large segments of natural nucleic acids of any type has become a target of biological synthesis. A series of recombinant techniques has been established to obtain natural DNA and RNA of almost any base sequence. An exceptionally important step has been taken by the development of the PCR, which nowadays has grown into a broad field of techniques giving also access to significant quantities of nucleic acids. In contrast to synthesis of larger DNA segments, oligonucleotide synthesis of natural DNA, RNA – and of course of all modified types of oligonucleotide – has remained a domain of chemical synthesis. Solid support-based techniques, by which – compared with the Merrifield technique in peptide synthesis – the oligomer is built up by adding building blocks step by step, have been shown to be the most efficient approach in oligonucleotide synthesis. Among a series of competitive methods for making the phosphorodiester bond, the so-called phosphoramidite method is by far prevailing in importance. Therefore, the most common techniques of ODN synthesis by the phosphoramidite method are described in the following section. No description will be given for the diester, triester and H-phosphonate methods, which can also be carried out with most of the commercially available DNA synthesizers, but have been replaced in most cases by phosphoramidite synthesis. Synthesis of modified nucleic acids usually proceeds using modified protocols of ODN synthesis. Therefore, the principle of synthesis will be described in detail for natural compounds. Synthetic aspects specifically required for modified oligonucleotides will be covered directly in the respective sections. Nowadays, building blocks of natural nucleotides as well as those of the most frequently used modifications are commercially available. Custom synthesis of oligonucleotides is offered by several companies at conditions, which frequently present the most economic solution to obtain standard oligonucleotides. A comparison with peptide synthesis reveals that ODN synthesis is on the one hand less complex with respect to the fact that a by far lower number of different building blocks is required. On the other hand, the multifunctionality of these building blocks requires rather sophisticated strategies, e.g., to achieve selective cleavage of protective groups. In contrast to biological synthesis, chemical synthesis of DNA and RNA oligomers usually proceeds in the direction from the 3´- to the 5´-end of the sugar moiety. This approach allows to make use of the higher reactivity of primary 5´-hydroxy groups in the coupling step. Coupling takes place with the phosphoramidite moiety attached to the 3´-hydroxy group, which reacts with the free 5´-hydroxy group of the oligomer, that is attached to the solid phase. During this reaction the 5´-hydroxy group of the new monomer remains protected with an acid-labile protective group, preferentially DMT. Functional groups of the nucleobases generally also undergo protection, usually by acylation to avoid side reactions.

Unmodified ODNs are by far more stable when compared with unmodified RNA, which is cleaved easily by almost ubiquitous enzymes. In addition, RNA synthesis requires an additional protective group at the 2´-hydroxy group. Thus, it is no wonder that automated RNA synthesis was first published only about 10 years ago by Ogilvie [33], who modified the synthetic phosphoramidite protocol previously established for ODN synthesis.

With oligonucleotide drugs reaching the level of clinical studies and even the market, automated DNA-synthesis protocols are supplemented by methods of large-scale block synthesis, giving access to large amounts of material as required for regular drug production. Statistically, by this approach a 16-mer oligonucleotide consisting of natural nucleotides may be synthesized in four consecutive reaction steps via the dimers, tetramers and octamers involving a maximum of eight dimerization reactions, four tetramer and two octamer formations.

9.5.1 Building Blocks

Synthesis of standard ODN building blocks proceeds from the nucleosides and comprises the following steps: protection of nucleobases, protection of the 5´-hydroxy group and attachment of the phosphoramidite moiety. A comprehensive review on protective groups for use in oligonucleotide synthesis is given in [34].

A remarkably efficient method for selective N-acylation of nucleosides (given in Fig. 4 for benzoyladenosine) is the transient protection method [35]. In a one-pot procedure, the func-

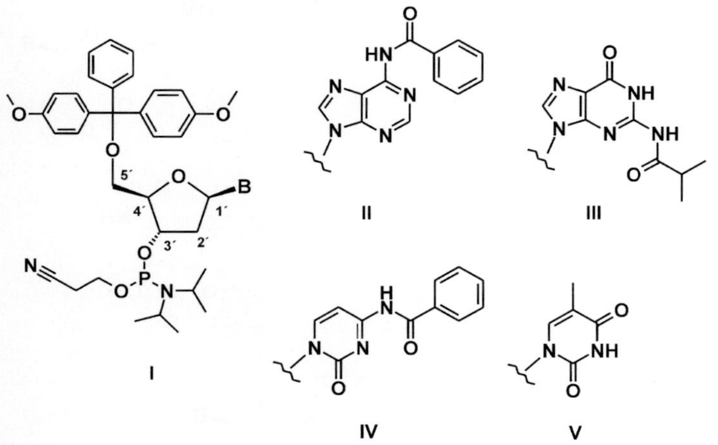

Figure 3. Protective group strategy in DNA-synthesis by the standard phosphoramidite method on solid support. I: Protected 5′-hydroxy group by the DMT (dimethoxytrityl) group, 3′-cyanoethyl-phosphoramidite; II, III, IV: Protection of nucleobases; V: Thymine remains unprotected.

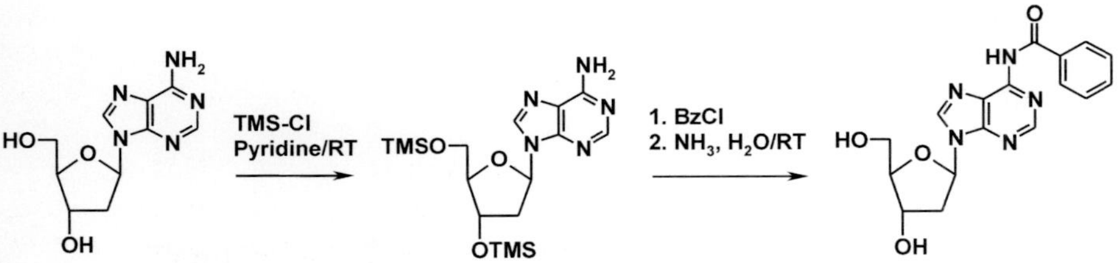

Figure 4. Transient protection method for N-acylation of nucleosides. Modified from [35].

tional groups of the nucleoside are converted to silyl ethers using trimethylsilyl chloride. Solubility of the nucleosides in organic solvents is markedly increased by silyl ether formation, resulting in far easier handling. After selective N-acylation trimethyl silyl ethers are cleaved by aqueous ammonia at room temperature. This technique may also be used to introduce other N-protective groups, e.g., trityl groups [36]. In section 5.4, selected acyl type protective groups are presented, which are cleaved particularly easily, thus providing strategy for mild protocols in ODN-synthesis.

9.5.2 Solid Support

Controlled pore glass (CPG) with a pore size of 500 Å is widely used in the synthesis of short oligomers up to 25-mer. Glass particles with wider pores (1000 Å) are used in the synthesis of longer oligomers. Polystyrene particles may be used as an alternative (Loading capacity: 20–30 μmol g^{-1}). As shown in Fig. 5, CPG glass is modified with aminopropyltriethoxysilane and passivated using trimethylchlorosilane/pyridine. The amino group is then reacted with

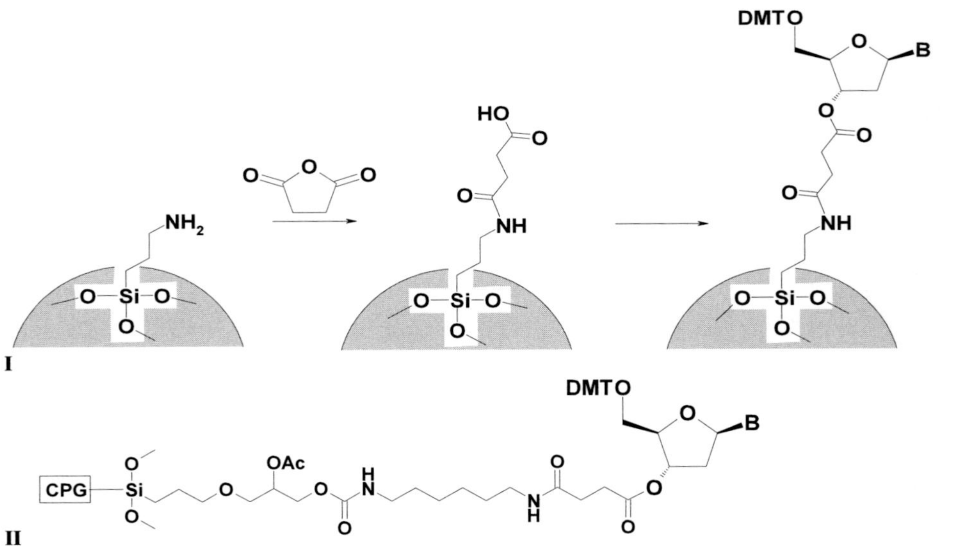

Figure 5. Activation of CPG-glass beads and linkage of starter monomer in standard phosphoramidite oligonucleotide synthesis.

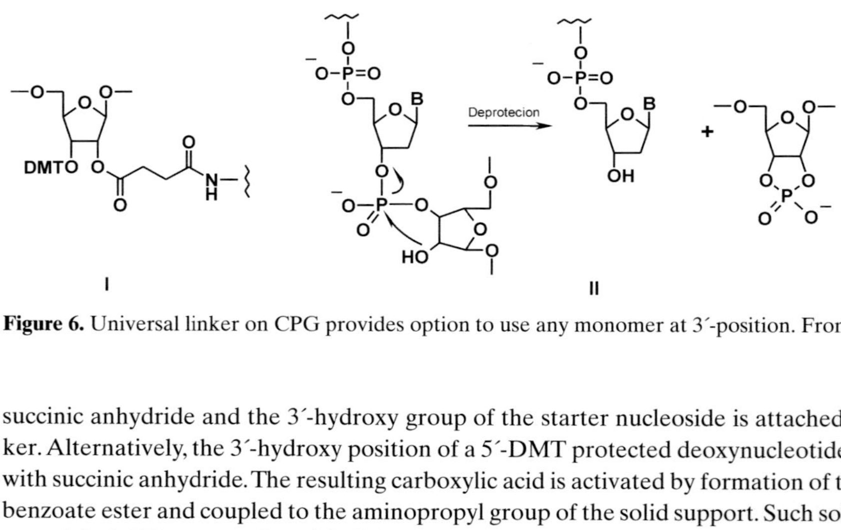

Figure 6. Universal linker on CPG provides option to use any monomer at 3′-position. From [37].

succinic anhydride and the 3′-hydroxy group of the starter nucleoside is attached to the linker. Alternatively, the 3′-hydroxy position of a 5′-DMT protected deoxynucleotide is reacted with succinic anhydride. The resulting carboxylic acid is activated by formation of the p-nitrobenzoate ester and coupled to the aminopropyl group of the solid support. Such solid support material of different quality with different starter nucleosides is commercially available.

Figure 6 illustrates the principle of action of a new sugar-type CPG-linker (I), which has recently become commercially available (license from Zeneca PLC). It can be used for any starting monomer (A,T,C,G), because – upon deprotection using standard cleavage procedures – it is eliminated as a cyclic sugar phosphate ester.

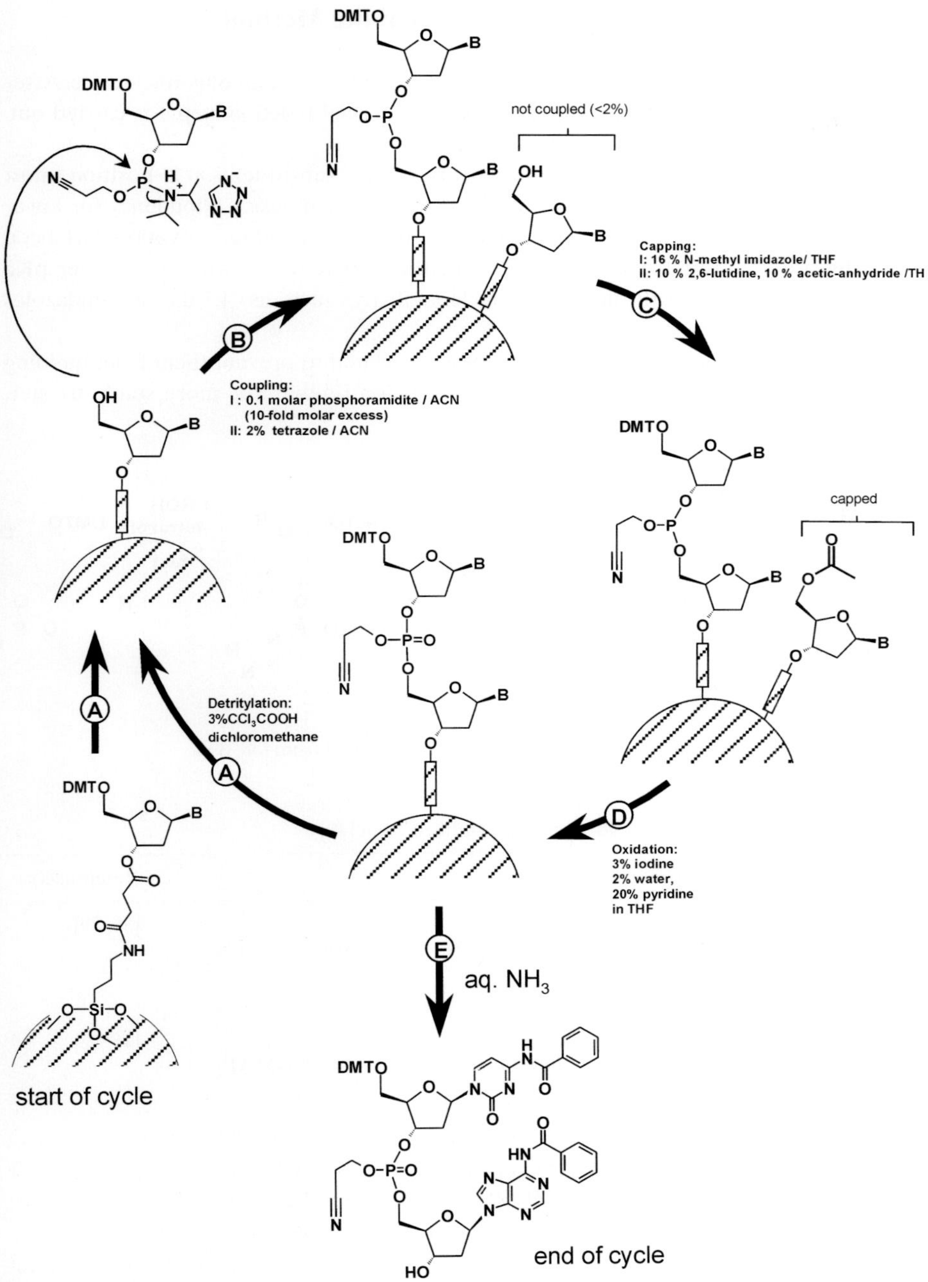

Figure 7. Synthesis cycle in standard oligonucleotide synthesis using the phosphoramidite method. A: 5´-OH-group is released by acid cleavage of the DMT-group; B: Substitution of the $N(iPr)_2$-group by the 5´-OH-group is catalyzed by 1*H*-tetrazole; C: Unreacted 5´-OH-functional groups are capped by catalyzed acetylation; D: Cyanoethyl-phosphite is oxidized by iodine; E: Oligomer is cleaved from spacer by concentrated ammonium hydroxide solution and kept at 55 °C for 8 h for complete deprotection.

9.5.3 Cycle of Synthesis via the Phosphoramidite Method

Figure 7 illustrates the different steps in solid support synthesis of an oligonucleotide. After detritylation, tetrazole-catalyzed coupling – the most crucial reaction step – is carried out. Figure 8 indicates the mechanism of tetrazole action.

With modified nucleotides, above all those bearing bulky substituents at 2´-position (most important the 2´-t-butyl(dimethyl)silyl-protection in ORN-synthesis), a tendency for lower coupling yields is observed. Improvement of 1*H*-tetrazole amidite activation has been achieved by use of 5-(4-nitrophenyl)- and 5-ethylthio-1*H*-tetrazole with their lower pKa values and compounds with increased nucleophilicity, such as 4,5-dicyanoimidazoles (Table 3).

After coupling, unreacted 5´-hydroxy groups are acylated to prevent them from building up truncated oligomer sequences. Subsequent oxidation renders the more stable triester.

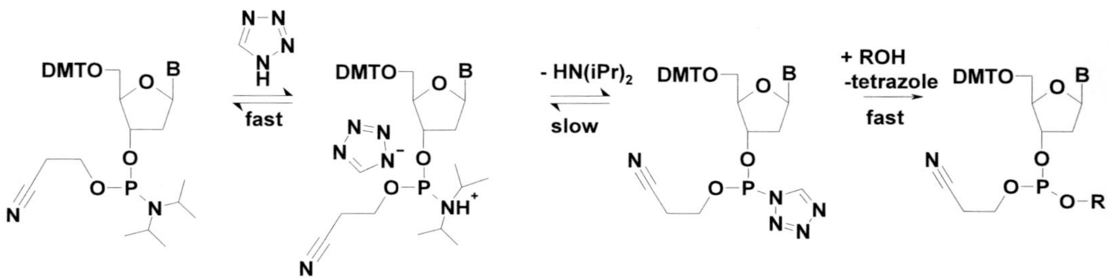

Figure 8. Proposed mechanism of activation with tetrazole. Modified from [38, 39].

Table 3. Activators of phosphoramidite coupling. Modified from [40]

Activator	Details	Reference(s)
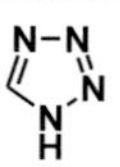	1*H*-tetrazole pKa: 4.8; Solubility: 0.5 M in acetonitrile	[38, 39]
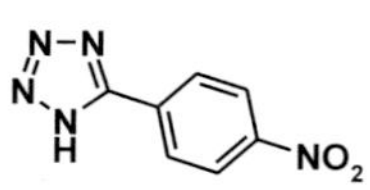	5-(4-nitrophenyl)-1*H*-tetrazole 0.12 M in acetonitrile in combination with DMAP in RNA-synthesis	[41, 42]
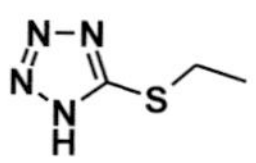	5-ethylthio-1*H*-tetrazole use in RNA synthesis	[43–45]
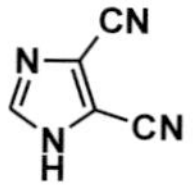	4,5-dicyanoimidazole pKa: 5.2; Solubility: 1.1 M in acetonitrile; used for 2´-amino-2´-deoxy-nucleotide coupling	[46, 47]

(With phosphorothioates it is recommended to carry out sulfurization before capping to avoid oxidation of P(III) by peroxides in THF.) Detritylation makes the 5′-hydroxy group accessible for the next steps. At the same time, oligonucleotide synthesizers are in a position to determine the coupling yield by measuring UV-absorption or conductivity of the waste solutions containing the cleaved DMT-cation. After finishing the desired number of cycles, the oligomer is cleaved from the solid support by use of aqueous ammonia.

Automated synthesis generally proceeds with high coupling efficiency. Almost quantitative yields are obtained, so that with standard oligonucleotides rather long oligomers can be synthesized. It must be taken into account that reagents and building blocks usually are applied in molar excess. Therefore, particularly in the synthesis of modified oligonucleotides, careful control of coupling efficiency from the beginning is required both to guarantee good oligomer yields and to avoid waste of material. Control of coupling times and selection of auxiliary reagents are the first measures to be taken in this respect.

9.5.4 Specific Measures for Efficient Deprotection

Deprotection involving concentrated ammonia is a critical step in the synthetic cycle. In particular, the use of (fluorescent) dyes linked to oligonucleotides or some other specific modifications is limited due to their lack of stability in the presence of concentrated ammonia solution, which is required for cleavage of the amides. Sometimes an improvement in deprotection yields may be achieved by an increase of reaction temperature from 55 °C to 80 °C, which reduces reaction time to 60 min [48]. A prerequisite for significantly milder cleavage protocols, however, is a change in protective groups at pyrimidine and purine bases.

Thus, the Beckmann Instruments company has developed a modified process called FAST-DNA-synthesis, by which a minor change in protective group strategy allows significantly more rapid cleavage of the oligomers from the CPG-column accompanied by cleavage of all protective groups [49]. As indicated in Fig. 9, N4-acetylcytidine is used instead of N4-benzoylcytidine. Aqueous methylamine is added to the ammonia solution (AMA). Work

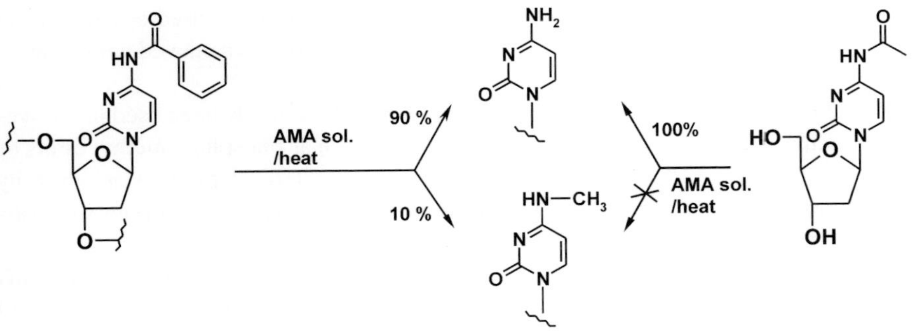

Figure 9. In contrast to N-benzoyluridine, N-acetyluridine phosphoramidites offer the key to fast cleavage and deprotection using AMA (20% methylamine and 15% ammonium hydroxide in water), since N-methyl uridine formation is not observed. No other bases have to be changed. Duration of cleavage is shortened to 5 min at 65 °C.

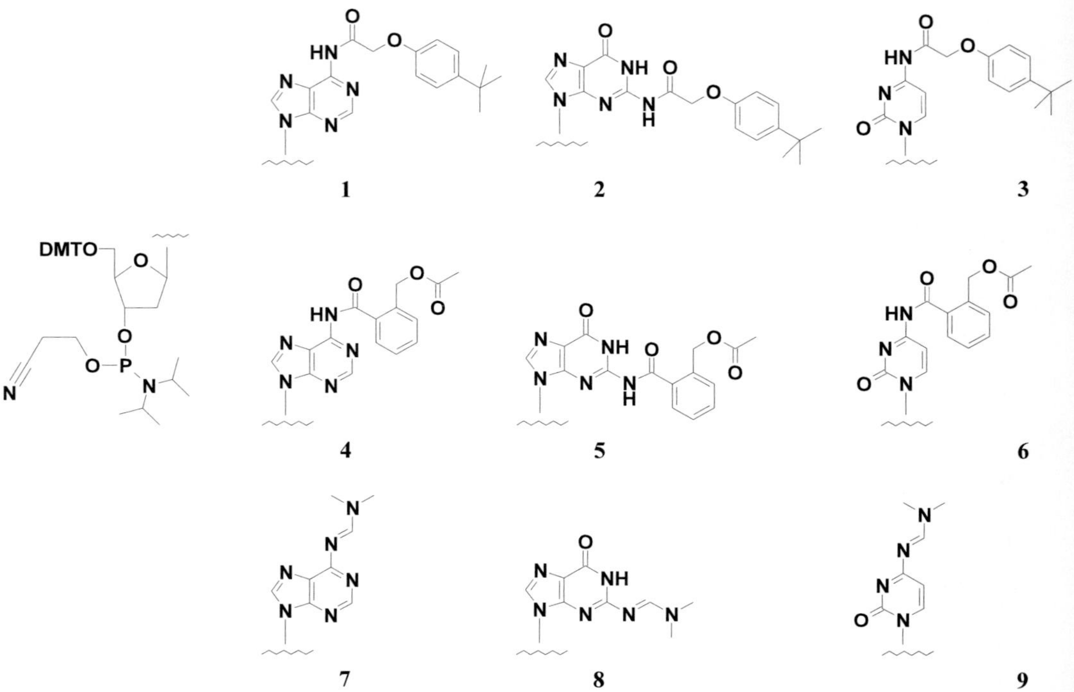

Figure 10. Improvement in mild deprotection protocol.

with these reagents requires some attention, because rubber fittings bulge and fluorescein, which is a frequent label with oligonucleotides, is not stable under cleavage conditions.

Another option for mild deprotection is the general use of labile N-protective groups, such as the phenoxyacetyl group. A further improved protective group is applied in the Expedite™* method, in which t-butylphenoxyacetyl-protective groups (Fig. 10) (**1**, **2**, **3**) are used for all bases except of thymine and t-butylphenoxyacetic anhydride is used as capping reagent. Cleavage conditions in this case are very mild. Deprotection is achieved with concentrated ammonia solution at 55 °C within 15 min [50].

The 2-(acetoxymethyl)benzoyl-protective group (**4**, **5**, **6**) has initially been used in the synthesis of particularly base-labile backbone modifications, such as phosphate methyl esters or methylphosphonates [51, 52], again thymine remains unprotected. Cleavage is achieved by elegant intramolecular reaction within 90 min at room temperature by use of potassium carbonate in dry methanol (Fig. 11).

Another established method for base protection is the use of dimethylformamidine (dmf) (**7**, **8**, **9**) protected phosphoramidites. In this case deprotection is achieved in concentrated ammonia solution at 55 °C within 60 min [53, 54].

* Expetide is a Trademark of the Millipore Corporation

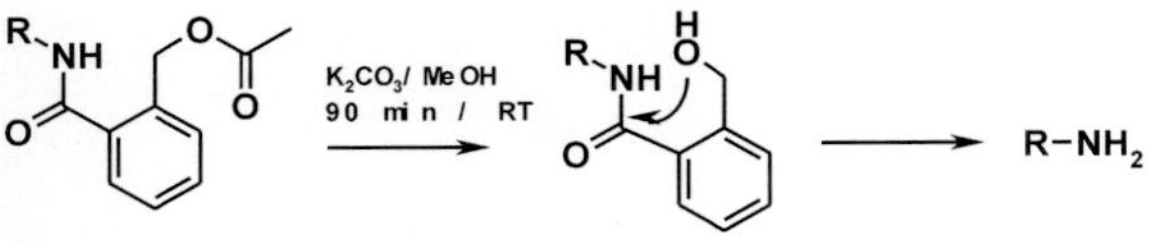

Figure 11. 2-(Acetoxymethyl)benzoyl protective group is easily removed by potassium carbonate in anhydrous methanol at RT.

9.5.5 Synthesis in "Large Scale"

Solid phase synthesis in 3–4 mmolar scale proceeds along the same protocols as small-scale synthesis. However, special synthesizers are used such as the Perkin-Elmer ABI 390 Z, Perspective Biosystems 880 DNA-Synthesizer and Oligo Pilot II synthesizer of Pharmacia Biotech. More efficient coupling is achieved by convection. Vortexing or permanent circular pumping allow use of only 1.5–2 molar equivalents of the phosphoramidite applied in a 0.1–0.2 molar solution in acetonitrile. As a rule, recovery of monomer solutions is feasible. Choice of narrow-pore (500 Å) "high loading" CPG beads with short-length oligonucleotides or polystyrene-based supports allow further increase in batch size. When using polystyrene beads, it must be considered that dichloroacetic acid used for deprotection is absorbed by the material. This acid has to be removed by repeated washing with DMF/ACN.

An important aspect in the synthesis of oligonucleotides in gram or even kilogram quantities is the limited availability of 2-deoxynucleosides. The main source of their supply is salmon milt (the sperm sac of the male). An estimated annual crop of 237 000 tons of salmon corresponds to 15 tons of ODNs in case of application of standard methods of isolation and synthesis. It is evident that increased introduction of oligonucleotide drugs into the market will prompt search for other sources and development work on synthesis of nucleosides [55].

9.5.6 5´-3´-Synthesis

With the 5´-hydroxy functional group of ribofuranosyl nucleosides being by far more reactive than the 3´-hydroxy group, oligonucleotides are generally built up from the 3´ to the 5´ end. For the study of hairpin structures, complementary sequences are synthesized linked with a polyoxyethylene-linker [56] or via a nucleoside [57]. In a comparable approach, stability of oligonucleotides against action of 3´-exonucleoases may be increased by terminal 3´,3´-coupling. In all these cases part of the oligonucleotide has to be built up from the 5´ to the 3´end. In this case the corresponding building blocks are 3´-dimethoxytritylated and have the phosphoramidite attached to the 5´-position.

9.6 Oligonucleotides with Modifications of the Nucleotide-Linkage

The half-life of ODNs in fetal calf serum is as short as 5 min [58]. Therefore, it is evident that unmodified ODNs cannot be expected to be directly applicable in antisense drug therapy. Nucleases, which cleave the phosphoric ester bond, are the most relevant ODN-degrading

Table 4. Phosphorus-containing backbone modifications. Modified from [59]

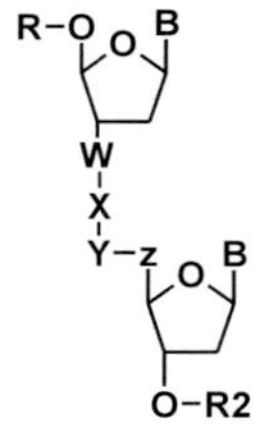

	-W-	-X-	-Y-	-Z-	Trivial	Reference
10	-O-	-P(O)S—	-O-	$-CH_2-$	*Phosphorothioates*	[60]
11	-O-	-P(S)S—	-O-	$-CH_2-$	*Phosphorodithioates*	[61]
12	-O-	$-P(O)CH_3-$	-O-	$-CH_2-$	*Methylphosphonates*	[62]
13	-O-	-P(O)OR-	-O-	$-CH_2-$	*Phosphoric acid triesters*	[63]
14	-O-	-P(O)NHR-	-O-	$-CH_2-$	*Phosphoramidates*	[64]
15	-NH-	-P(O)O—	-O-	$-CH_2-$	*Phosphoramidates*	[65]
16	-O-	$-P(O)BH_3$—	-O-	$-CH_2-$	*Phosphoroboronates*	[66]

enzymes. Therefore it is no surprise that modifications of the phosphate linkage were the first targets of antisense drug development. Figure 6 indicates such modifications, some of them constituting the basis for antisense drugs of the "first generation".

With respect to stereochemistry, substitution of one unbonded oxygen atom of the prochiral phosphorus linkage in an oligonucleotide by sulfur, nitrogen or borane creates a chiral center leading to diastereomeric mixtures of R_P and S_P compounds in synthesis. Since the number of diastereomers increases by 2^n, this sums up to the formation of about half a million diastereomeric compounds during the synthesis of a 20-mer. Nevertheless, antisense activity remains intact in such diastereomeric mixtures.

9.6.1 Phosphorothioates

Due to their nuclease stability and straightforward accessibility, oligonucleotides of the phosphorothioate type (SODN)(**10**) [67] were found rather early to be good modifications and have become the most frequently used class of modified oligonucleotides. SODNs are effective in nanomolar concentrations [68]. Drug/target duplexes are in a position to activate RNaseH. A major disadvantage is imperfect sequence specificity, particularly with high concentrations [68–70].

SODNs may be prepared without difficulty using the regular ODN phosphoramidite method. Several reagents used for sulfurization are shown in Table 5.

Apart from TETD, Beaucage's reagent (3*H*-1,2-benzodithiol-3-one-1,1-dioxide) [74] is used most frequently for sulfurization after coupling of the phosphoramidite. The latter is applied in 0.05 molar solution in anhydrous acetonitrile. Sulfurization proceeds within 0.5 min.

With respect to the dominance of the phosphoramidite method, the formation of phosphorothioates by the phosphite [75] and phosphotriester methods [76] is not outlined here. As mentioned previously, a new chiral center is generated with each sulfurization step of phos-

Table 5. Reagents for synthesis of phosphorothioates by the phosphoramidite method.

Reagent	Description	Reference
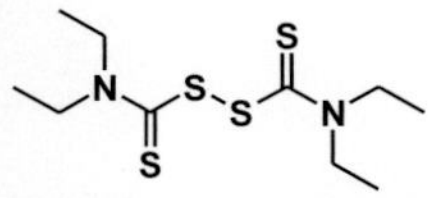 **Tetraethylthiuram disulfide**	TETD reagent: 0.5 M solution in anhydrous acetonitrile; Sulfurization time: 15 min; Sulfurization proceeds better in acetonitrile than in dichloromethane or THF.	[71]
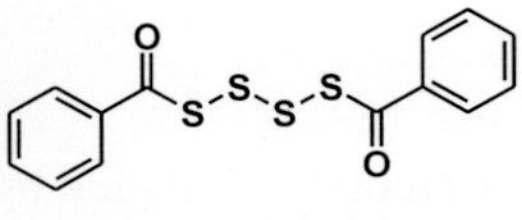 **Dibenzoyl tetrasulfide**	0.4 M solution in THF or dichloromethane (insoluble in acetonitrile); Sulfurization time: 1 min	[72]
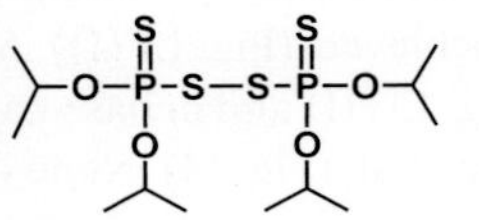 **Thioperoxydiphosphoric acid tetrakis (1-methylethyl) ester**	0.5 M solution in acetonitrile; Sulfurization time: 15 s. Used as antioxidant in rubber industry; Formation by oxidation of sodium O,O-diisopropylphosphorodithionate; Solution is not commercially available, it has to be prepared freshly before use.	[73]

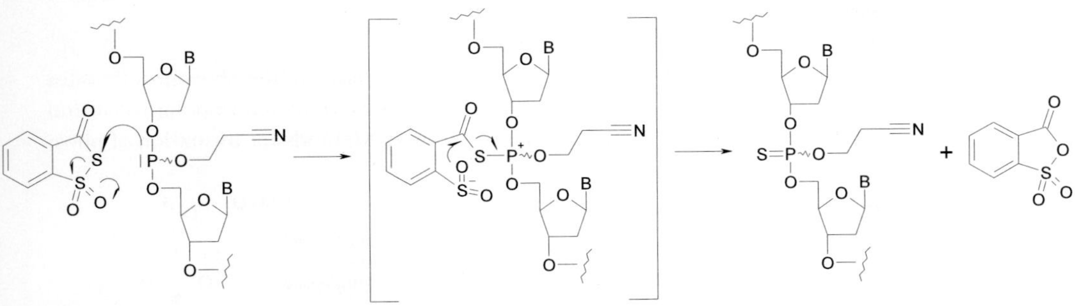

Figure 12. Preparation of phosphorothioate oligomers using Beaucage's reagent. Mechanism as proposed in [74].

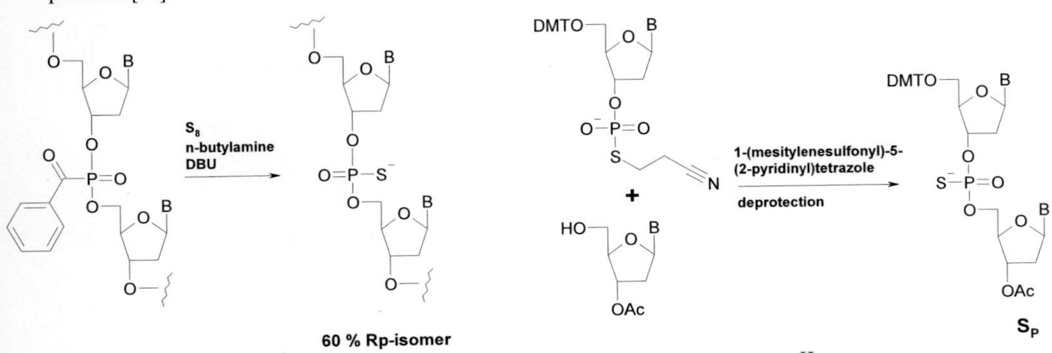

I II

Figure 13. Synthesis of chiral phosphorothioates. Modified from [77, 78].

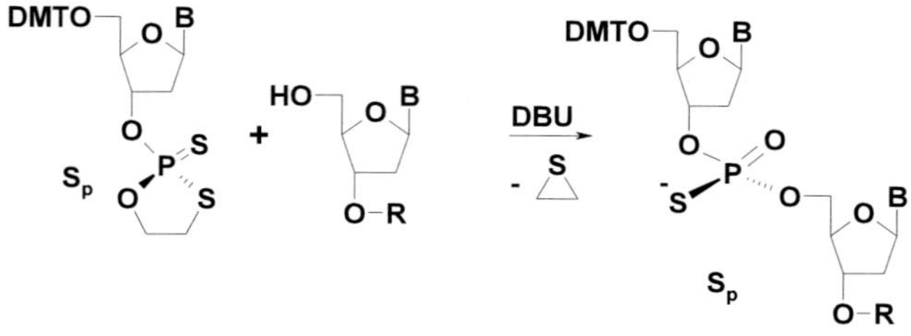

Figure 14. Synthesis of chiral phosphorothioates by ring opening of 1,3,2-oxathiaphospholanes. Modified from [79, 80].

phorothioate synthesis. Attempts have been made to overcome diastereomer formation by diastereoselective synthesis.

Thus, by use of aroylphosphonates and an alkylamine in the presence of DBU preferential formation of *R*p diastereomers of phosphorothioate dimers was achieved (Fig. 13 (I)). *S*p configuration prevails in the direct coupling method presented in Fig. 13 (II) and in base-catalyzed ring opening of 1,3,2-oxathiaphospholanes as studied by Stec et al. (Fig. 14). None of these methods reached broad application.

9.6.2 Phosphorodithioates

The phosphorodithioate linkage (**11**) overcomes the chirality issue of the phosphorothioates by replacing two oxygen atoms by sulfur. The linkage is isostructural and isopolar to natural phosphates and is stable towards enzymatic hydrolysis [81]. Methods for formation of phos-

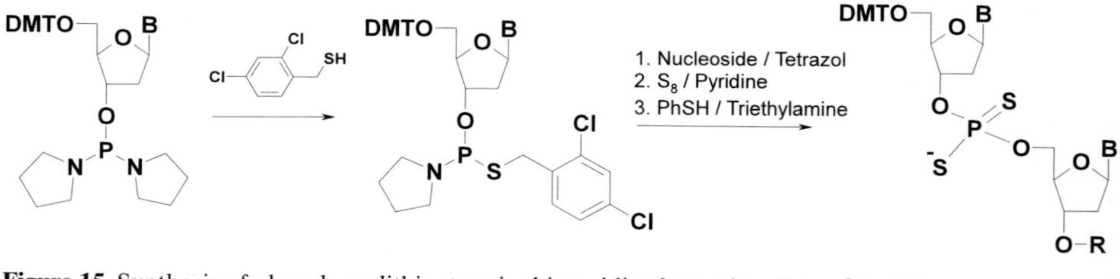

Figure 15. Synthesis of phosphorodithioates via thioamidite formation. From [76, 82].

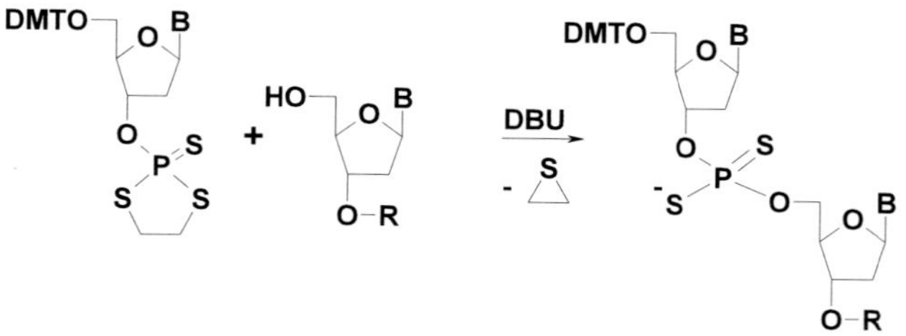

Figure 16. Synthesis of phosphorodithioates by ring opening of 2-thio-1,3,2-dithiaphospholane. From [80].

phorodithioates have been reviewed in [82]. Sulfurization via formation and reaction of phosphorthioamidites (Fig. 15) is one of the established methods.

Ring opening of 3´-*O*-2-thio-1,3,2-dithiaphospholane nucleosides – comparable with the formation of chiral phosphorothioates – may also be employed to obtain phosphorodithioates (Fig. 16). To date, none of the described building blocks is available commercially.

9.6.3 Methylphosphonates

The use of methylphosphonates (**12**) as antisense compounds has been reviewed by several authors [83, 84]. Very good nuclease resistance and a different mechanism of membrane penetration are the most remarkable features of this ODN modification [85]. The lack of anionic charge however renders the compounds relatively insoluble in water. Drug/target duplexes do not activate RNase-H. Therefore, high concentrations of up to 100 μM are required to achieve antisense activity. Nevertheless, methylphosphonates seem to be suited elements of chimeric oligonucleotides.

The method of choice for their synthesis employs methylphosphoramidite technology, which proceeds with high yields (97%) in each coupling step [86, 87] (Fig. 17 (I)). While T and dA monomers are soluble in acetonitrile, a 2:1 mixture of acetonitrile/dichloromethane is used as a solvent for dC and dG. With this method capping is carried out after the oxidation step. Since methylphosphonates are unstable under basic conditions, deprotection usually proceeds in two cycles. First, the support-bound oligomer is treated for one day at room temperature with a solution of hydrazine hydrate in pyridine/acetic acid buffer, followed by 6 h of treatment with a 1:1 mixture of ethylenediamine/ethanol at room temperature. Different protocols for one-pot procedures [88] or alternative protective group strategies (see section 5.4) have been published. Reduced T_m-values have been reported with methylphosphonates, probably due to their diasteromeric nature. Stereospecific synthesis has resulted in products exhibiting R_p configuration in up to 95% purity. Figure 17 (II) illustrates that stereospecific inversion at the phosphorus atom was achieved by activation of the 5´-hydroxy group with tert.-butylmagnesium chloride [80]. However, only tetramers have been synthesized by this method and there is little chance for its implementation in solid phase synthesis. Alkylphosphonates with S_p configuration have also been described [89].

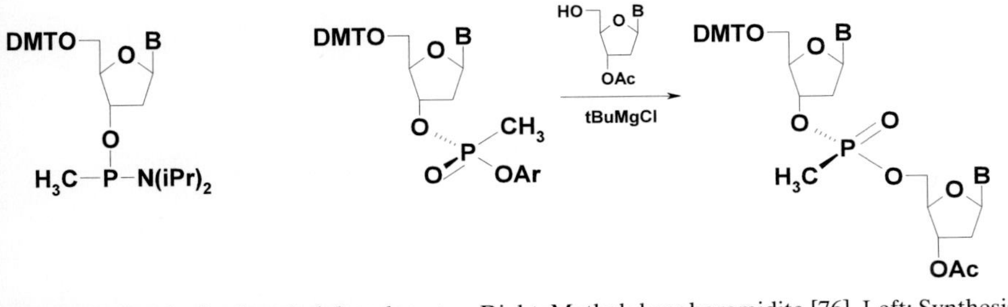

Figure 17. Synthesis of methylphosphonates. Right: Methylphosphoramidite [76]. Left: Synthesis of chiral methylphosphonates. From [80].

9.6.4 Phosphoric Acid Triesters

Triphosphates (**13**) may be formed using O-ethyl or O-isopropyl phosphoramidites. These esters do not undergo hydrolysis by concentrated aqueous ammonia solution applied for deprotection after synthesis. Another option for their synthesis involves use of 3´-O-phosphorodimorpholidites, which are subjected to alcoholysis prior to the final oxidation step [90]. By choice of specific phosphoramidites this technology may be used to insert conjugated functional groups into an oligonucleotide.

9.6.5 Phosphoramidates

Phosphoramidates (**14**) are a type of modification which – like triphosphates – offers further options for functionalization of oligonucleotides. They may be synthesized using phosphoramidite technology with oxidation of the phosphite triester with iodine in the presence of alkylamines [91]. t-Butylamine in methanol at 45 °C is used for their deprotection and cleavage from the support. The protocol allows modification of a single linkage for attachment of functional groups.

Alternative methods for the synthesis of phosphoramidates are presented in Table 6. Suitability for these techniques has been reviewed in [91].

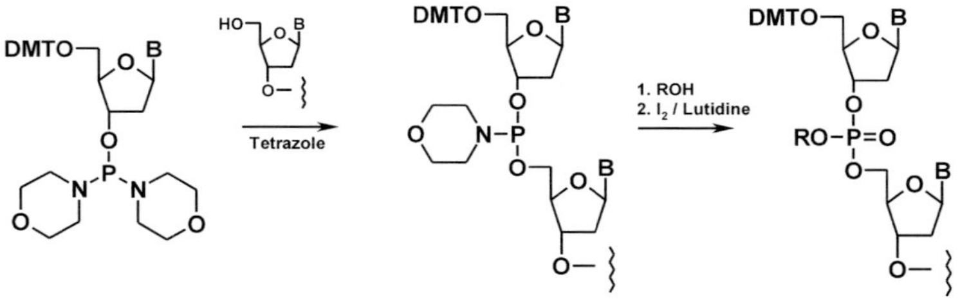

Figure 18. Formation of phosphate triesters by formation of phosphormorpholites. ROH: R = lower alkyl substituent. From [90].

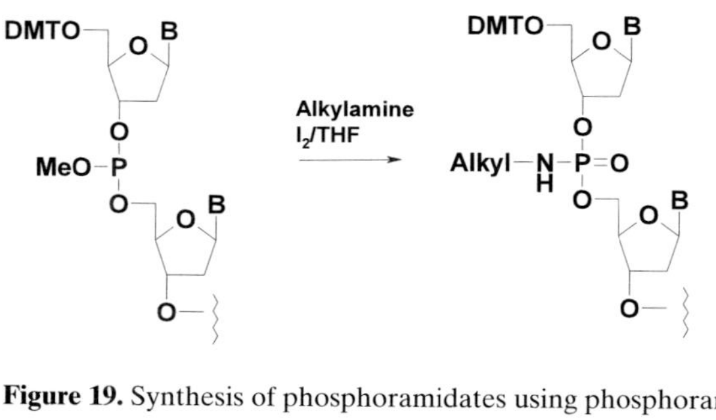

Figure 19. Synthesis of phosphoramidates using phosphoramidite technology. Modified from [76].

Table 6. Synthesis of phosphoramidates

Reaction	Description	Reference(s)
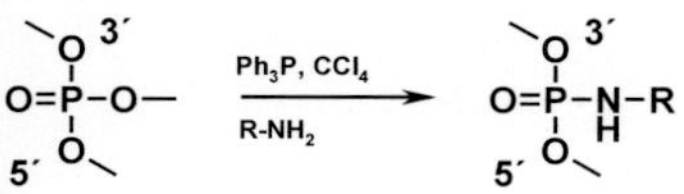	Usually gives moderate yields (24%); Method has been used for the formation of anilides	[92, 93]
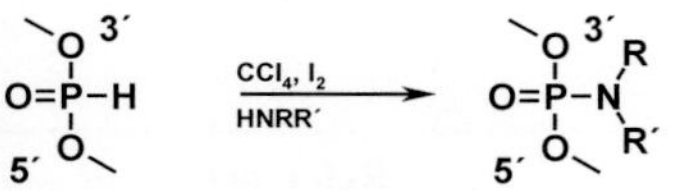	Preparation of all-phosphoramidate oligonucleotides (both primary and secondary)	[92, 93]

9.6.6 Phosphoroboronates

Phosphoroboronates (**16**) are a class of modified ODNs with particularly high nuclease resistance. Development in this field has been reviewed recently [96]. Phosphoroboronates may be synthesized starting from H-phosphonates easily accessible by synthesis of phosphormorpholidites as outlined in Fig. 20. Resistance towards snake venom phosphodiesterase (3´,5´-exonuclease) is due to the generated chiral phosphorus atom. *S*p boronated thymine dimers display hydrolysis at small rates, whereas the *R*p diastereomer is completely stable. Digestion by bovine spleen phosphodiesterase (3´,5´-exo) proceeds also very slowly, however without discrimination of the diastereomeric dimers [97]. The dimer (Fig. 20: A) (partition coefficient n-octanol/water: 1.6×10^{-3}) is more lipophilic than the corresponding phosphate dimer (partition coefficient: 8.7×10^{-5}). Incorporation of boronated dG in a 14-mer led to little deviation of the T_m-values compared with unmodified strands [97]. Direct PCR-sequencing might become a possible application of 2´-deoxynucleoside-5´-α-[P-borano]-triphosphates. DNA-sequences are amplified with boron nucleotides triphosphates and digested by exonucleases, leaving DNA fragments terminated with the backbone modification. Similar techniques with 5´-α-[P-thio]-triphosphates have not been satisfying because of uneven band intensity for the four terminal nucleotides [98, 99].

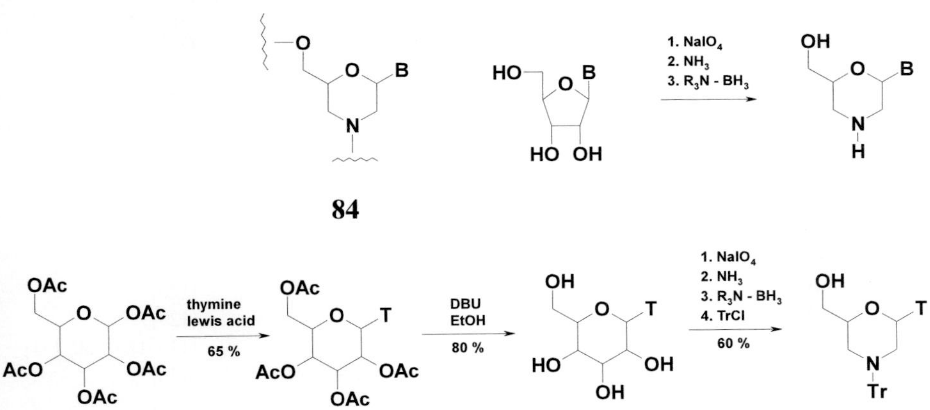

Figure 20. Synthesis of phosphoroboronates starting from H-phosphonates. From [66].

Table 7. Non-phosphorus-containing linkages. Linkages are sorted by composition and CIP-rules following 3´ to 5´ direction in ascending priority. (42: represents five chain members)

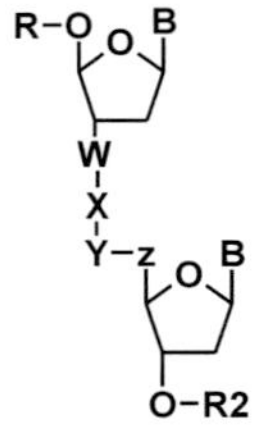

	-W-	-X-	-Y-	-Z-	Reference
Carbon-, silicon- and oxygen-containing linkages					
17	-CH_2-	-CH_2-	-CH_2-	-CH_2-	[100]
18	-CH_2-	-CH_2-	-CH_2-	-O-	[101]
19	-CH_2-	-CO-	-CH_2-	-CH_2-	[102, 103]
20	-CH_2-	-CO-	-O-	-CH_2-	[104, 105]
21	-O-	-CH_2-	-CH_2-	-CH_2-	[106]
22	-O-	-CH_2-	-CH=	=CH-	[106]
23	-O-	-CH_2-	-CH_2-	-O-	[101]
24	-O-	-CH_2-	-O-	-CH_2-	[76, 107, 108]
25	-O-	-CO-	-O-	-CH_2-	[109, 110]
26	-O-	-SiR_2-	-O-	-CH_2-	[76, 108, 111–113]
Nitrogen-containing linkages					
27	-CH_2-	-CH_2-	-NH-	-CO-	[114]
28	-CH_2-	-CH_2-	-N(CH_3)-	-CH_2-	[100]
29	-CH_2-	-CO-	-NH-	-CH_2-	[115–118]
30	-CH_2-	-CO-	-NCH_3-	-CH_2-	[117, 118]
31	-CH_2-	-CO-	-N(iPr)-	-CH_2-	[117, 118]
32	-CH_2-	-N(CH_3)-	-CH_2-	-CH_2-	[100]
33	-CH_2-	-NH-	-CO-	-CH_2-	[119]
34	-CH_2-	-N(CH_3)-	-N(CH_3)-	-CH_2-	[120]
35	-CH_2-	-NH-	-O-	-CH_2-	[121]
36	-CH_2-	-N(CH_3)-	-O-	-CH_2-	[76, 108, 120, 122]
37	-CO-	-NH-	-CH_2-	-CH_2-	[123]
38	-NCH_3-	-CO-	-CH_2-	-CH_2-	[124]
39	-NCH_3-	-CO-	-NH-	-CH_2-	[125]
40	-NCH_3-	-CO-	-O-	-CH_2-	[126]
41	-NH-	-CO	-CH_2-	-CH_2-	[124]
42	-NH-	-CO-CH_2-	-NH-	-CO-	[127]
43	-NH-	-CO-	-NH-	-CH_2-	[125]
44	-NH-	-CO-	-NCH_3-	-CH_2-	[125]
45	-NH-	-CO-	-O-	-CH_2-	[126]
46	-NH-	-(C=N-CN)-	-NH-	-CH_2-	[128]
47	-NH-	-(C=N-R)-	-NH-	-CH_2-	[129]
48	-O-	-CO-	-NH-	-CH_2-	[115, 126, 130, 131]
49	-O-	-CO-	-NCH_3-	-CH_2-	[126]
50	-O-	-CO-	-N(iPr)-	-CH_2-	[126]
51	-O-	-N(CH_3)-	-CH_2-	–	[132]

Table 7. Continued

	-W-	-X-	-Y-	-Z-	Reference
Sulfur-containing linkages					
52	-CH_2-	-CH_2-	-S-	-CH_2-	[133]
53	-CH_2-	-S-	-CH_2-	-CH_2-	[76, 108]
54	-CH_2-	-SO_2-	-CH_2-	–	[134]
55	-NH-	-SO_2-	-CH_2-	–	[134]
56	-O-	-CH_2-	-S-	-CH_2-	[107]
57	-O-	-SO_2-	-CH_2-	–	[134]
58	-S-	-CH_2-	-CH_2-	-CH_2-	[106]
59	-S-	-CH_2-	-CH=	=CH-	[106]
60	-S-	-CH_2-	-O-	-CH_2-	[76, 108]

9.6.7 Non-Phosphorus Linkages

Due to stereoelectronic effects, the phosphate diester group contributes significantly to nucleic acid conformation. Although replacement of this hydrolyzable moiety by stable structural elements would present a particularly helpful modification in drug synthesis, it is by no means self-explanatory that such modifications will retain their ability to undergo base pairing. Many efforts have been made during the past years to synthesize backbone-modified analogs of nucleic acids, frequently starting with modifications of natural nucleosides, sometimes by rearrangement or radical reactions, sometimes even from sugars. It would be beyond the scope of this article to discuss all these modifications. Instead, Table 7 provides some information on such compounds focusing on their synthesis and use.

Pairing properties of oligomers with amide, urea and carbamate linkages have been systematically studied by De Mesmaeker and co-workers [59]. As a rule, binding affinity to RNA is reduced with these compounds, the exceptions being N-methyl-hydroxylamine **35** and the thioformacetal linkers **56** with pairing properties comparable with RNA. Oligonucleotides with modification **29** and **33** display ΔT_m-values similar to those of wild-type DNA [59]. It is trivial to mention that most of the listed types of compounds are not subjected to nuclease degradation; indeed, an impact on future developments might be expected from some of these.

9.7 Modifications of the Nucleoside

9.7.1 2´-Modification of the Sugar Moiety

Substitution at the 2´-position of the sugar moiety has become one of the most widely used modifications. Above all, 2´-alkylation of riboses have been studied by several groups during the late 1980s. It seems worth mentioning that formally, 2´-substituted oligomers may be considered on the one hand to be alkylated ORNs, and on the other hand to be alkoxylated ODNs. Thus, prompted particularly by the discovery of the catalytic effect of ribozymes, the

Table 8. 2´-Alkyl-, 2´-O-alkyl and 2´-fluoro-modified nucleotides in oligonucleotides synthesis. Sorted by chain length and CIP rules in ascending order.

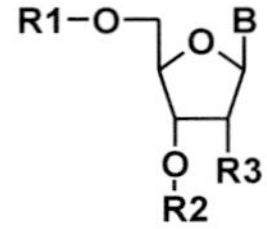

R3	*Base*: Reference
$-CH_3$	[137]
$-CH_2OH$	[137]
$-CH_2-CH=CH_2$	[137]
$-C_6H_5$	[137]
$-CH=CH-C_6H_5$	[137]
$-OCH_3$	*A*:[138]; *A, U, G, C*:[139–141]; *T*:[59]; *T, C*:[142]; I [135]
$-O-CH_2-CH_3$	*A, U, G, C, I:* [135, 139, 143–145]; *A*:[138]
$-O-CH_2-CH_2-CH_3$	*A*:[138]
$-O-CH(CH_3)_2$	*T*:[142]
$-O-CH_2-CH=CH_2$	*A, U, C, I*:[146]; *A*:[138]; *A, U, G, C*:[147]
$-O-CH_2-CH=C(CH_3)_2$	*A, U, C, I*:[146]
$-O-[CH_2]_3-CH_3$	*A, U, G, C*:[148]; *A*:[138]; *A*:[149]
$-O-[CH_2]_4-CH_3$	*A*:[138]
$-O-[CH_2]_6-CH_3$	*A*:[149]
$-O-CH_2-C_6H_5$	*A*:[138]
$-O-[CH_2]_9-CH_3$	*A*:[149]
$-O-[CH_2]_{10}-CH_3$	*A*:[138]
$-O-[CH_2]_{15}-CH_3$	*A*:[149]
$-O-CH_2-CH(OH)-CH_2OH$ (*R*)	*T*:[142]
$-O-CH_2-CH(CH_3)-O-CH_3$	*T*:[142]
$-O-CH_2-CH_2-O-CH_3$	*T*:[142]; *A*:[149]
$-O-[CH_2]_5-NHR$	*A*:[150]
$-O-[CH_2]_6-NH_2$	*A*:[151]
$-O-[CH_2-CH_2-O]_2-CH_3$	*T*:[142]; *A*:[149]
$-O-[CH_2-CH_2-O-]_3-CH_3$	*T*:[142]; *A*:[149]
$-O-[CH_2-CH_2-O-]_4-CH_3$	*A*:[149]
-F	*A, U, G, C*:[152]

authors group tried to stabilize RNA-oligomers by alkylation of the 2´-hydroxyl group in a cooperation with Birnstiel's group at the IMP in Vienna. Based on studies on stereoelectronic effects, a stabilizing effect of the substitution could be expected [135]. 2´-O-Methylated nucleosides are also structural elements of t-RNA ("rare" bases); their occurrence in nature has been reviewed [136]. Chimeric 2´-O-methylated ORNs are monomers in antisense drugs like GEM-92, GEM-132 and GEM-231 and are usually classified as "second generation" antisense compounds.

Different synthetic routes have been described for synthesis of 2´-modified nucleosides [152]. 2´-Fluoro nucleosides are usually prepared starting from arabinofuranosylnucleosides. 2,2´-Anhydroarabinofuranosyluracil **61** has been reported to yield 2´-deoxy-2´fluorouridine upon treatment with hydrogen fluoride. 2´-Amino-2-deoxyuridine may be easily prepared by similar rearrangements in large scale synthesis [153]. Synthesis of 2´-deoxy-2´-fluoropurine

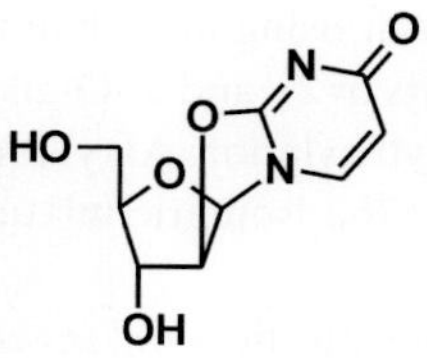
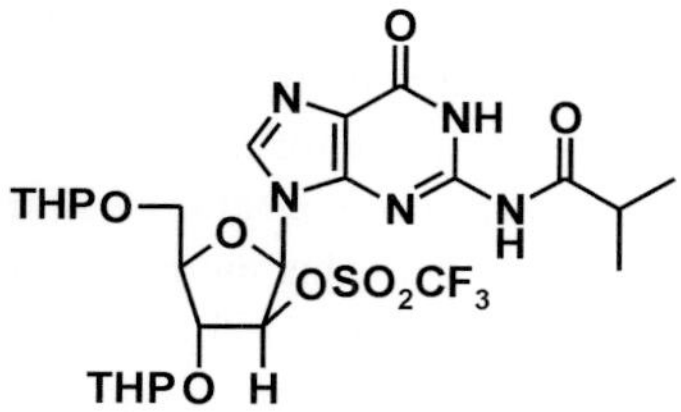
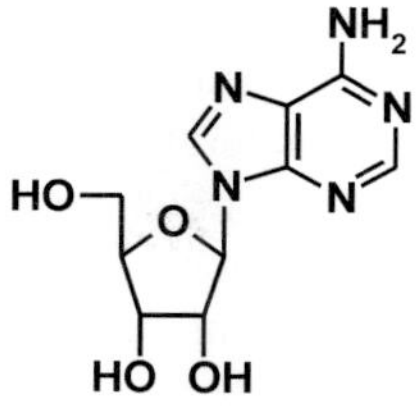

61 62 63

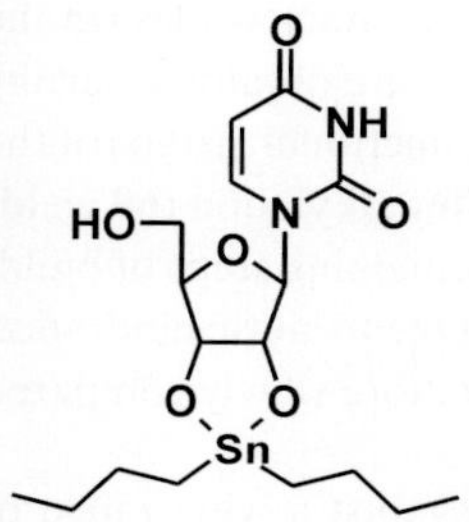
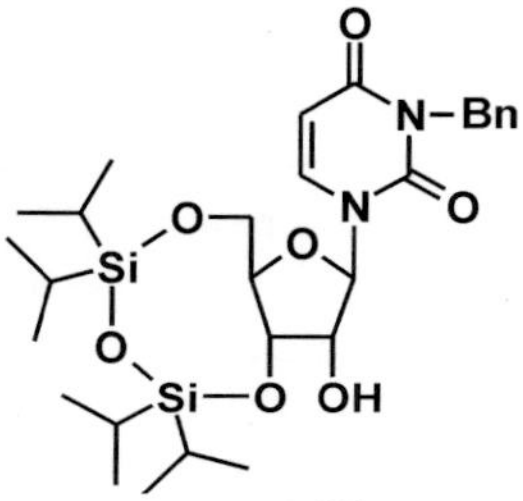
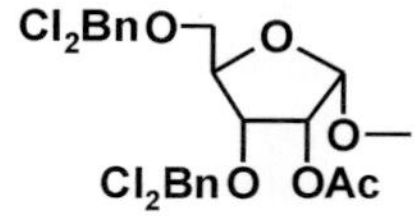

64 65 66

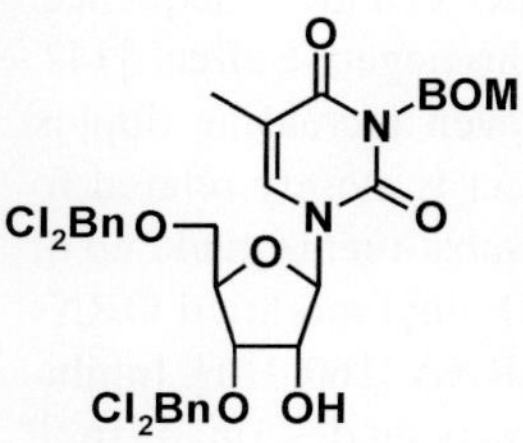
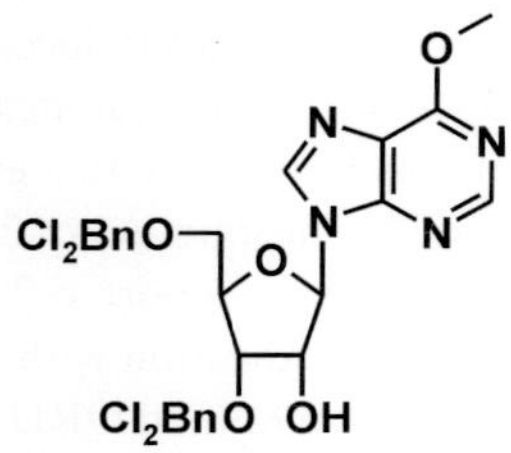
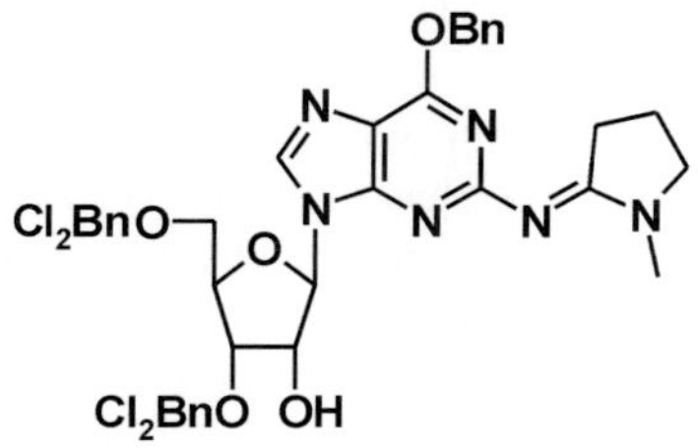

67 68 69

Figure 21. Synthetic strategies in 2´-derivatization of ribonucleosides. Compounds used in S_N2-displacement and 2´-alkylation reactions.

nucleosides was achieved by S_N2-displacement of 2´-β-O-triflyl groups of arabinofuranosyladenines and -guanines (**62**) [152].

Several strategies have been reported for the synthesis of 2´-O-alkylnucleotides. The long-known effect of selective 2´-methylation of adenosine (**63**) is due to the fact that the 2´-hydroxy group is more acidic than the 3´- and 5´-OH groups. Applying this reaction to other nucleosides, leads to the strategy of nucleotide building block synthesis starting with *direct alkylation* of nucleosides [135, 143]. Using strong bases like sodium hydride for deprotonation, DMF as a solvent and alkyl halides as alkylating agents leads to rather selective formation of 2´-O-alkyl nucleosides as main products. The NH-group of uridine must be protected before alkylation [138, 143–145, 149–151].

Direct methylation with diazomethane [154] or alkylation of organotin derivatives of uridine **64** lead to non 2´-O-selective alkylations. The organotin approach using alkyl halides and temperatures up to 130 °C leads to the formation of equal amounts of 2´- and 3´-O-alkyluridines, which may be separated by chromatography after dimethoxytritylation. Alkylation of the uracil moiety does not occur with this method. Purification of the isomeric mixture may be achieved after further derivatization [155].

Since direct alkylation is not perfectly selective and requires further purification, several attempts have been made to optimize synthesis of 2´-O-substituted nucleosides. In the most prominent approach, the 3´- and 5´-hydroxy groups are simultaneously capped by the tetraisopropyldisiloxanyl protective group [156]. Selective methylation of the 2´-hydroxy group occurs, but benzylation of uridine (**65**) or application of 6-chloroadenine nucleosides (in the synthesis of guanosine) is required. Higher yields of 2´-alkylnucleosides are obtained starting from ribose derivative **66** and ribosylation of nucleobases [142, 157]. Functional groups of the nucleobases like imids need to be protected (**67**) or are introduced after alkylation (**68** yields adenosine derivatives; **69** yields guanosine derivatives) [142]. In the remaining steps of building block synthesis, standard protective group strategies are applied. Due to steric hindrance, coupling of the corresponding phosphoramidites proceeds somewhat more slowly compared with common reagents (25 s vs. 9••••).

A major advantage of 2´-ODN modifications is their resistance against a wide range of RNA- and DNA-specific nucleases [139], which seems to be related to the steric hindrance exerted by 2´-subtituents. T_m-values have been reported to decrease with length of alkyl substituents [138, 139, 149] and increasing number of modified nucleotides in a given sequence. Replacement of alkyl groups by 2´-O-oxyethylene moieties reduces this negative effect. [142, 158, 159]. With oligonucleotides containing modified thymidines, even increasing duplex stability has been observed [142]. It may be assumed that this effect is closely related to favorable gauche effects leading to an extended conformation of 2´-substituents made up of oxyethylene units [59, 158, 159]. Several authors have reported 2´-O-alkyl modified ORNs not to be in a position to activate RNase-H in hybridization with mRNA [160, 161]. Inhibition of certain reverse transcriptases by 2´-O-alkyl modified ORD has been described [162].

9.7.2 Modified Bases

Due to the steric constraints of Watson–Crick base pairing, modifications of bases, which do not negatively affect base pairing properties are possible only to a limited extent. Some modifications are shown in Table 9.

Hydrophobic interactions exerted by substituents at position 5 of cytosine or uracil enhance stability of DNA/RNA-hybrids (**70**, **71**) [163]. Derivatives with longer alkyl chains reduce duplex stability [169], with aminoalkyl residues being an exemption [170]. Above all, ethinyl substitution at position C5 of pyrimidine bases results in significantly increased binding affinity, as shown with binding of modified poly(rU) and poly(rA) [164]. Also, substitution with propinyl showed the same effect (**73**, **74**), which was attributed to π-π-interactions of the alkinyl substituent with the nucleobase in the 5´-neighbourhood. Oligomers of this type form duplexes with RNA, which can activate RNase-H. Since there is only a minor influence on nuclease resistance, such modifications are preferentially applied in combination

Table 9. Base modification in oligomers. Modified from [59]

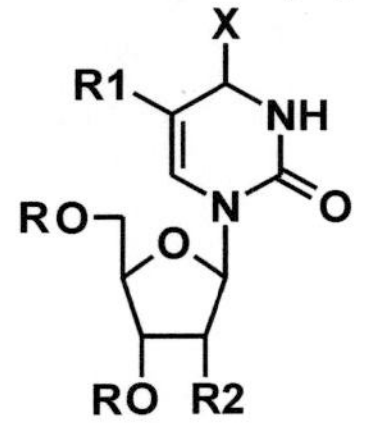

	-X	-R1	-R2	Reference
70	$-NH_2$	$-CH_3$	-H	[163]
71	$-NH_2$	-Br	-H	[163]
72	-OH	-C≡CH	-OH	[164]
73	-OH	$-C{\equiv}CCH_3$	-H	[165]
74	$-NH_2$	$-C{\equiv}CCH_3$	-H	[165]
75	-OH	$-C{\equiv}CCH_3$	$-OCH_2CHCH_2$	[166]
76	$-NH_2$	$-C{\equiv}CCH_3$	$-OCH_2CHCH_2$	[166]
77	-OH	-2-pyridyl	-H	[167, 168]
78	-OH	-2-thiophenyl	-H	[167, 168]
79	-OH	-2-thiazoyl	-H	[167, 168]
80	-OH	-2-imidazoyl	-H	[167, 168]

with phosphorothioate [171] or 2′-alkoxy modifications (**75**, **76**). Stabilizing effects of C-5-alkinyl substituents at uridine on triple helix formation have been studied by Dervan's group [172]. The favorable properties of this type of base modification had been observed previously in studies on 2′-O-allyl-RNA oligomers [173]. Suitability as a structural element of an antisense drug was tested with an oligomer directed against mRNA of HIV [174]. Adding this modification to 2′-O-methyluridine-phosphoramidites reduces the solubility of the phosphoramidites in acetonitrile. In this case, oligonucleotide synthesis is to be carried out using THF as solvent. It is somewhat surprising that even relatively drastic modifications of the heterocyclic base may lead to an increase in binding affinity. With a ΔT_m/mod. value of +1.6 °C, the thiazole derivative **79** exhibited a duplex stability comparable with that of propinyluridine [167]. An increase in transient temperature of 2 to 5 °C per monomer unit has been reported for the heterocycle phenoxazine (to act as a modified cytidine) **81** [168] using SODN and ODN oligomers [161]. Introduction of C-nucleosides into oligomers (**82**) led to

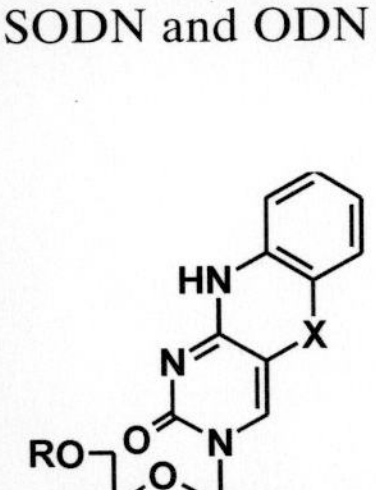

81

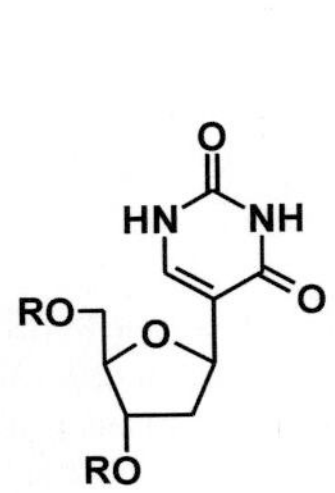

82

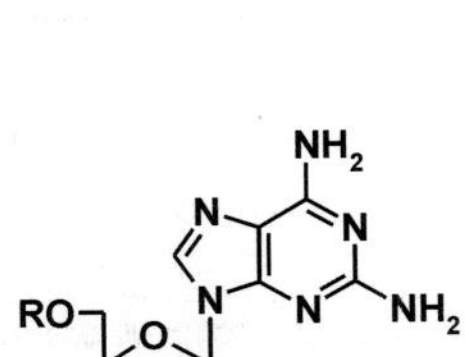

83

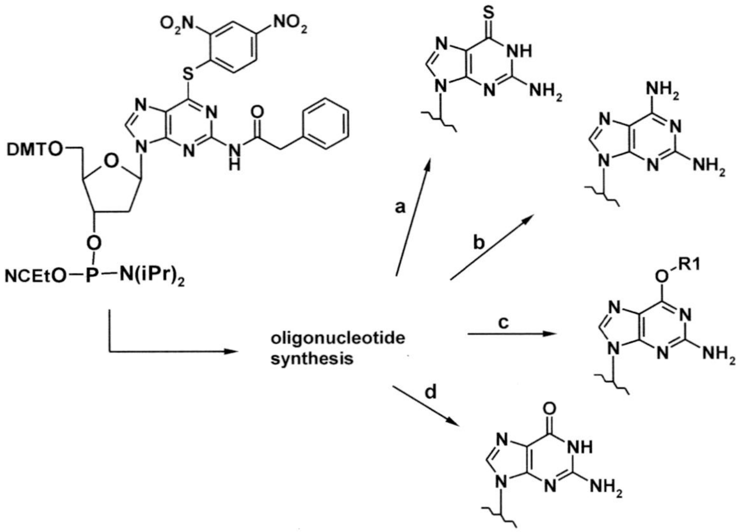

Figure 22. 6-(2,4-Dinitrophenyl)thio-N_2-(phenylacetyl)-2′-deoxyguanosine, a convertible purine nucleoside. Postsynthetic reactions of the oligonucleotide: (a) 10% mercaptoethanol in conc. ammonium hydroxide, 48 h, RT; (b) 65 mM tetramethylguanidine, 75 mM 2-nitrobenzaldoxime in conc. ammonium hydroxide, 48 h, RT; (c) for R1 = CH_3: 10% V/V DBU in anhydrous methanol, 48 h; (d) 0.5 M aqueous sodium hydroxide, 48 h, RT. Modified from [178].

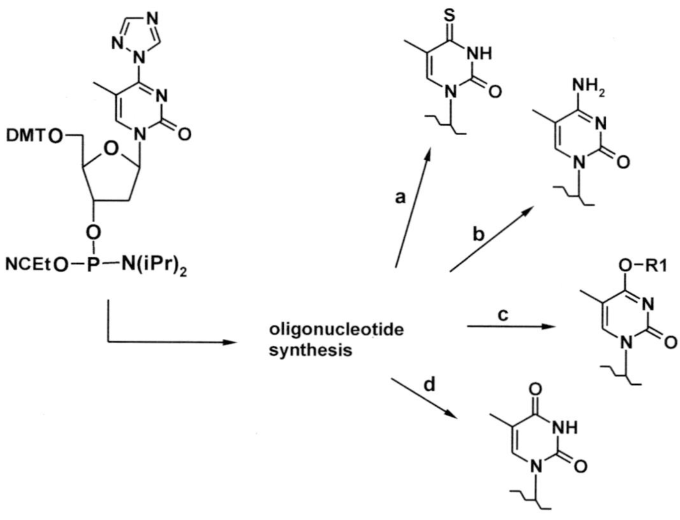

Figure 23. Convertible triazolyl pyrimidine nucleoside. Postsynthetic reactions of the oligonucleotide: (a) 10% thioacetic acid in acetonitrile on solid support, deprotection with 10% DBU then conc. ammonium hydroxide, 24 h, RT; (b) 10% 1,1-dimethylhydrazine in acetonitrile on support, 3 h, RT; 0.5 M aqueous sodium hydroxide, 16 h, RT; (c) for $R_1 = CH_3$, 10% V/V DBU in anhydrous methanol 24 h, RT; $R_1 = CH_2CH_3$, 10% V/V DBU in anhydrous ethanol, 48 h, RT; (d) 0.5 M aqueous sodium hydroxide, 24 h, RT. Modified from [179].

an increase in transient temperature of ΔT_m/mod. = +1.8 °C [175]. 2,6-Diaminopurines are capable of building up an additional hydrogen bridge with pyrimidine bases, thus increasing duplex stability of oligomer/RNA-adducts. However, in this case an alkylation of the ribose 2´-hydroxy group (**83**) is required to avoid depurinization (hydrolysis of the aminal) [176]. A comprehensive review on base modifications and their impact on antisense technology has been published [177].

As demonstrated in Figs 22 and 23, phosphoramidites of "convertible" nucleosides allow introduction of non-natural modifications of nucleobases after solid-phase synthesis.

9.8 Purification and Characterization of Oligonucleotides

ODN synthesis on polymer support by no means results in a perfectly clean product. Although purification of such oligomers is not easy in principle, a set of standard methods offered by commercial suppliers has been developed in the meantime, which allows rather straightforward product purification. A specific requirement with ODN purification is given by the fact that – in contrast to block synthesis and apart from other oligonucleotide impurities – the (n–1)-mer of the target product is the impurity to be expected as a most probable byproduct with ODN synthesis on polymer support. Due to its similarity to the end-product, leading to similar retention- and migration-times, problems may occur in purification. HPLC-purification of ODNs may preferably be carried out before cleavage of the DMT-protective group from the product (DMT-on), applying HPLC with reversed-phase (RP) stationary phase. Examples for separation of modified ODNs are given in Table 10. Diastereomeric mixtures as SODNs frequently exhibit only partial resolution.

The efficiency of separation depends to a certain degree on oligonucleotide composition. Thus, with G-rich oligonucleotides, an increase in buffer pH-value may be required. In such cases, a switch from RP-silica gel phases to RP-polystyrene phases is advisable. Retention times in RP-systems depend on changes in logP-values. OD-cartridges for reversed-phase purification are available commercially and protocols allow detritylation and desalting during wornup.

Table 10. HPLC-separation of modified ODNs: TEAA: triethylammonium acetate; TEAB: triethylammonium borate; ACN = acetonitrile.

Modified oligomer	Stationary phase	Eluent system	Reference
2´-O-Alky and 2´-O-polyethylene glycol-modified RNA	ET 250-8-4 Nucleosil® 100-5 C18	A: 0.1 M TEAA buffer pH 7. B: 20% 0.1 M TEAA buffer pH 7/80% ACN	[149]
Phosphorodithioates	ODS Hypersil® (5μ)	0–20% ACN in 0.1 M TEAB buffer in 20 min.	[180]
3´- and 5´-cholesteryl substituted ODN	RP-18	A: 50 mM TEAA, pH 7; B: ACN	[181]
Methylphosphonates	ODS Hypersil® (5μ)	A: 0.1 M TEAB buffer; B: ACN	[182]
Phosphoramidates	RP-C18 HP	A: 0.1 TEAB buffer, pH 7.0; B: ACN	[64]

9.8.1 Ion-Exchange Chromatography

Ion-exchange chromatography is a favorable method for removal of (n–1) and (n–2) sequences. At first, the DMT protective group is cleaved using 40–80% acetic acid at room temperature for 2 h. Choice of acetic acid concentration depends on the 5′-terminal nucleobase. Special care must be taken with adenine nucleotides, which depurinize easily. With SODNs, high acid concentrations may lead to an exchange of sulfur by oxygen.

Polystyrene-type anion exchangers are used in ODN purification. 1.5 M sodium chloride solution and pure aqueous phosphate buffer (pH = 6.8–8) are used. The content of the sodium chloride solution is increased from 10% to 100% during 2 h by use of gradient programs. Fractions obtained are of high purity with respect to the desired oligomer; however, inorganic salts must be removed by desalination. This can be achieved by RP-HPLC, size-exclusion chromatography or precipitation with isopropanol.

9.9 Analysis of Oligonucleotides

RP-HPLC and ion-exchange chromatography discussed in the previous section are the most important methods for both analysis and purification of ODNs. A variety of protocols and commercial columns has become available based on these methods.

9.9.1 Capillary Electrophoresis

A series of reviews has been published on the analysis of oligonucleotides using CE [183–185]. A comprehensive review on separation of mono, oligo- and polymeric nucleic acids by this method has also been published [186]. In contrast to RP-HPLC and ion-exchange chromatography, CE – which has been established as the third technique of choice – is a purely analytical method [187]. Ion-exchange HPLC fails particularly with SODNs [185]. RP-HPLC does not at present allow baseline separation of SODNs, which differ by only one base pair [188]. Thus, control of preparative purifications by RP-HPLC has become an important application of CGE. Table 11 illustrates the resolution efficiency of CGE when compared with other methods.

Table 11. Limits of resolution in detection of unmodified oligonucleotides using different analytical methods. *: Size limitations for baseline separation of ODNs differing by a single base-pair. Modified from [189].

<table>
<tr><th>Method</th><th>Plates/m</th><th>Baseline achieved*</th><th>Preparative</th></tr>
<tr><td>CGE</td><td>3×10^6 to 7×10^6</td><td>250</td><td>No</td></tr>
<tr><td>RP-HPLC</td><td rowspan="2">4×10^4 to 8×10^5</td><td>30</td><td>Yes</td></tr>
<tr><td>Ion-Exchange</td><td>40</td><td>Yes</td></tr>
</table>

* EM_{Rel} = (electrophoretic mobility of reference substance)/(electrophoretic mobility of sample compound).

In CE, analytes are subjected to counteracting forces EOF and EM (μ). EOF is generated by ionization of the silanol groups of the quartz capillary. The migration time is a result of the combination of EOF and EM, which with polyanionic oligonucleotides is directed towards the anode. Due to the limited durability of the capillaries internal standards are frequently used and separation is reported in values of relative electrophoretic mobility (EM_{Rel}). Basically, an electrophoretic separation of oligonucleotides may be carried out in a quartz capillary, applying the standard method without any form of molecular sieving effects. However, with decreasing differences in EM with increasing chain length short oligomers, such CZE is only suited for *short oligomers*, which has been demonstrated with $A_{[2]}$–$A_{[20]}$ oligomers [185].

$$EM = \frac{q}{E \times M_T}$$

EM = electrophoretic mobility; q = charge; E = electric field strength; and M = migration time

Application of gels has provided a suited approach for resolution of oligonucleotides by CE. With CGE, silanization of the surface of the quartz capillaries suppresses EOF. Only EM and the sieving effect exerted by the gel are responsible for separation of oligonucleotides in this case. Several examples of *antisense oligonucleotide* separation by CGE have been reported [190–192]. Sieving effects have been achieved by different approaches: either crosslinked polyacrylamide gels may be filled into the capillaries or entangled linear (not crosslinked) polyacrylamides may be attached to the capillary surface. Alternatively, polymers such as methylcellulose and hydroxymethylcellulose may act as sieving buffers [193]. Typical conditions for the separation of a mixture of six SODNs (16- to 21-mer) differing in length by one nucleotide are:

- Electrokinetic injection at –8 kV for 5 s
- Running voltage of –22 kV
- Gel matrix and running buffer consisting of 10% Micro-Gel® in 35 mM boric acid, 15% ethylene glycol (pH 9.0) with 50 cm × 100 μm I.D. capillaries [185].

Table 12. Separation of oligonucleotides by CGE (capillary gel electrophoresis)

Sample	Conditions	Reference
Mixture of $p(dA)_{40-60}$	Crosslinked polymer: 5% T; 5% C; Tris-borate buffer pH 8.85; 7 M urea;	[194]
Mixture of SODNs	Entangled polymer: 10% Micro-Gel, 35 mM Tris; 5.6 mM boric acid; pH 9.0; 15% ethylene glycol	[185]
Mixture of 2´-O-substituted ORN/ODN	Linear polymer: 44% (w/v) Tris; 56% (w/v) boric acid, 7 M urea	[149]

%T = [acrylamide (g) + crosslinking agent (g)]/100; %C = {[crosslinking agent (g)]/acrylamide (g) + [crosslinking agent (g)]} × 100.

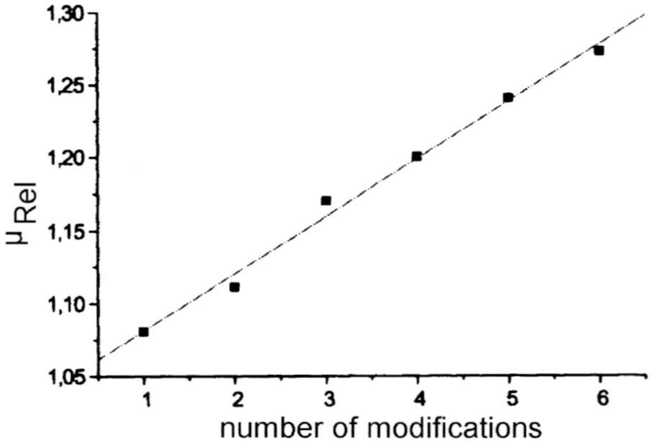

Figure 24. Change in relative electrophoretic mobility (μ_{Rel}) due to number of modified nucleotides in 12-mer (3´-dA$_{[Y]}$-A$_{[X]PEG}$-5´) [x] = number of modifications. PEG = 2´-O-(3,6,9,12,15-pentaoxahexadecyl). From [149].

Changes in EM are caused by an increase of molecule size with unchanged charge. An increasing number of 2´-oxyethylene-substituted RNA-nucleotides in a dA[12]-mer leads to a linear increase (Fig. 24) of EM$_{Rel}$ which is due to the sieving effects of the linear polyacrylamide gel applied. Other substituents at the 2´-position give similar effects. However, with 2´-alkyl-subsituents the relation is not linear [149].

Monomeric nucleosides and *nucleotides*, which appear as impurities in oligonucleotide synthesis, may be analyzed using the method of micellar electrokinetic chromatography (MEKC). With addition of SDS to the buffer solution, the surfactant concentration is above the critical micellar concentration. Movement of the micelles towards the anode against the EOF involves distribution processes, in which also nucleosides, nucleobases and nucleotides participate. Typical conditions are: 50 mM phosphate buffer; 40 mM SDS; pH 6.5; 20 kV (116 mA); 21 °C [186].

Separation of *DNA-fragments* requires conditions different from those of oligonucleotide separation. First, multivalent metal ions, which stabilize DNA tertiary structure, are complexed by the addition of EDTA salts. A different composition of gel is required to increase the size of the pores of the gel-matrix. Effects of gel composition on the resolution of small DNA fragments (25–250 bp) have been studied systematically. Polyacrylamide capillary gels were prepared with a fixed amount of crosslinking (5% C) and an acrylamide content of 2.5–6.0% T. Gels prepared by this method exhibited a pore size of 2.5 to 3.5 nm [195]. Capillaries with entangled linear gels are now commercially available, though details of their composition remain undisclosed by the manufacturers. Frequently, ethidium bromide is added to the gel system, resulting in better resolution of fragments in the range from 20 to 2000 bp [196]. Obviously this method is only suited for analysis of double-stranded DNA, which, as a rule, will migrate faster than single-stranded DNA of the same molecular size.

With *modified oligonucleotides* formation of secondary structures by base pairing may result in separate signals for one compound. This effect appears most frequently with G-rich sequences. To avoid this effect, gels and buffers in ODN-CE are generally prepared in the presence of 7 M urea or formamide. Other options for analysis of dG-rich ODNs include a pH value between 9 and 10 (dissociation of the nucleobases thymine and guanine), analysis at temperatures above 60 °C and heating of the samples to 95 °C for 5 min before the run [185].

9.9.2 Mass Spectrometry

MS does not only allow precise determination of the molecular mass of oligonucleotides with their high molecular weight, but does permit sequencing of nucleic acids when using extremely small amounts of sample material.

ESI-MS has been used to determine sequences of DNA- and RNA-fragments [197] and the mass of oligonucleotides [198]. The method of ionization leads to the formation of multiply charged molecules. A series of signals resulting from these masses allows determination of the molecular mass, if the number of charges is known. An overlap of the series may occur, if positively charged ions (e.g., sodium) are present at the same time. Therefore it is recommended that the metal ions be replaced by ammonium ions. This can be achieved either by ammonium acetate precipitation, by treatment with ammonium ion-loaded ion exchanger, or in some cases simply by dissolving the sample in ammonium hydroxide solution [199].

The principle of MALDI-TOF is based on the fact that a laser may effect desorption and ionization of a macromolecule imbedded in a crystalline matrix. This matrix is prepared by application of a mixture of the sample with an excess of an organic acid (sinapinic acid, nicotinic acid) to a steel plate and subsequent drying in vacuo. A cost-effective nitrogen laser (337 nm) allows use of 2,4,6-trihydroxyacetophenone as matrix material [200]. Pretreatment of the sample with an ammonium ion-loaded ion exchanger reduces signals of nucleic acids with bound metal ions [201]. The time difference between desorption by a short laser pulse and detection is proportional to the square root of the mass/charge ratio. Analysis of side products from ODN synthesis, which gives direct access to the synthesized sequence, is a particular application of MALDI-TOF mass spectroscopy. Sequence analysis of DNA-fragments up to 150 kDa has been successfully carried out [202]. Also sequence analysis of oligonucleotides containing non-natural nucleotides may be performed using MALDI-TOF in combination with snake venom phosphodiesterase 3´,5´-digestion. Detection of oxidized SODNs as impurities is also possible because the mass of the oligomer will decrease by 16 Dalton upon each oxidation event.

9.9.3 ^{31}P-NMR Spectroscopy

^{31}P-NMR spectroscopy is a valuable tool in the analysis of ODNs, and particularly in the analysis of their phosphorus-containing modifications. However, it is not used as a routine method in oligomer analysis, because the required material quantities (1–20 mg) are rather high. ^{31}P-NMR spectra allow quantitative interpretation via their integrals. Shift values are within the range of 50–60 ppm for phosphorothioates, 110 ppm for phosphorodithioates and 0 to –2 ppm for unmodified phosphodiesters. Further information on spectroscopy of phosphorus-containing compounds is available [203]. Application of ^{31}P-NMR in biological systems has also been described [204].

9.9.4 UV Spectroscopy

To estimate the concentration of an oligonucleotide solution*, the absorption is measured at 260 nm. Correct weights of solid oligonucleotides are difficult to obtain due to non-stoichiometric inclusion of water molecules into the solid. UV spectra of nucleic acids exhibit a pronounced hypochromicity. In double-stranded nucleic acids, base stacking with a distance of about 3.5 Å lowers UV-absorption by up to 20% due π-systems interaction when compared with single-stranded oligomers (Fig. 25).

It is this difference in UV absorption that may be used to determine the "transition temperature" (T_m) of double-stranded oligomers. The terms "melting temperature" and "melting point" are also used (not quite correctly) to describe this biophysical parameter. A typical procedure involves preparation of a buffer system (140 mM NaCl, 10 mM Na_2HPO_4 buffer pH 7.0, 1 mM EDTA), dissolution of equimolar amounts of single strands in 1 ml buffer (absorbance reading should not exceed 2.0) and degassing of the solution with nitrogen or argon. Alternatively, a buffer containing 0.15 M NaCl, 0.01 M Tris–HCl at pH 7.0 may be used [151]. The sample is heated to 90 °C for 5 min and allowed to equilibrate at the starting temperature for 15 min. The cuvette is sealed to stop evaporation during heating. Recording of the transition curve is performed at 260 nm (or 274 nm for GC-rich strands) by heating in steps of 0.75 °C min^{-1} and measuring absorbance every 0.5 °C. The transition temperature is obtained by differentiation of d(Aλ)/dT [199]. As a rule of thumb, increase of T_m by 3–5 °C

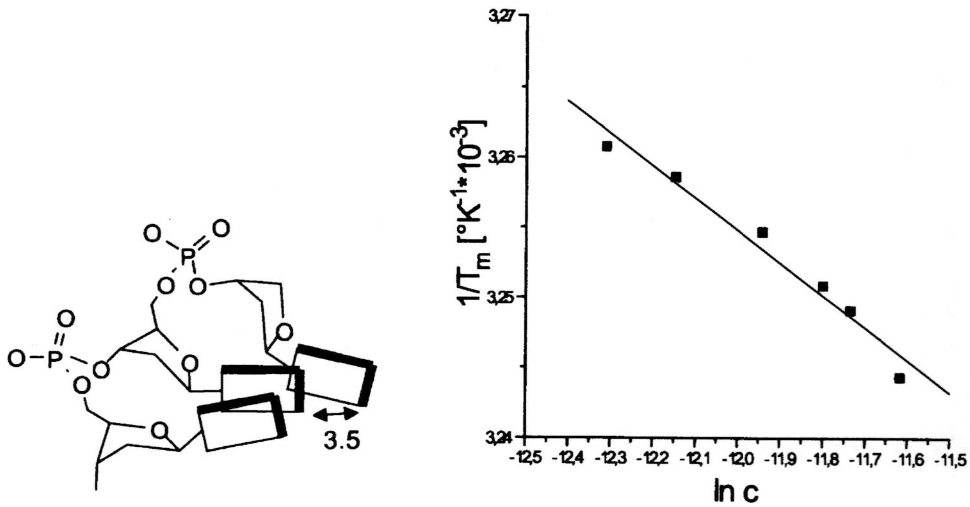

Figure 25. Left: Base stacking in double-stranded DNA (only one strand shown); Right: Determination of thermodynamic parameters of duplex formation: Oligonucleotides 3′-$dA_{[11]}$-$A_{[MEG]}$-5′ and $T_{[12]}$ were diluted starting from 9 µM to 4.5 µM in 0.15 M NaCl, 0.01 M Tris–HCl, pH 7.0, 25 °C and hybridization data were collected. From [149].

* Extinction coefficient of a given oligonucleotide (light at 260 nm wavelength) may be estimated using to following equation: [205]: $\varepsilon\ (260) = (A^*\ 15.34 + G^*\ 12.16 + T^*\ 8.7 + C^*\ 7.6) \cdot 0.9 \cdot 10^3\ [M^{-1}cm^{-1}]$. Concentration of the oligonucleotide: $c = OD_{(260)} / \varepsilon * l$.

equals a 10-fold increase of the association constant of a 15-mer oligonucleotide [206]. Measuring T_m at different concentrations provides data to calculate thermodynamic parameters of duplex formation. Models for formation of duplices have been described [207, 208]. Dilution of the sample has to be carried out with buffer to obtain constant ionic strength. $1/T_m$ is plotted versus ln(c). The slope should correspond to $R/\Delta H°$ and the intercept to $\Delta S°/\Delta H°$ (Fig. 25). It must be taken into account that single-stranded modified oligonucleotides may also show co-operative and reversible UV melting curves due to intramolecular hairpin formation. In this case, a 10- to 15-fold increase in concentration does not effect a change in T_m values. A recent example for this phenomenon and interpretation of such data is given for hairpin formation of α-anomeric oligomers in [209].

9.9.5 ORD Spectroscopy

A second and frequently used method to determine transition temperatures of oligonucleotides and, in general, to study nucleic acid conformation is ORD spectroscopy. Chiral chromophores exhibit differences in absorption of left and right polarized light ($\varepsilon_l \neq \varepsilon_r$). This difference is named circular dichroism, and depends on wavelength and path-length through the solution, as well as on the type and concentration of the chiral compound. When light of opposite direction of polarization passes a solution of a chiral compound, differences in absorption of the light beams are measured. The extent of this effect is expressed by the term of ellipticity, θ. As with UV-absorption, ellipticity exhibits a strong dependence on temperature due to loss of stacking effects. This allows determination of the transition temperature by heating a solution of complementary oligomers in a CD-spectrometer. With nucleic acids the effect of ellipticity is above all generated by the chirality center at the C1´-carbon atom of the sugar moiety, which exerts a significant effect on light absorption via the π-electron system of the nucleobases. Thus, CD spectroscopy allows, in addition to determination of the transition temperature, an approximate prediction on the conformation of a duplex, because the different polymorphic forms of nucleic acids exhibit different shapes in ORD (Fig. 26).

Usually a RNA : DNA duplex displays A-type polymorphism. However, if the amount of A : T-pairs increases B-type CD-spectra like Fig. 27 are found. The B-type duplex changes is shape dramatically with the length of the alkyl chain in 2´-position.

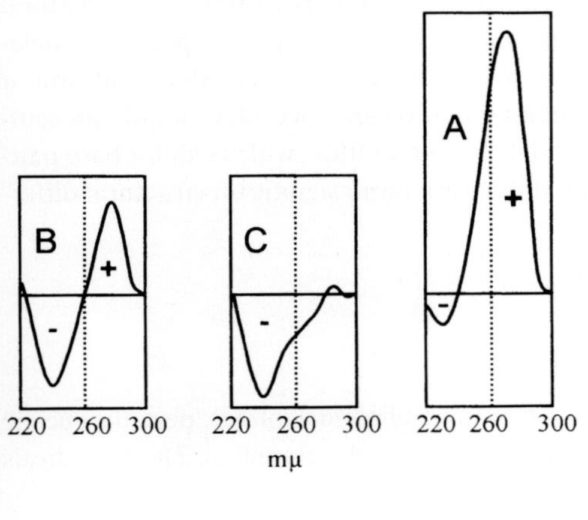

Figure 26. Different CD spectra in DNA-polymorphism: B-Form, C-Form and A-Form. From [210].

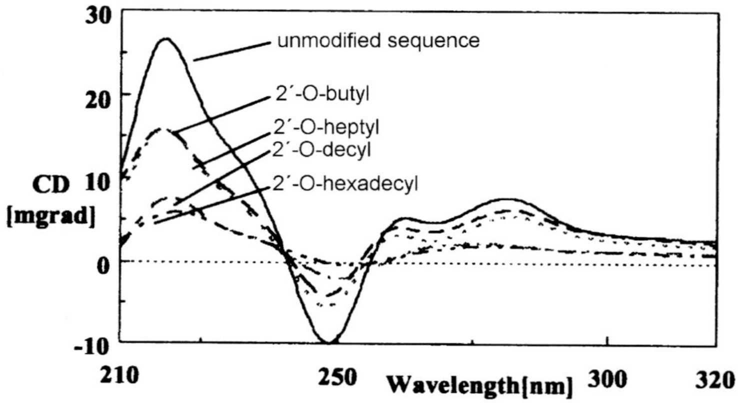

Figure 27. Influence of length of 2´-O-alkyl-subsituents on duplex-forming abilities. Oligonucleotide 5´-*a*AA AAA AAA AAA AAA-3´; *a*: 2´-O-alkyl-substituted adenosine; Conditions: Oligonucleotides were hybridized with $T_{[12]}$, 0.15 M NaCl, 0.01 M Tris, pH 7.0; From [149].

9.9.6 Crystal Structures of Oligonucleotides

X-ray structures of several oligonucleotides have been published and can be easily accessed. Methods for crystallization of oligonucleotides have been developed. These differ only slightly in solvents and conditions. Thus, sodium dimethylarsenate buffer containing magnesium chloride and spermine tetrahydrochloride and 2-methyl-2,4-pentadiol (MPD) as a precipitating agent are applied in almost 90% of the publications [25, 211, 212]. Proper conditions for crystal formation may also be found by using commercially available screening kits [213, 214].

9.10 Antisense Compounds Exhibiting "Major" Modifications

It has been mentioned previously that replacement of natural structures by non-natural bioisosteric elements is a general strategy in drug synthesis that may also be applied to nucleic acids. All drug candidates presently in the stage of clinical development exhibit only minor modifications of natural ODNs. Apart from the modified oligonucleotides already presented, further modifications have been reported and deserve attention, which exhibit base pairing with natural systems, although their SRUs exhibit even more significant structural differences from natural nucleic acids.

9.10.1 Peptide Nucleic Acids

With their structures only slightly reminiscent of natural oligonucleotides, peptide nucleic acids (PNAs) have had a significant impact on antisense drug development. Their synthesis and study has above all been connected with the work of Nielsen [215, 216]. PNA oligomers are achiral and neutral molecules, while hybridization with DNA and RNA and formation of triple helices have also been shown [217]. The oligomers are resistant to endo- and exonuclease-mediated degradation, as well as to protease digestion [218]. Uptake of PNAs into cells represents the most important hurdle against the use of the oligomers in antisense therapy. Microinjection of PNA into cells has led to selective transcriptional and translational arrest [219]. Efforts have been made to overcome poor absorption by linking PNA to carrier moieties [220].

The properties and synthesis of different types of PNA have been summarized recently [221, 222]. Synthesis is carried out following standard *t*Boc or Fmoc protocols as used in peptide synthesis. In the latter case, the benzhydryloxycarbonyl group is used to protect N-6 (adenine), N-2 (guanine) and N-4 (cytidine) functional groups. Synthesized oligomers should bear a glycine at the N-terminus to prevent intramolecular rearrangement at pH-values higher than 9.5. PNA reads nucleic acids in antiparallel direction, thus the N-terminus should be seen as the 5′-end of PNA. Since PNA is not charged, the resulting duplex with DNA exhibits increased T_m-values (~1 °C per bp). T_m-values will also increase with low ionic strength. As a rule of thumb, a 10-mer PNA/DNA-duplex will have a T_m of 50 °C and a 15-mer a T_m of 70 °C. Due to enhanced affinity, oligomer length should not exceed 18 base pairs. Since PNA tends to aggregate, the purine content of an oligomer should be kept between 25% and 50%. Not more than four to five consecutive purines and not more than three consecutive Gs should appear. Thymine and cytosine homopolymeric PNA binds to DNA by formation of triple helices (2 × PNA/DNA).

9.10.2 Morpholino Oligomers

Morpholino oligomers (**84**) constitute a class of a major structural modifications of natural ODNs. The five-membered sugar ring is replaced by a six-membered nitrogen-containing heterocycle. Nevertheless, morpholino oligomers have been found to exhibit good to excellent binding properties to RNA.

Table 13. PNA duplex characteristics; I: PNA/RNA and PNA/DNA duplexes show increased stability when compared with natural duplexes (Conditions: 10 mM phosphate buffer; 0.1 mM EDTA; 100 mM NaCl; pH 7.0). II: Dependence of T_m-values of PNA/DNA and natural duplexes on counterion strength. From [223]

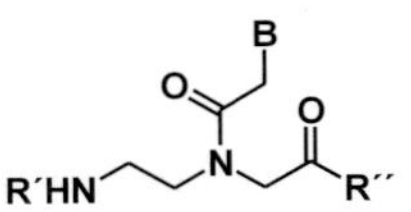

Duplex	T_m
15-mer PNA/DNA	69 °C
15-mer DNA/DNA	54 °C
15-mer PNA/RNA	72 °C
15-mer DNA/RNA	50 °C

I

NaCl conc.	DNA/DNA (T_m)	PNA/DNA (T_m)
0 mM	38 °C	72 °C
140 mM	56 °C	72 °C
1000 mM	65 °C	65 °C

II

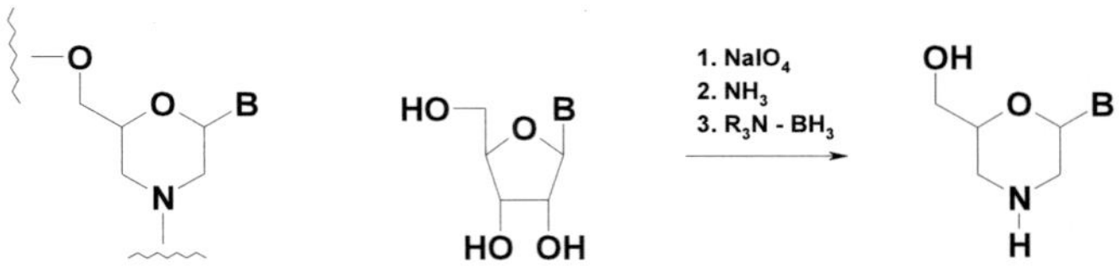

84

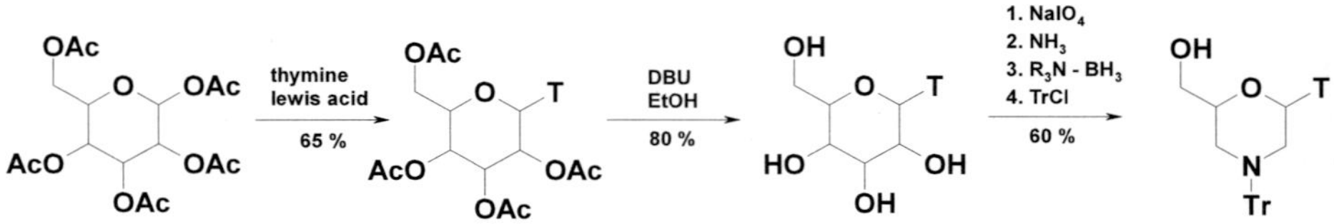

Figure 28. Morpholino oligomers: synthesis starting from hexopyranoses.

Particularly elegant is the ready accessibility of these compounds either from natural nucleosides or from natural sugars. At first, monomers were synthesized starting from ribonucleosides by periodate oxidation (Fig. 28). An alternative method for their synthesis has been developed, starting from hexopyranoses including a glycosylation step (Fig. 28) [224–227]. During this synthesis bases are protected using N4-benzoylcytosine, N6-benzoyladenine, N2-phenylacetylguanine (uracil remains unprotected). Morpholino dimers containing a rigid carbamate linkage exhibit little or no base stacking [228]. In the absence of specific solubilizing groups, morpholino oligomers containing such carbamate linkages are quite insoluble in aqueous solutions [229].

9.10.3 Modification at the Anomeric Center

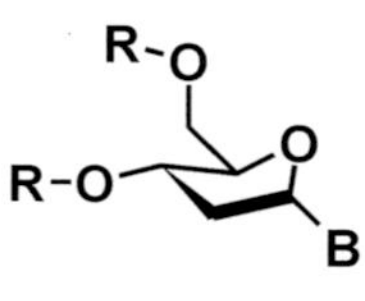

85

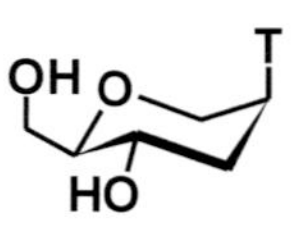

86

α-Anomeric oligonucleotides (**85**) represent a very elegant concept of antisense of antisense drug design [230–232]. It is clearly not self-explanatory that such oligomers are in a position to undergo base pairing, in which the antisense-oligomer is associated with the target-RNA not in an antiparallel but in a parallel manner. α-Anomers are not substrates for nucleases [233]. Nevertheless, lack of RNAse-activation affords high drug concentrations, if the "translation initiation site" or the "coding region" of mRNA are chosen as targets [68]. In contrast to these results, drug effects in micromolar concentrations could be shown, when the α-oligomer sequence was directed against the CAP-site of the mRNA [31]. Inhibition of translation could be shown with cell lysates [234]. An attempt has been made to demonstrate the strong pyrimidine/pyrimidine-base pairing exhibited by α-nucleosides making use of a 20-mer construct comprising 10 α-thymidines and 10 β-thymidines. Transition temperature was significantly reduced (T_m = 19.2 °C) compared with the homo-oligonucleotide (d[$^{\alpha}T_{20}$]–C], T_m = 52.2 °C). From the fact that transition temperatures did not depend on oligonucleotide concentration, it was deduced that intramolecular hairpin structures had formed [209]. Compounds like **86** formally are homonucleosides with a methylene group inserted between pyranose oxygen and anomeric sugar carbon atom. This modification, which abolishes the glycosidic nature of the molecule, renders it stable against deglycosylation. In spite of the major structural modification, a change in transition temperature of duplex formation with poly(rA) of only ΔT_m/mod. = –1.0 °C was found with a tridecamer of the nucleoside shown [235]. Replacement of the sugar oxygen of a nucleoside by a methylene group is a very old modification that is already well-established in nucleoside chemistry, before the advent of the antisense concept. As with the compounds with extended sugar rings, a major advantage of this type of modification is the stability against deglycosylation.

9.10.4 Three-Atom 2′-5′-Linkages

Oligonucleotides containing 3′-deoxyribonucleotides linked as 2′,5′-phosphates are "minor" modifications, and have been studied with respect to their ability to form duplices with ODN and ORN. Mixed sequences consisting of both 3′,5′- and 2′,5′-linked nucleotides did not hybridize with complementary strands (forming only triplex helices at high salt concentrations instead). This is not surprising, because – due to the four-atom 2′-5′-phosphate linkage – the number of connecting atoms in the structural repeating unit (SRU) increases from six to seven. Nevertheless, homologous 2′,5′-oligonucleotides displayed selective binding to RNA but not to DNA. Furthermore, it was found that these 2′,5′-oligonucleotides are not suited for self-recognition [236]. Similar results have been found for 2′,5′-linked RNA [237].

Non-phosphate three-atom 2′-5′-linked oligonucleotides constitute a significantly different modification, although at the first glance they resemble 2′-5′-phosphate-linked ODNs and ORNs: Like natural nucleic acids, they exhibit six connecting atoms in their SRU. Molec-

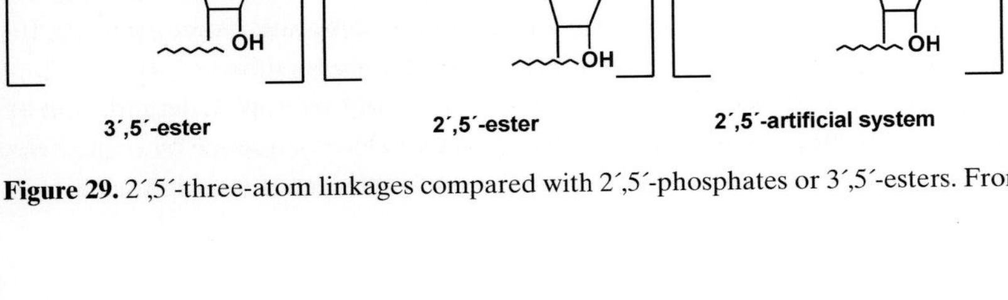

Figure 29. 2′,5′-three-atom linkages compared with 2′,5′-phosphates or 3′,5′-esters. From [238].

ular modeling experiments revealed that oligonucleotides made up of three-atom 2´-5´-linked units are structurally very well suited to undergo base pairing with natural nucleic acids [238]. It has been shown in experiments on triple helix formation that this type of a antisense compound is able to recognize natural nucleic acid sequences [36, 239–241] (Fig. 29).

9.11 The Triple-Helix Concept

Direct influence on DNA transcription via triple helix formation may be achieved both by use of oligonucleotides and by peptides. Starting from the observation that duplex strands of homopolymers of dA and dU undergo rearrangement under conditions of increased salt concentration [242], a series of triple helix-forming systems has been found following a common general concept. In triple-helix formation, a third strand binds into the major groove of a double strand. A planar triplet of bases is formed, in which hydrogen bridges of the third strand do not follow the Watson–Crick scheme, but undergo a type of base pairing, called (reversed) Hoogsteen base-pairing, which was first described by Hoogsteen with base pairing in crystals of complexes of adenine and thymine derivatives [243]. In Fig. 30, several possible base triplets are depicted. Orientation and number of hydrogen bridges depend on the sequence of the oligomers involved.

There are two options for formation of a T×T*A base triplet. In case A, Hoogsteen base-pairing takes place, with the second homopyrimidine strand oriented parallel to the purine strand of the duplex. In case B, the same pyrimidine strand may undergo reversed Hoogsteen base-pairing by taking an orientation antiparallel to the purine strand. The option for the third strand for both parallel and antiparallel arrangement is given for the homopolymers of T and G. It has been shown that oligopyrimidines exhibit a preference for parallel orientation to the homopurine strand of a DNA-duplex, while oligopurines bind in antiparallel orientation. With G- and T-rich oligomers particularly stable T×A*T and G×G*C Hoogsteen-hydrogen bonds in parallel orientation compete with the minor backbone torsion involved with antiparallel orientation. Depending on the sequence, an orientation antiparallel to the homopurine strand of the duplex has been found [244]. Use of such differences in preferred orientation of homopurine- and homopyrimidine-oligomers [245] provides an approach for sequence specific triple helix-forming antisense compounds.

Oligomers containing protonated cytidine (pKa of the protonated N3 is 4.3) may undergo both parallel and antiparallel orientation (C). The stability of such triple-helix systems, however, decreases with increase of the pH value. An arrangement involving two neighbored protonated cytidines is unfavorable due to electrostatic repulsion. Adenine polymers are an exception, because they undergo reversed Hoogsteen base-pairing with antiparallel orientation. In oligomers with parallel orientation adenine has to be replaced by inosine. It has to be considered in the design of triple-helix forming oligomers, that only in case of the triplets T×T*A and C(+)× G*C the C-1´ positions are located at the same site (isomorphism). Introduction of other bases into a mixed oligomer leads to backbone torsions.

Triple helix formation first of all stiffens the duplex, thus promoting duplex degradation by UV-rays, radicals or metal chelates. Impeded methylation of a DNA sequence by triple helix formation renders the non-methylated DNA sequence a good substrate for endonucleases. It

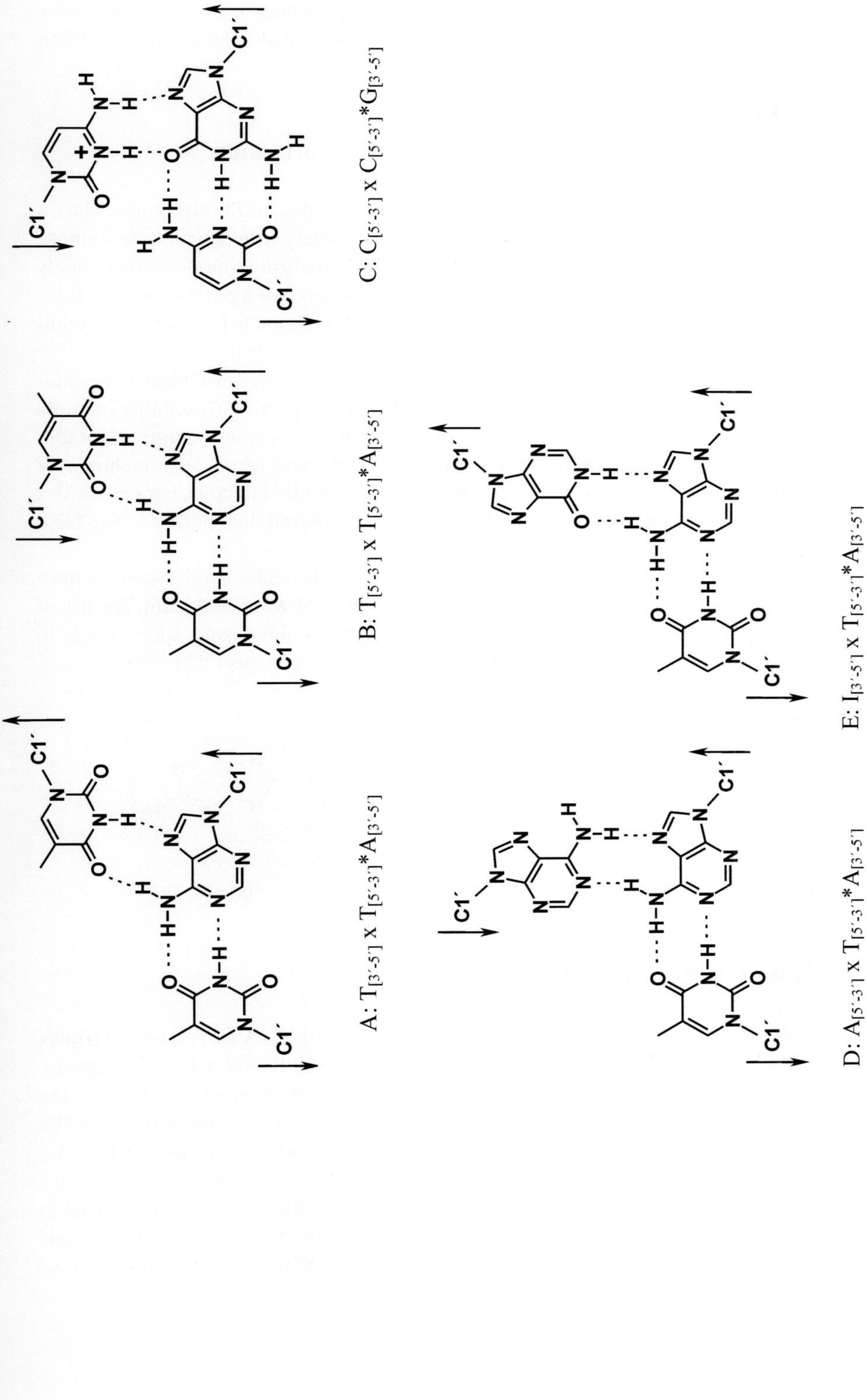

Figure 30. Examples of base triplets formed in triple helices; Mode of orientation and hydrogen bondings. Modified from [246].

is known that presence of Mg^{2+} ions or other cations, e.g., of polyamines, are prerequisites for triple helix formation [247]. Stabilization of triple helices has been also achieved by addition of benzopyridoindoles [248].

9.11.1 Modified Oligonucleotides in Triple Helix Formation

Tendency towards triple helix formation strongly depends on nucleotide structures. Thus, a decrease in tendency is observed with a change in the sugar moiety with deoxyribose < ribose < 2´-O-methylribose [249]. With phosphorothioates, triple helix formation depends strongly on the sequence. A modification next to the 5´-position in presence of a purine base increases stability of the triple helix, while a modification next to the 5´-position in presence of a pyrimidine base decreases stability [250].

With α-oligomers it has been found that homopyrimidine oligomers exhibited an orientation parallel or antiparallel to the homopurine strand of the duplex. With stability of triple helices significantly lower in comparison with corresponding β-oligomers, additional stabilization through conjugates is required [251]. With oligopyrimidine nucleotides, attachment of an intercalator at the 5´-end of an oligomer has been shown to be effective, because in this case the intercalator is in a position to intercalate with a 5´-pyrimidine–purine-5´ site [252, 253].

With the N-3 atom of cytidine unprotonated under physiological conditions, bases have been proposed to circumvent this disadvantage **87** [254], **88** [255], **89** and **90** [256]. The use of artificial nucleosides with bases such as 4-(3-benzamidophenyl)imidazole, which binds to T*A and C*G, has also been proposed in triple helix antisense concepts [257].

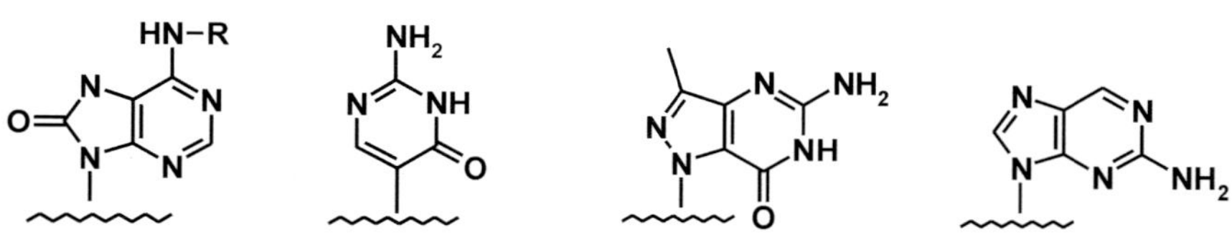

87 **88** **89** **90**

9.11.2 Minor Groove Recognition By Peptides

Peptides exhibiting sequence recognition in the minor groove of DNA correspond to triplex helix-forming oligonucleotides in a similar manner as do PNAs with duplex-forming oligonucleotides. Although an extensive coverage of this topic would be beyond the scope of the present article, this approach cannot be omitted when discussing triple helix formation. The concept has been found with natural products like netropsin and distamycin A, which are known to interact with DNA duplexes at sites of four or five successive A,T base pairs. Recognition is provided by the carboxamide NH groups of the peptides, which participate in hydrogen bonds with N-3 adenine and O-2 thymine atoms on the floor of the minor groove. The aromatic hydrogens of the N-methylpyrrole rings are anchored too deep in the minor

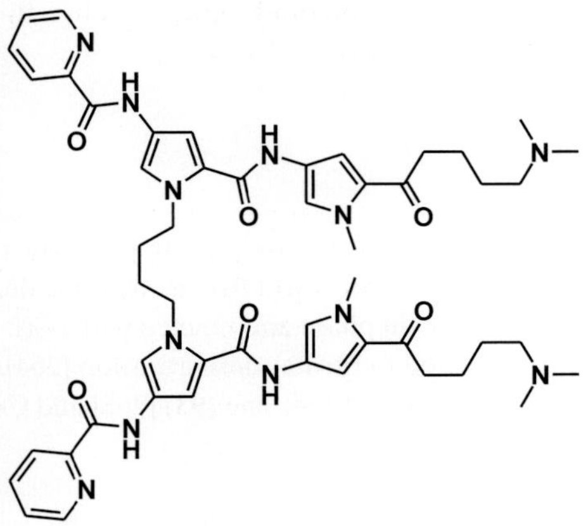

Figure 31. Minor groove-binding dimeric peptide. Symmetry reflects reading of the sequence 5´-TGTCA-3´ in duplex DNA. From [261].

groove to give room for the 2-amino group of the guanine of a G, C base pair. Shortly after it had been found that distamycin at high concentrations (2–4 mM) is capable of binding to the minor groove of 5´-AAATT-3´ as a dimer [258], details of the first synthetic analogs with different recognition patterns were published [259, 260]. As an example of this class of compounds a dimeric peptide is shown, that is capable of reading the sequence 5´-TGTCA-3´ with high affinity (Fig. 31).

9.12 Synthetic Methods for Ligand Attachment

Comprehensive compilatory work has been carried out by other authors to provide information about ligands covalently linked to oligonucleotides [262, 263]. Generally speaking, these attached functional moieties may either be *effectors*, which increase the antisense effect of the compound (increase of pharmacodynamic activity). They may be called *modulators*, if they add desirable properties to the drug (specific biological functions, improvement of pharmacokinetic properties). As there is a broad use of such oligonucleotide modifications to improve detection of oligonucleotides, a third group called *detectors* has also been defined (diagnostic tools).

One possibility of attaching ligands to oligonucleotides is to introduce properly protected amino or mercapto groups into the oligonucleotide to be deprotected and *derivatized after oligonucleotide synthesis*. Primary aminoalkyl substituents will react 100 times faster in nucleophilic reactions than will the amino groups of nucleobases. Thus, the primary amine may react with NHS esters to form amide bonds, or with isothiocyanates to form thiourea derivatives. The following examples have been selected to provide an overview on the positions at pyrimidine and purine nucleotides that can be used as points of attachment and on protective group strategies employed. The main advantage of this postsynthetic modification method is the possibility to use effector groups that are not stable under deprotection conditions of oligonucleotide synthesis.

Effector groups may also be attached by *direct ligand incorporation* during synthesis cycles. This is usually done at the 3´-or 5´-end of the oligonucleotide. A variety of such modifications is available commercially. The main advantage of this method is provided by efficient coupling. Examples employing this method will be discussed.

9.12.1 Attachment to Nucleobases

Base modification at the 4-position of pyrimidines leads to loss of base-pairing properties. The thiol moiety in **91** was deprotected after ODN synthesis using 1 M DBU in acetonitrile. The modified oligothymidylate was then cleaved from the solid phase and reacted with N-(2-chloroethylthio)phthalimide to yield **92**, which was subjected to further derivatization [264]. Purine base positions accessible for ligand attachment are C-8 of adenosine (**93**) [265] and C-2 of guanosine (**94**) [266].

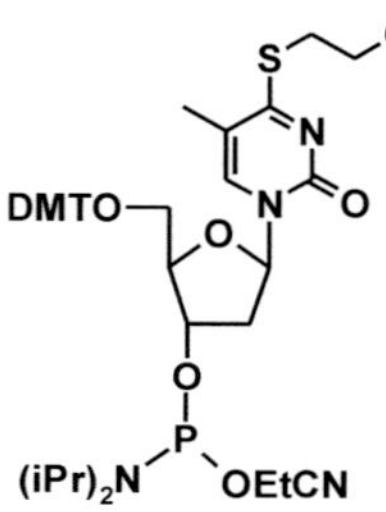

91

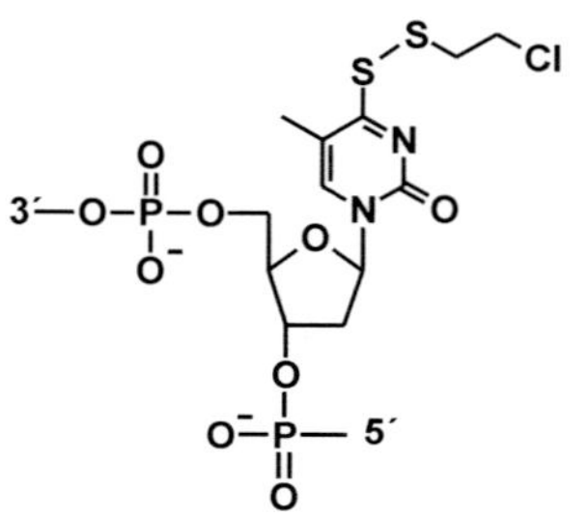

92

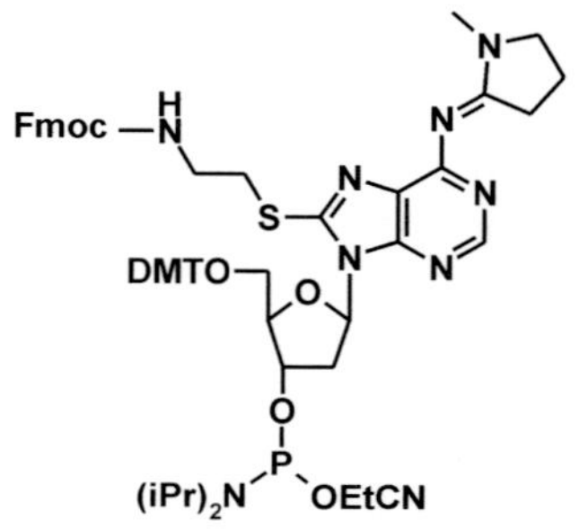

93

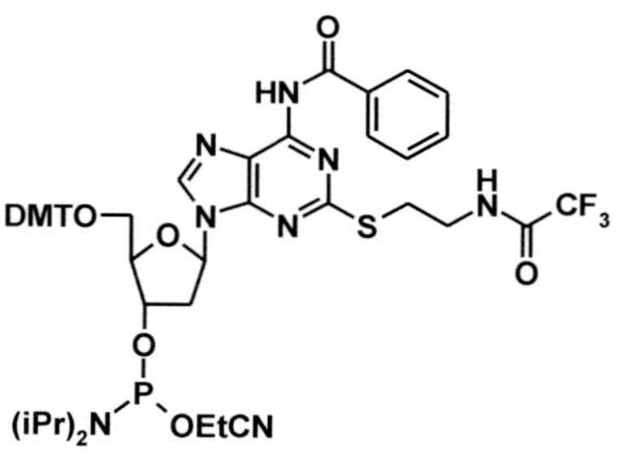

94

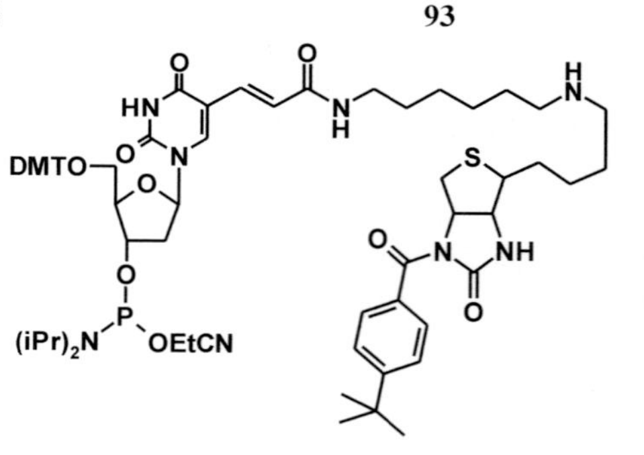

95

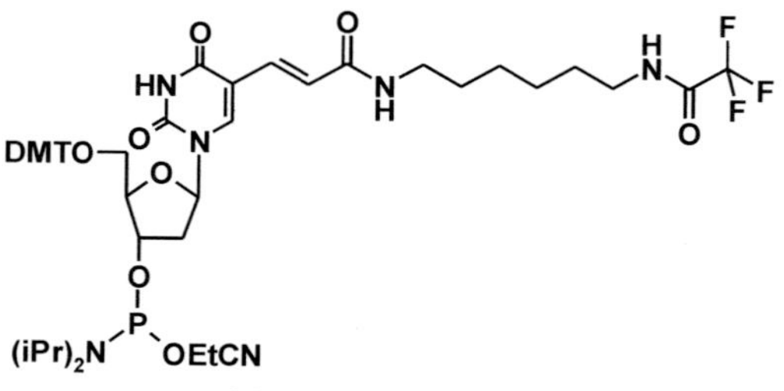

96

A particularly suited option to introduce functional moieties into oligonucleotides is given by the use of 5-alkyl pyrimidines as nucleobases (**95**, **96**). Oligonucleotides with such modifications still exhibit good hybridization properties [267, 268]. The trifluoroacetyl-protective group may be easily removed using ammonia solution. The free amino group is ready for derivatization [269, 270].

9.12.2 Attachment to the Sugar

After attachment of the building block **97** at the 5′-end, cleavage of the trifluoroacetyl group proceeded by use of aqueous ammonium hydroxide. Such 5′-aminated oligomers have been reacted with various fluorescent dyes [271]. 2′-O-aminoalkylated ribonucleotides allow post-synthetic derivatization after cleavage of the phthalimide protective group **98** [150].

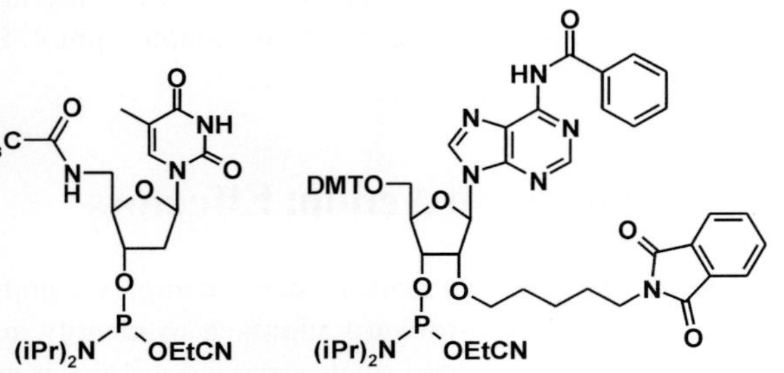

97 **98**

9.12.3 Attachment to the Phosphate Bridge

The use of special CPG-phosphoramidites with glycerol linkers has been proposed to test different 3′-modifications of an oligonucleotide of identical sequence. After solid-phase synthesis, glycerol esters are cleaved and oxidized. The resulting acid or aldehyde may be further re-

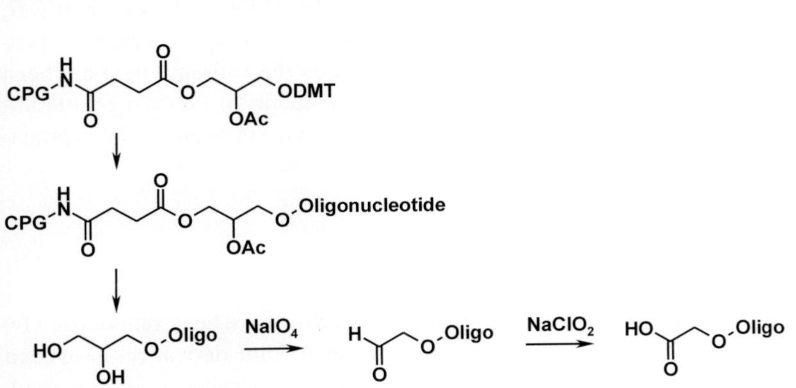

Figure 32. 3′-Modification by using glycerol succinic esters and subsequent oxidation.

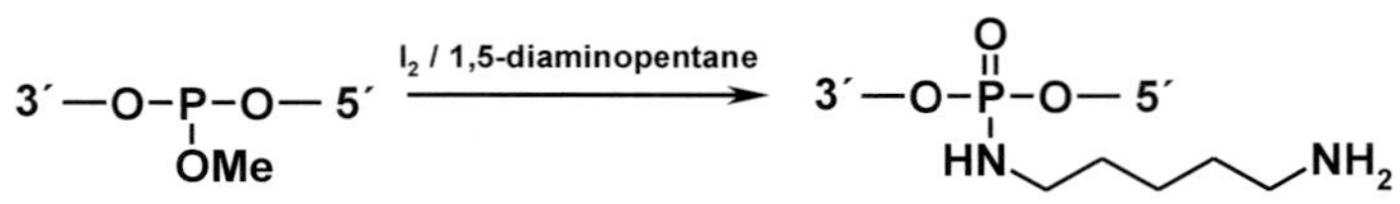

Figure 33. Preparation of phosphoramidates.

acted with amino groups containing functional moieties. A specific advantage of this method is given by the fact that only one CPG-building block has to be synthesized [272] (Fig. 32).

As pointed out in section 9.6.5, insertion of 3′-O-(N,N-diisopropylamino)methoxyphosphine nucleotide monomers and subsequent amidation using 1,5-diaminopentane/iodine leads to aminoalkylated phosphoramidates, which may act as linkers for functional moieties (Fig. 33).

However, such modifications are not suited in sequencing and amplification experiments in which the 3′-hydroxy group of the oligonucleotide has to remain free. Both 5-alkylpyrimidines and 5′- and 2′-modifications of sugars (2′-alkoxy-, 2′-deoxy-2′-alkylamino- and 2′-O-aminoalkyl-linkers) may be applied.

9.13 Ligands as Enhancers of Antisense Action: Effectors

Interaction between antisense and target RNAs generally leads to stable complexes under physiological conditions [273]. Nevertheless, the straightforward approach to identify and inhibit that part of a mRNA, which is present as an unfolded single-stranded mRNA, is not always successful. Enzymes and ribosomes reading the mRNA-sequence are in a position to strip off oligonucleotides from mRNA, without significantly impeding translation. An important approach to achieve efficient translational arrest by antisense-oligonucleotides is the attachment of intercalators to the oligomer to increase binding energy and to generate irreversible blockage of the target sequence. Such irreversibly bound effectors of antisense activity represent, however, by no means the most elegant strategy. Superior to target inhibition is target destruction, preferably including a catalytic mechanism. Metal chelators attached to an oligonucleotide represent this approach. A review also covering the field of ODN-intercalator ligands including this aspect has been published [274]. For the past 10 years however, the most elegant approaches have been those by which the antisense drug itself exerts or provokes the catalytic effect. The impact of ribozyme research on the antisense field has been already mentioned. A short survey on this field will given in section 16. Of even greater importance might be the observation (and the study) of the mechanism of enhanced degradation of double-stranded DNA:RNA by RNAse-H (see section 16.4).

9.13.1 Tricyclic Rings Linked to Oligonucleotides

Publications about oligonucleotides linked *to acridine derivatives* have been summarized by several authors [246, 275–277]. The intercalating properties of acridine derivatives have been shown by NMR-studies [278, 279]. 9-Amino-6-chloro-2-methoxyacridine is mostly used, which differs only slightly in size from the important intercalator drug, psoralen. Acridine-linked oligonucleotides are fluorescent, but in contrast to psoralen, crosslinks are not formed upon light treatment. In triple helix-forming oligomers, acridine substituents increase the stability of the complex by intercalation [280, 281]. Compounds **99** and **100** [282] or **101** [232] have been attached to the 5′-end of oligonucleotides by direct ligand incorporation [283, 284], probably resulting in increased duplex stability. A conjugate of this type has been successfully tested for inhibition of Ras-p21 synthesis [285]. Attachment of acridine substituents to the 3′-end or the phosphate linkage has also been described [286].

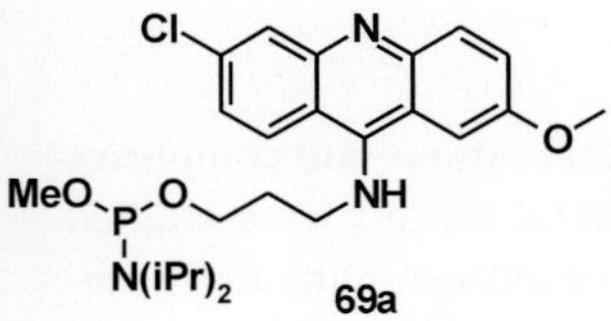

99

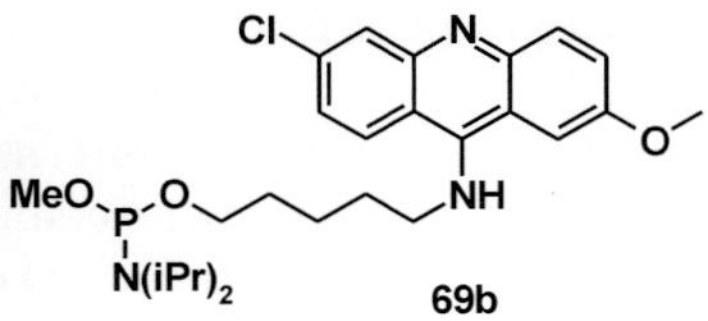

100

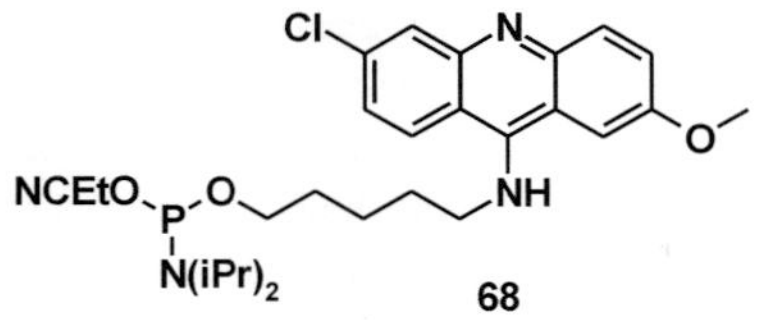

101

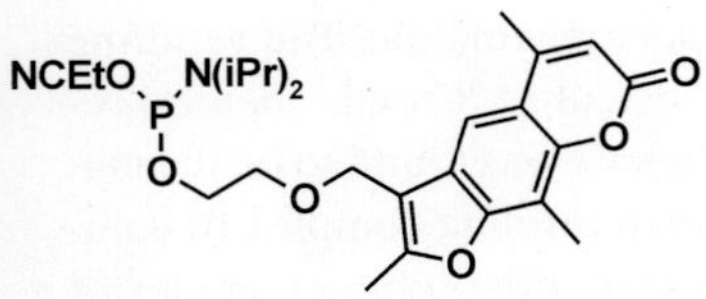

102

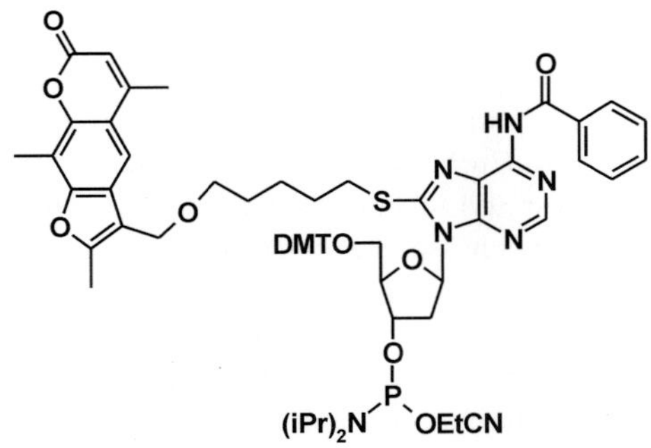

103

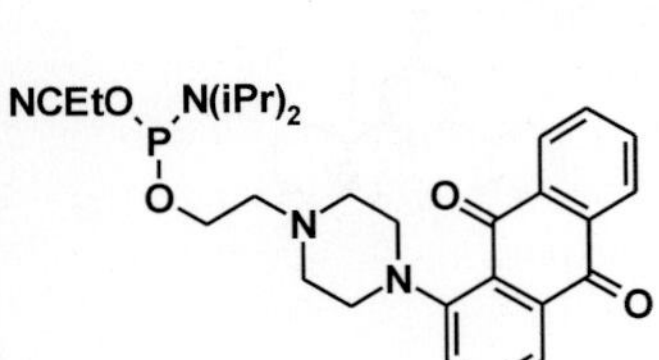

104

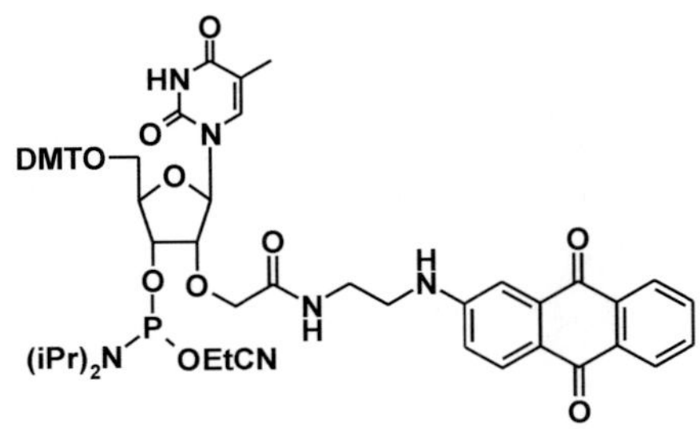

105

Attachment of *4,5′,8-trimethylpsoralene* to the C-8 position of the purine **103** has been described. The duplex of antisense and target strand was exposed to light of 345 nm wavelength for 50 min, resulting in over 90% crosslinking of strands [287]. Direct ligand incorporation has been achieved with compound **102** [288]. DNA–psoralen adducts are formed via light-induced cyclobutane formation with the furan and pyran ring of psoralen and pyrimidine (preferentially thymine) residues. Application of a linker made up of six methylene groups makes intercalation in triple helix formation feasible [289]. *Anthraquinoyl phosphoramidites* such as compound **104** have been suggested for incorporation into oligonucleotides [290]. As known from antitumor therapy, anthraquinones do not require photoactivation for radical generation. Their insertion into oligonucleotide sequences may also be achieved via phosphoramidite **105** [291].

9.13.2 Chelators Linked to Oligonucleotides

For some time it has been known that certain Lanthanide (III) metal cations and complexes catalyze the hydrolysis of RNA. In a pioneering study, several metal cations were studied with respect to catalytic turnover of RNA transesterification and complex stability in a hexadentate macrocyclic ligand similar to porphyrin. From these studies a positively charged *europium complex* was chosen for further development [292]. Thus, europium complexes like **106** were attached to oligonucleotides by amination at the 5´-end or incorporation of building block **96** and subsequent derivatization using the corresponding NHS-esters of the carboxylic acid **106** in amide formation [293]. Assays using complementary RNA indicated a concentration of 2.5 nM to be sufficient for RNA depletion. Since 2´,3´-cyclic ribonucleotides have been found in the assay solution after action, the authors suggested a hydrolytic mechanism with Eu(hexadentate) acting as a Lewis acid that facilitates intramolecular attack of the 2´-hydroxyl group to effect cleavage. *Bipyridyl-copper chelators* like phosphoramidite **108** have been used for 5´-phosphitylation of 3´-O-acetyl-2´-deoxythymidine. The resulting phosphate triester was incubated with Cu^{2+} ions and poly $rA_{[12–18]}$ at 31 °C under neutral reaction conditions for 24-48 h. Cyclic 2´,3´-adenosine monophosphate was found to be the major degradation product [294]. Attachment of the copper phenanthroline complex of compound **109** led to depletion of single-stranded RNA [295]. The terpyridyl copper complex of phosphoramidite **107** was shown to cleave RNA by reaching over the major groove of an A form DNA/RNA duplex to attack the opposite strand [296].

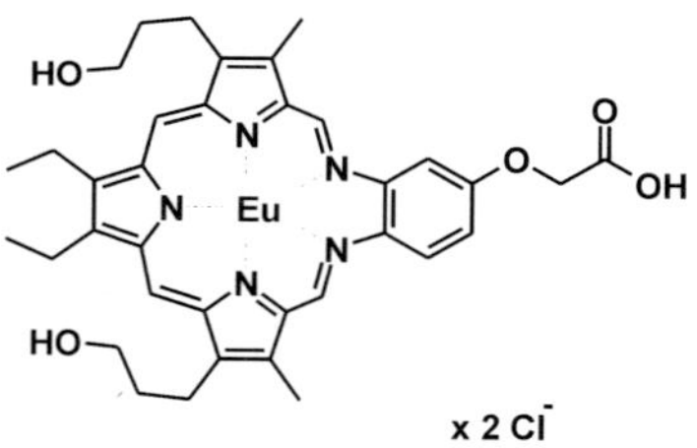

106

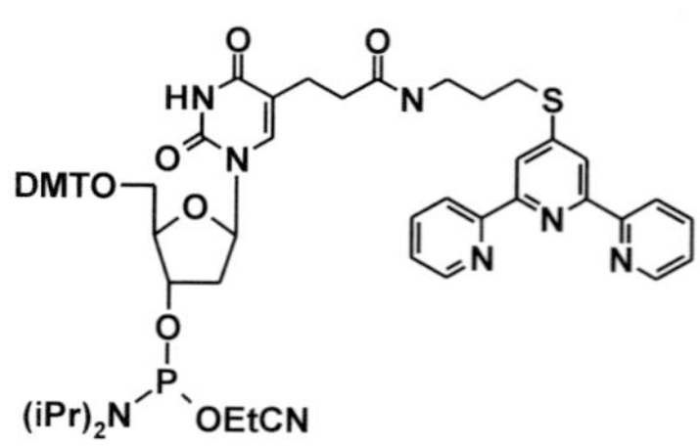

107

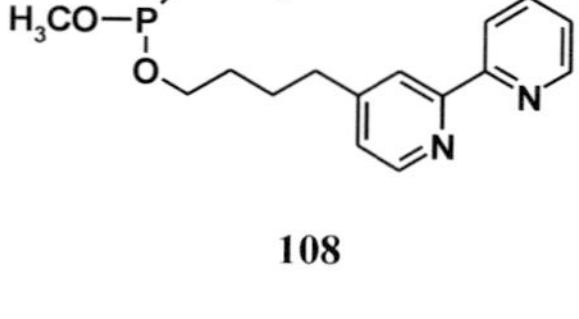

108

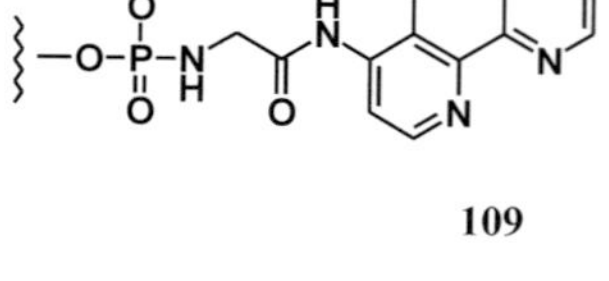

109

9.13.3 Zwitterionic Oligonucleotides

Linkers containing amino groups are frequently used to allow attachment of effectors, modulators and detectors (see section 12). The use of such amino linker modifications without attached functional moiety in oligonucleotides means adding a charge to the oligonucleotide, which is positive under physiological conditions. Duplex stability of duplex- and triplex-forming oligonucleotides is increased due to the decreased repulsion of the negatively charged phosphate backbones. A switch of negative charge to positive charge at specific linkages has been described with guanidine linkages (**110**) [129] and aminoalkylphosphonates (**111**) [297]. With regard to obtaining zwitterionic oligomer units, modifications of pyrimidine nucleobases at the 5-position (**112**) [170, 298], as well as attachment of alkylamino linkers to the N-6 position of guanine (**114**) [299] and N-4 of 5-methyl-dC (**113**) [300], have been de-

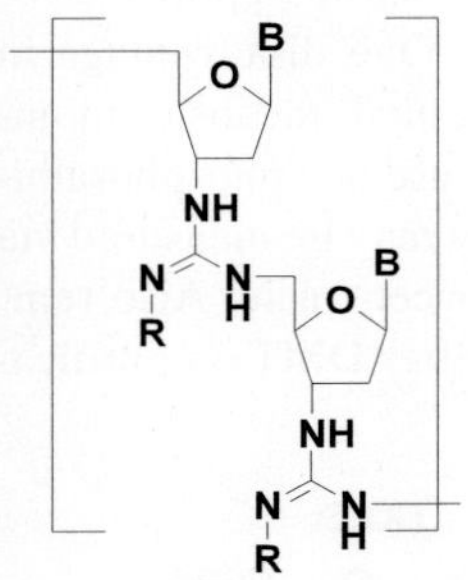

110

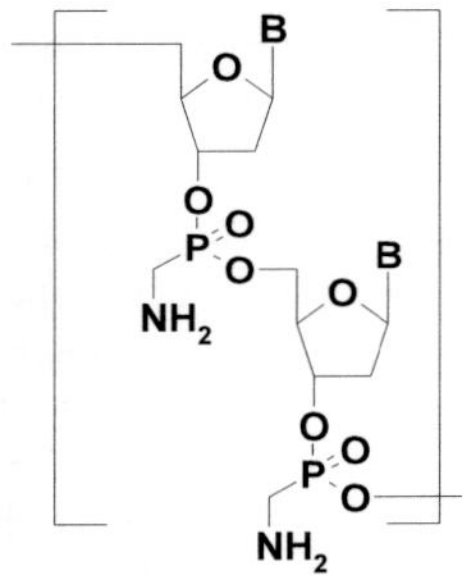

111

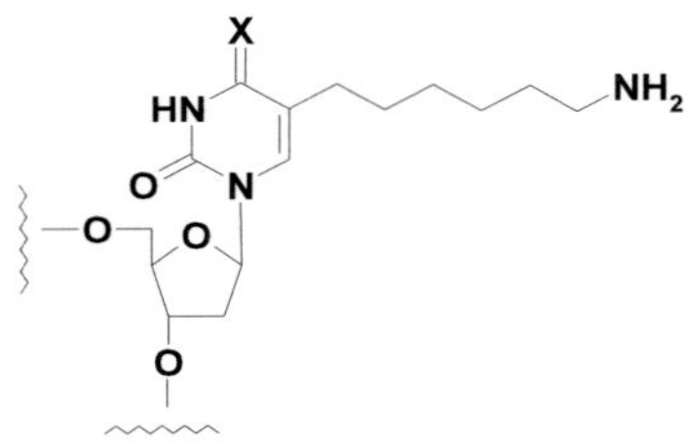

112

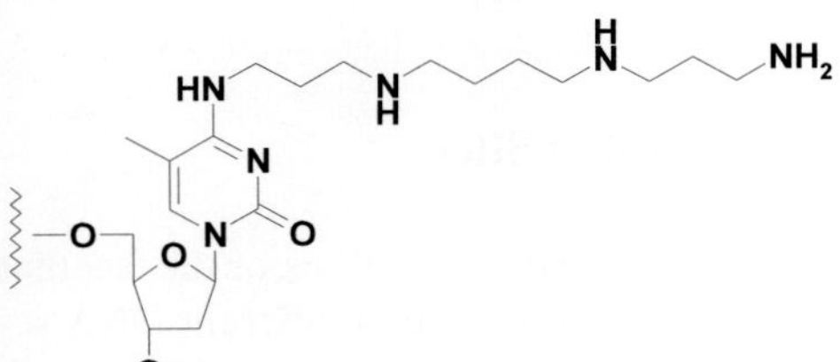

113

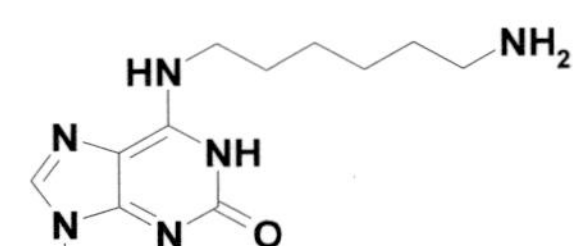

114

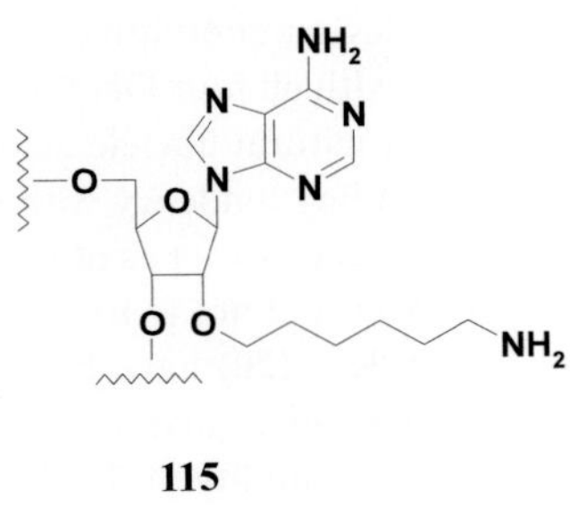

115

scribed. Molecular modeling studies on 2´-O-aminohexyl modification of adenosine (**115**) have shown that interstrand zwitterionic bond formation is feasible to enhance duplex stability. A set of oligonucleotides with improved hybridization properties has been synthesized [301, 302].

9.14 Modification of Biological Effects of Oligonucleotides: Modulators

9.14.1 Phosphorylation of Oligonucleotides

For use in cloning, gene construction and in ligase chain reactions, oligonucleotides have to be phosphorylated at the 5´-end. This can be achieved enzymatically by use of T4 polynucleotide kinase or chemically by applying phosphoramidite technology. One disadvantage in using biscyanoethyl phosphoramidites **116** [303] is the lack of a convenient means of monitoring the coupling reaction. This disadvantage has been overcome by use of a phosphoramidite synthesized from sulfonyldiethanol **117** [304]. Coupling efficiency may be measured via release of dimethoxytrityl cations upon deprotection using dichloroacetic acid. Also remnants of compound **118** have been cleaved from the oligonucleotide after DMT-on purification upon treatment with aqueous acid [305].

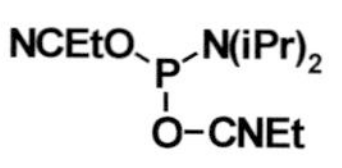

116

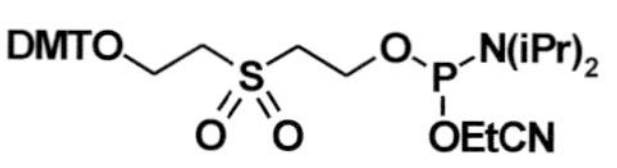

117

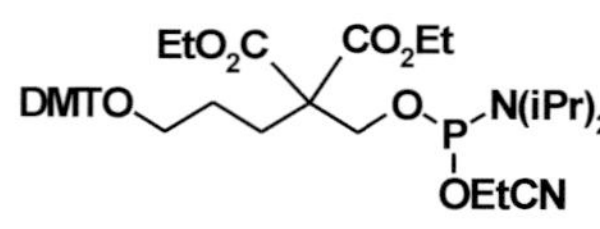

118

9.14.2 Base-Modified Nucleotides at Degenerated Sites

Imperfect gene–protein correlations result from incomplete peptide analysis, or the fact that due to degenerated codes a protein sequence may be coded by several different DNA sequences. Resulting difficulties in PCR amplification of a DNA-sequence coding for a specific protein may be overcome by placing a mixture of all four phosphoramidites at such undefined sites in oligonucleotide synthesis. Different reactivities of the phosphoramidites and increasing complexity of the reaction mixture with permutation of more than one base (4^n molecules) led at least to difficulties in purification. A rather elegant alternative is provided by the use of heterocycles, which on duplex formation interact with all four DNA/RNA-bases. Unfortunately, 2´-deoxy inosine (**119**), which also occurs in natural nucleic acids and is most frequently used for this purpose, exerts differences in binding tendency with the four possible bases of the counter strand. Indeed, it has a clear preference to act as dG [306]. In a search of heterocycles that are particularly prone to base stacking, but not to form hydrogen bridges to the nucleobases of the counter strand, 3-nitropyrroles (**120**) have been studied (based on the known fact that 4-nitroaniline is the smallest known intercalator). Along similar lines it was observed that 3-nitropyrrole decreases the overall stability of the duplex, the effect being more pronounced with sequences with central placement of the "universal" base [307–309]. Therefore, use of 5-nitroindoles (**121**) has been proposed. Depending on the position of the nitro group, duplex destabilization is reduced in the sequence: 3-nitropyrroles < 6-nitroindoles < 4-nitroindoles < 5-nitroindoles.

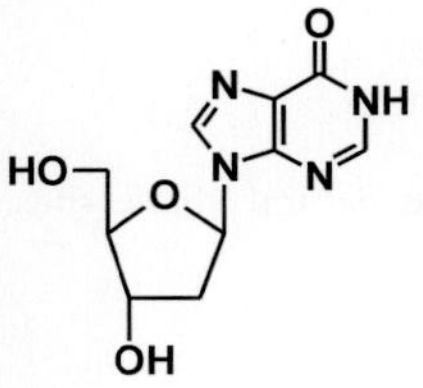

119

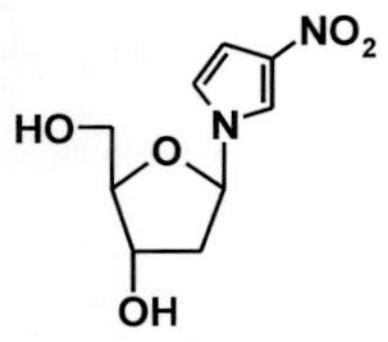

120

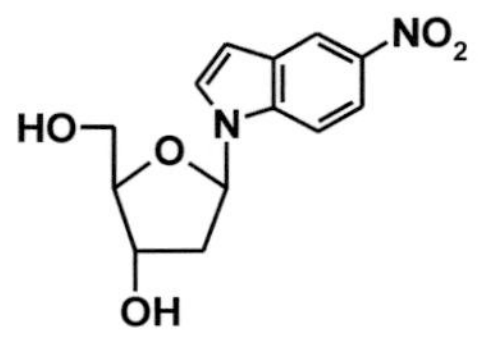

121

A further option is given by application of bases with "inexact" base-pairing properties. Due to tautomeric structures, nucleobase P forms hydrogen bridges with adenine and guanine, and nucleobase K with cytidine and thymine (Fig. 34). Thus, a mixture of phosphoramidites P and K may be used at positions of unknown bases instead of the mixture of the phorphoramidites dA, dG, dT, dC.

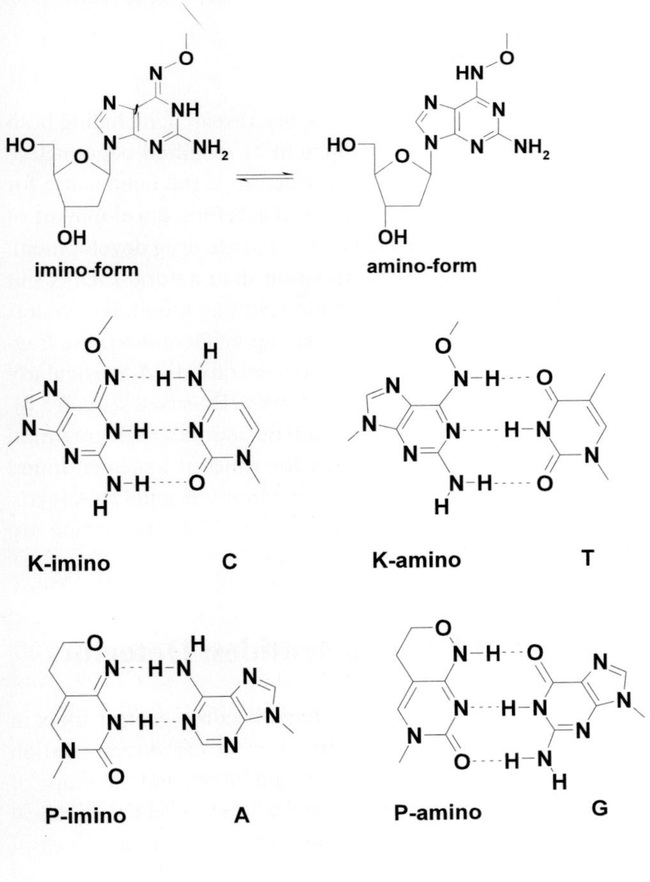

Figure 34. Left: Tautomerism of dK; Right: Binding sites of the tautomers of nucleobases P and K in oligonucleotides led to base mismatching during PCR.

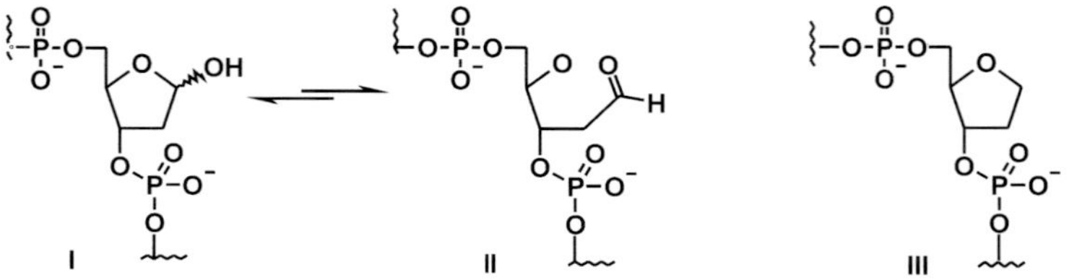

Figure 35. Naturally occurring abasic sites in DNA I, II: Unstable acetals in natural abasic sites. III: Tetrahydrofuranyl derivatives after oligonucleotide synthesis.

9.14.3 Natural Abasic DNA-Nucleotides and Tetrahydrofuran-Spacers

Hydrolytic cleavage of N,O-acetals due to enzymatic processes leads to nucleic acid chain elements without a nucleobase. Such positions are potential mutation sites. To study such mechanisms and to induce directed mutagenesis, oligonucleotides have been synthesized containing tetrahydrofuran building blocks introduced via corresponding phosphoramidites [310, 311] (Fig. 35).

9.14.4 Inhibitors of Enzymatic Degradation

Inhibition of enzymatic degradation is a central target of antisense research, including both strategies of nucleotide functionalization and direct development of modified oligonucleotides as described in the previous sections. Activity of 3´-exonucleases is the main cause for instability of natural oligonucleotides in parenteral application. Therefore, development of 3´-nuclease-resistant oligonucleotides has been a central aim of antisense drug development. Phosphorothioate oligomers are more stable towards degradation than natural ODNs, but they are also subject to degradation in blood. However with the resulting monomers undergoing much more rapid renal clearance compared with the intact oligonucleotides these fragments are not detected in blood in toxicologically relevant concentrations. A particularly simple method to prevent degradation is by linkage of 3´-ends of two ODNs (see section 5.6). Such 3´,3´-linked oligomers are degraded only slowly with natural nucleotides as degradation products [312]. Other target enzymes have been polymerases, for which it has been found that a spacer at the 3´-end of the oligonucleotide can induce their inhibition, and DNA (cytosine-5-)-methyltransferase. Thus, oligonucleotides containing 5-fluoro-2´-deoxy-uridine are known to be inhibitors of the latter enzyme [313].

9.15 Ligands for Localization of Oligonucleotides: Detectors

It was mentioned in the introduction that oligonucleotides have become central tools in physiological and pharmacological research. Although there is a strong direct relation between many of those applications and drug development, it would be beyond the scope of the article to discuss this matter in detail. Nevertheless, an overview is given on the chemical basis of functionalization of ODNs with "detectors", and this might be useful in the development of diagnostic tools.

9.15.1 Biotinylation of Oligonucleotides

Biotin undergoes specific strong interactions with proteins such as avidin or streptavidin in solution or tissue probes, which themselves can be linked to a reporter enzyme. Frequently, alkaline phosphatases or peroxidases are used for this purpose. To achieve optimum condi-

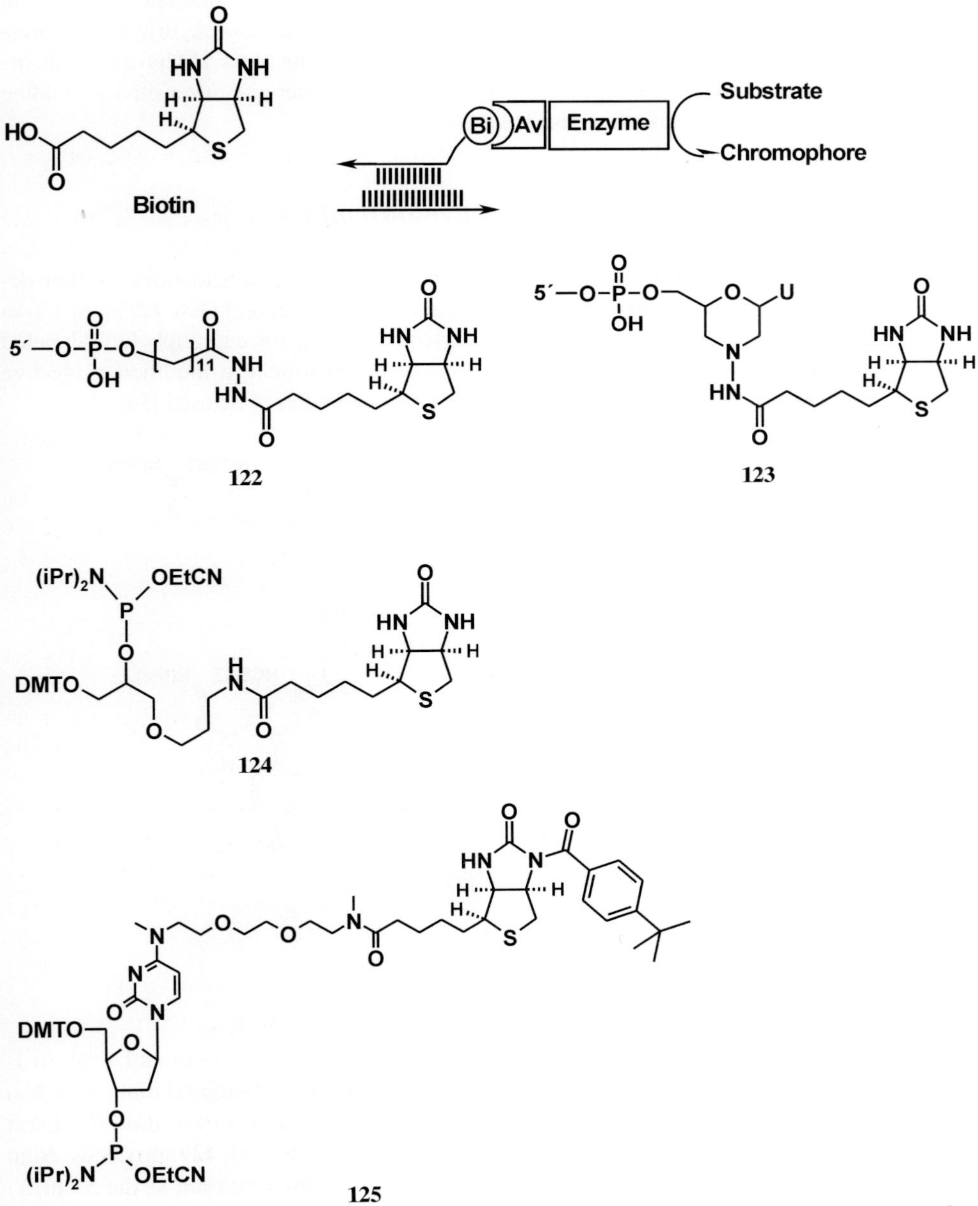

Figure 36. Biotin–avidin system for detection and purification of oligonucleotides. Modified from [277]; Examples for direct ligand incorporation.

tions, a spacer of 10 atoms length should be placed between the oligonucleotide and the biotin moiety. Biotin residues do not interfere with most of the assay techniques applied with DNA. Biotin residues at the 3´-end of an oligonucleotide have been reported to enhance stability against 3´-exonucleases [220] (Fig. 36). Techniques of biotinylation have been summarized [76]. Using the solid support method, an oligomer was phosphitylated at the 5´-end followed by reaction for the carboxylate spacer with biotin hydrazide to furnish **122** [314]. Starting from a 5´,5´-uridine residue, morpholino-derivative **123** was obtained via oxidation with sodium metaperiodate and reductive ring closure with sodium borohydride in the presence of biotin hydrazide [315]. Up to eight units of biotin building block **124** have been incorporated at the 5´-end. Depending on the assay used, a 10-fold increase of sensitivity could be achieved by such multiple attachments [316, 317].

9.15.2 2,4-Dinitrophenyl Groups in Oligonucleotides

2,4-Dinitrophenyl groups (DNP) are attached to the 5´-end of oligonucleotides to allow detection by monoclonal and polyclonal antibodies and suitable markers (**126**, **127**) [76]. Up to ten DNP groups have been inserted into oligonucleotides [318]. Since multiply DNP-labeled oligonucleotides are sensitive to ammonium hydroxide treatment, a modified protective group strategy, such as with dmf-protected phosphoramidites, has to be used [54].

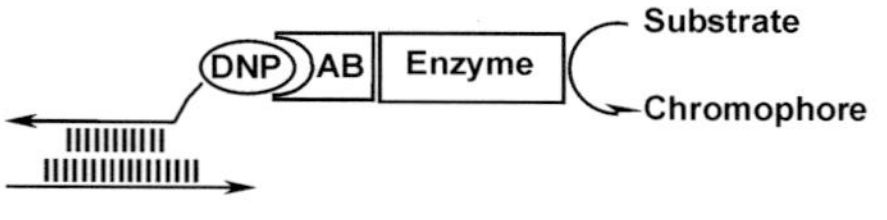

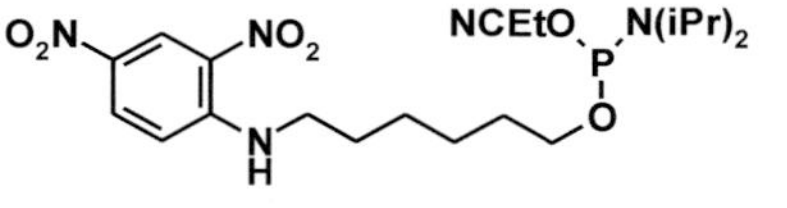

126

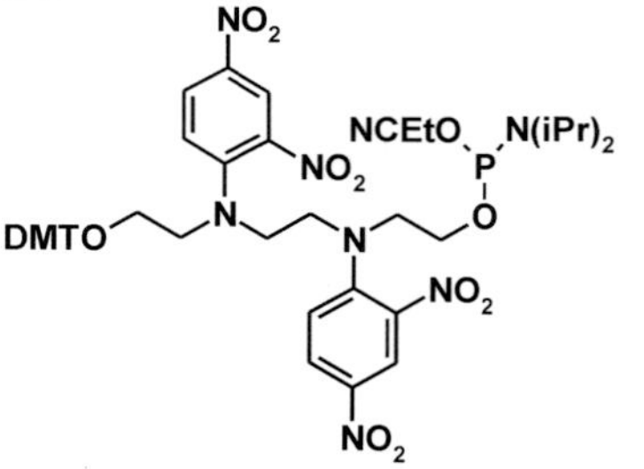

127

9.15.3 Fluorescein-Linked Oligonucleotides

Compound **128** has been synthesized by reaction of fluorescein methyl ester and 4-chloro-1-(4,4´-dimethoxytrityloxy)butane followed by detritylation and phosphitylation with bis-(N,N-diisopropylamino)methoxyphosphine. The phosphoramidite has been used for direct ligand incorporation at the 5´-end [319]. Synthesis of compound **129** proceeds from 5-carboxyfluorescein and isopropylidene glycerol (solketal). Incorporation at the 5´- or 3´-

end proceeds slowly and requires prolonged coupling times. Postsynthetic modification as been carried out using active esters of 4-acetylamino-1,8-napthalimido-N-caproic acid **130** as fluorescent tags (excitation at 360 nm and emission at 460 nm wavelength of light) [320].

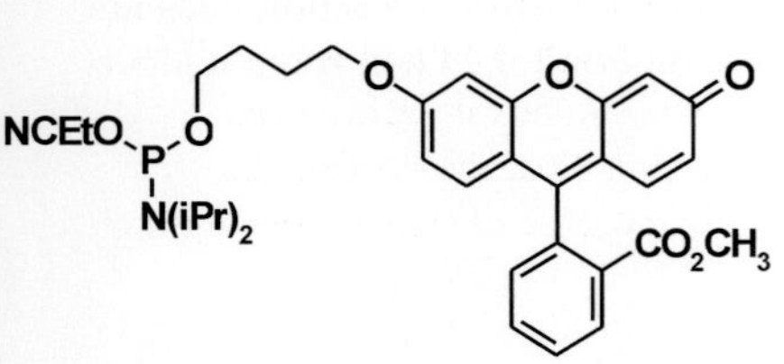

128

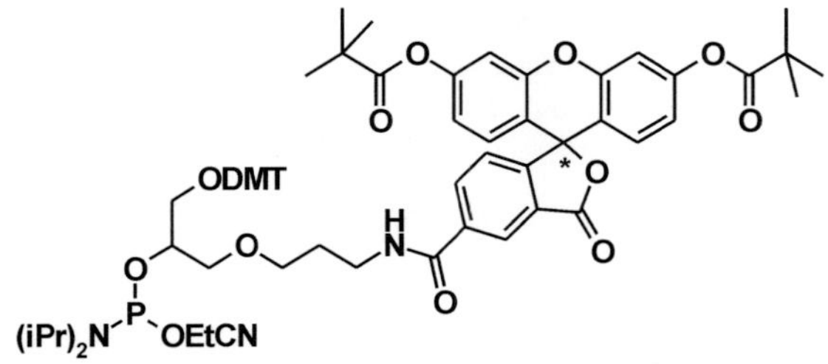

129

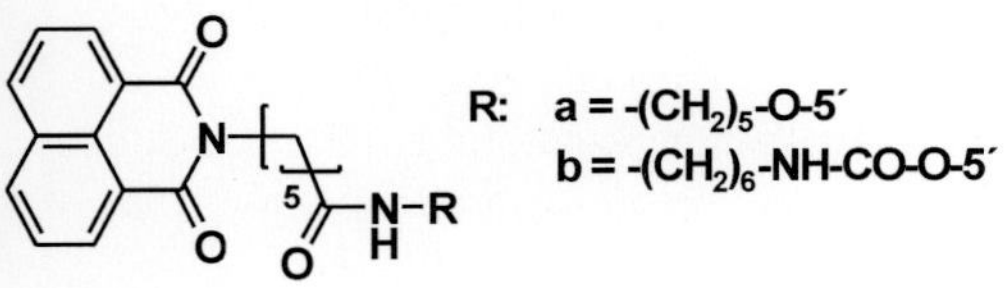

130

9.15.4 1,N^6-Ethenopurines as Fluorophores

1, N^6-Ethenoderivatives of adenosine (2′-deoxy adenosine) may be obtained as metabolites from the reaction of DNA with vinyl chloride, chloroacetaldehyde or chloroethylenoxide [321]. Excitation wavelength of such heterocycles is 300 nm with a maximum of emission at 412 nm. Phosphoramidite (**131**) and similar compounds have been incorporated into oligonucleotides [322–324].

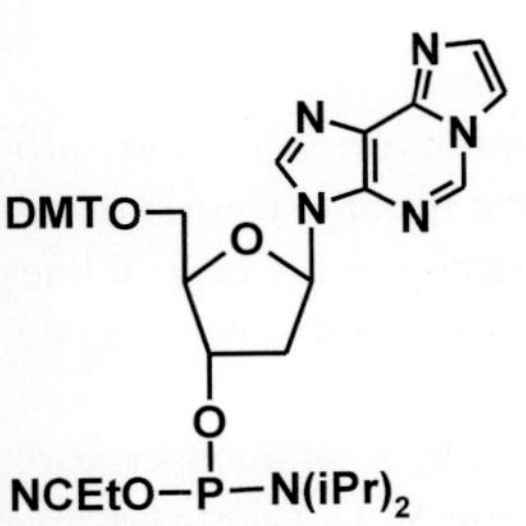

131

9.15.5 Halogenated Pyrimidines and Purines as Photo Crosslinkers

Crosslinking of DNA and proteins is a versatile method to study protein–DNA interactions. Among others, phosphoramidites of brominated or iodinated 2´-deoxycytidines [325], 2´-desoxuridines [326, 327] and recently also of 2´-deoxyadenosines **132** have been used [328]. Nucleophilic replacement of the bromine by an amino group during deprotection does not occur when reaction conditions of concentrated ammonia solution for 24 h at room temperature are used after oligonucleotide synthesis.

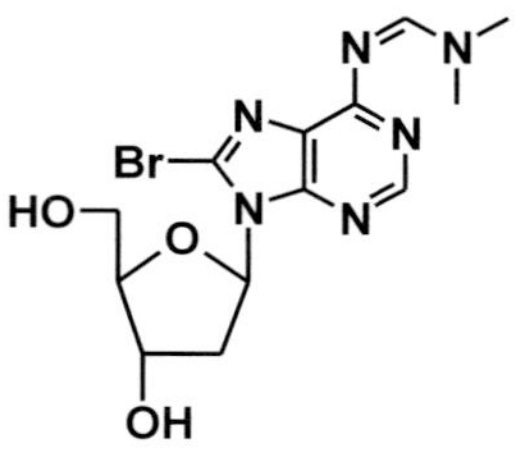

132

9.15.6 Applications of Radioactive Oligonucleotides

Radioactive [^{35}S]-phosphorothioates have been used for in vivo evaluation of chronic or cumulative toxicity of oligonucleotides. To achieve site-specific ^{35}S-labeling in phosphoramidite chemistry, either a mixture of [^{35}S]/carbon disulfide/pyridine/triethylamine [329] or ^{35}S-labeled Beaucage reagent [330] have been used in the sulfurization step. ^{3}H- and ^{14}C-labeled compounds have been tools in investigations of the metabolic fate of the nucleobases of an oligomer [331]. ^{3}H may be inserted at the 5´-end by formation of the 5´-carboxyaldehyde by Mofatt–Pfitzner reagent, followed by reduction with [^{3}H]-sodium borohydride in 2-propanol [332].

9.15.7 Immobilization and Array Techniques

With the increasing importance and complexity of gene-diagnosis, miniaturization and automation of test systems to achieve "high-throughput diagnostics" have become central development tasks. Immobilization and development of specific array techniques are central lines of nucleic acid research. Due to the potential direct and indirect impact on antisense drug development, this topic will be discussed in brief.

Immobilization of oligonucleotides *after synthesis* may be achieved by 5´-aminoalkylation [333] or by use of the biotin/streptavidin system [334, 335] (see section 15.1). Light-sensitive protective group strategies allow construction of DNA arrays for diagnostic and research purposes. By employing 5´-MeNPOC-phosphoramidites like **133** on glass substrates, which

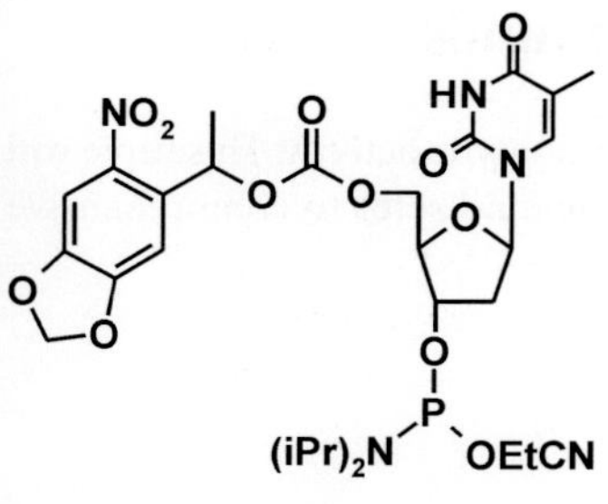

133

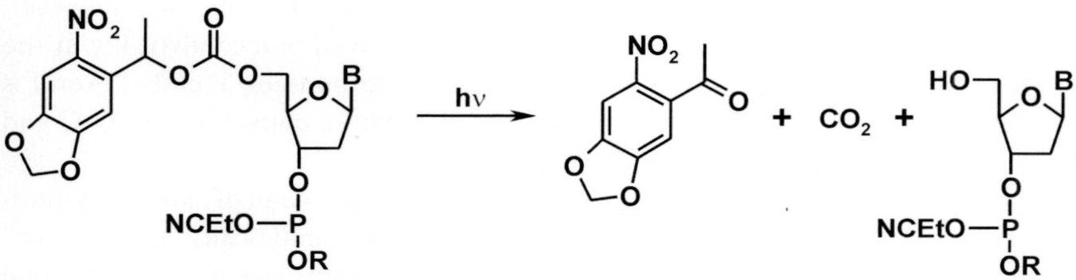

Figure 37. Light-sensitive phosphoramidites for array technology; Left: Phosphoramidite, thymine residue remains unprotected; Other bases: N6-phenoxyacetyl-dA, N2-isobutyroyl-dG, N4-isobutyroyl-dC; Right: Deprotection of 5´-terminus by light. From [336].

are modified with MeNPOC-protected hexaethyleneglycol linkers in the first step, structures of a resolution down to 10 × 10 µm may be generated (Fig. 37). By this method, arrays with 10^6 oligonucleotide probes per cm^2 may be synthesized in 40 reaction cycles. Deprotection proceeds by exposure to light over 10 half-lives (11–13 s at 25.5 mW cm^{-2}). However, yields obtained in each cycle are lower, when compared with standard phosphoramidite methods (92–96% versus 98%), leading to 10% of truncated sequences, which usually do not disturb sequential analysis carried out with these arrays. The average density of oligonucleotide loading is 10–30 pmol cm^{-2}. The oligomers are placed at a 40 Å distance, which might cause quenching effects when labeled with fluorescent dyes [336].

To broaden the applicability of these arrays towards genomic sequential analysis and diagnostics, efforts have been made to use standard phosphoramidite technology in combination with soluble polyimide underlayers (XU-218, CIBA-Geigy) and photoresistant epoxy resins (SU-8, SHELL). Using these techniques, resolutions down to 4–8 µm have been achieved [337, 338]. A comparatively simple approach to obtain oligonucleotide arrays useful in the search for potential antisense motifs has been described [339]. The method makes use of standard DNA synthesis. A square or round-shaped reaction chamber is moved over a glass surface for a distinct stretch after each cycle. During the cycle, the capping step is omitted, leaving a pattern of oligonucleotides that differ in length and sequence. These arrays can be used in connection with mRNA fragments bearing a radioactive label to scan effectively for sequences suited as antisense drug targets.

9.16 Catalytically Active Antisense Compounds

Ribozymes may be defined as oligonucleotides with intrinsic catalytic activity. This topic will be treated only briefly in this article, and interested readers should refer to comprehensive reviews on the subject [340–342].

9.16.1 Group I Introns

Group I introns are found in fungal mitochondria, chloroplasts and the genome of eubacteria. In vitro, they effect their own cleavage at the 5′-end. For in vivo cleavage, however, protein factors are required. Upon cleavage, an intron is removed autocatalytically at the 5′-splicing site (5′ss) and the 3′ss of the flanking exons. The corresponding 3′-end of exon 1 is attached to the 5′-end of exon 2. Cleavage occurs between the base pairs A-U and U-G and is initiated by a co-factor, providing a free 3′-OH group of rG.

As depicted in Fig. 38, the IGS is complementary to the 3′-terminal end of exon 1. By autocatalyic removal of the intron from exon 1 and exon 2, it remains catalytically "trans-active" [343], if an appropriate co-factor is present. Development of drugs by modification of group 1 introns appears to be difficult, because only two base pairs corresponding to the IGS are sufficient to achieve cleavable substrates.

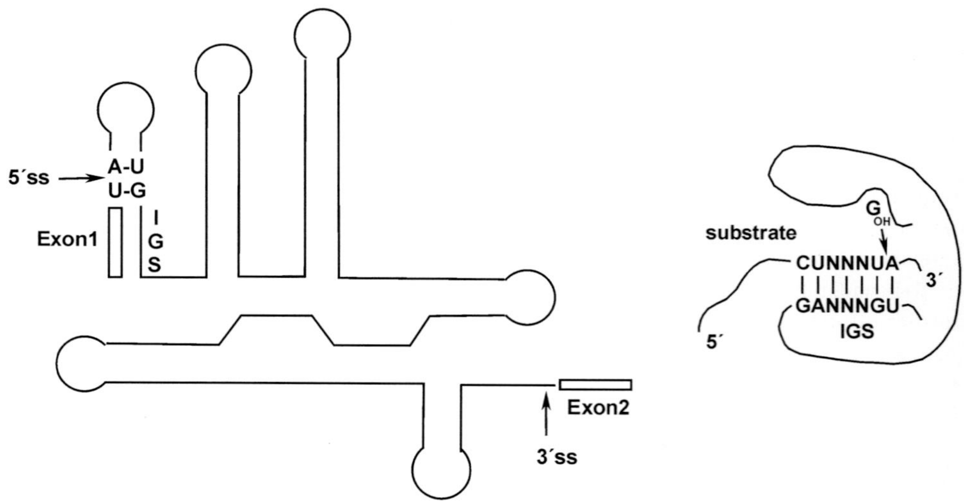

Figure 38. Left: Schematic structure of group 1 introns. Right: Modified group 1 intron exhibits trans-catalytic activity to substrates complementary to the IGS. Modified from [342].

9.16.2 RNase P

RNase P is an endonuclease, which is made up of a RNA and a protein unit and appears ubiquitously in the cytosol. It contributes significantly to maturing of tRNA precursors. As with group I introns, the phosphate diester is hydrolyzed to yield a 3´-hydroxy and a 5´-phosphate group. Activity of the endonuclease, including the protein unit, is higher than activity of the isolated RNA unit. Nevertheless, the intrinsic catalytic activity of the RNA unit renders it a real ribozyme of particular interest due to transactive cleavage. Recognition of the splicing site does not occur via Watson–Crick base-pairing, but involves specific tertiary structures. According to studies with the RNase-like enzyme, which in *E. coli* is made up from M1 RNA and the corresponding protein, the existence of a single hairpin is sufficient, if a single-stranded trinucleotide CCA appears at the 3´-side of the hairpin loop [344]. Bacterial RNase P is in a position to cleave target-mRNA, if a corresponding antisense-ODN is present with an external guiding sequence (EGS), including a NCCA-sequence at the 3´-proximal end of the EGS (see Fig. 39) [345]. With human RNase P, conditions are obviously more complex, requiring structures very similar to tRNA precursors [346].

9.16.3 Viroids and Virusoids (Hammerhead Ribozymes)

Viroids and virusoids are circular RNA, including hammerhead ribozyme structures. They represent the smallest known ribozyme motif. Their name is derived from the shape of their autocatalytical sequences, which resembles a hammerhead (see Fig. 39). Hammerhead ribozymes consist of three stem–loops joined by a 11-mer single-stranded domain. Cleavage occurs between two stem–loops with a cyclic 2´,3´-phosphate and a free 5´-hydroxy group being formed. It has been shown that similar effects may be achieved with two or three single-stranded ORNs [347–349]. Obviously, any RNA containing a 5´-XUN-3´(X being A, C, G or U; N being A, C, or U) sequence is cleavable at this site by a transactive hammerhead ribozyme (see Fig. 39) [342].

Hammerhead ribozymes have been the preferred class of ribozymes to be used for in vitro experiments. In such experiments, the antisense sequence consists of 6–8 bases flanking both sites of the catalytic center. A lower number of bases would lead to decreased binding properties, while an increased number of pairing bases would reduce the turnover by decrease of

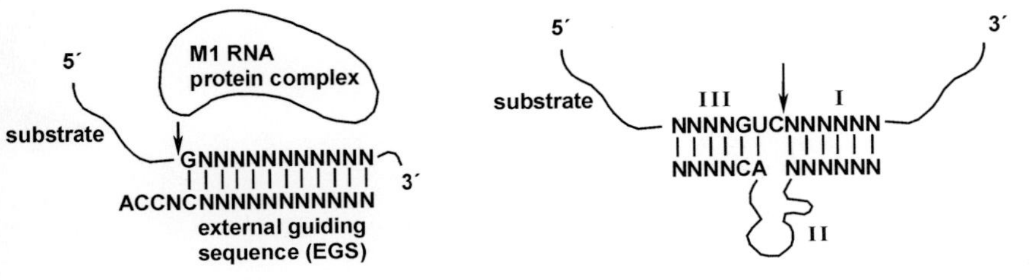

Figure 39. Left: Required assembly for trans-active RNA cleavage by RNase P-like M1 RNA protein complex in *E. coli*. Right: Assembly of two oligonucleotides forming a hammerhead-like ribozyme; (N variant bases). Modified from [342].

dissociation constants [350–352]. Cellular enzymes denaturing base pairs of dsRNA are activated by 15–20 base pairs [353]. At present, only empirical methods exist to establish whether a mRNA sequence including GUC or XUN triplets will undergo ribozyme cleavage. Stabilization of easily degradable ribozymes has been attempted many times. Among others, ribozymes with 2´-fluorocytidines, 2´-fluorouridines and 2´-aminouridines have been studied [354]. Modifications, however, must not affect the catalytical center. The topic has been painstakingly reviewed [352].

9.16.4 RNase H: Chimeric Oligonucleotides

RNase H was first described by Hausen and Stein [355]. While dsRNA is not recognized, the enzyme cleaves RNA in RNA/DNA-duplexes. The biological role of the enzyme is assumed to be cleavage of RNA primers in Okazaki fragments and digestion of ribonucleotides inserted into DNA by error [356]. Cleavage products are 5´-phosphates, and magnesium ions are involved in the proposed mechanism [357]. The order of cleavage rate is rA > U > rC >G [358].

Design of chimeric oligomer is a field of development which deserves specific attention (see table 18 for the structure of such molecules). One segment may improve drug uptake into the cell, while the other is in a position to activate RNase-H. The strategy has been recommended for both methyl phosphonates [359] and phosphorothioates [360, 361]. Recent chimeric antisense ODNs comprise 2´-O-ethyleneglycol-ORN-phosphorothioates/ODNs [362] and ORNs/2´-O-methyl-ORN [363]. A recent study on the selection of a modification which is best suited for an anti-cancer assay system might provide further information [364]. The present challenge in the design of such oligomers is to establish specifically effective targets of their interaction with mRNA metabolism. Thus, criteria for adapting oligonucleotides to act as inhibitors of splicing of "pre mRNA" have been already described [365].

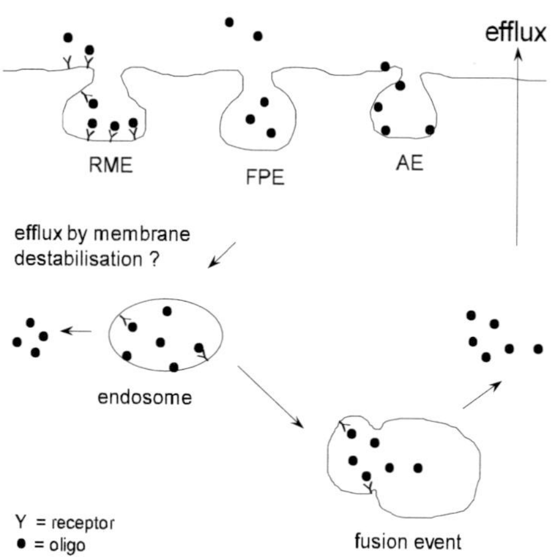

Figure 40. Uptake and intracellular fate of oligonucleotides. RME: receptor-mediated endocytosis. FPE: fluid-phase endocytosis (pinocytosis). AE: adsorptive endocytosis. Modified from [366].

9.17 Drug Delivery

9.17.1 Cellular Uptake and Intracellular Fate of Oligonucleotides

Diffusion of negatively charged ODNs through phospholipid membranes appears to be a slow process [367]. Endocytosis is assumed to be the predominant mechanism with uptake of ODNs into eucaryotic cells. In cultured cells, maximum uptake of ODNs occurs after 2 h of incubation, but this is reduced at lower temperatures and by incubation with known inhibitors of endocytosis such as deoxyglucose, cytochalasin B and sodium azide [368, 369]. Mechanisms involved may be absorptive endocytosis (Fig. 40: RME, AE) or fluid-phase endocytosis (Fig. 40: FPE; pinocytosis). Absorptive endocytosis has been proposed for phosphate diesters and phosphorothioates because of the high-affinity binding to cellular membranes and increased cellular uptake [370] in comparison with non-ionic compounds such as methylphosphonates and PNA, which show only poor delivery into cells. Recent advances in cell delivery have been summarized in review articles [371, 372]. Replacement of ODNs bound to cellular surface binding sites by non-labeled poly-dC phosphorothioate oligomers or with tritium-labeled Suramine may be used to determine K_d-values for ODN absorption [370, 373, 374]. Competition in uptake mediated by absorptive endocytosis has been found for SODNs, phosphorodithioates and ODNs. Various different proteins have been identified as binding sites in absorption of ODNs and SODNs to cellular membranes [375, 376]. A critical evaluation of the identified proteins seems to be required, because their ability to adsorb nucleic acids does not correspond with their main physiological function in the cellular membrane [366, 373]. The uptake of oligonucleotides in concentrations above those found for the K_d-values for absorptive endocytosis (22 nM for ODNs [370] and 1 μM for SODNs [374] in HL-60 cells) is most likely mediated by pinocytosis. Both pathways lead to internalization of the oligomer into the endosomal compartment, resulting in considerable "degradation stress". During incubation of cultured cells with SODNs, the steady state of uptake of oligomers and active transport out of the cytosol is reached within 6 h [373]. Rapid accumulation of the oligomers in the nucleus is achieved by direct injection of oligonucleotides into the cytosol [377].

For apolar modifications such as methylphosphonates, cellular uptake by passive diffusion through the cellular membrane has been proposed [378]. However, it has been shown that transport through phospholipid membranes proceeds slowly (efflux $t_{1/2} > 4$ days) and does differ only slightly from the diffusion of unmodified ODNs [367]. Cellular uptake of methylphosphonate oligomers labeled with fluorescent dyes was found to depend on temperature. The oligomers were found in endosomal compartments within the cells [379].

9.17.2 Oligonucleotide Ligands for Improved Drug Delivery

To overcome low cellular uptake of oligonucleotides, several delivery systems have been developed. Comprehensive summaries are available [220, 380], but it should be mentioned that most of these techniques have shown to improve oligonucleotide uptake using in vitro assay systems. Several attempts have been made to attach *hydrophobic groups* to oligomers. Phosphoramidites **134**, **135** and **136** have been incorporated at the 5′-end of a phosphorothioate oligomer, but they had little influence on the observed denaturation temperatures of duplexes [381]. Phosphoramidite **137** has been used for modification of oligonucleotides targeted against a loop-forming site of the influenza A/PR8/34 viral RNA and suppressed the development of virus in Madin–Darby canine kidney cells [382]. Along similar lines D,L-α-tocopherol phosphoramidite **139** has been attached to oligonucleotides for in vitro HIV testing [383].

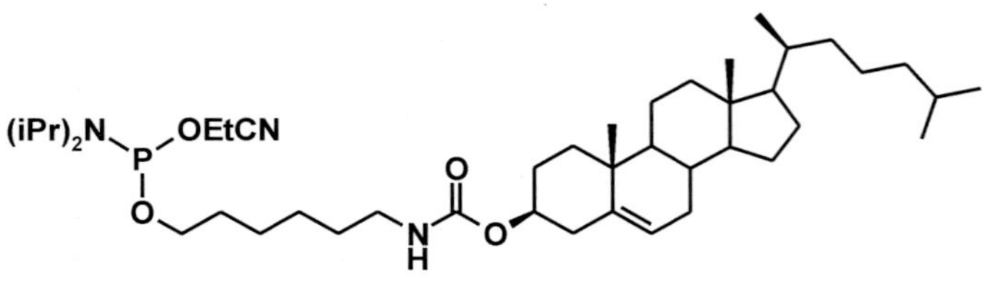

134

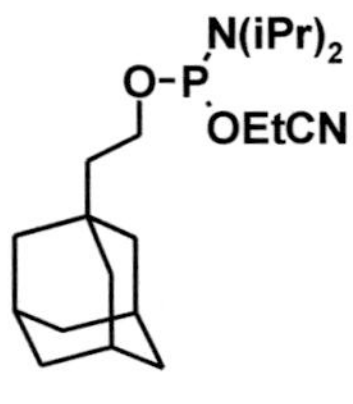

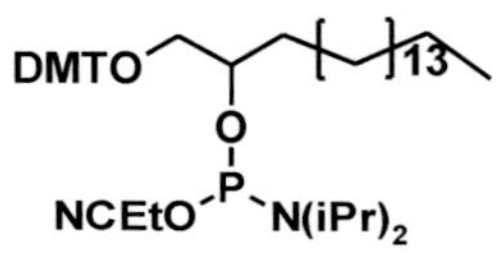

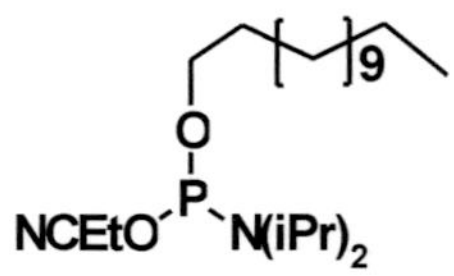

135 **136** **137**

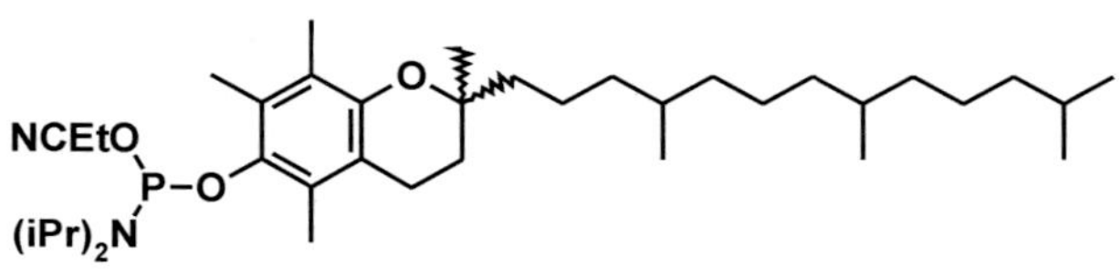

138

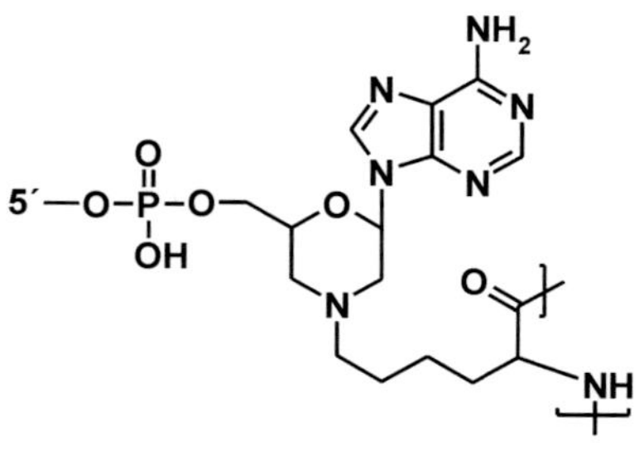

139

Modifications involving linkage of cholesterol have been the most prominent approach [135, 381, 384–388]. ODNs, which are linked to cholesterol at their 5′-end, exhibit increased nuclease stability. Their uptake into human cancer cells is increased by a factor of 30–100 when compared with natural ODNs [389]. In addition, they have been found to reduce

MDR 1 (multiple drug resistance) protein levels [390]. Incubating 5′-end cholesteryl-modified oligonucleotides with purified human LDL revealed co-operative uptake of oligonucleotides by LDL in a ratio 7.5:1. The oligonucleotides were not stripped from their binding site by heparin or suramin. The authors suggested that the cholesteryl moiety partitioned into the lipid fraction of the particle [391]. This increased uptake has been associated with binding to LDL or a LDL-receptor-mediated endocytosis [391]. However, the suitability of such conjugates for antisense therapy has been questioned [31]. Low solubility of cholesterol-modified phosphoramidite building blocks in acetonitrile has turned out to be a considerable problem, but may be overcome by adding 5% THF to the solvent. Most frequently, cholesterol is introduced with the CPG-starter-monomer at the 3′-end [392]. In the meantime, a variety of spacers is available commercially, including such cholesterol modifications. Cholic acid, also generally suited as lipophilic carrier, was converted into its N-hydroxysuccinimide ester or pentafluorophenylester and reacted with 5′- or 3′-amino-linkers of oligonucleotides. The compounds failed to show advantages in ICAM-1 assays, as long as no cationic lipids were added [181].

Polylysine-linked oligonucleotides may be synthesized by conjugation of one of the ε-amino group of polylysine (approx. 14 kDa) to a ribonucleotide at the 3′-end. Treatment with sodium periodate ($NaIO_4$) leads to oxidative C-C-cleavage of the ribose moiety. Reductive amination with sodium cyanoborohydride ($NaBH_4CN$) leads to the conjugated morpholino modification (**138**, [393]). Polylysine-modified natural ODNs are resistant to 3′-endonucleases, and endocytosis is also enhanced. Within the cell, ODNs are released by proteolytic cleavage of the polylysine moiety by cellular proteases [220]. In cultured cells poly-L-lysine-conjugated oligomers are 10–15 times more potent than unmodified ODNs [394, 395]. Unfortunately, poly-L-lysine has been found to be a non-specific cytotoxin when applied to tissue cultures [396].

9.17.3 Liposomes

In terms of structural chemistry, liposomes are supramolecular aggregates. A drug will also associate with the liposome by via non-covalent binding. With modified and unmodified ODNs, liposomes have become a particularly promising approach, aided by the fact that specific ODN-modifications may also be used specifically to influence drug–liposome interaction. Thus, base-modified cholesterol derivatives have been developed to act as specific anchor points for ODNs in liposomal formulations [397–408]. Application of pH-sensitive liposomes made from DOPE/OA [409] or DOPE/OA/CHOL [410] has also been suggested. As a first important effect of liposome formulation, protection of ODNs from extracellular nucleases has been observed [411–413]. Use of such formulations has proven to be a valid method for cellular internalization in cultured cell lines, an effect which has been reviewed by a number of authors [414, 415].

The use of *cationic lipids*, which are particularly effective in the transfection of large DNA fragments and antisense oligonucleotides, has been reported and summarized [416–420]. Thus, an effect has been reported for a 21-mer antisense ODN at the 3′-*tat* splice acceptor site of HIV-1 using distearoylphosphatidylethanolamine [421], for a 15-mer complementary to the 5′-region including the AUG codon of the *env* gene of a retrovirus [422] and for SODNs directed to the c-*myc* sequence in hematopoietic cells in culture [423]. Very often, a combination of DOTMA (**140**) and phosphatidylethanolamine, has been used based on the pioneering work of Felgner in 1987 [417]. It has become well known under its commercial name, Lipofectin®. Using DOTMA, a 1000-fold increase in potency of antisense SODN directed to the methionine initiation codon of human ICAM-1 has been achieved [424]. Destabilization of endosome membranes has been proposed as a mechanism of the increased efficiency obtained by application of these formulations [425]. Cellular uptake of ODNs should be optimal if the ratio of positiv charges of the cationic lipid and the negative charges of the ODN is ten to ODN. However, unfortunately these lipids are concomitantly cell toxic [424].

As an example (among many others) for a new development, a formulation of the cationic lipid GS 2888 **141** and DOPE (Cytofectin®) has become commercially available [426]. This seems to be the only protocol to permit facilitated oligonucleotide delivery by cationic lipids in up to 50% serum content in assay systems [427]. In future – based on knowledge about lipoproteins – liposome formulations may also comprise proteins or other moieties, which are able to undergo recognition of specific cell-surface epitopes.

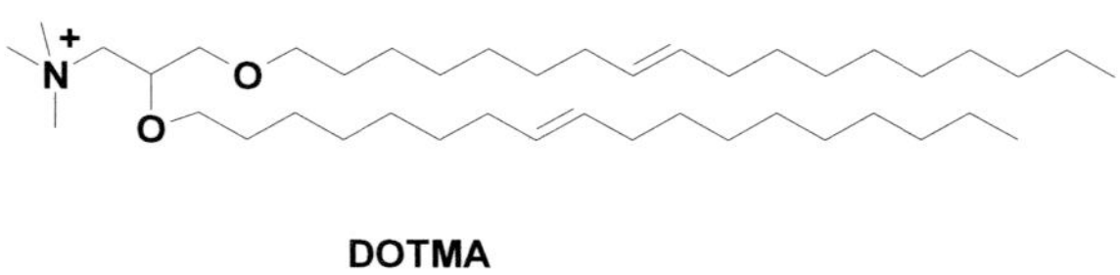

140

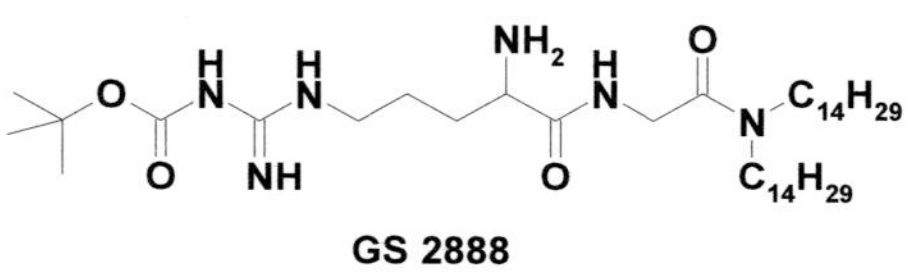

141

9.17.4 Nanoparticles

Nanoparticles are solid colloidal drug carriers that range in size from 10 to 1000 nm in diameter, exhibit characteristic features, and which have become an important tool in drug delivery. Their use as carriers for oligonucleotides has recently been reviewed [428]. Initial experiments included monitoring adsorption of ODN to cyanoacrylate nanoparticles [429, 430].

ODNs directed to a point mutation in codon 12 of the HA-*ras* mRNA adsorbed to such *nanoparticles* were injected into tumor tissue, and suppressed oncogene-mediated tumor proliferation at concentrations 100 times lower than the free oligonucleotides [435]. Because of the negative surface potential (ζ-potential) of polycyanoacrylate particles, a cationic copolymer such as DEAE-dextran (known as a transfection enhancer), or a hydrophobic cationic detergent such as CTAB, are usually used. In the meantime, base-modified nanoparticles of the methacrylate type have also been described [399–401, 404, 405, 407]. They have shown to increase plasma stability of ODNs and increase cellular uptake. Effects of toxicity were observed depending on the type of modification. As they are also used as car-

Table 14. Oligonucleotide nano- and microparticle preparations. Modified from [428]

Particle polymer	Antisense target	ODN structure	Reference
Gelatin	HLA-DRA, HIV 1 LTR	ssODN, dsODN	[431]
Methoxy poly(ethylene glycol)-poly(DL-lactic acid)	Ha-*ras* mRNA	ODN	[432]
Polyisobutylcyanoacrylate	$PdT_{[18]}$	ODN	[430]
	$PdT_{[16]}$	ODN	[433, 434]
Polyisohexylcyanoacrylate	$PdT_{[18]}$	ODN	[430]
	Ha-*ras*	ODN	[435]
	VSV	ODN	[436]
	Ha-*ras*	ODN linked to cholesterol	[437]
	$PdT_{[16]}$	ODN	[433]
Poly(D,L-lactic acid-co-glycolide)	HIV *tat*, c-*myc*	ODN, SODN, rib, hair	[438, 439]
	Rat tenascin mRNA	SODN	[440]
Polyhexylcyanoacrylate	HSV 1 IE 3,4	ODN, SODN	[398]

riers in gene transfer, nanoparticles have become an important target in the field of nucleic acid drug delivery. It may be expected that, in the near future, a series of different polymeric structures suited for formation of nanoparticles will be described.

9.17.5 Virus Capsids

Virus capsid-forming proteins have been established as particularly effective transfection systems in gene therapy, with adenovirus [441, 442] and polyoma virus-derived capsids [443] undergoing intensive study. It is evident that they also offer an option to be used as drug carriers for antisense oligonucleotides. Immune response, which is closely connected with application of this type of drug carrier, is one of the obstacles to be overcome.

9.18 Standards of Antisense Drug Development

Antisense research has been dominated by the hectic activities of biotech companies, who have each aimed to launch their drugs as early as possible. However, concern has been expressed about the significant lack of systematic studies in this field, as this might jeopardize the antisense concept as a whole (see section 21). This lack of systematic research refers a great deal also to data on the pharmacological properties of antisense oligonucleotides. As mentioned previously, the phosphorothioate oligonucleotide fomivirsen sodium, bearing the trade name Vitravene®* (ISIS Pharmaceutical), was approved by the FDA in August 1998. Vitravene®, used to treat the cytomegalovirus (CMV) retinitis often found in HIV patients, represents the first commercially available antisense oligonucleotide for drug therapy worldwide. The successful approval of this SODN drug has set at least the following standards, which must be met in future by every competitor:

* Vitravene is a registered trademark of the ISIS Novartis Pharm.

1. Synthesis and coupling of the oligomers must be efficient.
2. During application and distribution the oligomer drug and the encapsulated formulation, respectively, must be stable.
3. Non-specific toxicity must be minimized.
4. If SODNs are applied, the mode of application and therapeutic dosage must not cause activation of the complement, and dangerous accumulation in kidney, liver or spleen tissues must not occur in multiple-dose regimens of SODNs.
5. Degradation products must not show additional physiological effects. Specific attention must be attributed to degradation products of base-modified oligonucleotides.
6. Absorption and internalization of the oligomer or the encapsulated formulation into the cell must be efficient in order to achieve high intracellular concentrations. Affinity to cellular membranes must not prevent the compound from distributing freely into the cytosol once it has been liberated from the endosome.
7. The oligomer must bind selectively to the target mRNA or DNA.
8. Oligomers should catalyze depletion of target mRNA by cellular nucleases (e.g., RNase-H).

At the present stage, it seems difficult to attain all these requirements. At first glance, an increase in hydrophobicity and the choice of an apolar backbone seem to be valid strategies to increase the cellular uptake and stability of oligonucleotides. Methylphosphonates have been proved to use an alternative pathway in cellular uptake. Attempts have been made to compare the properties of modified oligonucleotides in view of their suitability for physiological studies [444]. To date, it is mostly SODNs that have been filed for clinical trials. From it can be seen that SODNs must be administered directly into the central compart-ment, a requirement that is due to their rapid degradation in the digestive system by 3´-exonucleases.

With respect to toxicity and side effects, only limited in vivo data are available. As most of the compounds in clinical testing are SODNs, it is clear that some of the non-specific toxicity data presented in Table 16 are closely related to this class of backbone-modified compounds.

Given intravenously they are rapidly bound to serum albumin and α-2-macro-globulin. Thus renal elimination is reduced and a half-life of 20–60 h is a chiered. Oral bioaviability does not esceed 5 % due to degradation, an exaption bling 2´-O-methylate chimeric ORN's. It can be seen from Fig. 41 that phosphorothioates doses which produce significant toxicity are considerably higher than those associated with pharmacological activity. Unwanted side effects that occur at dose levels of 0.006 to 6 mg kg^{-1} SODN are classified into three groups (Table 16) [446].

Table 15. SODN in clinical development and route of application [445]. Development of GEM 91 has been halted in clinical Phase II in 1997, due to decreased platelet count after 10 days of treatment.

Oligomer	Target-mRNA	Application
ISIS 2922 (Fomivirsen sodium)	CMV-retinitis	Intravitreal
ISIS 2302	ICAM (intracellular adhesion molecules)	Intravenous
ISIS 3521	PKC (protein kinase C)	Intravenous
ISIS 5132	c-*raf*	Intravenous
GEM 91	AIDS	Intravenous

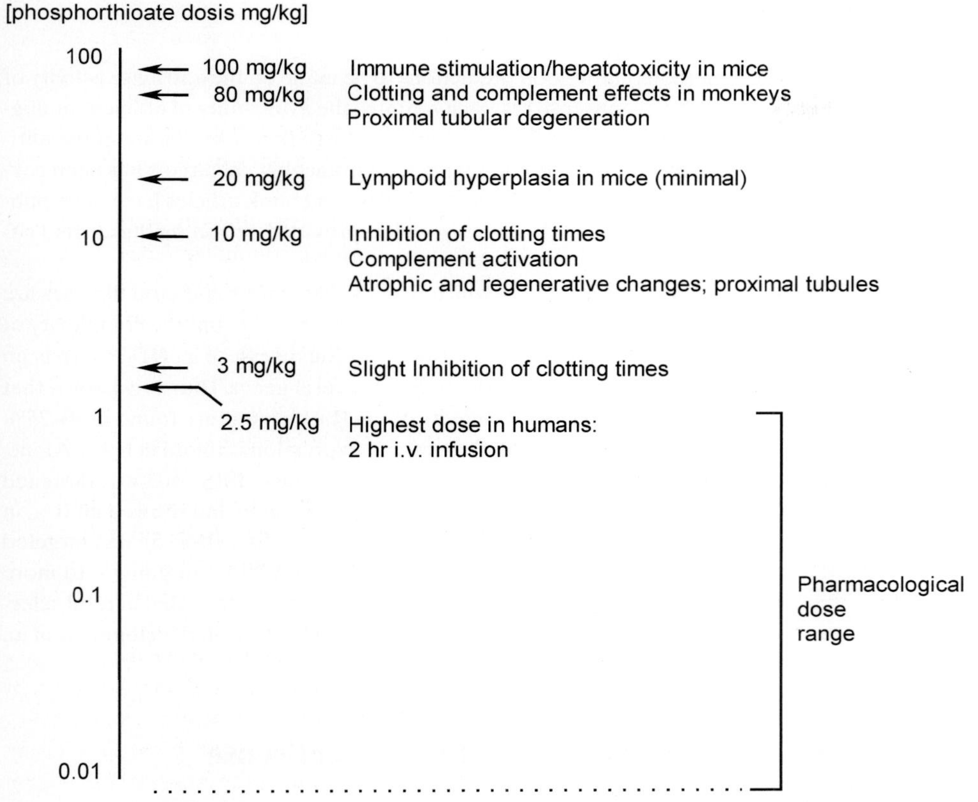

Figure 41. Dose-dependent toxicity of phosphorothioate antisense oligomers (Inhibitors of PKC-a (ISIS 3521/CGP 64126A) and C-*raf* kinase (ISIS 5132/CGP 69846A)). Modified from [446].

Table 16. Effects of high dose levels of phosphorothioate oligomers in animal testing.

Effects due to protein binding of phosphorothioates	Compared with ODNs, phosphorothioates display higher affinity to protein surfaces. • Binding sites on regulatory proteins of blood coagulation are occupied. (Increased clotting time can be treated with protamine sulfate injection) • Activation of complement system. Formation of C5a was observed during rapid injection of phosphorothioate oligomers in apes. Phosphorothioates therefore have to be administered at slow infusion rates.
Pharmacokinetic effects	Repeated dosage regimens lead to accumulation of oligomers in kidney due to active resorption of SODN. Very high concentrations are therefore found in kidneys.*
Sequence-dependent toxicity	Palindromic sequences and dinucleotide motifs containing CpG sequences in particular, have been shown to posses potent immunostimulatory properties in rodents. Similar effects do not seem to occur in primates.

* In multiple dosage regimes of 10 mg kg^{-1} and 80 mg kg^{-1} daily, concentrations of 1276 ± 292 μg g^{-1} and 3069 ± 1759 μg g^{-1} have been found in monkey kidneys [446].

9.19 Therapeutic Targets

Although biophysical methods allow good predictions to be made on the antisense activity of a compound, the complexity of antisense research requires the availability of efficient biological or pharmacological testing systems from the outset. Most articles appearing in the antisense field deal with such testing, but a broad scope of pharmacological targets has been covered in the meantime. A number of comprehensive reviews and book articles have been published; in the present article, only an outline survey will be provided, reflecting the current situation at the patent level.

In Fig. 42, a patent overview is presented, which reveals that cancer and viral diseases are still the dominating targets of antisense drug research. Taking into account the complexity of a gene structure, it is not surprising that even for one specific disease, e.g., HIV, there is an almost unlimited variety of different target sequences at several genes. Thus, it is known that mutations of the *ras*-gene, which lead to overactivity of the Ras-protein, are found in 10–25% of all human tumors. A survey on influencing Ras-protein expression is found in [447]. Alone, the ISIS company has taken several approaches to target this gene. ISIS 2503 was designed to hybridize to the H-*ras* mRNA in the initiation of translation region and showed an IC_{50} in the range of 45 nM when cationic lipids were added to cultured cells. ISIS 6957 was targeted to H-*ras* mRNA in the 5′-UTR and ISIS 9827 was active in both. Without going into more detail or discussing other gene targets, it should be mentioned at this stage that careful selection and choice of the gene-target site is a crucial step to be taken early in development of an antisense drug.

9.20 State and Tendencies of Industrial Antisense Drug Development

Basic patents covering most of the relevant oligonucleotide structures have been granted. As can be seen from Fig. 43, most of these patents were filed between in the first half of the 1990s, most of them claiming results of research on backbone and sugar modifications and conjugates. It is worth mentioning that, due to the oligomeric character of oligonucleotides

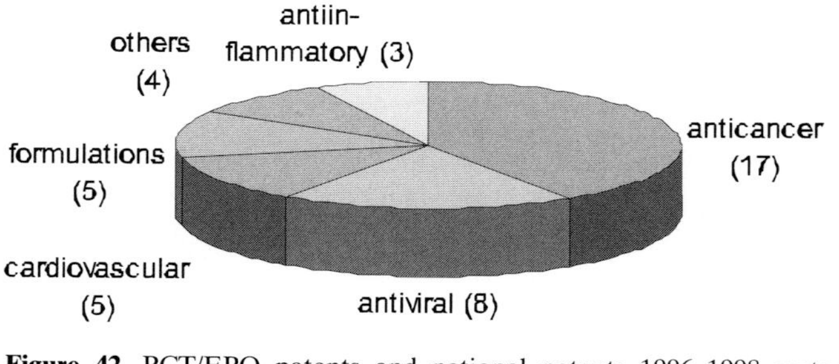

Figure 42. PCT/EPO patents and national patents 1996–1998 sorted by therapeutic applications. “Other” includes: Treatment of bacterial infections or wounds and oligonucleotides used as immune response modifiers.

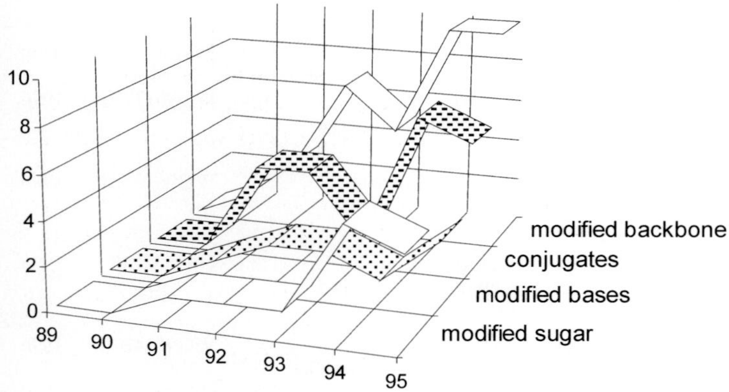

Figure 43. Development in antisense technology. Number of patents (PCT/EPO + US chemistry patents) in 1989-1995. Modified from [448].

and the broad scope of the applications, some of these patents cover billions of chemical compounds.

The update of patents, presented in Table 17, shows a still ongoing research activity that is aimed at novel antisense structures.

Of particular worth in mentioning is the ongoing activity of patenting in the field of catalytically active chimeric oligomers. Section 16.4 and Table 18 indicate the potential of this approach.

Table 17. PCT/EPO-patents related to medicinal chemistry of antisense oligonucleotides from 1996 to December 1998. (a) US-Patents of particular interest

Assignee (Licensee)	Type/Description	Patent	Year
Imanishi, Takeshi; Obika, Satoshi	Sugar modification Bridged artificial riboses, 3´,5´-linkage	WO9839352	1998
Imanishi, Takeshi	Sugar modification Bridged artificial riboses, 2´,5´-linkages	WO9822489	1998
Lynx Therapeutics Inc.	Sugar modification 2´-Fluoronucleotides (a)	US5684143	1997
Hoffmann-La Roche	Sugar modification 2´,6´-Amide-linked hexopyranosyl nucleotides	WO9714709	1997
Nippon Shinyaku Co., Ltd.,	Sugar modification 1´-Methyl riboses	WO9618640	1996
ICN Pharmaceuticals	Sugar modification	WO9614329	1996
Van Gelder, Russell N.; Von Zastrow, Mark E.; Barchas, Jack D.; Eberwine, James H.	Oligomer synthesis Antisense RNA production (a)	US5545522	1996

Table 17. Continued

Assignee (Licensee)	Type/Description	Patent	Year
Chiron Corporation	Oligonucleotide synthesis Screening method for antisense oligonucleotides development	WO9617955	1996
Hybridon Inc.	Oligonucleotide synthesis Synthesis of chiral phosphorothioates	WO9619572	1996
Dyad Pharmaceutical Corp.	Backbone modification 2′, 5′-Phosphate diesters bind to RNA (a)	US5532130	1996
Hybridon Inc.	Backbone modification Synthesis of radioactively labeled oligonucleotides	WO9614277	1996
Ciba-Geigy A.-G.	Conjugates Europium terpyridine complex for catalytic cleavage of nucleic acids	WO9607667	1996
Emory University, USA	Base, sugar, backbone modification Usage of carborane oligonucleotides in anticancer treatment	WO9614073	1996
Jewish Hospital of St. Louis, USA / Dartmouth College	Neurochemistry Regulation of expression of Ca^{2+}-channel	WO9613269	1995
Amgen Inc.	Backbone modification 5′-Dithiomodified oligonucleotides	WO9604295	1996
Lynx Therapeutics, Inc.	Oligonucleotide synthesis Oligonucleotide tags to identify synthesis products	WO9612039	1996
Hybridon, Inc.	Sugar modification 2′-O-Methyl-modified oligonucleotides	WO9612497	1996
Chiron Corporation	Base modification Non Watson–Crick paring oligonucleotides used in PCR	WO9606950	1996
Genta Inc.	Conjugates Oligonucleotides to cleave 5′-cap of complementary strands	WO9606621	1996
Stichting Rega VZW, Belg.	Sugar modifications 1,5-anhydrohexitol nucleosides linked via phosphodiester bridges	WO9605213	1996
Altman, Sidney; Fiedler, Paul; Levine, Robert A.; Wardlaw, Stephen C. USA	Conjugates Intercalating functional groups attached to oligonucleotides	WO9604788	1996

Table 18. Oligonucleotide-motif for chimeric oligonucleotides and related patents

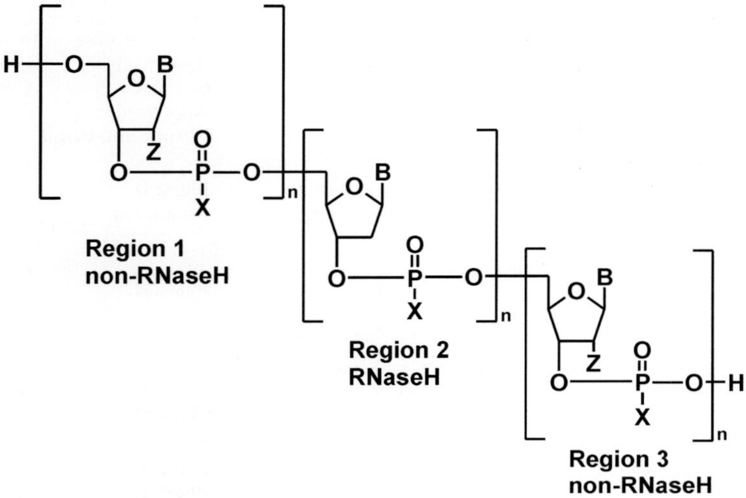

Assignee (Licensee)	Region 1&2		Region 2	Reference
	X =	Z =	X =	
Ajinimoto/ISIS	O	O-alkyl	O	[449]
Worcester (Hybridon)	Alkyl, NR_1R_2	H	S	[450]
Worcester (Hybridon)	Alkyl, NR_1R_2	H	O,S	[451]
Gilead	NR_1R_2	H	O,S	[452]
University of Iowa	S, NR_1R_2 O-alkyl, alkyl	H O	O	[453]
ISIS	S S S S Non-phosphorus linked	O-alkyl F α-sugars 2´,5´-linked	S S S S S	[454]
Worcester (Hybridon)	S	OH, O-alkyl	S	[455]
Genta	Chiral alkyl	H, O-alkyl	O, S	[456]
ISIS	Peptide nucleic acids		S	[457]

An increased tendency towards patents claiming antisense oligonucleotide application with specific mRNA-targets may be expected in future. According to expert opinion, such patents have a chance to be valid even in the case of previously patented cDNA-sequences. In this context, antisense-sequences and DNA-sequences may not be seen to be identical [448].

As shown in Table 19, there is at present a considerable number of antisense drugs in clinical development. The success of these drug candidates will certainly have a significant im-

Table 19. Antisense oligonucleotides in clinical trials (December 1998)

Product	Sequenz	Disease/Target	Status	Company
Fomivirsen sodium[1] Vitravene®	GCGTTTGCTCTTCTTC TTGCG	CMV retinitis, IE-2	Approved by FDA Approved in Europe and Brasil	ISIS/ Ciba-Vision
ISIS 2303[1]	GCCCAAGCTGGCATC CGTCA	ICAM-11,3´-UTR Crohn's disease, Kidney transplant rejection	 Phase II completed	ISIS[5]
ISI 3521[1]	GTTCTCGCTGGTGAG TTTCA	PKC-α, 3´-UTR NSCL cancer Breast cancer Ovarian cancer	 Phase II Phase II Phase II	ISIS/Novartis
ISIS 5132[1]	TCCCGCCTGTGACAT GCATT	*c-raf* Kinase Breast cancer Ovarian cancer	 Phase II Phase II	ISIS/Novartis
ISI 2503[1]	TCCGTCATCGCTCCTC AGGG	HA-*ras* oncogen All cancer Pancreatic cancer Colorectal cancer NSCL cancer	 Phase I Phase II Phase II Phase II	ISIS
ISIS 13312[2]	GC(ps)GTTTGC(ps)TC (ps)TTC(ps)TTC(ps)TTG CG	CMV retinitis, IE-2	Phase I–II	ISIS[5]
GEM 231[2]	GCGUGUCTCCTCACU GGC	RIa subunit of (cancer)	Phase II	Hybridon[2]
GEM 132[2]	UGGGGCTTACCTTGC GAACA	Cytomegalievirus (in AIDS patients)	Ohase I–II	Hybridon[2]
GEM 921[2]	UCGCACCCATCTCTC TCCUUC			Hybridon[2]
LR3280[1]	AACGTTGAGGGCAT	Stent Restenosis, c-myc protooncogene	Phase I	Lynx
LR4437[1]	GGACCCTCCTCCGGA GCC	Cancer, IGF-IR	Phase I	Lynx
LR3001[1]	TATGCTGTGCCGGGG TCTTCGGGC	Leukemia, proto-oncogene	Phase II	Lynx
GPI-2A[3]	G(ps)GTTC(ps)TTTT(ps) G(ps)TCC(ps)TTG(ps)TC (ps)T	AIDS, HIV-1 Gag	Phase I	Novopharm
G3139[1]	TCTCCCAGCGTGCGC CAT	Prostate cancer, bcl-1 protooncogene	Phase I/IIa	Genta

1: Phosphorothioate oligomers, **2**: Hybridon has licensed certain antisense compounds to OriGenix Technologies in Jan. 1999. The underline bases in the phosphororthioate oligomers GEM 92, GEM 132 and GEM 231 are 2´-O-methyl ribose derivatives. **3**: GPI-2A contains 7-phosphorothioate linkages, the rest of the oligomer are phosphordiesters. **4**: In ISIS 13312, the underlined bases are 2´-O-(methoxyethyl) sugar modifications and all U and C residues are 5-methyl derivatives. The oligomers contain 5 phosphorothioate linkages, the rest of the oligomer are phosphordiesters. **5**: Rights reacquired from Boehringer Ingelheim in Sep. 1999.

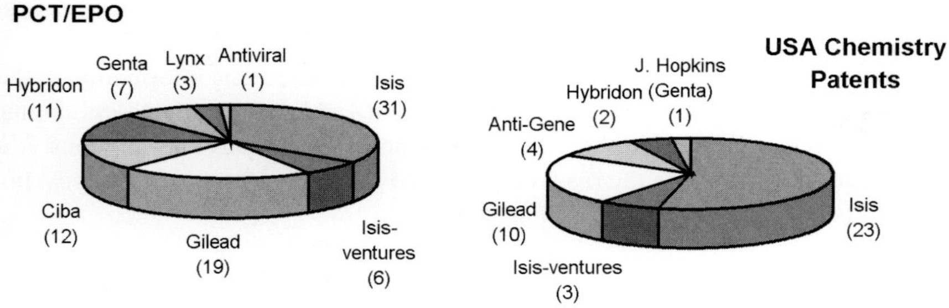

Figure 44. Licensees and PCT/EPO in the field of antisense-oligonucleotides; Licensees and US Chemistry Patents. Modified from [448] (1996).

pact on the future development of this field of therapeutics. It can be also seen that it is a characteristic feature of industrial antisense research that most of the projects are carried out by specialized companies, frequently in co-operation with large multinational companies..

This concentration on a small number of specialized companies is even more striking when looking at the situation concerning patent applications for base-, backbone-, and sugar-modifications of oligonucleotides as shown in Fig. 44. It should be mentioned, that according to a press release, ISIS Pharmaceuticals has purchased rights from Gilead Science Inc. antisense patents in December 1998. The patent of the J. Hopkins-Universität covers the methylphosphonates discovered by Ts'o and Miller. Further patents based on this technology are held by the company Genta.

9.21 Concluding Remarks

It can be seen from the previous sections that antisense drug development is a very "young" approach in drug research. Looking at published work – and particularly at patents – it may be stated that it took only about 10 years to promote this specific case of gene therapy from theory to the first filed drug. It is evident that antisense drugs of the first generation are by far not perfect. Many aspects of their pharmacokinetics or pharmacodynamics, which are trivial elements in established types of drugs, are not even understood [458]. It can be seen from this article that much of the vast amount of work in this field has been basic, and not related directly to drug research. Indeed, some of this work might even impact on antisense drug development in future. At the present stage, the task of checking the field with respect to the relevance of antisense drugs in medicinal chemistry is still challenging, and leaves "open space" for expected future development lines. At the same time, there is some danger not to consider duly aspects that are important in the near future. Thus, 2026 reports of "successful" antisense inhibition have been summarized. In more than 80% of these articles testing of a single oligomer was described [459]. This might be due to the fact, that as a rule of thumb only one in eight oligomers subjected to in vitro testing exhibits satisfying inhibition of translation [460, 461]. Thus, it has been hypothesized that only 12.5% of the primary experimental data have been published, and that certain sequence motifs such as G-quartets, that are known to display various cellular effects independently of their flanking sequences [462, 463], are over-represented in this subset [464–466].

Within this chapter, the authors have striven to obtain and validate the information in this field, concentrating on the chemistry of antisense compounds and giving more or less emphasis to different aspects according to their own experience, expectations and opinions. It is their hope that they may stimulate the readers' interested in one or the other of the most fascinating lines of research in this new field of drug therapy.

References

[1] Miescher, F, Hoppe-Seyler's Z. Physiol. Chem., 1871: p. 441–460.
[2] Kossel, A, Z. Physiol. Chem., 1883. **7**: p. 7–22.
[3] Kossel, A, Neumann,,A, Chem. Ber., 1893. **26**: p. 2753–2756.
[4] Kossel, A, Steudel, H, Z. Physiol. Chem., 1902/1903. **37**: p. 177–180.
[5] Kossel, A, Steudel, H, Z. Physiol. Chem., 1903. **38**: p. 49–59.
[6] Ascoli, A, Z. Physiol. Chem., 1900/1901: p. 161–164.
[7] Levene, P A, Jacobs, W A, Chem. Ber., 1909: p. 2102–2107.
[8] Levene, P A, Jacobs, W A, Chem. Ber., 1909: p. 2469–2473.
[9] Levene, P A, Jacobs, W A, Chem. Ber., 1909: p. 2474–2479.
[10] Levene, P A, Jacobs, W A, Chem. Ber., 1909: p. 2703–2707.
[11] Levene, P A, Mikeska, L A, Mori, T, J. Biological Chem., 1930. **85**: p. 785–787.
[12] Avery, O T, MacLeod, C M, McCathy, M, J. Exp. Med., 1944. **79**: p. 137–159.
[13] Watson, J D, Crick, F H C, Nature (London), 1953. **171**: p. 737–738.
[14] Franklin, R E, Gosling, R G, Nature (London), 1953. **171**: p. 740–741.
[15] Stokes, A R, Wilson, H R, Nature (London), 1953. **171**: p. 738–740.
[16] Belikova, A M, Zarytova, V F, Grineva, N I, Tetrahedron Lett., 1967. **37**: p. 3557–3562.
[17] Galileo, D S, Majors, J, Horwitz, A F, Sanes, J R, Neuron, 1992. **9**: p. 1117–1131.
[18] Colman, A J, Cell Sci., 1990. **97**: p. 399–409.
[19] Zamecnik, P C, Stephenson, M L, Proc. Natl. Acad. Sci. U.S.A., 1978. **75**: p. 280–284.
[20] Kruger, K, Grabowski, P J, Zaug, J A, Sands, J, Gottschling, D E, Cech, T R, Cell, 1982. **31**: p. 147–157.
[21] Bartel, D P, Szostak, J W, Science, 1993. **261**: p. 1411–1418.
[22] Haselhoff, J, Gerlach, W L, Nature (London), 1988. **334**: p. 585–591.
[23] Eschenmoser, A, Bolli, M, Micura, R, Pitsch, S, Helv. Chim. Acta, 1997. **80**: p. 1901.
[24] Egli, M, Angew. Chem., Int. Ed. Engl., 1996. **35**(17): p. 1895–1909.
[25] Saenger, W, *Principles of Nucleic Acid Structure*. 1984, New York: Springer Verlag. 556.
[26] Guschelbauer, W, Janowski, K, Nucleic Acids Res., 1980. **8**: p. 421–33.
[27] Helene, C, Toulme, J, Biochim. Biophys. Acta, 1990. **1049**: p. 99–125.
[28] Marcus-Sekura, C J, Anal. Biochem., 1988. **172**: p. 289–295.
[29] Andrea, D, Hepatology, 1996. **24**(6): p. 1517–1529.
[30] Nguyen, T T, Helene, C, Angew. Chem., 1993. **105**: p. 697–723.
[31] Boiziau, C, Toulme, J J, Biochimie, 1991. **73**: p. 1403–1408.
[32] Dolnick, B J, Pharmacol. Ther., 1997. **75**(3): p. 179–184.
[33] Usman, N, Ogilvie, K K, J. Am. Chem. Soc., 1987: p. 7845–7854.
[34] Beaucage, S L, Iyer, R P, Tetrahedron, 1992. **48**(12): p. 2223–2311.
[35] Ti, G S, Gaffney, B L, Jones, R A, J. Am. Chem. Soc., 1982. **104**: p. 1316–1319.
[36] Noe, C R, Windhab, N, Haberhauer, G, Arch. Pharm. (Weinheim, Ger.), 1995. **328**: p. 743–744.
[37] Scott, S, Hardy, P, Sheppard, R C, McLean, M J.. in *Innovations and Perspectives in Solid Phase Synthesis, 3rd International Symposium*. 1994: Mayflower Worldwide.
[38] Dahl, O, Nielsen, J, Dahl, B H, Nucleic Acids Res., 1987(15): p. 1729–1742.
[39] Berner, S, Muhlegger, K, Seliger, H, Nucleic Acids Res., 1989. **17**: p. 853–864.
[40] *DCI–A LOGICAL ALTERNATIVE ACTIVATOR*, in *The Glen Report*. 1997.
[41] Sproat, B S, Lamond, A I, Beijer, B, Neuner, P, Ryder, U, Nucleic Acid Res., 1989. **17**: p. 3373.
[42] Pon, R T, Tetrahedron Lett., 1987. **28**(32): p. 3643–3646.
[43] Sproat, B, et. al., Nucleosides & Nucleotides, 1995. **14**: p. 255–273.

[44] Tsou, D, Hampel, A, Vinayak, R, Nucleosides & Nucleotides, 1995. **14**: p. 1481–1492.
[45] Winscott, F, et al., Nucleic Acids Res., 1995. **23**: p. 2677–2684.
[46] Krotz, A H, Klopchin, P G, Walker, K L, Srivatsa, G S, Cole, D L, Ravikumar, V T, Tetrahedron Lett., 1997. **38**: p. 3875–3878.
[47] Vargeese, C, et al., Nucleic Acids Res., 1998. **26**: p. 1046–1050.
[48] Reynolds, T R, Buck, G A, Biotechniques, 1992. **12**: p. 518.
[49] Reddy, M P, Hanna, N, Farooqui, F, Tetrahedron Lett., 1994. **35**: p. 4311.
[50] Sinha, N D, Davis, P, Usman, N, Perez, J, Hodge, R, Kremsky, J, Casale, R, Biochimie, 1993. **75**: p. 13–23.
[51] Kuijpers, W H A, Huskens, J, Van Boeckel, C A A, Tetrahedron Lett., 1990. **31**: p. 6729–6732.
[52] Kuijpers, W H A, Kuyl-Yeheskiely, E, Van Boom, J H, Van Boeckel, C A A, Nucleic Acids Res., 1993. **21**: p. 3493–3500.
[53] McBride, L J, Kierzek, R, Beucage, S L, Caruthers, M H, J. Am. Chem. Soc., 1986. **108**: p. 2040.
[54] Vu, H, McCollum, C, Jacobson, K, Theisen, P, Vinayak, R, Spiess, E, Andrus, A, Tetrahedron Lett., 1990. **31**: p. 7269–7272.
[55] Toce, J A. *Large-Scale Isolation of Deoxynucleosides: Scaling Up to Ton Quantities*. in *Oligonucleotide Synthesis*. 1998. The Fairmont Hotel, San Francisco, California: Camebridge Healthtech Institute.
[56] Durand, M, Peloille, S, Thuong, N T, Maurizot, J C, Biochemistry, 1992. **31**: p. 9197–9204.
[57] Sande, J H v d, Ramsing, N B, Germann, M W, Elhorst, W, Kalisch, B W, Kitzing, E V, Pon, R T, Clegg, R C, Jovin, T M, Science, 1988. **241**: p. 551–557.
[58] Akhtar, S, Kole, R, Juliano, R L, Life Sci., 1991. **49**: p. 1793–1801.
[59] De Mesmaeker, A, Haener, R, Martin, P, Moser, H E, Acc. Chem. Res., 1995. **28**(9): p. 366–374.
[60] Matsukura, M, Shinozuka, K, Zon, G, Mitsuya, H, Reitz, M, Cohen, J S, Broder, S, Proc. Natl. Acad. Sci. U.S.A., 1987. **84**: p. 7706–7710.
[61] Marshall, W S, Caruthers, M H, Science, 1993. **259**: p. 1564–1570.
[62] Miller, P S, Yano, J, Yano, E, Carroll, C, Jayaraman, K, Ts ´o, P O P, Biochemistry, 1979. **18**: p. 5134–5143.
[63] Miller, P S, Fang, K N, Kondo, N S, Ts'o, P O P, J. Am. Chem. Soc., 1971. **93**: p. 6657–6665.
[64] Letsinger, R L, Singman, C N, Histand, G, Salunkhe, M, J. Am. Chem. Soc., 1988. **110**.
[65] Gryaznov, S, Chen, J K, J. Am. Chem. Soc., 1994. **116**: p. 3143–3144.
[66] Sergueev, e a, Nucleosides & Nucleotides, 1997. **16**(7–9): p. 1538–1553.
[67] Stein, C A, Subasinghe, C, Shinokuza, K, Cohen, J S, Nucleic Acids Res., 1988. **16**: p. 3209–3221.
[68] Cazenave, C, Stein, C A, Loreau, N, Thuong, N T, Neckers, L M, Subasinghe, C, Helene, C, Cohen, J S, Toulme, J J, Nucleic Acids Res., 1989. **17**: p. 4255–4273.
[69] Storey, A, Oates, D, Banks, L, Crawford, L, Crook, T, Nucleic Acids Res., 1991. **20**.
[70] Ho, P T C, Ishiguro, K, Wickstrom, E, Sartorelli, A C, Nucleic Acids Res., 1991. **1**: p. 329–342.
[71] Vu, H, Hirschbein, B L, Tetrahedron Lett., 1991. **32**(26): p. 3005–3008.
[72] Vama Rao, M, Reese, C B, Zhengyun, Z, Tetrahedron Lett., 1992. **33**: p. 4839.
[73] Stec, W J, Uznanski, B, Wilk, A, Hirschbein, B L, Fearon, K L, Bergot, B J, Tetrahedron Lett., 1993. **34**(33): p. 5317–5320.
[74] Beaucage, e a, J. Org. Chem., 1990. **55**: p. 4693–4699.
[75] Matsukura, M, Zon, G, Shinozuka, K, Stein, C A, Mitsuya, H, Cohen, J S, Broder, S, Gene, 1988. **12**: p. 343.
[76] Varma, R S, Synlett, 1993: p. 621–637.
[77] Fuji, M, Ozaki, K, Kume, A, Sekine, M, Hata, T, Tetrahedron Lett., 1986. **26**: p. 935.
[78] Cosstick, R, Williams, D M, Nucleic Acids Res., 1987. **15**: p. 992.
[79] Stec, e a, J. Am. Chem. Soc., 1992. **114**: p. 10197–10202.
[80] Okruszek, A, Biotechnologia, 1994(4): p. 16–39.
[81] Beaton, G, Dellinge, D, Marshall, W S, Caruthers, M H, *Synthesis of oligonucleotide phosphorothioates*, in *Oligonucleotides and Analogues: A Practical Approach*, F. Eckstein, Editor. 1991, IRL press: Oxford. p. 109–136.
[82] Caruthers, M H, Marshall, W S, Science, 1993. **259**: p. 1564.
[83] Uhlmann, E, Peyman, A, Chem. Rev. (Washington, D. C.), 1990. **90**: p. 544–584.
[84] Miller, P S, Cushman, C D, Lewis, J I, *Synthesis of oligo-2´-deoxyribonucleoside methylphosphonates*, in *Oligonucleotides and Analogues: A Practical Approach*, F. Eckstein, Editor. 1991, IRL press: Oxford. p. 137–154.
[85] Ts'o, P O P, Antisense Res. Dev., 1991. **1**: p. 273–276.
[86] Agrawal, S, Goodchild, J, Tetrahedron Lett., 1987. **28**: p. 3539.
[87] Loschner, T, Engels, J W, Nucleosides & Nucleotides, 1988. **7**: p. 729.

[88] Hogrefe, R I, Vaghefi, M M, Reynolds, M A, Young, K M, Arnold, L J, Nucleic Acids Res., 1993. **21**: p. 2031–2038.
[89] Samstag, W, Engels, J W, Angew. Chem., Int. Ed. Engl., 1992. **31**: p. 1386.
[90] Stec, W J, Uznanski, B, Wilk, W A, Tetrahedron Lett., 1987. **28**: p. 3401.
[91] Jäger, A, Levy, M J, Hecht, S M, Biochemistry, 1988. **27**: p. 7237–7246.
[92] Appel, R, Angew. Chem., Int. Ed. Engl., 1975. **14**: p. 801–811.
[93] Stec, W J, Acc. Chem. Res., 1983. **16**: p. 411–411.
[94] Froehler, B, Ng, P, Matteucci, M, Nucleic Acids Res., 1988. **16**: p. 4831.
[95] Letsinger, R L, Singman, C N, Histand, G, Salunkhe, M, J. Am. Chem. Soc., 1988. **110**: p. 4470.
[96] Schinazi, R F, Lesnikowski, Z J, Nucleosides & Nucleotides, 1998. **17**(1–3): p. 635–647.
[97] Li, H, Porter, K, Huang, F, Shaw, B R, Nucleic Acids Res., 1995. **23**: p. 4495.
[98] Nakamaye, K L, Gish, G, Eckstein, F, Vosberg, H-P, Nucleic Acids Res., 1988. **16**: p. 9947.
[99] Olsen, D B, Eckstein, F, Nucleic Acids Res., 1989. **17**: p. 9613.
[100] De Mesmaeker, A, Waldner, A, Sanghvi, Y S, Lebreton, J, Bioorg. Med. Chem. Lett., 1994. **4**: p. 395–398.
[101] Teng, K, Cook, P D, J. Org. Chem., 1994. **59**: p. 278–280.
[102] De Mesmaeker, A, Waldner, A, Fritsch, V, Romain, M, Eur. J. Med. Chem., 1995. **30**(Suppl., Proceedings of the 13th International Symposium on Medical Chemistry): p. 479s–484s.
[103] Lebreton, J, De Mesmaeker, A, Waldner, A, Synlett, 1994: p. 54–56.
[104] Jones, A S, MacCoss, M, Walker, R T, Biochim. Biophys. Acta, 1973. **365**: p. 365.
[105] Bleaney, R C, Jones, A S, Walker, R T, Nucleic Acids Res., 1975. **2**: p. 699.
[106] Matteucci, M D, Xiaodong, C, Tetrahedron Lett., 1994. **35**(15): p. 2325–2328.
[107] Jones, R J, Lin, K Y, Milligan, J F, Wadwani, S, Matteucci, M D, J. Org. Chem., 1993. **58**: p. 2983–2991.
[108] Sanghvi, Y S, Cook, P D, *Towards Second-Generation Synthetic Backbones for Antisense Oligonucleosides*, in *Nucleosides and Nucleotides as Antitumor and Antiviral Agents*, C.K. Chu, Baker, D. C:, Editor. 1993, Plenum Press: New York. p. 311–23.
[109] Mertes, M P, Coats, E A, J. Med. Chem., 1969. **12**: p. 154.
[110] Tittensor, J R, J. Chem. Soc. (C), 1971: p. 2656.
[111] Ogilvie, K K, Cormier, J F, Tetrahedron Lett., 1985. **26**: p. 4159.
[112] Cormier, J F, Ogilvie, K K, Nucleic Acids Res., 1988. **16**: p. 4583.
[113] Seliger, H, Feger, G, Nucleosides & Nucleotides, 1987. **6**: p. 483.
[114] De Mesmaeker, A, Lebreton, J, Waldner, A, Fritsch, V, Wolf, R M, Freier, S M, Synlett, 1993: p. 733–736.
[115] Gait, M J, Jones, A S, Walker, R T, J. Chem. Soc., Perkin Trans. I, 1974: p. 1684.
[116] Gait, M J, Jones, N D, Shephard, M J, Walker, R T, J. Chem. Soc., Perkin Trans. 1, 1979: p. 1684.
[117] De Mesmaeker, A, Lebreton, J, Waldner, A, Fritsch, V, Wolf, R M, Freier, S M, Angew. Chem. Int. Ed. Engl., 1994. **33**: p. 226.
[118] Lebreton, J, Waldner, A, Lesueur, C, De Mesmaeker, A, Synlett, 1994: p. 137.
[119] Lebreton, J, Waldner, A, Fritsch, V, Wolf, R M, De Mesmaeker, A, Tetrahedron Lett., 1994. **35**: p. 5225.
[120] Sanghvi, Y S, Swayze, E E,Peoc'h, D, Balkrishen, B, Stuart, D, Nucleosides & Nucleotides, 1997. **16**(7–9): p. 907–916.
[121] Vasseur, J-J, Debart, F, Sanghvi, Y S, Cook, P D, Tetrahedron Lett., 1992. **33**: p. 2645.
[122] Vasseur, J-J, Debart, F, Sanghvi, Y S, Cook, P D, J. Am. Chem. Soc., 1992. **114**: p. 4006–4007.
[123] De Mesmaeker, A, Lebreton, J, Waldner, A, Fritsch, V, Wolf, R M, Bioorg. Med. Chem. Lett., 1994. **4**: p. 873.
[124] Lebreton, J, De Mesmaeker, A, Waldner, A, Fritsch, V, Wolf, R M, Freier, S M, Tetrahedron Lett., 1993. **34**: p. 6383–6386.
[125] Waldner, A, De Mesmaker, A, Lebreton, J, Synlett, 1994: p. 57–61.
[126] Waldner, A, De Mesmeaker, A, Lebreton, J, Bioorg. Med. Chem. Lett., 1994. **4**: p. 405–408.
[127] Varma, R S, Singh, A K, Tetrahedron Lett., 1992. **33**: p. 2307.
[128] Pannecouque, C, Wigerinck, P, Van Aerschot, A, Herdewijn, P, Tetrahedron Lett., 1992. **33**: p. 7609.
[129] Dempcy, R O, Browne, K A, Bruice, T C, J. Am. Chem. Soc., 1995. **117**: p. 6140–6141.
[130] Mungall, W S, Kaiser, J K, J. Org. Chem., 1977. **42**: p. 703.
[131] Cull, J M, Carlson, D V, Weith, M L, Tetrahedron Lett., 1987. **28**: p. 745.
[132] Haly, B, Bellon, L, Mohan, V, Sangvhi, Y, Nucleosides & Nucleotides, 1996. **15**: p. 1383–1395.
[133] Kawai, S H, Wang, D, Giannaris, P A, Damha, M T, Just, G, Nucleic Acids Res., 1993. **21**: p. 1473.

[134] Reynolds, R C, Crooks, P A, Maddry, J A, Akhtar, M S, Montgomery, J A, Secrist, J A, J. Org. Chem., 1992. **57**: p. 2983.

[135] Birnstiel, M, Brunar, H, Cotten, M, Holzner, A, Issakides, G, Knollmüller, M, Noe, C R, *2´-O-Alkyl-oligoribonucleotide, Verfahren zu deren Herstellung und deren Verwendung als Antisense-Oligonucleotide*, in *Dtsch Pat. Anm. (Chem. Abstr. 1990, 114, 229211)*. 1992: Germany.

[136] Maden, B E, Corbett, M E, Heeney, P A, Pugh, K, Ajuh, P M, Biochimie, 1995. **77**(1–2): p. 22–29.

[137] Schmidt, C, Bevierre, M-O, De Mesmaeker, A, Altmann, K-H, Bioorg. Med. Chem. Lett., 1994. **4**: p. 1967–1974.

[138] Lesnik, E A, Guinosso, C J, Kawasaki, A M, Sasmor, H, Zounes, M, Cummins, L L, Ecker, D J, Cook, P D, Freier, S M, Biochemistry, 1993(32): p. 7832–7838.

[139] Lamond, A I, Sproat, B S, FEBS Lett., 1993. **325**(1,2): p. 123–127.

[140] Inoue, H, Hayase, Y, Asaka, M, Imura, A, Iwai, S, Miura, K, Ohtsuka, E, Nucleic Acids Res. Symposium Series,. **No. 16**: p. 165–168.

[141] Sproat, B S, Lamond, A I, Beijer, B, Neuner, P, Ryder, U, Nucleic Acids Res., 1989. **17**: p. 3373–3386.

[142] Martin, P, Helv. Chim. Acta, 1995. **78**: p. 486–504.

[143] Holzner, A, *An der 2´-Position modifizierte Nucleoside: Herstellung und Anwendung in der Oligonucleotidsynthese*,. 1991, TU Vienna.

[144] Cotten, M, Oberhauser, B, Brunar, H, Holzner, A, Issakides, G, Noe, C R, Schaffner, G, Wagner, E, Birnstiel, M L, Nucleic Acids Res., 1991. **19**: p. 2629–2635.

[145] Wagner, E, Oberhauser, B, Holzner, A, Brunar, H, Issakides, G, Schaffner, G, Cotten, M, Knollmüller, M, Noe, C R, Nucleic Acids Res., 1991. **19**: p. 5965–5971.

[146] Iribarren, A M, Sproat, B S, Neuner, P, Silston, I, Ryder, U, Lamond, A I, Proc. Natl. Acad. Sci. U.S.A., 1990. **87**: p. 7747–7751.

[147] Sproat, B S, Iribarren, A M, Guimil Garcia, R, Beijer, B, Nucleic Acids Res., 1991. **19**: p. 733–738.

[148] Sproat, B S, Lamond, A I, Guimil Garcia, R, Bejier, B, Pieles, U, Douglas, M, Bohmann, K, Carmo-Fonseca, M, Weston, S, O`Loughlin, S, Nucleic Acids Res. Symposium Series, 1991. **No. 24**: p. 59–62.

[149] Werner, D, *Synthese von 2´-O-Alkyl- und Oxylethylen-Adenosin: Synthese von Oligonucleotiden und Untersuchung der Paarungseigenschaften*. 1997, Aachen: Shaker Verlag.

[150] Manoharan, M, Guinosso, C J, Cook, P D, Tetrahedron Lett., 1991. **32**: p. 7171–7174.

[151] Brunar, H, in *Inst. f. Org. Chemie*. 1993, TU: Vienna.

[152] Kawasaki, A M, Casper, M D, Freier, S M, Lesnik, E A, Zounes, M C, Cummins, L L, Gonzalez, C, Cook, P D, J. Med. Chem., 1993. **36**: p. 831–841.

[153] McGee, e a, J. Org. Chem., 1996. **61**: p. 781–785.

[154] Issakides, G, *Seltene Ribonucleotide als Bausteine für die Synthese von "Antisense-Ribonucleotides"*,. 1991, TU: Vienna.

[155] Wagner, D, Verheyden, J P H, Moffat, J G, J. Org. Chem., 1974. **39**: p. 24–30.

[156] Markiewicz, W T, J. Chem. Research (S), 1979: p. 24–25.

[157] Niedballa, U, Vorbrüggen, H, J. Org. Chem., 1974. **39**: p. 3654, 3660, 3664, 3668, 3672.

[158] Werner, D, Brunar, H, Haberhauer, G, Noe, C R. *2´-Modified Oligonucleotides: Synthesis and Biophysical Analysis*. in *2nd European Congress of Pharmaceutical Sciences*. 1994. Berlin.

[159] Werner, D, Brunar, H, Haberhauer, G, Noe, C R, Eur. J. Pharm. Sci., 1994. **2**: p. 117–194.

[160] Larrouy, B, Boiziau, C, Sproat, B, Toulme, J J, Nucleic Acids Res., 1995. **23**(17): p. 3434–3440.

[161] Wagner, R W, Flanagan, W M, Grant, D, Lin, K-Y, Matteucci, M D, Nature Biotech., 1999. **17**: p. 48–52.

[162] Boiziau, C, Larrouy, B, Sproat, B S, Toulme, J J, Nucleic Acids Res., 1995. **23**(1): p. 64–71.

[163] Sanghvi, Y S, Hoke, G D, Freier, S M, Zounes, M C, Gonzales, C, Cummins, L, Sasmor, H, Cook, P D, Nucleic Acids Res., 1993. **21**: p. 3197–3203.

[164] Biala, E, Jones, A S, Walker, R T, Tetrahedron, 1980. **36**: p. 155–158.

[165] Froehler, B C, Wadwani, S, Terhorst, T J, Gerrad, S R, Tetrahedron Lett., 1992. **33**: p. 5307–5310.

[166] Froehler, B C, Jones, R J, Cao, X, Terhorst, T J, Tetrahedron Lett., 1993. **34**: p. 1003–1006.

[167] Gutierrez, A J, Terhorts, T J, Matteucci, M D, Froehler, B C, J. Am. Chem. Soc., 1994. **116**: p. 5540–5544.

[168] Lin, K-Y, Jones, R J, Matteuchi, M J, J. Am. Chem. Soc., 1995. **117**: p. 3873–3874.

[169] Sági, J, Szemzö, A, Ébinger, K, Szabolcs, A, Sági, G, Ruff, É, Ötvös, L, Tetrahedron Lett., 1993. **34**: p. 2191–2194.

[170] Hashimoto, H, Nelson, M G, Switzer, C, J. Am. Chem. Soc., 1993. **115**: p. 7128–7134.

[171] Wagner, R, Matteuci, M D, Lewis, J G, Gutierrez, A J, Moulds, C, Froehler, B C, Science, 1993. **260**: p. 1510–1513.

[172] Colocci, N, Dervan, P B, J. Am. Chem. Soc., 1994. **116**: p. 785–786.
[173] Froehler, B C, Jones, R J, Cao, X D, Terhorst, T J, Tetrahedron Lett., 1993. **34**: p. 1003–1006.
[174] Fenster, S D, Wagner, R W, Froehler, B C, Chin, D J, Biochemistry, 1994. **33**: p. 8391–8398.
[175] Hall, K B, McLaughlin, L W, Biochemistry, 1991. **30**: p. 1795–1801.
[176] Lamm, G M, Blencowe, B J, Sproat, B S, Iribarren, A M, Ryder, U, Lamond, A I, Nucleic Acids Res., 1991. **19**: p. 3193–3198.
[177] Sanghvi, Y S, *Heterocyclic Base Modifications in Nucleic Acids and their Applications in Antisense Oligonucleotides*, in *Antisense Research und Applikations*, S.T. Crooke, Lebleu, B., Editor. 1993, CRC Press Inc.: Boca Raton Fl. p. 273–288.
[178] Xu, X Z, Zheng, Q, Swann, P F, Tetrahedron, 1992. **48**: p. 1729–1740.
[179] Xu, X Z, Zheng, Q, Swann, P F, J. Org. Chem., 1992. **57**: p. 3841.
[180] Okruszek, A, Sierzchala, A, Sochacki, M, Stec, W J, Tetrahedron Lett., 1992. **33**(49): p. 7585–7588.
[181] Manoharan, M, Johnson, L K, Bennett, C F, Vickers, T A, Ecker, D J, Cowsert, L M, Freier, S M, Cook, P D, Bioorg. Med. Chem. Lett., 1994. **4**(8): p. 1053–1060.
[182] Lesnikowski, Z I, Jaworska, M, Stec, J, Nucleic Acids Res., 1990. **18**(8): p. 2109.
[183] Guttman, A, *Capillary Gel Electrophoresis*, in *Methods in Molecular Biology*, K. Altria, Editor, Humana Press Inc.: Totowa, NJ. p. 157.
[184] Schwartz, H E, Ulfelder, K, *Analysis of Bases, Nucleosides, and (Oligo)nucleotides by Capillary Electrophoresis*, in *Methods in Molecular Biology*, K. Altria, Editor, Humana Press Inc.: Totowa, NJ. p. 227.
[185] DeDionisio, L A, Loyd, D H, J. Chromatogr. A., 1996. **735**: p. 191–208.
[186] Singhal, R P, Xian, J, *Separation of Nucleic Acids and Oligonucleotides and Sequencing of DNA Fragments by Capillary Electrophoresis*, in *Pharmaceutical and Biomedical Applications of Capillary Electrophoresis*, S.M. Lunte, Radzik, D. M., Editor. 1996, Elsevier Science LTD. p. 387–424.
[187] Cohen, A S, Najarian, D R, Paulus, A, Guttman, A, Smith, J A, Karger, B L, Proc. Natl. Acad. Sci. U. S. A., 1988. **85**: p. 9660–9663.
[188] Cohen, A S, Vilenchik, M, Dudley, J L, Gemborys, M W, Bourque, A J,, J. Chromatogr., 1993. **638**: p. 293–301.
[189] Baba, Y T, Matsuura, T, Wakamoto, K, Tsuhako, M, J. Chromatogr., 1991. **558**: p. 273–284.
[190] Bourque, A J, Cohen, A S, J. Chromatogr. B, 1994. **662**: p. 343–349.
[191] Srivatsa, G S, Batt, M, Schuette, J, Carlson, R H, Fitchett, J, Lee, C, Cole, D L, J. Chromatogr., 1994. **680**: p. 469–477.
[192] Vilenchik, M, Belenky, A, Cohen, A S, J. Chromatogr., 1994. **663**: p. 105–193.
[193] Zhu, M, Hansen, D L, Burd, S, Gannon, F, J. Chromatogr., 1989. **480**: p. 311–319.
[194] Bruin, G J M, et al., J. Chromatogr., 1991. **559**: p. 163–181.
[195] Heider, D N, Cohen, A S, Karger, B L, J. Chromatogr., 1990. **516**: p. 3–48.
[196] Guttman, A, Cooke, N, Anal. Chem., 1991. **63**: p. 2038–2042.
[197] Little, D P, Thannhauser T W, McLafferty, F W, Proc. Natl. Acad. Sci. U. S. A., 1995. **92**: p. 2318–2322.
[198] Portier, N, Drosslaaer, A V, Cordier, Y, Roch, O, Bischoff, R, Nucleic Acids Res., 1994. **22**: p. 3895–3903.
[199] Engels, W, Schweitzer, M, *Analysis of Oligonucleotides*, in *Antisense – From Technology to Therapy*, R. Schliengensiepen, Schliengensiepen, K.-H., Brysch, W., Editor. 1997, Blackwell Science Ltd.: Berlin- Vienna. p. 78.
[200] Pieles, U, Zürcher, W, Schär, M, Moser, H E, Nucleic Acids Res., 1993. **21**(14): p. 3191–3196.
[201] Nordhoff, E, Ingendoh, A, Cramer, R, Overberg, A, Stahl, B, Karas, M, Hillenkamp, F, Crain, P F, Rapid Commun. Mass Spectrom., 1992. **6**: p. 771–776.
[202] Kirpekar, E, Nordhoff, E, Kristiansen, K, Roepstorff, P, Lezius, A, Hahner, S, Karas, M, Hillenkamp, F, Nucleic Acids. Res., 1994. **22**: p. 3866–3870.
[203] Gorenstein, D G, Prog. Nucl. Magn. Reson. Spectros., 1983. **16**: p. 1–98.
[204] O'Neill, I K, Richards, C P, *Biological 31-P NMR Spectroskopy*. Annual reports on NMR spectroscopy. Vol. 10A. 1980, London: Academic Press INC. Ldt. 133.
[205] Brown, T, Brown, D J S, in *Oligonucleotides and analogues. A practical approach*, F. In Eckstein, Editor. 1991, IRL Press: Oxford.
[206] Freier, S M, Lima, W F, Sanghvi, Y S, Vickers, T, Zounes, M, Cook, P D, Ecker, D J, *Gene Regulation: Biology of Antisense RNA and DNA*, ed. R.P. Erickson, Izant, J. G. 1992, New York: Raven Press Ltd. 95–107.
[207] Martin, F H, Uhlenbeck, O C, Doty, P, J. Mol. Biol., 1971. **57**: p. 201–215.
[208] Marky, L A, Breslauer, K J, Biopolymers, 1987. **26**: p. 201–215.

[209] Neidlein, U, Leumann, C, Tetrahedron Lett., 1992. **33**(52): p. 8057–8060.
[210] Ivanov, V I, Minchenkova, L E, Schyolkina, A K, Poletayev, A I, Biopolymers, 1973. **12**: p. 89–100.
[211] Wang, A-H, Gao, Y-G, in *Methods*, J.N. Abelson, Simon, M. I., Editor. 1990, Academic Press: San Diego. p. 91–99.
[212] Dock-Bregeon, A C, Moras, D, in *Crystallization of Nucleic Acids and Proteins*, A. Ducruix, Giege, R., Editor. 1992, Oxford University Press: Oxford. p. 145–174.
[213] Jancarik, J, Kim, S-H, J. Appl. Crystallogr., 1991. **24**: p. 409–411.
[214] *Hampton Research Crystallization News*,. 1995.
[215] Nielsen, P E, Egholm, M, Berg, R H, Buchardt, O, Science, 1991. **254**: p. 1497–1500.
[216] Nielsen, P E, Egholm, M, Buchardt, O, Bioconjugate Chem., 1994. **5**: p. 3–7.
[217] Demidov, V, Frank-Kamenetskii, M D, Egholm, M, Buchardt, O, Nielsen, P E, Nucleic Acids Res., 1993. **21**: p. 2103–2107.
[218] Demidov, V V, Potaman, V N, Frank-Kamenetskii, M D, Egholm, M, Buchard, O, Sonnichsen, S H, Nielsen, P E, Biochem. Pharmacol, 1994. **48**: p. 1310–1313.
[219] Haney, J C, Peffer, N J, Bisi, J E, Thomson, S A, Cadilla, R, Josey, J A, Ricca, D J, Hassman, C F, Bonham, M A, Au, K G, Carter, S G, Bruckenstein, D A, Boyd, A L, Noble, S A., Babiss, L E, Science, 1992. **258**: p. 1481–1485.
[220] Boado, e a, Adv. Drug Delivery Rev., 1995. **15**: p. 73–107.
[221] Hyrub, B, Nielsen, P E, Bioorg. Med. Chem., 1997. **4**(1): p. 5–23.
[222] Nielsen, P E, Biophys. Chem., 1997. **68**(1–3): p. 103–108.
[223] Egholm, M, Buchardt, O, Christensen, L, Beherns, C, Freier, S M, Driver, D A, Berg, R H, Kim, S K, Norden, B, Nielsen, P E, Nature (London), 1993. **365**: p. 566–568.
[224] Sumerton, J, Weller, D, Nucleosides & Nucleotides, 1997. **16**((7–9)): p. 889–898.
[225] Taylor, M F, Paulauskis, J D, Weller, D D, Kobzik, L, Cytokine, 1997. **9**(9): p. 672–681.
[226] Stein, D, Foster, E, Huang, S-B, Weller, D, Summerton, J, Antisense Nucleic Acid Drug Dev., 1997. **7**(3): p. 151–157.
[227] Hudziak, R M, Barofsky, E, Barofsky, D F, Weller, D L, Huang, S-B, Weller, D D, Antisense Nucleic Acid Drug Dev., 1996. **6**(5): p. 267–272.
[228] Kang, H, Chou, P, Johnson, C, Weller, D, Huang, S, Summerton, J, Biopolymers, 1992. **32**: p. 1351–1363.
[229] Stirchak, E P, Summerton, J E, Weller, D D, J. Org. Chem., 1987. **52**: p. 4202–4206.
[230] Gagnor, C, Bertrand, J R, Thenet, S, Lemaitre, M, Morvan, F, Rayner, B, Malvy, C, Lebleu, B, Imbach, J L, Paoletti, C, Nucleic Acids Res., 1987. **15**: p. 10419–10436.
[231] Morvan, F, Rayner, E, Imbach, J L, Chang, D K, Lown, J S, Nucleic Acids Res., 1986. **14**: p. 5019–5035.
[232] Thuong, N T, Chassignol, M, Tetrahedron Lett., 1988. **29**: p. 5905–5908.
[233] Morvan, F, Rayner, B, Imbach, J-L, Thenet, S, Bertrand, J-R, Paoletti, J, Malvy, C, Paoletti, C, Nucleic Acids Res., 1987. **15**: p. 3421–3437.
[234] Morvan, F, Rayner, B, Imbach, J-L, Anti-Cancer Drug Des., 1991. **6**: p. 521–529.
[235] Herdewijn, P, De Winter, H, Doboszewski, B, Verheggen, I, Augustyns, K, Hendrix, C, Saison-Behmoaras, T, De Ranter, C, Van Aerschot, A, *Hexapyranosyl-Like Oligonucleotides*, in *Carbohydrate Modifications in Antisense Research*, Sanghvi, Y S, Cook, P D, Editor. 1994, American Chemical Society: Washington DC. p. 80–99.
[236] Sheppard, T L, Breslow, R C, J. Am. Chem. Soc., 1996. **118**: p. 9810–9811.
[237] Giannaris, P A, Damha, M J, Nucleic Acids Res., 1993. **21**: p. 4742–4749.
[238] Haberhauer, G, *Struktur und Herstellung von Rückgrat-modifizierten Nucleotiden*,. 1994, TU: Vienna.
[239] Noe, C R, Buchstaller, H-P, Haberhauer, G, Holzner, A, Monatsh. Chem., 1997. **128**: p. 301–316.
[240] Noe, C R, Haberhauer, G, Kaufhold, L, *Nucleinsäure-Polyester-Polyamide*, in *Österr. Pat. Anm.* 1995: Austria.
[241] Noe, C R, Haberhauer, G, Kaufhold, L, *Nukleinsäure-Polyester-Polyamide*, in *PCT Anm. (Chem. Abstr. 1996, 125, 329291)*. 1996.
[242] Blake, R D, Massoulie, J, Fresco, J R, J. Mol. Biol, 1967. **30**: p. 291–308.
[243] Hoogsteen, K, Acta Crystallogr., 1963. **16**: p. 907–916.
[244] Giovannangeli, C, Rougee, M, Montenay-Garestier, T, Thuong, N T, Helene, C, Proc. Natl. Acad. Sci., 1992. **89**: p. 8631–8635.
[245] Beal, P A, Dervan, P B, J. Am. Chem. Soc., 1992. **114**: p. 4976–4982.
[246] Thuong, N T, Helene, C, Angew. Chem., 1993. **105**: p. 697–723.
[247] Rougee, M, Faucon, B, Mergyn, J L, Barcelo, F, Giovannangeli, C, Garestier, T, Helene, C, Biochemistry, 1992. **31**: p. 9269–9278.

[248] Mergyn, J L, Duval-Valentin, G, Nguyen, C H L, Perrouault, L, Faucon, B, Rougee, M, Montenay-Garestier, T, Bisagni, E, Helene, C, Science, 1992. **256**: p. 1681–1684.
[249] Shimizu, M, Konishi, A, Shimada, Y, Inoue, H, Ohtsuka, E, FEBS Lett., 1992. **302**: p. 155–158.
[250] Latimer, L J P, Hampel, K, Lee, J S, Nucleic Acids Res., 1989: p. 1549–1561.
[251] Sun, J S, Giovannangeli, C, Francois, R, Kurfürst, R, Montenay-Garestier, T, Asseline, U, Saison-Behmoaras, T, Thoung, N T, Helene, C, Proc. Natl. Acad. Sci. U. S. A., 1991. **88**: p. 6023–6027.
[252] Sun, J S, Francois, J C, Montenay-Garestier, T, Saison-Behmoaras, T, Roig, V, Thoung, N T, Helene, C, Proc. Natl. Acad. Sci. U. S. A., 1989. **86**: p. 9198–9202.
[253] Collier, D A, Thuong, N T, Helene, C, J. Am. Chem. Soc., 1991. **113**: p. 1457–1458.
[254] Krawczyk, S H, Milligan, J F, Wadwani, S, Moulds, C, Froehler, B C, Matteucci, M D, Proc. Natl. Acad. Sci. U. S. A., 1992. **89**: p. 3761–3764.
[255] Ono, A, Ts´o, P O P, Kan, L S, J. Org. Chem., 1991. **57**: p. 3225–3230.
[256] Koh, J S, Dervan, P B, J. Am. Chem. Soc., 1992. **114**: p. 1470–1478.
[257] Griffin, L C, Kiesling, L L, Beal, P A, Gillepsie, P, Dervan, P B, J. Am. Chem. Soc., 1992. **114**: p. 7976–7982.
[258] Pelton, J G, Wemmer, D E, Proc. Natl. Acad. Sci. U. S. A., 1989. **86**: p. 5723–5727.
[259] Wade, W S, Mrksich, M, Dervan, P B, J. Am. Chem. Soc,, 1992. **114**: p. 8783–8794.
[260] Mrksich, M, Wade, W S, Dwyer, T J, Geierstanger, B H, Wemmer, D E, Dervan, P B, Proc. Natl. Acad. Sci., 1992. **89**: p. 7586–7590.
[261] Mrksich, M, Dervan, P B, J. Am. Chem. Soc., 1993. **115**: p. 9892–9899.
[262] Goodchild, J, Bioconjugate Chem., 1990. **1**(3): p. 165–87.
[263] Beaucage, S L, Iyer, R P, Tetrahedron, 1993. **49**(10): p. 1925–1963.
[264] Coleman, R S, Siedlecki, J M, J. Am. Chem. Soc., 1992. **114**: p. 9229–9230.
[265] Roduit, J-P, Shaw, J, Chollet, A, Chollet, A, Nucleosides & Nucleotides, 1987. **6**: p. 349–352.
[266] Kido, K, Inoue, H, Ohtsuka, E, Nucleic Acids. Res., 1992. **20**: p. 1339–1344.
[267] Brumbaugh, J A, Middendorf, L R, Grone, D L, Ruth, J L, Proc. Natl. Acad. Sci., 1988. **85**: p. 5610–5614.
[268] Telser, J, Cruickshank, K A, Morrison, L E, Netzel, T L, J. Am. Chem. Soc., 1989. **111**: p. 6966–6976.
[269] Jablonski, E, Moomaw, E W, Tullis, R H, Ruth, J L, Nucleic Acids Res., 1986. **14**: p. 6115.
[270] Farmar, J G, Castaneda, M, Biotechniques, 1991. **11**: p. 588–589.
[271] Sanger, F, Nicklen, S, Coulson, A R, Proc. Natl. Acad. Sci. U. S. A., 1977. **74**: p. 5463–5467.
[272] Urata, H, Akagi, M, Tetrahedron Lett., 1993. **34**: p. 4015–4018.
[273] Zeiler, B N, Simons, R W, Cold Spring Harbor Monogr. Ser., 1998. **35**: p. 437–464.
[274] Asseline, U, Thuong, N T, Helene, C, New J. Chem., 1997. **21**(1): p. 5–17.
[275] Thuong, N T, Asseline, U, Montenay-Garestier, T, in *Oligodeoxynucleotides, Antisense Inhibitors of Gene Expression*, Cohen, J S, Editor. 1989, Macmillan: London. p. 25–51.
[276] Helene, C, Toulme, J J, in *Oligodeoxynucleotides, Antisense Inhibitors of Gene Expression*, J.S. Cohen, Editor. 1989, Macmillan: London. p. 137–172.
[277] English, U, Gauss, D H, Angew. Chem., 1991. **103**: p. 629–646.
[278] Lancelot, G, Guesnet, J-L, Asseline, U, Thuong, N T, Biochemistry, 1988. **27**: p. 1265.
[279] Lancelot, G, Thuong, N T, Biochemistry, 1986. **25**: p. 5357.
[280] Birg, F, Praseuth, D, Zerial, A, Thuong, N T, Asseline, U, Le Doan, T, Helene, C, Nucleic Acids Res., 1990. **18**: p. 2901.
[281] Sun, J-S, Francois, J-C, Montenay-Garestier, T,Saison-Behmoaras, T, Roig, V, Thuong, N T, Helene, C, Proc. Natl. Acad. Sci. U. S. A., 1989. **86**: p. 9198.
[282] Stein, C A, Mori, K, Loke, S, Subashinghe, C, Shinozua, K, Cohen, J S, Neckers, L, Gene, 1988. **72**: p. 333–342.
[283] Collier, D A, Mergny, J -L, Thuong, N T, Helene, C, Nucleic Acids Res., 1991. **19**: p. 4219–4224.
[284] Collier, D A, Thuong, N T, Helene, C, J. Am. Chem. Soc., 1991. **113**: p. 1457–1458.
[285] Saison-Behmoaras, T, Tocque, B, Rey, I, Chassignol, M, Thuong, N Y, Helene, C, EMBO J., 1991. **10**: p. 1111–1118.
[286] Asseline, U, Thuong, N T, Nucleosides & Nucleotides, 1988. **7**: p. 431.
[287] Pieles, U, Sproat, B S, Neuner, P, Cramer, F, Nucleic Acids Res., 1989. **17**: p. 8967–8978.
[288] Pieles, U, Englisch, U, Nucleic Acids Res., 1989. **17**: p. 285.
[289] Tagasugi, M, Guendouz, A, Chassignol, M, Decout, J L, Lhomme, J, Thuong, N T, Helene, C, Proc. Natl. Acad. Sci. U.S.A., 1991. **88**: p. 5602–5606.
[290] Mori, K, Subasinghe, C, Cohen, J S, FEBS Lett., 1989. **249**: p. 213–218.
[291] Keller, T H, Häner, R, Nucleic Acids Res., 1993. **21**(19): p. 4499–4505.

[292] Morrow, J R, Buttrey, L A, Shelton, V M, Berback, K A, J. Am. Chem. Soc., 1992. **114**: p. 1903–1905.

[293] Magda, O, Miller, R A, Sessler, J L, Iverson, B L, J. Am. Chem. Soc., 1994. **116**: p. 7439–7440.

[294] Modak, A S, Gard, J K, Merriman, M C, Winkeler, K A, Bashkin, J K, Stern, M K, J. Am. Chem. Soc., 1991. **113**: p. 283–291.

[295] Chen, C H B, Sigman, D S, J. Am. Chem. Soc., 1988. **110**: p. 6570.

[296] Bashkin, J K, Frolova, E I, Sampath. S, J. Am. Chem. Soc., 1994. **116**: p. 5981–5982.

[297] Fathi, R, Huang, Q, Syi, J-L, Delany, W, Cook, A F, Bioconjugate Chem., 1994. **5**: p. 47–57.

[298] Nara, H, Ono, A, Matsuda, A, Bioconjug. Chem., 1995. **6**(1): p. 54–61.

[299] Tor, Y, Dervan, P B, J. Am. Chem. Soc., 1993. **115**: p. 4461–4467.

[300] Barawkar, D A, Rajeev, K G, Kumar, V A, Ganesh, K N, Nucleic Acids Res., 1996. **24**(7): p. 1229–1237.

[301] Noe, C R, Brunar, H, *Modifizierte Oligonucleotide, Verfahren zu ihrer Herstellung und Verwendung als Wirkstoffe*, in *Österr. Pat. Anm.* 1993: Austria.

[302] Noe, C R, Brunar, H, *Modifizierte Oligonucleotide, Verfahren zu ihrer Herstellung und Verwendung als Wirkstoffe*, in *PCT Anm. (Chem. Abstr. 1996, 124, 30273)*. 1995.

[303] Uhlmann, E, Engels, J, Tetrahedron Lett., 1986. **27**: p. 1023–1026.

[304] Horn, T, Urdea, M S, Tetrahedron Lett., 1986. **27**: p. 4705–4708.

[305] Guzaev, A, Salo, H, Azhayev, A, Lonnberg, H, Tetrahedron, 1995. **1995**: p. 9375–9384.

[306] Kong Thoo Lin, P, Brown, D M, Nucleic Acids Res., 1989. **20**: p. 5149–5152.

[307] Nichols, R, Andrews, P C, Zhang, P, Bergstrom, D E, Nature (London), 1994. **369**: p. 492–493.

[308] Bergstrom, D E, Zhang, P, Toma, P H, Andrews, P C, Nichols, R, J. Am. Chem. Soc., 1995. **117**: p. 1201–1209.

[309] Loakes, D, Brown, D M, Nucleic Acids Res., 1994. **22**: p. 4039–4043.

[310] Takeshita, M, Chang, C N, Johnson, F, Will, S, Grollman, A P, J. Biol. Chem., 1987. **262**: p. 10171–10179.

[311] Kalnik, M W, Chang, C N, Grollman, A P, Patel, D J, Biochemistry, 1988. **27**: p. 924–931.

[312] Ortigao, J F R, Rosch, H, Selter, H, Frohlich, A, Lorenz, A, Montenarh, M, Seliger, H, Antisense Res. Dev., 1992. **2**: p. 129–146.

[313] Hanck, T, Schmidt, S, Fritz, H J, Nucleic Acids Res., 1993. **21**(2): p. 303–309.

[314] Kremsky, J N, Wooters, J L, Dougherty, J P, Meyers, R E, Collins, M, Brown, E L, Nucleic Acids. Res., 1987. **18**: p. 2891–2909.

[315] Agrawal, S, Christodoulou, C, Gait, M, Nucleic Acids Res., 1986. **14**: p. 6227–6245.

[316] Misiura, K, Durrant, I, Evans, M R, Gait, M J, Nucleic Acids Res., 1990. **18**: p. 4245–4354.

[317] Misiura, K, Durrant, I, Evans, M R, Gait, M J, Nucleosides & Nucleotides, 1991. **10**: p. 671–672.

[318] Will, D W, Pritchard, C E, Brown, T, Carbohydr. Res., 1991. **216**: p. 315–322.

[319] Schubert, F, Ahlert, K, Cech, D, Rosenthal, A, Nucleic Acids Res., 1990. **18**: p. 3427.

[320] Dubey, K K, Singh, R K, Misra, K, Neurochem. Int., 1997. **31**(3): p. 405–412.

[321] Barbin, A, Bartsch, H, Leconte, P, Radman, M, Nucleic Acids Res., 1981. **9**: p. 375.

[322] Hamanichi, J. Org. Chem., 1994. **59**: p. 1525–1531.

[323] Secrist, J A, Barrio, J R, Leonard, N J, Weber, G, Biochemistry, 1972. **11**: p. 3499–3506.

[324] Babic, I, Andrew, S E, Jirik, F R, Mutat. Res., 1996. **372**(1): p. 87–96.

[325] Hammond, P W, Lively, T N, Chech, T R, Mol. Cell. Biol., 1997. **17**(1): p. 296–308.

[326] Stump, W T, Hall, K B, RNA: A Publication of the RNA Society, 1995. **1**(1): p. 55–63.

[327] Green, L S, Jellinek, D, Jenison, R., Ostman, A, Heldin, C H, Janjic, N, Biochemistry, 1996. **35**(45): p. 14413–14424.

[328] Liu, J, Verdine, G L, Tetrahedron Lett., 1992. **33**: p. 4265–4268.

[329] Zhang, R, Iyer, R P, Yu, D, Tan, W, Zhang, X, Lu, Z, Zhao, H, Agrawal, S, J. Pharmacol. Exp Ther., 1996. **278**: p. 971–979.

[330] Iyer, R P, Tan, W, Yu, D, Agrawal, S, Tetrahedron Lett., 1994. **35**: p. 9521.

[331] Agrawal, S, Tan, W, Jiang, Z, Yu, D, Iyer, P, *Synthetic Methods for the Radioisotopic Labelling of Oligonucleotides*, in *Antisense – From Technology to Therapy*, R. Schliengensiepen, Schliengensiepen, K.-H., Brysch, W., Editor. 1997, Blackwell Science Ltd.: Berlin- Vienna. p. 60–77.

[332] Tan, W, Iyer, R R, Yu, D, Agrawal, S, Tetrahedron Lett., 1995. **36**: p. 3631.

[333] Zhang, Y, Coyne, M Y, Will, S G, Levenson, C H, Kawasaki, E S, Nucleic Acids. Res., 1991. **19**: p. 3929–3933.

[334] Agrawal, S, Tetrahedron Lett., 1989. **30**: p. 7025–7028.

[335] Gildea, B D, Coull, J M, Köster, H, Tetrahedron Lett., 1990. **31**: p. 7095–7098.

[336] McGall, G H, Barone, A D, Diggelmann, M, Fodor, S P A, Gentalen, E, Ngo, N, J. Am. Chem. Soc., 1997. **119**(22): p. 5081–5090.

[337] McGall, G H, Labadie, J, Brock, P, Wallraff, G, Nguyen, T, Hinsberg, W, Proc. Natl. Acad. Sci. U. S. A., 1996. **93**: p. 13555–13560.
[338] Anderson, R C, McGall, G, Lipshutz, R J, Top. Curr. Chem., 1998. **194**: p. 117–129.
[339] Southern, E M, Case-Green, S C, Elder, J K, Johnson, M, Mir, K U, Wang, L, Williams, J C, Nucleic Acids Res., 1994. **22**(8): p. 1368–1373.
[340] Arndt, G, Rank, G H, Genome, 1997. **40**(6): p. 785–797.
[341] Lavrovsky, Y, Chen, S, Roy, A K, Biochem. Mol. Med., 1997. **62**(1): p. 11–22.
[342] Castanotto, D, Rossi, J J, Sarver, N, Adv. Pharmacol., 1994. **25**: p. 289–317.
[343] Zaug, A J, Been, M D, Cech, T R, Nature (London), 1986. **324**: p. 429–433.
[344] McClain, W H, Guerrier-Takada, C, Altman, S, Science, 1987. **238**: p. 527–530.
[345] Foster, A C, Altman, S, Science, 1990. **249**: p. 783–786.
[346] Yuan, Y, Hwang, E-S, Altman, A, Proc. Natl. Acad. Sci. U.S.A., 1992. **89**: p. 8006–8010.
[347] Uhlenbeck, O C, Nature (London), 1987. **328**(596–600).
[348] Haseloff, J, Gerlach, W L, Nature (London), 1988. **334**: p. 585–591.
[349] Koizumi, M, Iwai, S, Ohtsuka, E, FEBS Lett., 1998. **239**: p. 285–288.
[350] Goodchild, J, Kohli, V, Arch. Biochem. Biophys., 1991. **284**: p. 386–391.
[351] Bertrand, E, Grange, R, Pictet, R, *Trans-acting Hammerhead Ribozymes in Vivo. Present limits and Future Directions*, in *Gene Regulation Biology of Antisense RNA and DNA*, R.P. Erickson, Izant, J G, Editor. 1992, Raven: New York. p. 71–81.
[352] Heidenreich, O, Pieken, W, Eckstein, F, FASEB J., 1993. **7**: p. 90–96.
[353] Nishikura, K, Yoo, C, Kim, U, Murray, J M, Estes, P A, Cash, F E, Liebhaber, S A, EMBO J., 1991. **10**: p. 3523–3532.
[354] Pieken, W, Olsen, D B, Benseler, F, Aurup, H, Eckstein, F, Science, 1991. **253**: p. 314–317.
[355] Hausen, P, Stein, H, Eur. J. Biochem., 1970. **14**: p. 278–283.
[356] Kitani, T, Yoda, K, Ogawa, T, Okazaki, T, J. Mol. Biol., 1985. **184**: p. 45–52.
[357] Nakamura, H, Oda, Y, Iwai, S, Inoue, H, Ohtsuka, E, Kanaya, S, Kimura, S, Katsuda, C, Katayanagi, K, Morikawa, K, Miyashiro, H, Ikehara, M, Proc. Natl. Acad. Sci. U. S. A, 1991. **88**: p. 11535–11539.
[358] Eder, P S, Walder, R Y, Walder, J A, Biochimie, 1993. **75**: p. 123–26.
[359] Giles, R V, Spiller, D G, Tidd, D M, Anti-Cancer Drug Des., 1993.
[360] Gosh, M K, Cohen, J S, Anti-Cancer Drug Design, 1993. **8**: p. 15–32.
[361] Monia, B P, Lesnik, E A, Gonzales, C, Lima, W F, McGee, D, Guinosso, C J, Kawasaki, A M, Cook, P D, Freier, S M, J. Biol. Chem., 1993: p. 14514–14522.
[362] Tereshko, V, Portmann, S, Tay, E C, Martin, P, Natt, F, Altmann, K H, Egli, M, Biochemistry, 1998. **37**(30): p. 10626–10634.
[363] Zagorowska, I, Kuusela, S, Lonnberg, H, Nucleic Acids Res., 1998. **26**(14): p. 3392–3396.
[364] Giles, R V, Spiller, D G, Grzybowski, J, Clark, R E, Nicklin, P, Tidd, DM, Nucleic Acids Res., 1998. **26**(7): p. 1567–1575.
[365] Hodges, D, Crooke, S T, Mol. Pharmacol., 1995. **48**: p. 905–918.
[366] Akhtar, S, Shoji, Y, Juliano, R L, Raven Press Ser. Mol. Cell. Biol., 1992. **1**(Gene Regil.): p. 133–145.
[367] Akhtar, S, Basu, S, Wickstrom, E, Juliano, R L, Nucleic Acids Res., 1991. **19**: p. 5551–5559.
[368] Yabukov, L A, Deeva, E A, Zarytova, V F, Ivanova, E M, Ryte, A S, Yurchenko, L V, Vlassov, V V, Proc. Natl. Acad. Sci. U.S.A., 1989. **86**: p. 6454.
[369] Vlassov, V V, Yabukov, L A, in *Prospects for Antisense Nucleic Acid Therapy of Cancer and Aids*, Crook, S T, Lebleu, B., Editor. 1993, CRC Press: Boca Raton.
[370] Stein, C A, Tonkinson, J L, Zhang, L-M, Yabukov, L, Gervasoni, J, Taub, R, Rotenberg, S A, Biochemistry, 1993. **32**: p. 4855.
[371] Vlassov, V V, Vlassova, I E, Pautova, L V, Prog. Nucleic Acid Res. Mol. Biol., 1997. **57**: p. 95–143.
[372] Rojanasakul, Y, Adv. Drug Delivery Rev., 1996. **18**(2): p. 115–31.
[373] Stein, C A, et al., Perspect. Drug Discovery Des., 1996. **4**(Antisense Therapeutics): p. 41–50.
[374] Stein, C A, Khan, T, Khaled, Z, Tonkinson, J, Clin. Cancer Res., 1995. **1**: p. 509.
[375] Loke, S L, Stein, C A, Zhang, X H, Mori, K, Nakanishi, M, Subasinghe, C, Cohen, J S, Neckers, L M, Proc. Natl. Acad. Sci. U.S.A., 1989. **86**: p. 3474–3478.
[376] Benett, R M, Gabor, G T, Morritt, M J, J. Clin. Invest., 1985. **76**: p. 2182–2190.
[377] Chin, D J, Green, G A, Zon, G, Szoka, F C, Straubinger, R M, New. Biol., 1990. **2**: p. 1091.
[378] Miller, P S, McParland, K B, Jayaraman, K, Ts´o, P O P, Biochemistry, 1981. **20**: p. 1874–1880.
[379] Shoji, Y, Akhtar, S, Periasamy, A, Herman, B, Juliano, R L, Nucleic Acid Res., 1991. **19**: p. 5543–5550.

[380] Rojanasakul, Y, Adv. Drug Delivery Rev., 1996. **18**(2): p. 115–131.
[381] Mackellar, C, Graham, D, Will, D W, Burgess, S, Brown, T, Nucleic Acids Res., 1992. **20**: p. 3411–3417.
[382] Kabanov, A V, Vinogradov, S V, Ovcharenko, A V, Krivonos, A V, Melik-Nubarov, N S, Kiselev, V I, Severin, E S, FEBS Lett., 1990. **259**: p. 327–330.
[383] Will, D W, Brown, T, Tetrahedron Lett., 1992. **33**: p. 2729–2732.
[384] Beucage, S L, Iyer, R P, Tetrahedron, 1993. **49**: p. 1925–1963.
[385] Oberhauser, B, Wagner, E, Nucleic Acids Res., 1992. **20**: p. 533–538.
[386] Stein, C A, Pal, R, Devico, A L, Hoke, G, Mumbauer, S, Kinstler, O, Sarngadharan, M G, Letsinger, R L, Biochemistry, 1991. **30**: p. 2439–2444.
[387] Rumo, E T, de Vrueh, R L, Sliedregt, L A, Biessen, E A, van Berkel, T J, Bijsterbosch, M K, Bioconjug. Chem., 1998. **9**(3): p. 341–349.
[388] Corrias, M V, Guarnaccia, F, Ponzoni, M, Journal of Neuro-Oncology, 1997. **31**(1–2): p. 171–180.
[389] Boutorine, A S, Boiziau, C, Le Doan, R, Toulme, J J, Helene, C, Biochimie, 1992. **74**: p. 485–489.
[390] Alahari, S K, Dean, N M, Fisher, M H, Delong, R, Manoharan, M, Tivel, K L, Juliano, R L, Mol. Pharmacol., 1996. **50**(4): p. 808–819.
[391] Krieg, A M, Tonkinson, J, Matson, S, Zahio, Q, Saxon, M, Zhang, L M, Bhanja, U, Yakubov, L, Stein, C A, Proc. Natl. Acad. Sci. U.S.A., 1993. **90**: p. 1048–1052.
[392] Marasco, C J, Angelino, N J, Paul, B, Dolnik, B J, Tetrahedron Lett., 1994. **35**: p. 3029–3032.
[393] Bisbal. C, S, M, Lemaitre, M, Bayard, B, Salehzada, T, Lebleu, B, Perree, T D, Blackburn, M G, Biochemistry, 1987. **26**: p. 5172.
[394] Degols, G, Leonetti, J-P, Benkirane, M, Devaux, C, Lebleu, B, Antisense Res. Dev., 1992. **2**: p. 293–301.
[395] Stevenson, M, Iversen, P L, J. Gen. Virol., 1989. **70**(70): p. 2673–2682.
[396] Morgan, D M L, Clover, J, Pearson, J D, J. Cell. Sci., 1988. **91**: p. 231–238.
[397] Zimmer, A, Zobel, H-P, Werner, D, Noe, C R, Kreuter, J, Proc. Int. Symp. Controlled Release Bioact. Mater., 1997. **24th**: p. 679–680.
[398] Zobel, H-P, Kreuter, J, Werner, D, Noe, C R, Kümel, G, Zimmer, A, Antisense Nucleic Acid Drug Dev., 1997. **7**: p. 483–493.
[399] Noe, C R, Kreuter, J, Zimmer, A, Zobel, H-P, Stienecker, F, Atmaca-Abdel Aziz, S, *Partikuläres Arzneistoff-Trägermaterial mit basischen Modifikationen*, in *Österr. Pat. Anm.* 1995: Austria.
[400] Noe, C R, Kreuter, J, Zimmer, A, Zobel, H-P, Stienecker, F, Atmaca-Abdel Aziz, S, *Partikuläres Arzneistoff-Trägermaterial mit basischen Modifikationen*, in *PCT Anm. (Chem. Abstr. 1996, 125, 257219)*. 1996.
[401] Noe, C R, Kreuter, J, Stieneker, F, Zimmer, A, Zobel, H-P, *Partikuläres Arzneistoff-Trägermaterial mit kurzkettigen basischen Modifikationen*, in *Österr. Pat. Anm.* 1996: Austria.
[402] Noe, C R, Kreuter, J, Stieneker, F, Zimmer, A, Zobel, H-P, *Partikuläres Arzneistoff-Trägermaterial mit kurzkettigen basischen Modifikationen*, in *PCT Anm. (Chem. Abstr. 1998, 129, 27887)*. 1996.
[403] Zobel, H-P, Atmaca, A, Werner, D, Noe, C R, Stieneker, F, Kreuter, J, Zimmer, A, *Preparation of Antisense Nucleic Acid loaded Nanoparticles and Liposomes.* Proceed. 2nd Antisense Nucleic Acids, Garmisch-Partenkirchen, 1995: p. 77.
[404] Zobel, H-P, Kreuter, J, Werner, D, Noe, C R, Zimmer, A, Europ. J. Pharm. Biopharm., Proceed. 42th Annual APV Congress, 1996. **13**: p. 47.
[405] Zimmer, A, Rühl, D, Müller, R-H, Atmaca Abdel-Azis, S, Noe, C R, Pharm. Res. (Supplement), 1996. **13**(9): p. S-386.
[406] Zimmer, A, Rühl, D, Müller, R-H, Atmaca Abdel-Aziz, S, Noe, C R, Pharm. Res. (Supplement), 1996. **13**(9): p. S-388.
[407] Zobel, H-P, Kreuter, J, Werner, D, Noe, C R, Zimmer, A, Proceed. Intern. Symp. Control. Rel. Bioact. Mater, 24th CRS-Symposium Stockholm, 1997: p. 679–680.
[408] Zimmer, A, Zobel, H-P, Kümel, G, Offensberger, W-B, Noe, C R, Kreuter, J. *Viral test systems for Oligonucleotide loaded colloidal drug carriers.* in *2nd APG/APV World Meeting*. 1998. Paris.
[409] Couvreur, P, Fattal, E, Malvy, C, Dubernet, C, J. Liposome Res., 1997. **7**(1): p. 1–18.
[410] Ropert, C, Lavignon, M, Duberner, C, Couvreur, P, Malvy, C, Biochem. Biophys. Res. Commun., 1992. **183**: p. 879.
[411] Iversen, P, Anti-Cancer Drug Des., 1991. **6**: p. 531–538.
[412] Thierry, A D, Nucleic Acids Res., 1992. **20**: p. 5691–5698.
[413] Bennett, C F, Chiang, M C, Shoemaker, J E E, Mirabelli, K, Mol. Pharmacol., 1992. **41**: p. 1023–1033.

[414] Clarenc, J P, Degols, G, Leonetti, J P, Milhaud, P, Lebleu, B, Anti-Cancer Drug Des., 1993. **8**(1): p. 81–94.
[415] Litzinger, D C, J. Liposome Res., 1997. **7**(1): p. 51–61.
[416] Bennett, C F, Chiang, M-Y, Chan, H, Grimm, S, J. Liposome Res., 1993. **3**(1): p. 85–102.
[417] Felgner, P L, Gadek, T R, Holm, M, Roman, R, Chan, H W, Wenz, M, Northrup, J P, Ringold, G M, Danielson, M, Proc. Natl. Acad. Sci. U. S. A., 1987. **84**: p. 7413.
[418] Leventis, R, Silvius, J R, Biophys. Acta, 1990. **1024**: p. 124.
[419] Behr, J-P, Demeneix, B, Loeffler, J-P, Perez, M J, Proc. Natl. Acad. Sci., 1989. **86**: p. 6982.
[420] Gao, X, Huang, L, Biochem. Biophys. Res. Commun., 1991. **179**: p. 280.
[421] Sullivan, S M, Gieseler, R K H, Lenzner, S, Ruppert. J, Gabrysiak, T G, Peters, J H, Cox, G, Richter, L, Martin, W J, Scolaro, M J, Antisense Res. Dev., 1992. **2**: p. 187–197.
[422] Ropert, C, Okruszek, A, Couvreur, P, Malvy, C, Antisense Res. Dev., 1994. **4**: p. 207–210.
[423] Loke, S L, Stein, C, Zhang, X, Arigan, M, Cohen, J S, Neckers, L, Curr. Top. Microbiol. Immunol., 1988. **141**: p. 282.
[424] Bennett, C F, Chiang, M-Y, Chan, H, Shoemaker, J E E, Mirabelli, C K, Mol. Pharmacol., 1992. **41**: p. 1023–1033.
[425] Gewirtz, A M, Stein, C A, Glazer, P M, Proc. Natl. Acad. Sci. U. S. A., 1996. **93**(8): p. 3161–3163.
[426] Lewis, J G, Lin, K -Y, Kothavale, A, Flagan, W M, Matteucci, M D, DePrince, R B, Mook, J R A, Hendren, R W, Wagner, R W, Proc. Natl. Acad. U. S. A., 1996. **93**: p. 3176–3181.
[427] Lewis, J G, Lin, K Y, Kothavale, A, Flanagan, W M, Matteucci, M D, DePrince, R B, Mook, R A, Hendren, R W, Wagner, R W, Proc. Natl. Acad. Sci. U. S. A., 1996. **93**: p. 3176–3181.
[428] Zimmer, A, Methods: A Companion to Methods in Enzymology, 1999. **18**: p. in press.
[429] Bertling, W M, Gareis, M, Paspaleeva, V, Zimmer, A, Kreuter, J, Nurnberg, E, Harrer, P, Biotechnol. Appl. Biochem., 1991. **13**: p. 390–405.
[430] Chavany, C, LeDoan, T, Couvreur, P, Puisieux, F, Helene, C, Pharmacol. Res., 1992. **9**: p. 441–449.
[431] Cortesi, R, Esposito, E, Menegatti, E, Gambari, R, Nastruzzi, C, Int. J. Pharm., 1994. **105**: p. 181–186.
[432] Emile, C, Bazile, D, Herman, F, Helene, C, Veillard, M, Drug Deliv., 1996. **3**: p. 187–195.
[433] Fattal, E, Vauthier, C, Aynie, I, Nakada, Y, Lambert, G, Malvy, C, Couvreur, P, J. Controlled Release, 1998. **53**: p. 137–143.
[434] Nakada, Y, Fattal, E, Foulquier, M, Couvreur, P, Pharm. Res., 1996. **13**: p. 38–43.
[435] Schwab, G, Chavany, C, Duroux, I, Goubin, G, Lebeau, J, Helene, C, Saison-Behmoaras, T, Proc. Natl. Acad. Sci. U. S. A., 1994. **91**: p. 10460–10464.
[436] Chavany, C, Saison-Behmoaras, T, Le Doan, T, Puisieux, F, Couvreur, P, Helene, C, Pharm. Res., 1994. **11**: p. 1370–1378.
[437] Godard, G, Boutorine, A S, Saison-Behmoaras, E, Helene, C, Eur. J. Biochem., 1995. **232**: p. 404–410.
[438] Lewis, K J, Irwin, W J, Akhtar, S, J. Controlled Release, 1995. **37**: p. 173–183.
[439] Akhtar, S, Lewis, K J, Int. J. Pharm., 1997. **151**: p. 57–67.
[440] Cleek, R L, Rege, A A, Denner, L A, Eskin, S G, Mikos, A G, J. Biomed. Mater. Res., 1997. **35**: p. 525–530.
[441] Nguyen, D M, Wiehle, S A, Koch, P E, Branch, C, Yen, N, Roth, J A, Cristiano, R J, Cancer Gene Ther., 1997. **4**(3): p. 191–198.
[442] Nguyen, D M, Wiehle, S A, Koch, P E, Branch, C, Yen, N, Roth, J A, Cristiano, R J, Cancer Gene Ther., 1997. **4**(3): p. 183–190.
[443] Forstova, J, Krauzewicz, N, Sandig, V, Elliot, J, Palkova, Z, Strauss, M, Griffin, B E, Hum. Gene Ther., 1995. **6**(3): p. 297–306.
[444] Baertschi, A J, Mol. Cell. Endocrinol., 1994. **101**(1–2): p. R15–R24.
[445] Agrawal, S, Trends Biotechnol., 1996. **14**(10): p. 376–387.
[446] Henry, S P, Monteith, D, Levin, A A, Anti-Cancer Drug Des., 1997. **12**: p. 395–408.
[447] Cowsert, L M, Anti-Cancer Drug Des., 1997. **12**(5): p. 359–371.
[448] Crooke, S T, Berstein, L S, Boswell, H, Expert Opin. Ther. Pat., 1996. **6**(9): p. 855–870.
[449] AJINOMOTO,. 1991, AJINOMOTO.
[450] FOUNDATION, W,. 1992.
[451] FOUNDATION, W,. 1993.
[452] GILEAD,. 1993.
[453] RESEARCH, U O I,. 1996.
[454] ISIS PHARMACEUTICALS, I,. 1993.
[455] HYBRIDON,. 1994.
[456] GENTA,. 1995.

[457] ISIS PHARMACEUTICALS, I,. 1995.
[458] Noe, C R, Med. Forsch., 1996. **9**: p. 139–147.
[459] Tu, G, Cao, Q, Zhou, F, Israel, Y J, J. Biol. Chem., 1998. **273**: p. 25125–25131.
[460] Dean, N, McKay, R, Condon, T, Bennett, C J, J. Biol. Chem., 1994. **269**: p. 16416–16424.
[461] Monia, B, Johnston, J, Ecker, D, Zounes, M, Lima, W, Freier, S J, J. Biol. Chem., 1992. **267**: p. 19954–19962.
[462] Benimetskaya, L, Berton, M, Kolbanovsky, Benimetsky, S, Stein, C A, Nucleic Acids Res., 1997. **25**: p. 2648–2656.
[463] Smetsers, T, Boezeman, J, Mensink, C, Antisense Nucleic Acid Drug Dev., 1996. **6**: p. 63–67.
[464] Stein, C A, Nature Biotech., 1999. **17**: p. 209.
[465] Bennett, C F, Biochem. Pharmacol., 1998. **55**(1): p. 9–19.
[466] Monia, B, Nat. Med., 1999. **5**: p. 127.

Index